COMPENDIUM IN ASTRONOMY

PROFESSOR JOHN XANTHAKIS

FELLOW OF THE NATIONAL ACADEMY OF ATHENS
SECRETARY OF THE PROCEEDINGS OF THE ACADEMY
EMERITUS PROFESSOR OF ASTRONOMY
AT THE UNIVERSITY OF THESSALONIKI

COMPENDIUM IN ASTRONOMY

*A Volume Dedicated to Professor John Xanthakis
on the Occasion of Completing Twenty-five Years of Scientific Activities
as Fellow of the National Academy of Athens*

Edited by

ELIAS G. MARIOLOPOULOS
*Fellow, National Academy of Athens
President, National Observatory of Athens*

PERICLES S. THEOCARIS
*Vice-President, National Academy of Athens
Rector, National Technical University of Athens*

and

L. N. MAVRIDIS
Professor, University of Thessaloniki

D. REIDEL PUBLISHING COMPANY

DORDRECHT : HOLLAND / BOSTON : U.S.A.

LONDON : ENGLAND

Library of Congress Cataloging in Publication Data
Main entry under title:

Compendium in astronomy.

Includes indexes.
1.   Astronomy—Addresses, essays, lectures.
2.   Xanthakis, John N., 1904-   —Addresses, essays, lectures.   I.   Xanthakis,
John N., 1904-   .   II.   Mariolopoulos, Elias G., 1900-   .   III.   Theocaris,
Pericles S., 1921-   .   IV.   Mavridis, L. N.   V.   Series.
QB51.C686   520   81-19949
ISBN-13: 978-94-009-7768-6   e-ISBN-13: 978-94-009-7766-2   AACR2
DOI: 10.1007/978-94-009-7766-2

Published by D. Reidel Publishing Company,
P.O. Box 17, 3300 AA Dordrecht, Holland.

Sold and distributed in the U.S.A. and Canada
by Kluwer Boston Inc.,
190 Old Derby Street, Hingham, MA 02043, U.S.A.

In all other countries, sold and distributed
by Kluwer Academic Publishers Group,
P.O. Box 322, 3300 AH Dordrecht, Holland.

D. Reidel Publishing Company is a member of the Kluwer Group.

PREFACE

When we first approached some colleagues all over the world to sound them about a volume dedicated to Professor John Xanthakis on the occasion of completing twenty-five years of scientific activities as fellow of the National Academy of Athens, any possible doubts as to the feasibility of the project were quickly dispelled by their warm and encouraging response. In a short time 50 authors from 15 countries, coming from a wide range of Professor Xanthakis' immediate colleagues, pupils and friends joined to produce the 36 contributions included in this volume.

Some of those who where originally approached found themselves unable to contribute, because of the time-limit necessarily imposed. Happily, they were only few in number, and we should like to record our gratitude to them for their good wishes for the success of the venture. Their warm words were among the many sources of inspiring encouragement extended to us.

This response can easily be understood, if one takes into account the career, scientific achievements and personality of Professor Xanthakis, which made him a widely known and highly estimated member of the international scientific community. His research work, which opened new perspectives in the fields of Solar Physics and Solar-Terrestrial Relations, his pioneer work as the first Professor of Astronomy at the University of Thessaloniki, his long and creative scientific activity as Fellow, Secretary of the Proceedings, and President of the National Academy of Athens, where he founded, as early as twenty years ago, the Research Center for Astronomy and Applied Mathematics, as well as his continuous efforts in his capacity as President of the Greek National Committees for Astronomy, for Mathematics and for Space Research, to ensure the harmonious co-operation of the Greek scientists in the international scientific programs, had already, long ago, created the proper climate for this enthousiastic response to our request.

It is our pleasant duty to extend our warmest thanks to all those who contributed to the preparation of the present volume. First of all to the authors of the papers for their excellent co-operation. Then to the publishers, the D. Reidel Publishing Company, for their unfailing helpfulness and great care, which they devoted to the production and publication of this volume, especially to Mrs. N.M. Pols-v.d. Heijden, who always gave every possible consideration to our suggestions. Finally, to Miss J. Karrinti, who assisted us in many ways during the preparation of the volume, and Mrs. E. Kolokotroni and Miss M. Stamatelou, who typed the camera-ready manuscript.

v

E. G. Mariolopoulos et al. (eds.), Compendium in Astronomy, v–vi.
Copyright © 1982 by D. Reidel Publishing Company.

We should like to add a final personal footnote to our Preface. Circumstances have given us the opportunity of long-continued contact with Professor Xanthakis' inspiring personality. It is an experience, which we cherish more deeply than we can readily express, and which we consider one of the greatest gifts of our life.

August 1981   E.G. MARIOLOPOULOS
                                                      P.S. THEOCARIS
                                                      L.N. MAVRIDIS

PROFESSOR JOHN XANTHAKIS
BRIEF BIOGRAPHICAL NOTE AND LIST OF PUBLICATIONS

BRIEF BIOGRAPHICAL NOTE

Born in Gythion, Greece, November 21, 1904.
Graduate in Mathematics, University of Athens (1925); PhD in Mathematics,
University of Athens (1930).
Astronomer and Chief Assistant, University of Athens (1929-31).
Research Associate, Observatoire de Strasbourg (1931-34).
Professor of Mathematics, Military Academy of Athens (1938-39).
Professor of Astronomy, University of Thessaloniki (1939-55).
Fellow, National Academy of Athens (1955-now); President National Acad-
emy of Athens (1964); Secretary of the Proceedings, National Academy of
Athens (1966-now).
Founder and Supervisor of the Research Center for Astronomy and Applied
Mathematics, National Academy of Athens (1959-now).
President, Greek National Committee for Astronomy (1957-now).
President, Greek National Committee for Mathematics (1957-80).
President, Greek National Committee on Space Research (1964-now).
President, Administrative Council, Greek Oceanographic Institute (1960-
67).
President, Greek Mathematical Society (1965-73).
Served twice as Minister of Agriculture (1963,1964).
Published 6 text-books in Greek language, 2 books and 78 original papers.
His research work refers mainly to the Mathematical Analysis, Positional
Astronomy, Solar Physics and Solar-Terrestrial Relations.

BOOKS

1. Solar Physics, Proceedings of a NATO Advanced Study Institute Confer-
    ence, Athens, September 1965, (ed. by J. Xanthakis), Interscience
    Publishers a Division of John Wiley and Sons, London, New York,
    Sydney, 1967.
2. Solar Activity and Related Interplanetary and Terrestrial Phenomena,
    Proceedings of the First European Astronomical Meeting, Athens,
    September 4-9, 1972, Vol. I, (ed. by J. Xanthakis), Springer-Verlag,
    Berlin, Heidelberg, New York, 1973.

TEXT-BOOKS (in Greek)

1. General Mathematics, University of Thessaloniki Press, Thessaloniki,
    1947.

E. G. Mariolopoulos et al. (eds.), Compendium in Astronomy, vii–xii.

2. Probability Calculus, University of Thessaloniki Press, Thessaloniki, 1951.
3. Astronomy, Vol. I: Spherical Astronomy, University of Thessaloniki Press, Thessaloniki, 1955.
4. Astronomy, Vol. II: Celestial Mechanics, University of Thessaloniki Press, Thessaloniki, 1952.
5. Astronomy, Vol. IIIa: Earth, Moon, Sun. University of Thessaloniki Press, Thessaloniki, 1954.
6. Astronomy, Vol. IV: Stellar Astronomy, University of Thessaloniki Press, Thessaloniki, 1951.

ORIGINAL PAPERS

1.. Sur la théorie des anomalies des équations différentielles du 1er ordre, Proc. Acad. Athens, 5, 53, 1930.
2.. Contribution à la théorie des singularités des équations différentielles du premier ordre.PhD Dissertation, University of Athens, 1930.
3. Sur les singularités des équations différentielles du premier ordre. Ann. Fac. Sci. Univ. Toulouse, 1932.
4. Sur la variation des longitudes géographiques, Proc. Acad. Athens 7, 255, 1932.
5. Sur le terme Z de M. Kimura, Proc. Acad. Athens, 8, 32, 1933.
6. Sur les déplacements apparents de l' étoile polaire, Compt. Rend. Acad. Sci. Paris 196, 1649, 1933.
7. Sur les fluctuations du méridien de Greenwich, Bull. Astron. Obs. Paris 8, Fasc. III, 123, 1934.
8. Sur quelques erreurs systématiques des ascensions droites des étoiles circumpolaires, Proc. Acad. Athens 9, 136, 1934.
9. Sur la variation diurne des azimuts, Proc. Acad. Athens 11, 464, 1936.
10. Observations de l' éclipse totale de soleil du 19 Juin 1936. Mem. Natl. Obs. Athens, Ser. I. Astron., No. 1, 20, 1937.
11. Sur les causes de la variation des azimuts, Proc. Acad. Athens 12, 478, 1937.
12. Sur la variation d'azimut de la ligne des mires méridiennes à l'Observatoire de Strasbourg, Compt. Rend. Acad. Sci. Paris 206, 171, 1938.
13. Sur une relation remarquable entre les températures moyennes mensuelles de l'air à Alexandrie, Athènes, Rome, Paris, Proc. Acad. Athens 13, 424, 1938.
14. A New Method for the Determination of the Logistic Curve. Application to the Population of Greece, Proc. Acad. Athens 14, 412, 1939.
15. Sur les temperatures moyennes mensuelles de l'air, Volume dedicated to the thirty-fifth anniversary from the appointment of Professor N. Criticos (1907-1942), p. 150, Athens, 1943.
16. Relation between the Mean Monthly Air Temperatures in the Temperate Zones, University of Thessaloniki Press, Thessaloniki, 1948.
17. Relation Between the Mean Monthly Air Temperatures in the Temperate Zones, Bull. Am. Meteorol. Soc. 29, 550, 1948.
18. Statistical Study of the Results of the Entrance Examinations to the Schools of the University of Thessaloniki during the Academic Years 1946-47, 1947-48, 1948-49, Bull. Univ. Thessaloniki, 1949.

19. Sur une relation entre les valeurs moyennes mensuelles de la radia-
    tion solaire observée en 12 stations de l' hémisphère nord, Proc.
    Acad. Athens 26, 208, 1951.
20. Sur la variation d' azimut du cercle meridien de l' Observatoire
    National d' Athènes, Bull. Hellenic Army Geographical Service 12,
    1, 1952.
21. A Correlation Between the Coefficient of Continentality and the Excen-
    tricity of the Ellipse of the Mean Monthly Air Temperatures in the
    Temperate Zones, Proc. Acad. Athens 27, 101, 1952.
22. Justification théorique d' une relation empirique entre les valeurs
    moyennes mensuelles de la température de l' air et de la radiation
    solaire, Proc. Acad. Athens 27, 168, 1952.
23. New Relations between the Mean Monthly Air Temperatures, University
    of Thessaloniki Press, Thessaloniki, 1953.
24. Sur la distribution symétrique des températures moyennes mensuelles
    de l' air, Proc. Acad. Athens 28, 47, 1953.
25. Les températures saisonnières de l'air pendant les périodes de l'
    action solaire, Proc. Acad. Athens 28, 91, 1953.
26. Le problème des déplacements apparents de l' étoile Polaire, Proc.
    Acad. Athens 28, 274, 1953.
27. Expression de la radiation solaire en fonction de la longitude du
    Soleil en 11 stations de l' hémisphère nord, Proc. Acad. Athens 28,
    299, 1953.
28. Study of the Mean Monthly Air Temperatures during the Successive
    Sunspot Cycles, Publ. Astron. Dept. Univ. Thessaloniki No. 1, 1955.
29. Sur les variations des températures saisonnières de l' air à six
    stations de l' Europe Centrale et du Nord-Ouest pendant les 150
    dernières années, Proc. Acad. Athens 30, 115, 1955.
30. Sur une corrélation importante entre l' action solaire et l' atmo-
    sphère inférieure, Proc. Acad. Athens 30, 125, 1955.
31. Study of the Mean Monthly Air Temperatures during the Successive
    Sunspot Cycles, Archiv Meteorol. Geophys. Bioklimatol. Ser. A
    9, 54, 1956.
32. Sur une classification des cycles des taches solaires et des facules,
    Astron. Zh. 34, 391, 1957.
33. Current Solar Studies. Inaugural Presentation in the Academy of
    Athens, Proc. Acad. Athens 32, 202, 1957.
34. Les aires des taches solaires et des facules en fonction du temps
    d' ascension, Geofis. Pura e Appl. Milano 39, 1, 1958.
35. The Areas of Sunspots in the Two Sun Hemispheres, Trans. Acad. Athens
    24, 1, 1958.
36. L' expression de l' activité solaire en fonction du temps d' ascension
    Ann. Astrophys. 22, 855, 1959.
37. Les aires des taches solaires vers le Nord et vers le Sud de l'
    equateur du Soleil en fonctions du temps d' ascension. Compt. Rend.
    Acad. Sci. Paris 249, 1315, 1959.
38. L' asymétrie N-S dans l' activité solaire. Compt. Rend. Acad. Sci.
    Paris 249, 1458, 1959.
39. The Sunspot Areas and the Relative Sunspot Numbers. Geofis. Pura e
    Appl. Milano 46, 11, 1960.

40. Sur les maximums singulièrement élevés de l' activité solaire. Compt. Rend. Acad. Sci. Paris 253, 1311, 1961.
41. The Sunspot Areas and the Wolf Numbers. A Study of the Analytical Relations Given by J. Xanthakis and J. Mergentaler. Mem. Soc. Astron. Ital. XXXIII, 291, 1962. (in collaboration with G. Banos).
42. Les relations analytiques des nombres des groupes de taches solaires et de nombres relatifs. Ann. Astrophys. 25, 342, 1962.
43. Two Interplanetary Phenomena of 468 B.C., Trans. Acad. Athens 24, No. 4, pp. 24-45, 50-53, 1963. (Marinatos, S., comments by J. Xanthakis).
44. A Study of the Sunspot Magnetic Field Strengths, Mem. Soc. Astron. Ital. XXXVI, 25, 1965.
45. The Departures of the Maxima of Solar Activity from the Parabolic Law. Publicazioni del Comitato Nazionale per le Manifestazioni Celebrative del IV Centenario della Nascita di Galileo Galilei, Tomo 2: Atti del Convegno sulle Macchie Solari, pp. 63-67, 1966.
46. Probable Value of the Time of Rise for the Sunspot Cycle No. 20. Nature 210, 1242, 1966.
47. The Relative Sunspot Numbers and the Time of Rise. Bull. Astron.Inst. Czech. 17, 215, 1966.
48. The Probable Mean Values of the Different Indices of Solar Activity During the Sunspot Cycle No. 20 (1964-1975). Proc. Acad. Athens 41, 384, 1967.
49. The Different Indices of Solar Activity and the Time of Rise, in Solar Physics (ed. by J. Xanthakis), Interscience Publishers, London, N. York, Sydney, p. 157, 1967.
50. Probable Values of the Time of Rise for the Forthcoming Sunspot Cycles Nature, 215, 1046, 1967.
51. On a Relation Between the Indices of Solar Activity in the Photosphere and the Corona. Solar Phys. 10, 168, 1969.
52. Relations entre l' indice des aires, le nombre des éruptions à protons (proton flares), la variation du radio-diamètre de la couronne solaire et l' intensité des rayons cosmiques. Comp. Rend. Acad. Sci. Paris, 271, 1009, 1970.
53. On a Relation Between the Indices of Solar Activity in the Photosphere and the Corona, Proc. Acad. Athens 44, 153, 1970.
54. Relations Between the Areas Index and Different Phenomena in the Chromosphere, the Corona and the Interplanetary Space, in Physics of the Solar Corona (ed. by C. Macris), D. Reidel, Dordrecht-Holland, p. 179, 1971.
55. Solar Activity and Precipitation, in Solar Activity and Related Interplanetary and Terrestrial Phenomena (ed. by J. Xanthakis), Springer-Verlag, Berlin p. 19, 1973.
56. Solar Activity and Precipitation Within the Zones of Latitude $0°-40°N$. Proc. Acad. Athens 49, 187, 1974 (in collaboration with C. Poulakos and B. Tritakis).
57. Remarks on High Latitude Air Temperature Ranges and Solar Activity, Proc. Acad. Athens 50, 118, 1975 (in collaboration with Ch. S. Zerefos).
58. Probable Periodical Variations in the Frequency of the Etesian Winds, in the Special Volume in Memoriam "Demetrios Eginitis", (ed. by C. Carapiperis, D. Kotsakis, and C. Macris), Athens p. 305, 1975.

59. Solar Activity and a Global Survey of the Precipitation. Trans. Acad. Athens 37, 1975.
60. Analytical Expression of the Mean Annual Variation of the Precipitation Within Various Latitude Zones of the Earth. Proc. Acad. Athens 51, 600, 1976 (in collaboration with B. Tritakis).
61. Analytical Expression of the Mean Annual Variation of the Precipitation Within Various Latitude Zones of the Earth. J. Interdiscipl. Cycle Res. 8, 226, 1977 (in collaboration with B. Tritakis).
62. Preliminary Results on the Discovery of Two Possible Galaxies in Cepheus. Proc. Acad. Athens 52, 230, 1977 (in collaboration with C. Poulakos).
63. A Forecast of Solar Activity for the 21st Solar Cycle. Solar Phys. 56, 467, 1978 (in collaboration with C. Poulakos).
64. A Microdensitometer Tracing of a New Elliptical Galaxy in Cepheus. Proc. Acad. Athens 52, 620, 1977 (in collaboration with C. Poulakos).
65. A New Index of Solar Activity. Proc. Acad. Athens 53, 286, 1978 (in collaboration with C. Poulakos).
66. Possible Sun-Weather Correlation. Nature 275, 775, 1978.
67. Possible Sun-Weather Correlation II. Nature 280, 254, 1979.
68. Prediction of the Radio Emission Indices of the Sun in the Frequency Range 1000 MHz ⩽ f ⩽ 3750 MHz , in Solar Terrestrial Predictions Proceedings Vol. 3: Solar Activity Predictions (ed. by R.F. Donnelly), U.S. Dept. of Commerce, NOAA, Environmental Research Laboratories, p. 75, 1980 (in collaboration with C. Poulakos).
69. The Mean Annual Variation of the Precipitation and the Pressure in the Zone 0°-10° of the Northern Hemisphere , in Low Latitude Aeronomical Processes COSPAR Symp. Ser. Vol. 8 (ed. by A.P. Mitra) Pergamon Press, Oxford, N. York p. 245, 1980 (in collaboration with C. Poulakos and B. Tritakis).
70. On the Variation of the Annual Mean Sea-Level Pressure in Latitude Zones of the Northern Hemisphere, in Solar-Terrestrial Predictions Proceedings Vol. 4: Prediction of Terrestrial Effects of Solar Activity (ed. by R.F. Donnelly), Space Environment Laboratory, Boulder, Colorado, F-63, 1980.
71. Influence of Solar Proton Event on Upper Stratospheric Temperatures. Proc. Acad. Athens 55, 362, 1980 (in collaboration with C. Zerefos, S. Sehra, C. Repapis, C. Poulakos).
72. Cosmic-Ray Intensity Related to Solar and Terrestrial Activity Indices in Solar Cycle No. 20. Astrophys. Space Sci., 74, 303, 1981. (in collaboration with H. Mavromichalaki and B. Petropoulos).
73. Effects du champ magnetique solaire sur l'intensite semi annuelle de la raie verte coronale. Soleil et climat, Toulouse, 30 Septembre-3 Octobre 1980. CNES, CNRS, DGRST, p. 101, 1980 (in collaboration with B. Petropoulos and H. Mavromichalaki).
74. The Evolution and the Secondary Maximum of the Corona Line Intensity. Solar Phys. (in press) (in collaboration with B. Petropoulos, H. Mavromichalaki).
75. A Study of the Inferred I.J.F. Polarity Periodicities. J.I.C.R. (in press) (in collaboration with B. Tritakis and C. Zerefos).
76. Rocket Sonde-Derived Stratospheric Temperature Changes Associated with a Major Solar Cosmic Ray Event. Astrophys. Space Sci. 74, 475,

1981 (in collaboration with C.A. Zerefos, P. Sehra, C. Repapis and C. Poulakos).

77. Possible Periodicities of the Annually Released Global Seismic Energy (M ⩾ 7.9) during the Period 1898 – 1971. Tectonophysics (in press).

78. Possible Periodicities of the Annually Released Planetary Seismic Energy (M ⩾ 7.8) during the Period 1898-1977. Proc. Acad. Athens (in press).

TABLE OF CONTENTS

PART I
HISTORY AND PHILOSOPHY

EPISTEME AND GNOSIS: A TENSION IN EUROPEAN THOUGHT

Alan H. Batten,
Dominion Astrophysical Observatory,
Herzberg Institute of Astrophysics,
Victoria, B.C., Canada

It can be plausibly argued that science is a Greek invention, so I trust that it will not seem inappropriate to honour a modern Greek scientist by discussing the nature of science and the role played by science in European (or Western) culture, and thus to pay tribute to a member of the modern Academy of Athens by considering some of the questions that would have interested the members of the original Academy in that city.

My own first visit to Greece and my meeting with Academician Xanthakis were connected with the commemoration of Aristarchos of Samos. We honour Aristarchos now for his suggestion that the Sun and not the Earth is the centre of the planetary system, but some at least of his contemporaries thought that he should have been charged with impiety for making this very suggestion. If he had been so charged, he would have been in good company: only 150 years or so earlier, Anaxagoras had been charged by the Athenians with the same offence for teaching that the Sun is a fiery stone "bigger than the Peloponnese". The early atomists were widely regarded as atheists. Even Aristotle found it expedient to return to his own town "lest the Athenians sin a second time against philosophy", while the victim of their first sin, Socrates, still serves as the prime example of a man persecuted - ostensibly, at least, on religious grounds - for doing nothing worse than thinking for himself and asking questions.

The mature Socrates was certainly not a scientist in anything like the modern sense of the term. At his trial, according to Plato, Socrates explicitly disavowed an interest in the natural world and dissociated himself from the teachings of Anaxagoras. Even Socrates may have stretched a point when his life was at stake, however. In the Phaedo, Plato has Socrates admitting to a youthful interest in physics and astronomy. Aristophanes caricatured Socrates as an absent-minded professor of science - a caricature that must have contained some truth if it was to have any point at all. Arguing along these lines, Taylor (1933) has plausibly argued that the young Socrates was indeed what we should now call

3

E. G. Mariolopoulos et al. (eds.), Compendium in Astronomy, 3–9.
Copyright © 1982 by D. Reidel Publishing Company.

a scientist, and that the famous Delphic oracle characterizing him as
the wisest man in the world was based on the reputation he acquired in
that way. The Socrates we know, whose prime interests were in understanding the nature of abstractions like "justice", or in finding the limits
to his own knowledge and that of others, was, according to Taylor,
Socrates in his later years. This reconstruction of Socrates' life is,
of course, conjectural, but even if it is entirely wrong, the spirit of
Socrates' relentless questioning, regardless of the cost to himself or
others, is exactly that of modern science, and we scientists of today
can rightly claim him as a spiritual ancestor whether or not, as a young
man, he engaged in pursuits essentially similar to ours.

We may draw several parallels between the lives of Socrates and
Galileo. Both men were fiercely independent thinkers with a ready wit
that they did not hesitate to use to make their opponents appear ridicu-
lous. Their ultimate fates might indeed have been different if each of
them could have resisted the temptation to use his wit in that way. We
who laugh, comfortably separated from them by the centuries or millennia,
at the sallies each could thrust at his opponents, would be the first to
squirm and smart if we met either of them in the flesh and were rash
enough to argue. Neither Socrates nor Galileo seemed to care that by
behaving in this way they were bound to make enemies. Thus, each was
brought to trial in old age, neither of them for a real offence, but
rather because each had made the establishment of his day look ridicu-
lous. The formal indictment of each included what amounted to a charge
of heresy.

Much has been written about the trial of Galileo, and I do not expect in this short essay to add anything new. In bygone days he was seen as
a valiant hero, a scientific David fighting an ecclesiastical Goliath,
and all our sympathies were with David. We live in an age that does not
like heroes, and now the fashion is to emphasize Galileo's faults. They
were many and obvious, and certainly contributed to his downfall, but I
doubt if, on their account, we can dismiss the whole episode as a conflict of personalities. There was a real issue between the authoritarian
Church and this pioneer of the modern scientific way of thinking. If
there had been no Galileo, someone else would have had to fight the battle. A "new trial" for Galileo may be very welcome if it sets the record
straight, but unless it is coupled with a recognition of this fact, it
will largely miss the point. The battle was not <u>about</u> whether the Earth
goes round the Sun or <u>vice versa</u> - that was only the field on which it
was fought. The battle was about something much more important: the nature of knowledge and how man acquires it.

Like most people, I have a certain affection for my native language,
which happens to be English, but I have to acknowledge that it has one
severe defect - a paucity of words for the different kinds of knowledge.
Virtually every other European language, for example, has different words
for "knowledge of facts" and "knowledge of people". I want to make a
different distinction, in the present context, and having no English
words with which to make it, I shall fall back on two Greek ones. I apol-

ogize to my Greek readers if I am misusing their language, but I believe
that my usage will be consistent with the way in which words derived from
these two Greek roots are used in other languages. I shall equate the
Greek word "episteme" with rational knowledge, particularly scientific
knowledge, based on critical analysis of and logical deduction from experience, while I shall use "gnosis" to mean knowledge believed to be
obtained by some intuitive faculty or special revelation.

I do not think anyone will quarrel with the identification of
"episteme" with the Latin "scientia" and modern European words derived
from it. I now suggest that all religion is essentially based on claims
of "gnosis". Some care is needed here: the word "gnostic" already exists
as a technical term for various heretical Christian or syncretistic sects
now dead and largely forgotten. Enough people remember them, however,
for there to be some risk of misunderstanding. The meaning of even this
technical sense of the word, however, is not dissimilar to the meaning
I wish to give it. Gnosticism was precisely a claim, on the part of its
adherents, to a special source of revealed knowledge, and there is a
residuum of this claim even in the most orthodox forms of Christianity.
This was clearly recognized by T.H. Huxley (1889) in the essay in which
he coined the word "agnostic" to describe his attitude to religious
questions. He did not simply mean that he did not know whether or not
God exists: he meant, as he said quite explicitly, to deny the possibility of knowledge about religious matters at all. More recently, Roszak
(1976) has opposed "gnosis" and "science" but the distinction he draws
between them is rather different from that proposed here.

Perhaps "episteme" and "gnosis" represent not so much two different
kinds of knowledge as two different routes to the same knowledge. Certainly mankind has sought knowledge in at least these two different ways from
time immemorial. The Greece that produced the ancient philosophers also
produced the Eleusinian mysteries. The two roads are very different, and
there is little mutual understanding between the wayfarers on each. The
road to episteme is by critical and original thought, by careful observation, and by the rejection of all claims to authority. The seeker after
episteme sees his attitude as one of humility because he will reject a
cherished theory if the ultimate test of observation shows it to be wrong
(or, at least, he likes to think he will). His refusal to accept authoritarian statements, however, makes him seem arrogant to those who do not
share his standards. Submission to the authority of a teacher or guru,
on the other hand, is the hallmark of the seeker of "gnosis". Some
Christians still speak of "making one's submission to the Church"; the
Buddha taught his disciples to aim for the complete cessation of all
desire; the very name of Islam means "surrender". Some religious persons
will indeed surrender their own critical judgment in favour of a church
or teacher, even when confronted with apparently paradoxical inspired
revelations. Such a person again sees his attitude as one of humility,
but so often the very strength of his own convictions makes him seem
frighteningly arrogant to those who do not share them.

The seeker after "gnosis" may see some areas of inquiry as tabu.

There are some matters that mortals ought not to seek know. This attitude is epitomised in that very old story that represents the eating of
the fruit of the tree of knowledge of good and evil as an act of defi-
ance, in punishment for which the first humans were evicted from Paradise.
By contrast, nothing is sacred to the seekers after "episteme". Xenophanes
being scathing in his criticisms of the Olympian gods in the sixth century B.C. (see e.g. Kirk and Raven, 1957) is one in spirit with Galileo
daring to question the immobility of the Earth in the seventeeth century A.D., or with Darwin another two hundred years later asserting man's
kinship with the apes, or even with Bertrand Russell (1943) writing his
grossly unfair but gloriously witty "Outline of Intellectual Rubbish".
The spirit of all of these was summed up in Tennyson's (1842) poem
"Ulysses" (he should, of course, have called it "Odysseus") celebrating
the hero's determination on his last voyage.
          To sail beyond the sunset, and the baths
          Of all the western stars, until I die
and ending with the line
          To strive, to seek, to find, and not to yield.
This line became famous as the epitaph of Captain Scott and his companions on his ill-fated journey to the South Pole, but it can equally apply
to mental exploration, and many research institutes would happily accept
it as an appropriate motto.

Amongst the many interwoven themes to be found in the history of
Galileo's trial, this conflict between the ways of "gnosis" and "episteme"
is one of the most basic. In one sense it does not matter very much
whether the Earth goes round the Sun or not. It was, however, a part of
the faith of a whole culture that the Earth is immobile. Not only did
Galileo challenge this item of faith, but he defied the right of the
"gnostic" authorities to rule on the matter at all. His sin was precisely to ask questions that ought not to be asked - and that at a time when
the whole fabric of the culture was being ripped apart anyway. Clearly
he had to be silenced: it was indeed expedient that one man should recant
for the people. It is this underlying conflict between those who insist
on the right to question everything and those who wish to hedge around
certain areas as closed to inquiry that must be resolved if particular
conflicts between science and religion (trivial in themselves) are to be
seen in their true perspective. A "new trial" for Galileo will not guarantee the freedom of those who honestly believe it to teach Darwinian
evolution, or help the plight of scientists with the courage to speak
out against totalitarian regimes under which they live.

Naturally, the actual course of any particular conflict will depend
on the personalities involved: Galileo recanted but Socrates did not.
The Athenians at least allowed Socrates to administer the instrument of
his own death with dignity. European civilization advanced in the next
two thousand years and devised means of depriving its victims of all human dignity before executing them. Galileo, at least formally, was threat-
ened with torture. I doubt, however, if it was that threat that fright-
ened him into submission. He believed, as Socrates did not, that his
accusers had power over his soul after death. To die excommunicate was

to take a risk that Socrates could not envisage and most of us moderns have to make a special effort to understand. Even Galileo's judges conceded that in all matters not related to the heliocentric theory he "answered like a good Catholic". In the context of his time that can only mean that he accepted the claims of the Roman Church to be the final arbiter of truth and to possess the keys of Heaven and Hell. His special agony was to learn by the road of "episteme" an exciting new scientific "truth" and yet be constrained to deny it by the demands of the "gnosis" in which he had been brought up.

To claim that either side was right or wrong is to oversimplify the issue. No scientist, or any other person, is a perfect seeker after "episteme". We do not always reject our pet theories when observations contradict them; we do not always present the evidence with complete fairness. Scientists are human too and sometimes succumb to that very human trait of arguing from authority. Roszak (1969) has had his quite legitimate fun talking about the priesthood of science. There are ecclesiastics who argue much more carefully and temperately than many scientists do. Cardinal Bellarmine, by suspending judgment, appeared a paragon of moderation compared to Galileo. The difference is in the ideals we set ourselves. The scientist who argues from authority or ignores a relevant fact knows, in his better moments, that he has been untrue to his profession. Religious people, on the other hand, will often persist in faith despite apparently overwhelming contradictory evidence, and see this as a positive virtue. These two attitudes are bound to conflict. It is not that one is right and the other wrong, or even that one is better. Perhaps both are necessary parts of human experience, but a tension must always exist between them. It may, like many other tensions be a sign of life, but it is always there - usually latent, sometimes, when all the circumstances conspire together, erupting into the tragedy of a Socrates or Galileo.

This perpetual latent conflict is symbolised in that powerful story of Renaissance Europe, the legend of Faust. A reaction of ordinary people to all new learning, the story had attained enormous popularity by the time of the scientific awakening of the sixteenth and early seventeenth centuries. Faust obtained great powers by selling his soul to the Devil. The parallel with modern science and the technological powers it gives us is obvious. The first attempt to make a serious drama out of the folk tale was made by the sixteenth-century English dramatist Christopher Marlowe. In one scene of Marlowe's play, Faust demonstrates his supernatural (and diabolical) powers by serving fresh fruit out of season. We take for granted our own ability to do that every day, and I doubt if even the most devoted adherent to the counter culture would regard the deep-freeze as an invention of the Devil. To the medieval mind, however, it would have been so incomprehensible as, very likely, to appear to be just that. Faust is a type of the inquiring rational mind, of the seeker after "episteme", but the human race is half afraid of its own capacity for knowledge and so the powers he obtained are popularly explained as the result of a contract with the Devil. This fear of knowledge, this belief in the diabolical origin of the power it brings, surfaces every time a Socrates, a Galileo, or even a John Scopes is brought to trial.

It was over two centuries after Marlowe that a greater poet, Goethe, raised the Faust story to new heights of expression. Even he waited till the end of his life before publishing the second part of the drama with its final twist. Faust was not damned but saved, and the only reason given for saving him was that (like Odysseus) he strove with all his might. Faust never gave up his relentless quest for knowledge and self-knowledge and this, Goethe seems to be saying, was more important even than a deliberate compact with the Devil. Goethe was no friend of the science of his day: he abhorred the mechanistic direction it was taking as a result of Newton's influence and it is well known that he considered Newton's theory of light and colour to be totally misguided. Of all people, he might have been tempted to join the counter-culture of his day. He did not join it, however, he did not "opt out", but used his great talents in government service and seems, at the end of Faust, to be saying that scientists must go along their chosen route to the perhaps bitter end.

One wonders how Goethe would react today. Again our primeval fear of our own knowledge and the powers it gives us has surfaced. We are afraid of the power we have let loose from within the atom, which we cannot fully control. We are afraid of the power that may shortly be ours to alter the inheritance of those yet unborn. The fundamental question is being asked again: are there some things too dangerous for us to know, that we should not even seek to know? The instinctive answer of scientists to this question is "no", but in recent years even scientists have begun to doubt their instincts. Thus, in many countries recently we had a moratorium on experiments in genetic recombination while their possible dangers were assessed. This was a wise and cautious move, but it is no suprise that it has been determined that the dangers were exaggerated or that experiments have begun. An indefinite moratorium on any kind of scientific investigation is impracticable. Such a moratorium could not be universally enforced; even if it could, we cannot collectively forget what we have learned. We have eaten the fruit of the tree of knowledge, we have been expelled from Paradise, and the route by which we left is barred against us. Our only way back is like Faust to strive with all our might, or like Tennyson's Odysseus

> To follow knowledge like a sinking star,
> Beyond the utmost bound of human thought.

So we today feel the same tensions that Socrates, Anaxagoras and many others felt in their relationships with their contemporaries and compatriots. As perhaps Galileo did, individuals often feel the tension within themselves, for none of us are pure seekers after "episteme" nor yet content to be mere recipients of "gnosis". Perhaps the tension is an inevitable part of the search for knowledge through the use of reason and the senses, and is as much a part of the inheritance that Greece has bequeathed to the rest of the world as art and science, philosophy and democracy. If so, conflicts between science and religion are from time to time inevitable. This is not because some particular item of belief is contradicted by some particular discovery: history has shown that it

is easy enough to reinterpret a text whose literal meaning is demon-
strably absurd. It is rather because the attitudes of mind of so many of
the adherents of the two systems are very different and bound to clash.
There are exceptions: Eddington (1929) writing on this same subject said
"You will understand the true spirit neither of science nor religion
unless seeking is placed in the forefront". Men of his breadth of vision,
however, are unfortunately rare; too often religion is presented not as
a search, but as a package of beliefs to be swallowed whole. Therein lie
the seeds of all the conflicts from the banishment of Anaxagoras to the
present-day fight of American fundamentalists against the teaching of
evolution in secondary schools, and to unknown and unnumbered individuals
and causes yet to come.

REFERENCES

Eddington, A.S.: 1929, Science and the Unseen World (Swarthmore Lecture),
     George Allen and Unwin, London, p. 54.
Goethe, J.W.: 1842, Faust, Part 2, Act 5, Scene 7.
Huxley, T.H.: 1889, Agnosticism reprinted in Selections from the Essays
     (ed. by A. Castell) AHM Publishing Corp., Northbrook, Illinois, p.69.
Kirk, G.S., and Raven, J.E.: 1957, The Presocratic Philosophers, Cambridge
     Univ. Press, Cambridge, p. 163.
Marlowe, C.: 1604, The Tragical History of Dr. Faustus, Act 4, Scene 7.
Plato c380 B.C., The Apology of Socrates and Phaedo transl. by H.
     Tredennick in The Last Days of Socrates, Penguin Books, Harmondsworth,
     pp. 57 and 153.
Roszak, T.: 1969, The Making of a Counter Culture, Doubleday, Anchor
     Books, New York, p. 263.
Roszak, T.: 1976, in Science and Its Public: The Changing Relationship
     (ed. by G. Holton and W.A. Blanpied), D. Reidel, Dordrecht-Holland,
     p. 17.
Russell, B.: 1943, An Outline of Intellectual Rubbish reprinted in The
     Basic Writings of Bertrand Russell 1961 (ed. by R.E. Enger and L.E.
     Denonn), Simon and Schuster, New York, p. 73.
Taylor, A.E.: 1933, Socrates (the Man and his Thought), reprinted 1953
     Doubleday, Anchor Books, New York.
Tennyson, A.: 1842, Ulysses.

A NEW, RATIONAL ENDEAVOUR FOR UNDERSTANDING THE ERATOSTHENES NUMERICAL
RESULT OF THE EARTH MERIDIAN MEASUREMENT

Massimo Cimino
Institute of Astronomy, University of Rome
Rome, Italy

ABSTRACT

    Hultsch's discovery of the 157,5 m value for the stadium Eratosthenes
must have used, instead of the traditional olympic stadium of 185 m, in
his famous determination of the length of the earth's meridian, seems to
have convinced the most modern historians as to the particular excellence
of the result, contained within an error of but 1,5% on the real value.
However, setting the problem on the mere basis of ascertained historical
and geographical data and on the application of a more correct logic as
regards modern data, leads one back again to the more traditional conclusions. Yet at the same time, it also allows room for granting recognition to the remarkable critical capacities of the Alexandria scholar
in the utilization of observation data.

    Lastly, a spontaneous hypothesis is advanced for the consideration
of Hultsch's stadium as an independent entity from the problem under consideration, and which could have been used by Eratosthenes in a given
preliminary elaboration of his work.

BASES FOR A RATIONAL APPROACH TO THE PROBLEM

1. The Eratosthenes Measurement and Modern Criticism

    The length of the terrestrial meridian obtained by Eratosthenes of
Cyrene (276-195 BC) was highly renowned, and still is today the world
over, both because of the simplicity of its concepts and its especial
elegance of method. Yet at the same time, due to the limited amount of
the data having reached us directly and the contradictions found in the
indirect ones, any judgment on the exactness of the results of the measurement cannot but continue to be controversial.

    And, in truth, the two essential historical data – viz. the differ-
ence of geographical latitude between Alexandria and Syene (now Aswan)
of 7° 12', equal to 1/50 of the entire circumference, observed by Erato-

11

E. G. Mariolopoulos et al. (eds.), Compendium in Astronomy, 11–21.
Copyright © 1982 by D. Reidel Publishing Company.

sthenes, and the distance between the two towns (erroneously supposed to be situated on the same meridian), which he assumed as 5.000 stadia – do not alone suffice to formulate a definite judgment [1]. This is true so long as the metric equivalent of the adopted stadium is not specified. It is therefore no wonder that research on this equivalent becomes ever more complicated, and that it extends to other analogous problems resulting in numerous, and often controversial proposals of equivalence between the ancient and the modern measuring standards.

For our present purpose, however, it is not necessary to enter into this dispute in depth, but rather to consider two such standards alone, i.e. the olympic one, known as Alexandrine or geographic, of 184,8 m largely used in antiquity and considered in the past to be the stadium used by Eratosthenes, and the stadium of 157,5 m discovered by F. Hultsch around the middle of the last century and now largely accepted by modern historians [2] . The two values differ considerably one from the other, but in a numerical discussion, this is an advantage. By using the olympic stadium, the length of the meridian arc between the two terrestrial parallels of Alexandria and Syene produces the result of 5.000 x 0,1848 = =924 kilometres, as against the 800 km of reality, with a difference of 124 km and an error (referred to the true value) of 15,5 %; whereas by using Hultsch's, we would obtain 5.000 x 0,1575 = 787,5 km, with a difference of less than 12,5 km only, and an error of 1,56 % [3] . Therefore, for the length of the whole meridian circle, we would obtain (by multiplying the values of the meridian arc by 50) respectively 46.200 km and 39.375 km, as against the 800 x 50 = 40.000 km of reality [4].

2. A more Rational Criterion for a Comparison with Modern Data

The preference given by modern historiography to the result obtained with the 157,5 m stadium, while confirming Eratosthenes as a scholar of manifold cultural interests and as a scientist of uncommon ability as an observer, also leaves in us, on another side, a certain sense of perplexity : how to understand his being able to reach such a precise result, within a 1,5 % range, when he certainly must have committed a far greater error, though it be due to a reason unbeknown to him. According to historical tradition, the value of 5.000 stadia refers to the distance between Alexandria and Syene, and not to a meridian arc (Fig.); the two cities differ about 3° in longitude, and contrary to what is generally thought, this difference can radically change the terms of the comparison with modern data and lead to different conclusions.

It is quite clear that in order to arrive at a more correct judgment, it is not sufficient to take into consideration the total error only, but bear in mind that two other distinct errors meet up together: the first due to having followed a wrong path in measuring the distance, the other, to having made possible technical errors in measuring it once the path was chosen.

It is to this last error alone that we should refer when formulating a rational judgment on Eratosthenes as an observer.

The simultaneous examination of the two errors can enlighten us on the choice of the stadia.

3. Error of Path and Error of Observation

In this and in the next paragraphs, will be set out the simple but rational criteria we shall follow in attempting to make a correct analysis of the Eratosthenes measurement.

We invite the reader to refer to the Figure and attached Table of Errors.

In the Figure are represented schematically some of the hypothetical paths we call the m-, g-, a-, z-, b- path (Table column(1)). We suppose we know their lengths exactly, expressed in kilometres or in each of the two stadia of 157,5 m or of 184,8 m (columns (2), (3), (4)). The differences $\nabla_i = g_m - g_i$ of each g_i with the g_m length of the arc of meridian m, and the quotients $\delta_i = \nabla_i / g_m$ are given in columns (5) and (6). These represent the absolute and relative errors we would make in choosing the generical path i instead of m, which we have called errors of path. Obviously, the error of path on the meridian itself (i≡m) is zero.

Let us now suppose that we wish to control the values g_i experimentally, and for measuring them use both the stadia, and that we always obtain the same value of 5.000 stadia for any path and with both stadia; it should be evident that, but for the two exceptions we shall examine in No. 9, the results of our controls will definitely be wrong. The differences $\Delta_i = g_i - 5.000$ and the quotients $\varepsilon_i = \Delta_i / g_i$ (where the g_i are expressed in stadia) are indicated in columns (7), (8), (9) and (10). These represent the absolute and the relative errors originating from all the technical operations conducted in measuring the paths [5] . Therefore they are errors of observation.

4. The Total Error

The δ_i errors are referred to the true value g_m of the arc of the meridian; the ε_i to the true value g_i of the paths. In order to calculate the total error σ referred to the true value g_m, it is necessary to first also refer the ε_i to the meridian by multiplying them by $h_i = g_i / g_m$ and then adding them to the δ_i. In this way, we obtain $\sigma = \delta_i + h_i \varepsilon_i$. It should be noted that σ does not depend on the i-paths, but only on the value of the stadium (columns (12), (13)). On the meridian, there is $\delta_m = 0$ and $h_m = 1$, and therefore $\sigma = \varepsilon_m$, which depends on g_m alone expressed in stadia and on the measured value 5.000.

The values of σ enable us to measure the true value λ of the whole terrestrial meridian. This is :

$$\lambda = 5.000 \times 50 \times u + \lambda \cdot \sigma$$

where the quantity $\lambda \cdot \sigma$ is the total absolute error; for the two stadia

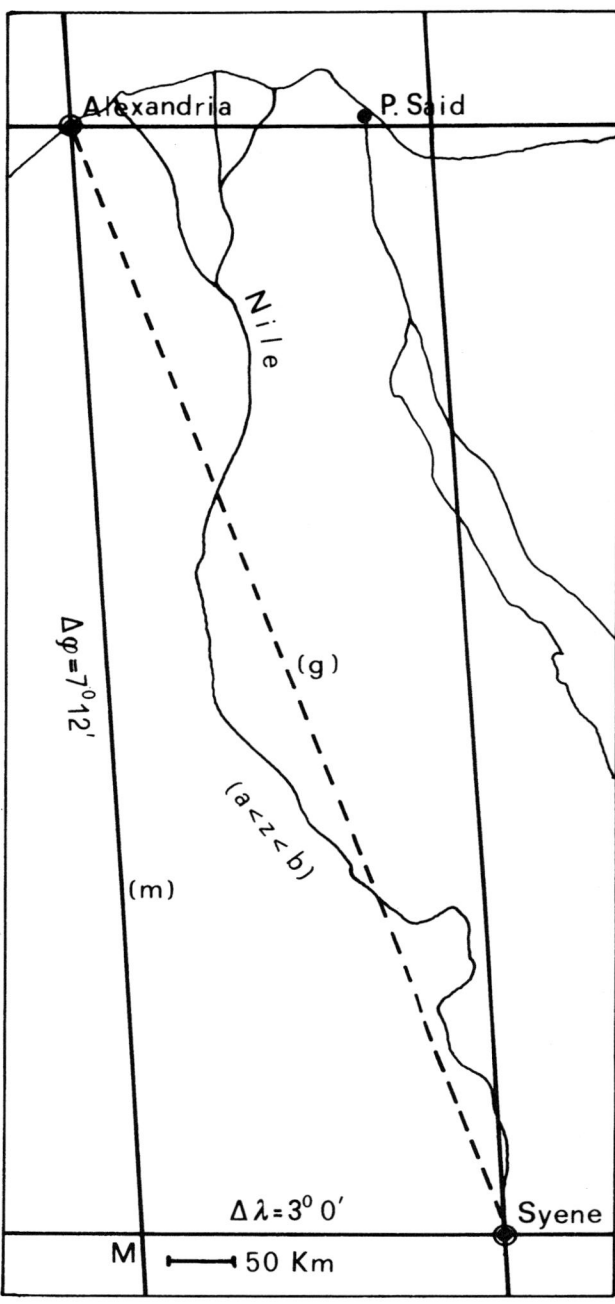

Fig. AM: meridian arc   AS: geodetic arc   Δλ: difference of longitude
Δφ: difference of latitude   (m),(g),(a),(z),(b) = paths

Table of Errors

Path i	Path length g_i			Path-error ∇_i, δ_i		Observation-error Δ_i, ε_i				h_i	Total error σ	
	km	Stadia 157,5	Stadia 184,8	km	%	Stadia 157,5	Stadia 184,8	157,5 %	184,8 %		157,5	184,8
(1)	(2)	(3)	(4)	(5)	(6)	(7)	(8)	(9)	(10)	(11)	(12)	(13)
m	800,0	5079	4329	—	—	+ 79	−671	+ 1,56	−15,50	1,000	+1,56	−15,50
g	844,0	5359	4567	− 44	− 5,50	+359	−433	+ 6,69	− 9,48	1,055	+1,56	−15,50
a (z)	900,0	5714	4870	−100	−12,50	+714	−130	+12,50	− 2,67	1,125	+1,56	−15,50
b	924,0	5867	5000	−124	−15,50	+867	0	+14,77	0,00	1,155	+1,56	−15,50
	980,0	6222	5303	−180	−22,50	+1222	+303	+19,64	+ 5,71	1,225	+1,56	−15,50

→ probable paths ←

Definitions:

$$\nabla_i = g_m - g_i$$
$$\delta_i = \frac{\nabla_i}{g_m}$$
$$g_m = g_i + g_m \cdot \delta_i$$

$$\Delta_i = g_i - 5.000$$
$$\varepsilon_i = \frac{\Delta_i}{g_i}$$
$$g_i = 5.000 + g_i \cdot \varepsilon_i$$

$$h_i = \frac{g_i}{g_m}$$

$$\sigma = \delta_i + h_i \cdot \varepsilon_i$$

u_1=0,1848 km and u_2=0,1575 km we obtain the numerical relations :

$$40.000 = 46.200 - 6.200$$
$$40.000 = 39.375 + 625$$

on which current historical criticism is based.

A RATIONAL COMPARISON OF THE NUMERICAL RESULT OF THE ERATOSTHENES MEASUREMENT WITH MODERN DATA

5. The "Weighted Mean" of Eratosthenes

Indeed, Eratosthenes never tracked a path on the ground, nor person- ally executed any measurement. If this be so, what sense can it have to speak of paths we do not know, and of relative errors of observation ? We have no wish to formulate hypotheses or work on our fantasy. Our aim is merely to think out anew, but logically and realistically, what ex- actly Eratosthenes had to do.

Tradition speaks rather clearly : Eratosthenes, judging Alexandria and Syene to be on the same meridian circle, gathered information on the distance between the two towns, and from this deduced the 5.000 stadia reported by the historians. It would appear obvious that he disposed of a certain number of data and, in modern times, he would have drawn from them a mean value in order to assume it as the most probable value; then, from the differences between any single datum and the mean value, he would have been able to draw information on the precision of the data itself. Should Eratosthenes have done all of this,he would have antici- pated by twenty centuries the fundamental principles on which the modern Theory of Errors of observation rests, viz. one of the fundamental pillars of the modern experimental method. But Eratosthenes did not. He elabor- ated what today we call a "weighted mean value", following his own par- ticular criteria (which, of course, we ignore), and without taking into account the differences of each single datum from this mean, viz. the errors of observation.

Nevertheless, we shall demonstrate that it is possible to take a few steps along the path for recognizing such errors; for, if it is true that we cannot reconstruct the system of errors, yet it is possible to determine with sufficient certitude an interval in which these errors must necessarily fall.

6. The Interval of the Lengths of the Paths and the Intervals of the Errors

This possibility depends upon the fact that we can determine an interval for the lengths of the paths between Alexandria and Syene which are of interest to our problem.

Indeed, such an interval has an inferior limit (minimum length) in the geodetic arc between the two towns [6] , of about 844 km , whereas for the superior limit, it would appear reasonable to choose the length of the caravan road which follows the river Nile close to its banks for roughly 980 km (and is therefore one of the less rectified) [7] .

We call this road the b-path, whereas the g-path is the geodetic one. The latter can reasonably be substituted by an a-path, which represents a sufficiently rectified caravan road (by cutting a medium bend North of Syene) of about 900 km, a little less than half way between the g- and b-paths.

If we then assume as the probable interval of the lengths that between the lengths of the a- and the b-paths, then two intervals of probable errors can be found in our Table : one for the error of path, the other for the error of observation. If we suppose the measurements to be made with the stadium of 184,8 m, then the interval of observation error is comprised between -2,7% and +5,7%; whereas for the stadium of 157,5 m, it is comprised between +12,5% and +19,6%. On the contrary, the interval of the path error is the same for both stadia (as is obvious, since the error of path does not depend on the measurements operation); such an interval is comprised between -12,5% and -22,5%; (columns (10), (9), (6) respectively).

7. A more Rational Meaning of Eratosthenes' Numerical Result

We are now able to formulate a more rational and meaningful judgment of Eratosthenes' measurement. Our conclusions are as follows :

(a) Eratosthenes elaborated the results of measurements made by others, of the distance between Alexandria and Syene; he obtained the value of 5.000 stadia, which represents (from a logical point of view) a "weighted mean value", according to his criteria which are unknown to us;

(b) the stadium to which he referred, in the last analysis, is the olympic one of 184,8 m;

(c) the length he found for the entire maximum circle was 250.000 stadia, which corresponds to 46.200 km, with an error (in its true value) of 6.200 km in excess, equal to 15,5%;

(d) only a minor part of this error, between about -3% and +6% (column (10)), is to be attributed to errors of observation by the operators, viz. a maximum of 2.400 km; the rest, about 3.800 km, must be attributed to an error of path, which in those times, it was impossible to identify;

(e) the Eratosthenes' scientific personality emerges as magnified by the seriousness of the criteria he followed in selecting and elaborating the experimental data.

In my opinion, this is the most rational and credible conclusion, which should find its place in a critical history of scientific progress, rather than the "summary" judgment which, by ignoring the internal structure of the total error, ends up actually completely reversing the result.

8. The Measurement with the 157,5 m Stadium

An analogous judgment on the measurement with the 157,5 m stadium, would in fact be as follows :

(a) and (b) are the same, except for the change of the stadium with the 157,5 m one;

(c) the length of the entire meridian circle of 250.000 stadia now corresponds to 39.375 km, with an exceptionally small error (for the period) of but 625 km in minus, equal to 1,5% of the true value;

(d) since the measured path was not the meridian arc, a number of path errors (which at the time were not identified) must necessarily have occurred, which (for the probable paths followed by the operators) are between -12,5% and -22,5% (column (6)). On such paths, the operators' technical ability was rather mediocre, with observation errors of between +12,5% and +19,6% (column (9));

(e) is it possible to accept that a measure - which posterity will recognize as extremely close to the true value - might have been obtained with technical operations such as those in measuring the distances ?

To what is such a spectacular compensation of errors of two quite distinct kinds [path and observation errors : columns (6), (9), (12)] due ?

9. A Mathematical Explanation; the z-Path

In our Table, we carry data concerning an ideal path, which we call the z-path. On such a path, we suppose that the error of observation of the length be zero. The error, therefore, will be of path alone, and equal to the total error (the same as for every other path). With the 184,8 m stadium, such an error is -15,5 %.

The z-path depends on three fundamental data alone : the 800,0 km of the meridian arc; the 5.000 stadia of the historical tradition; the 184,8 m of the metric equivalent of the olympic stadium. On the z-path, the Eratosthenes "weighted mean"is (ideally) a true value (and such it was for Eratosthenes himself).

The z-path is amongst those that can really be followed (longer than the g-path). On the paths around it (amongst the longest and the shortest ones), there always exists an exchange of the algebric sign of the observation errors, while only a partial compensation (on each path) may occur between the path-error and the observation-error, which is never complete (total error different from zero; columns (6), (10), (13)).

A z-type path can be defined for every other choice of stadium differing from the 184,8 m one. Should we look for a path relative to the 157,5 m stadium we would find a z-path of 787,5 km, of 12,5 km (1,5%) less than the meridian arc. But what is important is that the measuring of such a path cannot be considered as real (within the framework of our problem), since its length is less than the length of the geodetic arc between the two cities.

The approximate coincidence of the z- and the meridian arcs explains why a nearly complete compensation of the path and observation errors occurs on any other path (total error +1,56%).

The exceptional character of the high precision in the final measuring result depends on how much the z-path differs from the meridian arc, viz. on how close the value of the stadium is to an ideal value of 160,0 m [8] , and not on the ability of the technical operators. On the contrary, when measuring real paths, the operators must have inevitably committed considerable observation errors in order to compensate for path errors which they cannot know.

From this follows the paradoxical aspect of the hypothesis that the value of 5.000, fruit of real measurements of paths, might be referred to the 157,5 m stadium.

10. An Hypothesis on a Probable Utilization of the 157,5 m Stadium

Are we then to definitively refute the hypothesis that Eratosthenes may have helped himself with the 157,5 m stadium in elaborating his measurement ? Certainly not. The question is merely one of understanding in which phase of his work such a stadium could have been used.

The answer is very simple and to us appears sufficiently logical and credible. Actually, it would seem highly reasonable to posit that while gathering information on the distances between various places along the Nile, Eratosthenes may have found it more practical to reduce them preliminarily to one single length standard, given that along the river course the measurements may have been made with various local standards. For such a preliminary operation, he might have chosen the 157,5 m stadium, as well as any other.

If what we have demonstrated in No. 6 concerning the entity of the errors of observation contained within 5-6% on the probable paths [9] be accepted, Eratosthenes would have obtained his "weighted mean" expressed in stadia of 157,5 m. He would have found a mean value to the order of 5.870, which he would have subsequently reduced (probably with some rounding-off) to the 5.000 olympic stadia (see Table, path z). In this way, we again find the result of No. 5.

On the basis of this hypothesis, the Hultsch stadium would no longer

be in cause as regards its dependence or not on the problem under consideration. At the same time, its utilization would not lead to the aston-
ishing conclusions of No. 8.

As to the reduction to the olympic stadium this would appear almost
obvious : the geographic interests of the scholar-librarian of Alexandria
are too well known not to imagine that he was obliged to reduce the most
disparate measures reaching him from all over the world to one single
universal standard. Therefore he would also have reduced the measure
along the Nile to this stadium; a fundamental experimental basis for his
great geographic work.

COMMENTS

[1] These are the values of Eratosthenes' measurements which Cleomedes
(II century BC) carries in his Theory of Circular Motions of Celestial
Bodies, in which he also gives the value of 250.000 stadia for the whole
meridian circle. The values carried by Cleomedes are those which were
generally accepted by the whole of antiquity, and particularly by Hippar-
chus (c. 150 BC), in his Geography. Only later does Strabo (I cent. BC-
I cent. AD) carry a measuring by Poseidonius (I century BC) from Rhodes
to Alexandria, which allocates 180.000 stadia to the meridian circle;
this being the value picked up by Marinos of Tyre (II century AD) in his
Geography, and by Ptolemy (II century AD) in his Almagesto. Cf. F. Enri-
ques-G. de Santillana, Storia del pensiero scientifico, Zanichelli, Bologna 1932, pp. 279-281; P. Tannery, Recherches sur l'Histoire de l'astronomie ancienne, Gauthier-Villars, Paris, 1893. For a more extensive bibliography and more up-to-date information, cf. the recent book by G.
Dragoni, Eratostene e l'apogeo della scienza greca, CLUEB, Bologna, 1979,
pp. 173-74 and p. 224 (Nota 33).

[2] F. Hultsch was able to set the value of 157,5 m for the stadium used
by Eratosthenes, following the fortunate discovery of a statement by Pli-
nius setting an equivalence between the "schoenus" (an Egyptian standard
equal to 40 Eratosthenes stadia), and the Roman mile. According to a number of historians, following this discovery all doubts as to the value of
Eratosthenes' measurement could henceforth represent nothing but a useless
repetition of old books. Cf. G. Dragoni, loc.cit., p. 145 (Nota 10),
and p. 184.

[3] In historical papers, three significant digits are more than sufficient for setting out facts; in the Table we are publishing four to allow
for the safer internal checking of numerical calculus.

[4] We are assuming 800,0 km as the true value of the meridian arc of
the Earth spheroid, for the mean value 27° 30' of the latitudes between
the two towns. When multiplying by 50, we obtain the true length of the
meridian circle of Earth when supposed spherical.

[5] When speaking of technical operations, we mean any operation suscep-

tible of leading to a conclusive datum. Therefore, in addition to technical errors true and proper, one must also take into account errors of calculus, evaluation, numerical approximation, rounding-off, etc., which we will therefore include in the more general denomination of error of observation.

[6] The length of the geodetic arc between the two cities on the earth's sphere was determined by trigonometric calculus, both latitudes and the difference in longitude being known; but the usual nautical tables are also sufficient for such determinations.

The effective tracing on the ground of a geodetic arc and its measurement is not of elementary simplicity, since it is based on the principles of the modern differential geometry of curves on surfaces, and requires continuous controls on horizontal directions (azimuth) and longitude differences. In Eratosthenes' times, these operations were extremely difficult, especially due to the lack of portable precision clocks to keep the time.

[7] The length of the Nile banks between Alexandria and Syene and that of a number of caravan routes which follow them have been taken from a good geographic map (Mollweide's equivalent projection, origin 20° East of Greenwich, 1cm = 50 km scale, sufficient for the purpose; reading error 0,5 mm, equivalent to 2,5 km). The required corrections have been made (to the order of -7,5%) to the relevant lengths.

It can reasonably be accepted that the caravan routes cut through the river's small bends, i.e. those that can be ascertained by direct observation, in a region to the order of 10-15 km. The large bends with a basis to the order of 300 km could not, on the other hand, easily be corrected, since for this purpose the continuous checking of the azimuth is required. Proof that these large bends were not observed can be found in the very historical tradition which sets the course of the Nile as roughly parallel to a meridian.

[8] Precisely, the value of a hypothetical stadium, the z-path of which be the meridian arc of 800,0 km, would be given by 800.000/5.000=160,0 m/stadium.

[9] Errors of this order in the ground or road length measurements in ancient Egypt would appear to be easily acceptable within ranges of between 10-15 km (topographic field), with the use of primitive ancient instruments, but used by persons familiar with their trade. We can consider as such the land office officials and the "itinerum mensores" whose job it was to measure the road lengths. Thus the error of the sum of this measurement to obtain the distances over longer roads must have been of the same order. This is true only when considering the measurement of the road lengths but not their rectification by vast scale cutting [geographic field; cf. Note (7)] the bends.

ASTRONOMICAL ACTIVITY IN THE PERIOD 1600 TO 1880

G. Teleki
Astronomical Observatory
Belgrade, Yugoslavia

ABSTRACT

Proceeding from the data presented in the Houzeau-Lancaster Bibliography (1882) the growth of the total sum of information in Astronomy from 1600 to 1880 is analysed and estimated that the time interval, within which the number of data is doubled, were about 25 years. It is inferred, taking thereby into account Teleki's (1980) analysis of the number of catalogues of star positions within the same interval, that the growth of the total sum of the information in Astronomy as a whole has been more rapid - by about a quarter - than that relating to the catalogues alone. The ratio of the sum of papers and the total number of catalogues within this period is doubled in 69 years span.

1. In Houzeau-Lancaster Bibliography (1882) a statistical survey is given of the number of paper dealing with Astronomy published in the interval from 1600 to 1880. In a previous papers of ours (Teleki, 1980) we analysed the growth of the number of catalogues of star positions in the past, the period 1600 to 1880 inclusive.

It may, in our view, be useful to make comparison of the two groups of information by the methods of the Science of Science. This, first of all, because the determination of star positions has been one of the leading preoccupations of astronomers throughout the past centuries and it might, therefore, be of interest to "detach" this activity from that in Astronomy in general. Secondly, the period 1600 to 1880 witnessed an all-out blossom of sciences, including Astronomy, so it is useful to compare the scientific expansion of Astronomy as a whole with that of one of its parts.

2. The analysis of the number of the papers published in the interval 1600 to 1880 is accomplished in the same way as has been proceeded in our earlier paper (Teleki, 1980).

In Fig. 1 the number of the papers is represented according to ten-

23

E. G. Mariolopoulos et al. (eds.), Compendium in Astronomy, 23–27.
Copyright © 1982 by D. Reidel Publishing Company.

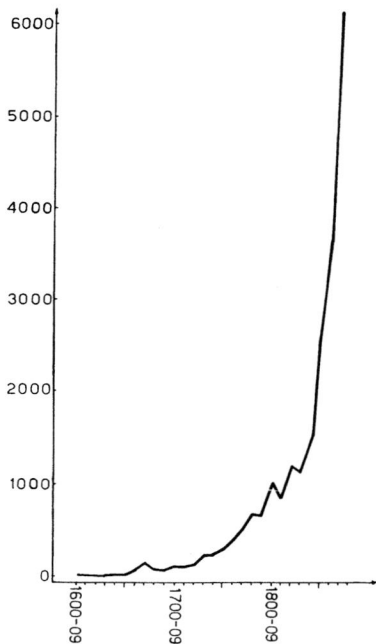

Fig. 1 : The number of the papers published according to decades in the
period 1600-1880 (on the basis of the Houzeau-Lancaster Bibliography (1882)).

years intervals. The exponential growth is clearly apparent. The variations in the curve in the interval 1790 to 1839 are, no doubt, due to
the political upheavals prevailing at that time in Europe.

        In Table 1 times T are given within which the number of the papers
i.e. the sum of papers, respectively, are doubled. The number of the
papers within decades are separately analysed as was the sum of all the
papers published within a given decade and in the period preceding it.

Table 1
The time T of doubling (in years) of the number and the sum of the published papers for the variants I, II and III.

	I	II	III
Number of papers	20	27	27
Sum of papers	20	22	32

        The computations were performed according to three variants. The
values T of the I variant denote those obtained from the mean values of
the ratios of data within adjacent decades. The values T obtained from

the data of the first and the last decades alone are denoted by II. Finally, the notation III is related to the period 1710 to 1879, which was analysed in the similar manner as I.

As seen in Table 1, both the number and the sum total of the astronomical papers is doubled in the period 1600 to 1880 in about 20 to 32 years - approximately in 25 years.

3. It has been stated in our previous paper (1980) that the number of catalogues, in the period 1710 to 1909, has been doubled every 34 years on the average. If the period 1710 to 1879 is taken separately, then the time T of doubling is about 35 years - the same tendency has therefore been preserved as in the previous case. On comparing this information with the one from Table 1 (variant III, 27 years), one sees that the general growth of astronomical information is more rapid - by about one quarter - than in the case of the catalogues.

If the number of catalogues within a given decade is taken apart from the mass of information on papers, we get practically the same results as presented in Table 1.

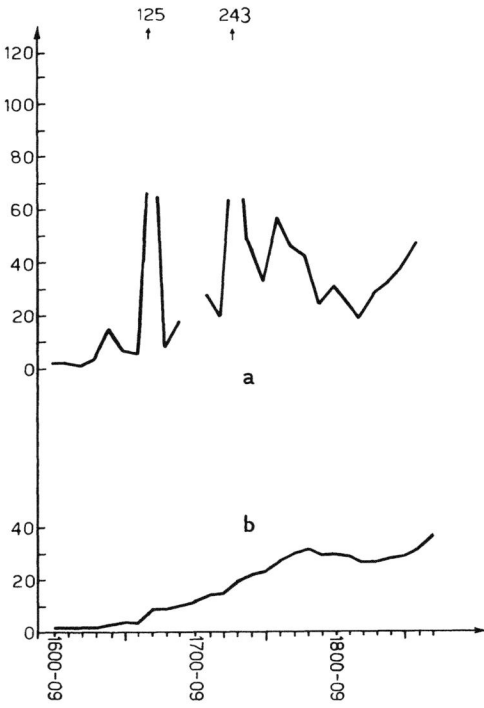

Fig. 2 : The graph a represents the ratio of the number of papers and the number of catalogues published according to decades. The graph b shows the ratio of the sum total of all papers and the sum total of all the catalogues published since 1600, according to decades.

4. Two graphs are plotted in Fig. 2. The upper graph, denoted by
a, illustrates the ratio between the number of papers and that of catalogues published by decades, while the lower one (denoted by b) the ratio
between the sum total of all papers and sum total of all the catalogues
since 1600, according to ten-years intervals.

The graph a displays a discontinuity (1700 to 1709) and two extreme
jumps (1670 to 1679 and 1730 to 1739). This is a result of the fact that
the number of the catalogues published during these decades has been low
or nil (that is : 0, 1 and 1). The general trend testifies of the fact
that since the thirties of the 17th century the astronomical development
in general has been more rapid regarding catalogues, and that trend of
"detachment" being preserved until about 1760 to 1769. Thereafter a
relative decline sets in to finally give place to a new rise from 1820-
1829 on.

A faster growth of the astronomical than that of the catalogues information is evident from the graph b, related to the summary data on
both subjects. Following a relative drop at the end of the 18th and the
beginning of the 19th centuries, there occurs a steep growth as a result,
in the first place, of the acquiring of new knowledge of the Sun (fifth
decade), advent of the photography (1851) and spectral analysis (1859-
1862). We feel justified to make extrapolation and to state that the
growth has been continued even after 1880.

By analysing the growth of the curve b we realize that in the period
1600-1880, speaking generally, the ratio between the sum total of the
papers and the sum total of catalogues is doubled within a period T of
about 69 years.

5. These inquiries pertain solely to the growth of the information
volume but not to the knowledge growth as such, as the pertinent elements
are lacking. Nevertheless, the following estimate will be attempted. Let
it be assumed that the astronomical knowledge during the period 1600-1880
has been growing with the same intensity as the total mass of information,
that is, with the period of 25 years. We revert in this connection to
Fig. 2 of our previous paper (Teleki, 1980) in which it can be seen that
the catalogue accuracy growth since 1601 (Brahe's catalogue) to 1890
(Cape catalogue) in time has been doubling in 18 years span. By equating
the accuracy growth with the knowledge growth - although this is not quite
correct - we arrive at the conclusion that the growth of astronomical
knowledge and that of the star positions has been progressing at about
the same rate within the period 1600-1880. This, of course, is but a
rough estimate.

One is led to the conclusion, on the ground of all that has been
said in the present paper, that the astronomical activity and development
in the period 1600-1880 has not been considerably impeded by the elaborating of catalogues, even though the working out of the catalogues has been
the principal preoccupation of many astronomers, absorbing much of their
active time. We are led to this conclusion by Fig. 2 where a clear "de-

tachment" of the general astronomical activity of the one connected with the catalogue constructing is noticeable after the middle of 17th century.

REFERENCES

Houzeau, J.C. and Lancaster, A. : 1882, Bibliographie générale de l' astronomie (ed. by X. Havermans), Bruxelles, 2, LXXI.
Teleki, G. : 1980, Proc. Symp. "Aristarchos of Samos", Athens.

PART II
DYNAMICS

THE BEHAVIOUR OF ADIABATIC INVARIANTS NEAR RESONANCES

B. Barbanis
Department of Astronomy, University of Thessaloniki
Thessaloniki, Greece

ABSTRACT

We investigate numerically the changes of the adiabatic invariants in some resonant cases of time dependent one-dimensional dynamical systems. We find that at each main resonance of the Mathieu equation there is a "resonant" adiabatic invariant which is much better conserved than other forms of adiabatic invariants. When the orbits are stable the resonant adiabatic invariant does not change appreciably. However, in the case of unstable orbits the variation of this adiabatic invariant becomes progressively larger as the time increases. Then we study the adiabatic invariant in the case of a simple pendulum with varying length. The changes of the adiabatic invariant are of the same order when the motion remains either oscillating or changes from oscillating to rotating.

1. INTRODUCTION

It is well known that an adiabatic invariant is a quantity which is approximately constant during a slow variation of the parameter of a mechanical system. Several investigators (Kulsrud, 1957; Gardner, 1959; Lenard, 1959; Vandervoort, 1961) have examined the constancy of adiabatic invariants in various cases. If the adiabatic invariants are expressed as power series in a small parameter it was found that under certain, quite general, conditions they are constant to all orders of the small parameter, namely they are formal integrals of motion (Kulsrud, 1957; Lenard, 1959; Kruskal, 1962).

Contopoulos (1966a) examined the similarities and differences between the "third" intergal and the adiabatic invariants. Using the simple one-dimensional time dependent Hamiltonian

$$H = \frac{1}{2} (\omega_1^2 x^2 + \dot{x}^2) - \epsilon x^2 \sin\omega t \qquad (1)$$

he found in second approximation the "third" integral Φ and applying Gardner's method (1959) the adiabatic invariant J. Comparing the values

31

E. G. Mariolopoulos et al. (eds.), Compendium in Astronomy, 31–42.

of Φ and J in some orbits calculated numerically he found that J is better conserved than Φ for small values of ω. However, for ω/ω_1 approaching
unity Φ is better conserved than J. This result is a manifestation of
the fact that the action J is not an adiabatic invariant if there is a
resonance between the frequency of the perturbation and the eigenfrequen-
cy of the system.

In the present paper we study the behaviour of the adiabatic invariants in the case of the Hamiltonian (1) for ω/ω_1 approaching 1 and 2,
and in the case of a simple pendulum whose length ℓ decreases slowly in
time.

In the first case we calculate besides the usual adiabatic invariant
a "resonant" adiabatic invariant appropriate for each particular resonance and we compare their variations near each resonance. In the case
of the pendulum if its length is continuously decreasing the amplitude
of oscillations gradually increases and finally the pendulum makes complete revolutions. It is known that the Hamiltonian is not conserved, but
the adiabatic invariant $J = 1/2\pi \oint pdq$ is well conserved if the variation
of ℓ is slow, so that $Td\ell/dt \ll \ell$. This "slowness condition" cannot be
satisfied when the motion of the pendulum changes from oscillating to
circular. This limiting case corresponds to an infinite period T and this
condition is violated. Our purpose is to examine how well conserved the
usual adiabatic invariant is in transition cases of this type of practical interest.

The systems above are both one dimensional time-dependent systems
and may be easier to handle from an analytical point of view. However,
there is an essential difference between them. In the case of the
Hamiltonian (1) there are regions of unstable orbits extending to infinity while the motion of the pendulum is always bounded. Furthermore, the
near-commensurabilities among the mean motions of the satellites and
planets and the spin-orbit resonant coupling of Mercury can be modelized
by a pendulum like system (see, for instance, Goldreich, 1966; Goldreich
and Peale, 1966; Peale, 1976). Yoder (1979) also considered a rigid pendulum which is subject to both a time-dependent periodic torque and a
constant applied torque. This system has been studied by Sinclair (1972)
and approximates the orbit-orbit interaction in certain cases.

2. ADIABATIC INVARIANTS AT THE MAIN RESONANCES OF THE MATHIEU EQUATION

The equation of motion of the Hamiltonian (1) is

$$\ddot{x} = - x(\omega_1^2 - 2\varepsilon\sin\omega t), \tag{2}$$

where dots mean time derivatives. Setting $y = x$, $\omega t = 2z$, $a = 4\omega_1^2/\omega^2$,
$q = 4\varepsilon/\omega^2$ this equation is reduced to

$$\frac{d^2y}{dz^2} + (a - 2q\sin2z)y = 0, \tag{3}$$

namely the well-known Mathieu equation. The resonances of the above system are given by the formula $\omega^2 = 4\omega_1^2/n^2$ where n is an integer. The main resonances correspond to n = 1 or 2 namely for $\omega = 2\omega_1$ or $\omega = \omega_1$.

Near a resonance we will construct a "resonant" adiabatic invariant as follows. We consider the time t as a new "dependent" variable and introduce the extended Hamiltonian $\tilde{H} = H + h$ in which h is canonically conjugate to t (Brouwer and Clemence, 1961). Introducing now the canonical transformation

$$x = (2I_1/\omega_1)^{\frac{1}{2}} \cos\vartheta_1, \qquad \dot{x} = -(2I_1\omega_1)^{\frac{1}{2}} \sin\vartheta_1 ,$$

$$t = \vartheta_2/\omega , \qquad h = \omega I_2 , \qquad (4)$$

we write

$$\tilde{H} = \omega_1 I_1 + \omega I_2 - \frac{\varepsilon I_1}{\omega_1} \{\sin\vartheta_2 + \frac{1}{2}\sin(\vartheta_2 + 2\vartheta_1) + \frac{1}{2}\sin(\vartheta_2 - 2\vartheta_1)\} = \text{const.} \qquad (5)$$

Then in the case of a particular resonance we apply the usual treatment. E.g., let us consider the resonance $\omega = 2\omega_1$. We can eliminate all trigonometric terms except the "resonant" term that contains $(\vartheta_2 - 2\vartheta_1)$ by means of a "determining function" (Brouwer and Clemence, 1961).

$$W = \vartheta_1 I_1^* + \vartheta_2 I_2^* - \frac{\varepsilon I_1^*}{\omega_1} \{ \frac{\cos\vartheta_2}{\omega} + \frac{\cos(\vartheta_2 + 2\vartheta_1)}{2(\omega + 2\omega_1)} \} . \qquad (6)$$

This defines new canonical variables through the relations

$$\vartheta_1^* = \frac{\partial W}{\partial I_1^*} = \vartheta_1 - \frac{\varepsilon}{\omega_1} \{ \frac{\cos\vartheta_2}{\omega} + \frac{\cos(\vartheta_2 + 2\vartheta_1)}{2(\omega + 2\omega_1)} \} ,$$

$$\vartheta_2^* = \frac{\partial W}{\partial I_2^*} = \vartheta_2 ,$$

$$I_1 = \frac{\partial W}{\partial \vartheta_1} = I_1^* + \frac{\varepsilon I_1^*}{\omega_1} \frac{\sin(\vartheta_2 + 2\vartheta_1)}{(\omega + 2\omega_1)} ,$$

$$I_2 = \frac{\partial W}{\partial \vartheta_2} = I_2^* + \frac{\varepsilon I_1^*}{\omega_1} \{ \frac{\sin\vartheta_2}{\omega} + \frac{\sin(\vartheta_2 + 2\vartheta_1)}{2(\omega + 2\omega_1)} \} .$$

In first approximation in ε, we find

$$\tilde{H} = \omega_1 I_1^* + \omega I_2^* - \frac{\varepsilon I_1^*}{2\omega_1} \sin(\vartheta_2^* - 2\vartheta_1^*) \qquad (7)$$

and if we set

$$J_1 = I_1^* + 2I_2^* , \qquad \psi_1 = \vartheta_1^* ,$$

$$J_2 = I_2^* \quad , \qquad \psi_2 = \vartheta_2^* - 2\vartheta_1^* ,$$

we derive

$$\tilde{H} = \omega_1 J_1 + (\omega - 2\omega_1)J_2 - \frac{\varepsilon(J_1 - 2J_2)}{2\omega_1} \sin\psi_2. \tag{8}$$

As ψ_1 is an ignorable coordinate, J_1 is an integral of motion, in this approximation, which can be written

$$J_{\omega=2\omega_1} = J_1 = I_1 + 2I_2 - \frac{2\varepsilon I_1}{\omega_1} \{ \frac{\sin\vartheta_2}{\omega} + \frac{\sin(\vartheta_2+2\vartheta_1)}{\omega + 2\omega_1} \} . \tag{9}$$

Following a similar procedure we find, after some algebra, that the appropriate adiabatic invariant near the resonance $\omega = \omega_1$, in second approximation in ε, is

$$J_{\omega=\omega_1} = I_1 + I_2 - \frac{\varepsilon I_1}{\omega_1} \{ \frac{\sin\vartheta_2}{\omega} + \frac{3\sin(\vartheta_2+2\vartheta_1)}{2(\omega + 2\omega_1)} - \frac{\sin(\vartheta_2-2\vartheta_1)}{2(\omega - 2\omega_1)} \} -$$

$$- \frac{\varepsilon^2 I_1}{\omega_1^2} [- \frac{1}{\omega^2-4\omega_1^2} (1 + 2\cos2\vartheta_1 + \cos4\vartheta_1 - \frac{\omega_1}{\omega} \cos2\vartheta_2) -$$

$$- \frac{1}{2(\omega+2\omega_1)} \{ \frac{2\cos2(\vartheta_2+\vartheta_1)}{\omega+\omega_1} + \frac{3\cos2(\vartheta_2+2\vartheta_1)}{2(\omega+2\omega_1)} \} - \frac{\cos2(\vartheta_2-2\vartheta_1)}{4(\omega-2\omega_1)^2}]. \tag{10}$$

We calculated numerically, using the Runge-Kutta method, many orbits for 100 time units with a step $\Delta t = 0.01$ time units, $\omega_1 = 1.0$, $x_0 = 0$ and $\dot{x}_0 = 0.1$. During the computation of an orbit we calculated the maximum and minimum values of $\tilde{H}$, $J_{\omega=\omega_1}$, $J_{\omega=2\omega_1}$, and of the usual adiabatic invariant, namely the action

$$J = \frac{1}{2\pi} \oint \dot{x}dx = \frac{1}{2\pi} \int_{t_0}^{t_0+T} \dot{x}^2 dt .$$

T is the time elapsed between two successive zeros of x and t_0 is the value of t when $x = 0$. The accuracy of the computations is checked up by the constancy of $\tilde{H}$. During each motion at least six significant figures of H remain unchanged.

Table I gives the relative changes of J, $J_{\omega=\omega_1}$, $J_{\omega=2\omega_1}$ in some calculated orbits for 100 time units when $\omega_1 = 1.0$, $\varepsilon = 0.1$, $x_0 = 0$, $\dot{x}_0 = 0.1$. We notice that when ω varies from 0 to 1.4, J is well conserved, except of the values around the resonance $\omega=\omega_1$, namely for $0.95<\omega<1.05$. $J_{\omega=\omega_1}$ is much better conserved when ω is in the range (0.5, 1.4) including the resonant region, while $J_{\omega=2\omega_1}$ is better conserved for $1.2 \ll \omega \ll 1.85$. In both cases their changes are smaller than 5%.

Table I

Relative variations of the adiabatic invariants J, $J_{\omega=\omega_1}$, $J_{\omega=2\omega_1}$ when $\omega_1=1.0$, $\varepsilon=0.1$, $x_o=0.0$, $\dot{x}_o=0.1$. Each orbit has been calculated for 100 time units.

ω	0.2	0.5	0.8	0.95	1.0	1.05	1.2	1.4	1.6	1.8	1.9
$\Delta J/J$	0.004	0.026	0.038	0.19	1.1	0.21	0.07	0.17	0.41	1.3	81.5
$(\Delta J/J)_{\omega=\omega_1}$	0.114	0.043	0.033	0.034	0.04	0.035	0.034	0.05	0.09	0.34	30.5
$(\Delta J/J)_{\omega=2\omega_1}$	0.21	0.10	0.07	0.22	1.2	0.21	0.052	0.03	0.02	0.017	0.38

However, the constancy of $J_{\omega=\omega_1}$ near or at the resonance becomes worse as the time t increases. In the region $0.992 \ll \omega \ll 1.001$ the changes of $J_{\omega=\omega_1}$ become progressively larger, with the time, although they are always much smaller than the changes of J. This happens because in this region the orbits are unstable. In fact, the resonances of equation (1) correspond to the integral values of $a=4\omega_1^2/\omega^2$, where a is one of the parameters of the Mathieu equation. It is well known that on the plane $(a, q=4\varepsilon/\omega^2)$ near each resonance there is a region, defined by the characteristic curves, where all orbits are unstable (McLachlan, 1964). If we calculate the characteristic values a_2 and b_2, using the first terms of the series

$$a_2 = 4 + \frac{5}{12} q^2 - \frac{763}{13824} q^4 + \dots ,$$

$$b_2 = 4 - \frac{1}{12} q^2 + \frac{5}{13824} q^4 - \dots ,$$

we find that $b_2 < a < a_2$ for $0.992 \ll \omega \ll 1.001$, i.e., these orbits are unstable.

Table 2 gives the relative changes of $J_{\omega=\omega_1}$ for time intervals which are multiples of 100 time units for four orbits near or at the resonance

$\omega=\omega_1$ and the corresponding changes of $J_{\omega=2\omega_1}$ for two orbits near $\omega=2\omega_1$.
We see that in the stable orbits there is not any appreciable change of
the appropriate adiabatic invariant, while in the unstable orbits the
changes become large with the time. Therefore even the suitable "resonant"
adiabatic invariant fails to be conserved near its own resonance. This
is the consequence of the instability of orbits near each resonance.

It is well known that in a weakly perturbed dynamical system of two
degrees of freedom there are regions of phase space in which there is an
isolating integral besides the energy (Hénon and Heiles, 1964;Barbanis,
1966). This integral fails near the resonances of the system but in each
resonance an appropriate form of a third integral can be constructed in
general (Contopoulos, 1966b).

Table II

Relative variation of J near the two main resonances of equation (1)
for various time intervals.

		time units					
ω	stability	0-100	0-200	0-300	0-400	0-500	0-600
		$(\Delta J/J)_{\omega=\omega_1}$					
0.990	stable	0.0334	0.0342	0.0358	0.0368	0.0380	
0.995	unstable	0.044	0.091	0.216	0.511	1.3	3.4
1.0	unstable	0.041	0.071	0.150	0.301	0.637	1.4
1.01	stable	0.0346	0.0348	0.0350	0.0350	0.0350	
		$(\Delta J/J)_{\omega=2\omega_1}$					
1.89	stable	0.0482	0.0482	0.0482	0.0482		
1.90	unstable	0.336	1.9				

This integral allows us to find the periodic orbits of the system. The
non periodic orbits follow the stable periodic orbits. However, in the
Mathieu equation all resonant periodic orbits are unstable. For this
reason all adiabatic invariants fail near these resonances. Therefore
the non-constancy of the adiabatic invariant near resonances of the
Mathieu equation is a general characteristic independent of the appropriate adiabatic invariant used.

3. NUMERICAL RESULTS FOR A PENDULUM OF VARYING LENGTH

Consider a simple pendulum whose length ℓ is decreased slowly during a finite time interval. If the potential energy is zero in the horizontal plane then at angle ϑ is

$$V = -mg\ell\cos\vartheta$$

and the kinetic energy

$$T = \frac{m}{2}\ (\dot{\ell}^2+\ell^2\dot{\vartheta}^2).$$

From the Lagrangian of the system we find the equation of motion

$$\ell\ddot{\vartheta} + 2\dot{\ell}\dot{\vartheta} + g\sin\vartheta = 0. \tag{11}$$

The total energy, per unit mass, is

$$E' = \frac{1}{2}(\dot{\ell}^2+\ell^2\dot{\vartheta}^2) - g\ell\cos\vartheta\ , \tag{12}$$

and the action

$$J' = \frac{1}{2\pi}\ \oint p_\vartheta d\vartheta = \frac{1}{2\pi}\oint \ell^2\dot{\vartheta}d\vartheta\ , \tag{13}$$

where the integration extends over one period.

Suppose that the length is varying linearly in time, namely

$$\ell = \ell_o\ (1-\beta t)\ , \tag{14}$$

with $\beta>0$, while the variation takes place during the time interval $(0,t_f)$. Before $t = 0$ and after $t = t_f$ the length remains constant. The total decrease of the length is $\ell_o\beta t_f$ and the "slowness condition" $T\dot{\ell} \ll \ell_o$ in this case becomes $\beta T \ll 1$. In our calculations we take T as the "initial" period. If, however, T is the "instantaneous" period of a pendulum of length ℓ, then $T \to \infty$ as we approach the transition from oscillatory to circular motion. Therefore in transitions of this type the "slowness condition" is violated.

Substituting ℓ in equations (11), (12) and (13) we obtain

$$\ddot{\vartheta} = (-\frac{g}{\ell_o}\ \sin\vartheta + 2\beta\dot{\vartheta})/(1-\beta t)\ , \tag{15}$$

$$E = E'/\ell_o^2 = \frac{1}{2}\ \{\beta^2+(1-\beta t)^2\dot{\vartheta}^2\} - \frac{g}{\ell_o}(1-\beta t)\cos\vartheta\ , \tag{16}$$

$$J = J'/\ell_o^2 = \frac{1}{2\pi}\ \int_{t_o}^{t_o+T} (1-\beta t)^2\dot{\vartheta}^2 dt\ , \tag{17}$$

where T is the time elapsed between two successive zeros of $\dot{\vartheta}$.

In all calculated cases we get $g/\ell_0 = 1-\beta t_f$, so that for $t \geqslant t_f$ the equation of motion becomes $\ddot{\vartheta} = -\sin\vartheta$. The initial conditions are always $\vartheta = \vartheta_0$, $\dot{\vartheta} = 0$.

For $\beta \neq 0$ the total energy is not conserved. During the time interval $(0, t_f)$ the total energy increases on the average although in some particular steps it decreases. The adiabatic invariant, calculated by formula (17), is generally increasing in successive integration steps by a small amount. The question is: How well conserved is the adiabatic invariant in the case when the oscillating motion of the pendulum becomes a circular rotation? This investigation is of interest because the usual "slowness condition", under which the constancy of the adiabatic invariants is proved, is now violated.

The transition from oscillating to circular motion depends on the quantity βt_f. To a given initial angle ϑ_0 corresponds a certain value $(\beta t_f)_{osc.}$, so that for any $\beta t_f \leqslant (\beta t_f)_{osc.}$ the motion of the pendulum is always oscillating for whatever value of t_f. Namely all orbits in the phase space $(\vartheta, \dot{\vartheta})$ circulate around the origin (Fig. 1). For βt_f larger than $(\beta t_f)_{osc.}$ there are some values of t_f for which the oscillating motion is changed to circular motion around the point of suspension of the pendulum. Then the motion in the phase space $(\vartheta, \dot{\vartheta})$ is progressive i.e., ϑ after some time increases (or decreases) continuously (Fig.2). We call an orbit like that of Fig. 1 "trapped" and like that of Fig. 2 "untrapped".

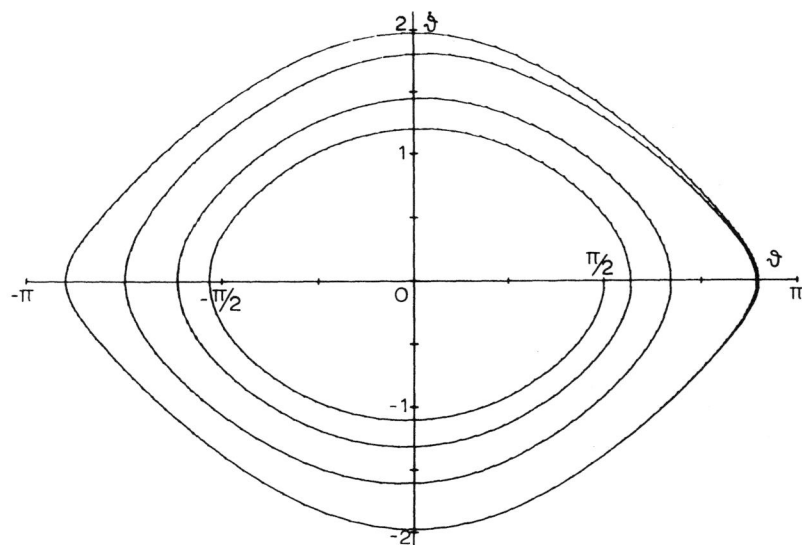

Fig. 1. A trapped orbit in the phase space $(\vartheta, \dot{\vartheta})$ corresponding to the oscillating motion of a pendulum with varying length when $\vartheta_0=90°$, $\dot{\vartheta}_0=0$, $\beta t_f=0.44$ and $t_f=32.0$.

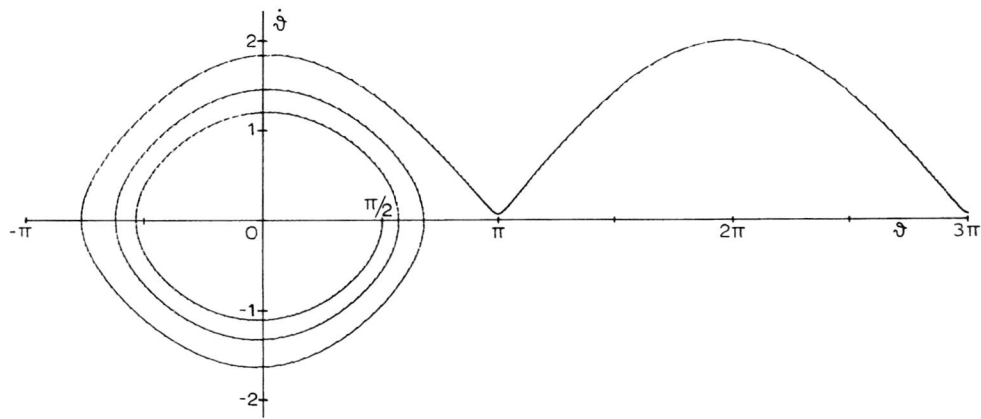

Fig. 2. The oscillating motion of a pendulum with varying length is
changed to a circular motion around the point of suspension and the orbit
in the phase space $(\vartheta, \dot\vartheta)$ from trapped becomes untrapped. Initial data:
$\vartheta_o = 90^\circ$, $\dot\vartheta_o = 0$, $\beta t_f = 0.44$ and $t_f = 31.0$

We have experimented with some values of ϑ_o but we studied in detail
the case $\vartheta_o = 90^\circ$. When the motion of the pendulum becomes circular it can
be direct or retrograde, namely the angle ϑ is continuously increasing
or decreasing. The results are similar and we confine our discussion to
the direct cases.

Table III represents a typical example of the changes of the energy
and of the adiabatic invariant. This Table gives some trapped and direct
untrapped orbits, as indicated in the second column, when $\vartheta_o = 90^\circ$ and
$\beta t_f = 0.44$. In the examples of Table III, ℓ takes its final constant value
during the eighth period of motion. We designate these examples as the
8-period groups of orbits. The third column gives the maximum value of
ϑ in the case of a trapped orbit or the minimum value of $\dot\vartheta$ in the case
of an untrapped orbit. The next column gives the maximum energy E_{max}.
As the initial energy in the case $\vartheta_o = 90^\circ$ is zero, E_{max} gives the corresponding change of the energy. ΔJ is the difference between the maximum
and the minimum values of the adiabatic invariant in each case.

The initial value of the adiabatic invariant is $J_{in.} = 0.40361$, namely
it is the value of J of a pendulum of constant length when $\vartheta_o = 90^\circ$ and
$g/\ell = 1 - \beta t_f = 0.56$. Each value of J is calculated by equation (17) between
two successive zeros of $\dot\vartheta$. We call a "step of integration" each integration
according to equation (17). In the case of a trapped orbit the final value
of J corresponds to these successive zeros of $\dot\vartheta$ that occur at times after
the time t_f when the pendulum reaches its final constant length. When an
orbit becomes untrapped from trapped the corresponding value of J is calculated at times when $\dot\vartheta = 0$ and then when $\dot\vartheta$ takes its next minimum value.

The last "step of integration" is calculated at times when ϑ takes successive minimum values.

In the first steps of integration, in each case of Table III, there is a small increase of J so that the increase of J at the end of the seventh period is less than 5×10^{-4}. However, the changes of J which occur in the last two steps of integration are relatively large and can be sometimes one or two orders of magnitude larger than the previous change of

Table III

A typical example of the changes of energy and adiabatic invariant in an 8-period group of orbits when $\vartheta_o = 90°$ and $\beta t_f = 0.44$. $E_{max.}$ is the maximum energy and ΔJ the total increase of J in each case.

t_f	kind of orbit	max.ϑ ($\dot\vartheta = 0$) min.$\dot\vartheta$	$E_{max.}$	$\Delta J \times 10^3$
77.8	trap.	3.03	0.3202	9.4
78.6	"	2.94	0.3174	14.2
79.0	"	3.00	0.3160	12.5
79.2	untrap.	0.05	0.3152	9.6
79.4	"	0.19	0.3191	7.9
80.2	"	0.31	0.3284	13.5
81.0	"	0.27	0.3264	10.0
81.4	"	0.24	0.3252	7.1
82.2	"	0.15	0.3226	2.9
82.6	"	0.06	0.3213	5.7
82.8	trap.	3.07	0.3206	7.7

J. The total increase of J is given in the last column of Table III. We notice that the adiabatic invariant is better conserved than the total energy in all cases. The maximum change of J is 0.014, about 3% of $J_{in.}$ This change is an order of magnitude smaller than the change of the energy when an orbit is either trapped or untrapped.

Table IV gives the maxima and minima of $E_{fin.}$ and ΔJ for various groups of orbits in the case $\vartheta_o = 90°$ and $\beta t_f = 0.44$ as well as the range of values of β in each group. We see that $(E_{fin})_{max}$ decreases and $(E_{fin})_{min}$ increases from group to group. In each group the value of J_1, namely the result of the first step of integration, decreases with increasing t_f but its changes are small; i.e., in the 8-period group this change is only 0.00001. Table IV gives the smallest value of J_1 in each group. J_1 tends to be equal to J_{in} with decreasing β. The variation of adiabatic invariant becomes progressively smaller with the increasing number of periods; its constancy is very well conserved when the number of periods is larger than eight.

Table   IV

Maxima and minima of $E_{fin.}$ and ΔJ for various groups of orbits in the case of $\vartheta_o = 90^o$ and $\beta t_f = 0.44$

Number of periods	$\beta \times 10^2$		$E_{fin.}$		J_1	$\Delta J \times 10^3$	
	from	to	max.	min.		max.	min.
3	1.67	1.41	0.3471	0.2943	0.40408	31.3	7.1
5	0.94	0.85	0.3352	0.3031	0.40377	20.5	4.4
8	0.57	0.53	0.3284	0.3076	0.40367	14.2	2.9
12	0.370	0.355	0.3246	0.3102	0.40364	10.4	2.1
16	0.274	0.267	0.3226	0.3115	0.40363	8.5	1.5
20	0.218	0.213	0.3214	0.3122	0.40362	7.3	1.2

We have found similar results when ϑ_o takes the values 60°, 120°, 150°, 170°. In all tested cases the adiabatic invariant is much better conserved than the total energy. However, the constancy of J depends on ϑ_o, so that J is better conserved as ϑ_o becomes larger.

4. CONCLUSIONS

The paper studies the behaviour of adiabatic invariants:
a) at the main resonances of the Mathieu equation b) in the case of the transition of a pendulum from oscillating to circulating motion.
a) The non-constancy of the adiabatic invariants at the main resonances is shown by solving numerically the Mathieu equation. These resonances are connected with regions of instability where all orbits, including the resonant periodic orbits are unstable. The consequence of this instability is that even the resonant adiabatic invariant fails to be conserved near its own resonance. Outside the resonant region the suitable resonant adiabatic invariant is much better conserved than the usual one.
b) In the case of a pendulum the interesting result is that the degree of constancy of the adiabatic invariant is practically the same when either the pendulum remains oscillating or its motion becomes rotating. The changes of J depend on both the rate of variation of the length of the pendulum and on its initial position. These results justify the use of adiabatic invariant in transition cases of practical interest.

Acknowledgments

The author would like to thank Professor G. Contopoulos for useful discussions on the subject of this paper and the staff of the computing center of the University of Thessaloniki for kind cooperation.

REFERENCES

Barbanis, B.: 1966, Astron. J. 71, 415.
Brouwer, D., Clemence, G.: 1961, Methods of Celestial Mechanics, Academic
    Press, New York, p. 531.
Contopoulos, G.: 1966a, J. Math. Physics 7, 788.
Contopoulos, G.: 1966b, Astrophys. J. Suppl. 13, 503.
Gardner, C.: 1959, Phys. Rev. 115, 791.
Goldreich, P.: 1966, Astron. J. 71, 1.
Goldreich, P., Peale, S.: 1966, Astron. J. 71, 425.
Hénon, M., Heiles, C.: 1964, Astron. J. 69, 73.
Kruskal, M.: 1962, J. Math. Physics, 3, 806.
Kulsrud, R.: 1957, Phys. Rev. 106, 205.
Lenard, A.: 1959, Ann. Physics 6, 261.
McLachlan, N.W.: 1964, Theory and Application of Mathieu Functions, Dover
    Publ. New York, pp. 16, 40.
Peale, S.: 1976, Ann. Rev. Astron. Astrophys. 14, 215.
Sinclair, A.: 1972, Monthly Notices Roy. Astron. Soc. 160, 169.
Vandervoort, P.: 1961, Ann. Physics 12, 436.
Yoder, C.: 1979, Celes. Mech., 19, 3.

RADIATION PRESSURE EFFECT ON DUST SURROUNDING THE PLEIADES BRIGHT STARS

P. Bouvier
Geneva Observatory

ABSTRACT

The action of the radiation pressure from a B-star on a dust grain
is reexamined in the celestial mechanics approximation of a two-body
problem. The trajectories of the grains belonging to a dust cloud incident
on a B-star build a one-parameter family of hyperbolae. The envelope of
these curves is a paraboloid of revolution having the star at its focus,
and representing an insuperable barrier for the dust approaching the
star. Such a paraboloid was also met by Radzievskii and Dagaev in a different context. When we turn to the reflection nebulae associated with
the bright stars of the Pleiades cluster in order to detect some possible parabolic profiles, we are faced with several observational difficulties; nevertheless, in the case of the Merope nebula, we do notice a
distinctly parabolic boundary on its eastern side and this confirms indirectly the existence of dust in the close neighbourhood of star 23 Tau.

1. INTRODUCTION

Interstellar matter associated with the Pleiades star cluster exhibits two distinct features; first, bright nebulosities of somewhat striated appearance around each of the brightest stars and secondly, ray-like
parallel filaments lying in east-west direction, apparently uncorrelated
with the bright stars, and the general picture suggested (Arny, 1977) is
that of an interstellar cloud travelling through the cluster from east
to west. When the cloud approaches one of the bright stars, which are
all of the B-type, the dust particles will be decelerated and deviated
by the stellar radiation pressure, while the gas component of the cloud,
mainly unaffected by the radiation field, shall tend to drag the dust
grains around the star, thus forming a thin dust shell. Any dust clump
within the cloud will undergo, in the vicinity of a B star, a shearing
stress and become elongated in a way recalling the formation of cirrus
clouds in the Earth's atmosphere, through the shearing of ice crystal
clumps by differential wind speeds.

E. G. Mariolopoulos et al. (eds.), Compendium in Astronomy, 43–54.
Copyright © 1982 by D. Reidel Publishing Company.

We shall not deal here with the ray-like filaments, possibly originating, according to Arny (1977), in matter stripped off from circumstel-
lar envelopes associated with low-luminosity stars. Our intention is to
reexamine the effect of radiation pressure, due to a B star, on the dust
grains of a passing cloud, in the two-body framework of celestial mechanics (section 2), thus neglecting first the drag of the grains by the gas.
In section 3, we find that every such B star tunneling through an interstellar cloud appears to be shielded from the dust by a shell of paraboloidal form, a result pointed out about 10 years ago by Radzievski and
Dagaev (1969) who proceeded however along very different lines.

We then comment, in section 4, on the possible influence of the gas
dragging the dust around the star, before turning to the expected bright
nebulosity profile, projected on the sky, in the neighbourhood of the
illuminating star (section 5) and concluding with the example of Merope,
where the associated nebulosity tends indeed to exhibit a parabolic profile, characteristic of the radiation pressure effects.

2. RADIATION PRESSURE ACTING ON A DUST GRAIN

Let M and L be respectively the mass and luminosity of an isolated
B star S on which a dust cloud arrives at constant velocity V and consider (Fig. 1) a dust grain of the cloud at distance r from S, coming from
infinity with an impact parameter D.

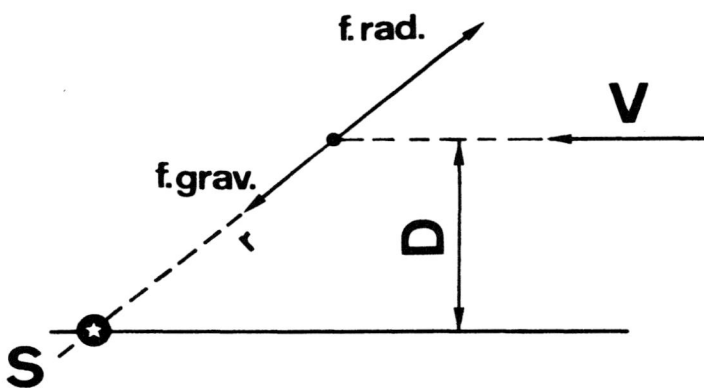

Fig. 1.

The grain will experience a repulsive force due to the radiation
pressure P_{rad} of the star S and we may write, per unit volume:

$$-\frac{dP_{rad}}{dr} = \frac{\kappa}{c} F \tag{1}$$

here κ is a weighted mean extinction coefficient of the dust and the energy flux is equal to

$$F = \frac{L}{4\pi r^2}$$ (2)

when we neglect the extinction of the radiation before it reaches the grain, which is justified for a cloud of sufficiently low density. The monochromatic coefficient κ_λ is related to the cross-section for the radiative interaction of the grains and to the dust grain density n_d through the product $\kappa_\lambda = \sigma_\lambda n_d$.

Taking the mean value κ of κ_λ, weighted by the fluxes, we get straightforwardly the force acting on a single grain

$$f_{rad} = \frac{1}{4\pi c r^2} \int_0^\infty \sigma_\lambda L_\lambda d\lambda$$ (3)

$L_\lambda \Delta\lambda$ is given by the spectral type of S in the interval $\Delta\lambda$ around λ and σ_λ is usually written in the form $\sigma = \pi r_d^2 q_\lambda$, r_d being the radius of a dust grain, assumed spherical, while q_λ is an efficiency factor depending on r_d/λ in the framework of the Mie theory (Wicramasinghe, 1967). Adding to f_{rad} the gravitational attraction of the star on the grain of mass m_d, namely

$$f_{grav} = -G \frac{M m_d}{r^2} ,$$ (4)

the equation of motion of the grain is that of the two-body problem in celestial mechanics

$$\frac{d^2 \vec{r}}{dt^2} = k \frac{\vec{r}}{r^3}$$ (5)

where

$$k = -GM + \frac{1}{4\pi c m_d} \int_0^\infty \sigma_\lambda L_\lambda d\lambda$$ (6)

It leads to a hyperbolic orbit of which S is the internal focus if $k < 0$ or the external focus when $k > 0$, which shall be the case here. The conservation law for angular momentum reads, with D as impact parameter,

$$r^2 \dot\theta = VD = h \ (const.)$$ (7)

and the radial motion of the grain is described by the equation

$$\ddot{r} - r\dot\theta^2 + \frac{k}{r^2} = \frac{h^2}{r^3} + \frac{k}{r^2}$$ (8)

the solution of which yields here a hyperbolic orbit of external focus
S (Fig. 2).

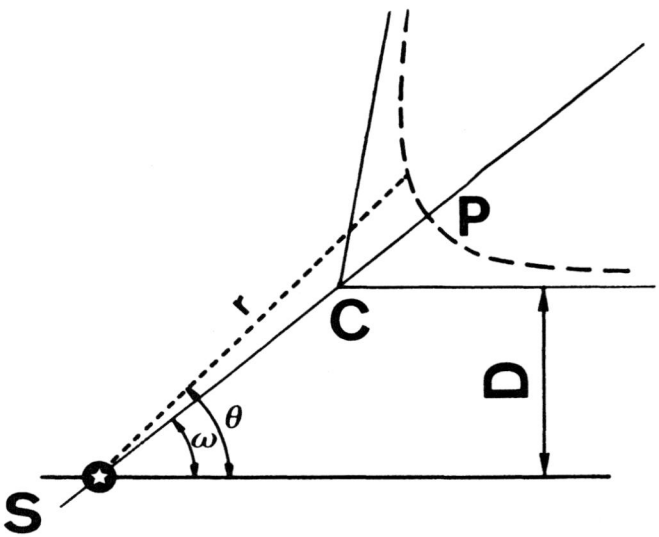

Fig. 2.

In standard notation we thus write

$$r = \frac{p}{e \cos(\theta-\omega)-1}$$ (9)

where

$$p = (h/k)^2, \qquad e^2 = 1 + (hV/k)^2$$ (10)

and ω is the polar angle of the hyperbola axis SCP with the direction
of V, which is also an asymptotic direction of the hyperbola (Fig. 2).

We therefore obtain a one-parameter family of hyperbolae, the parameter being precisely the angle ω of any hyperbola axis with the
asymptotic V-direction common to all the hyperbolae (Fig. 3). Consequently, if a is the semi-major axis and c = CS, we have cos ω = a/c and a is
related to p by the usual relation

$$p = a(e^2 - 1)$$ (11)

so that

$$a = \frac{k}{V^2}$$

Since for a given type of grain, k is constant, we see that all the hyperbolae of the family have the same semi-major axis. Moreover, the periastron distance r_p = SP is equal to

$$r_p = c + a = a (1 + 1/\cos \omega) \tag{13}$$

a relation showing that the locus of the periastra P, as ω (or D) varies, is the conchoid C of the straight line perpendicular to V passing at distance a from S and carrying the centers of all the hyperbolae.

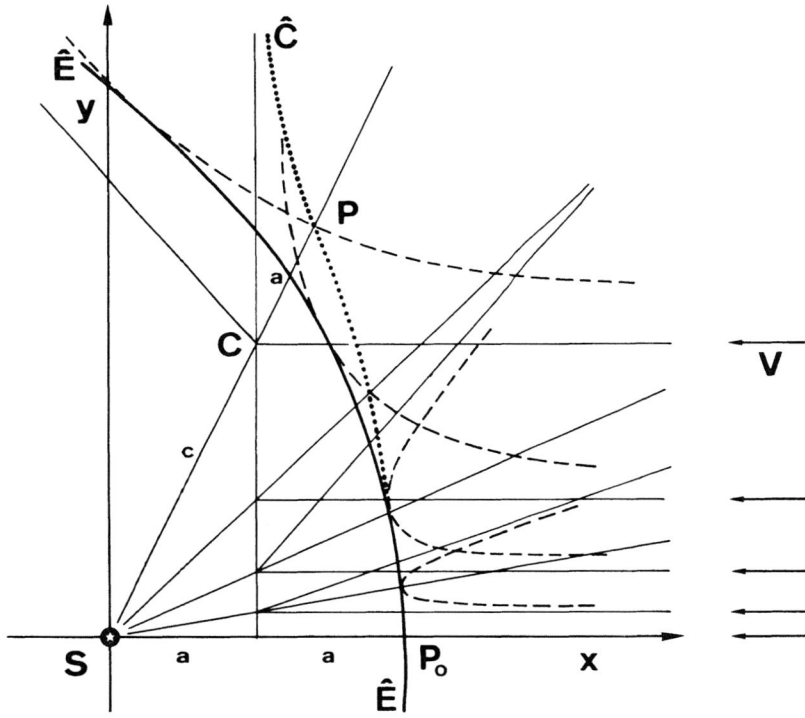

Fig. 3. Meridional profile of the paraboloidal shield of star S against dust incoming with velocity V. <u>dashed curves</u>: hyperbolic trajectories of dust grains. <u>dotted curve</u> $\hat{C}$: conchoid of axis x = a, this axis carries the centres C of the hyperbolae, while $\hat{C}$ is the locus of the vertices of the hyperbolae. <u>continuous curve</u> $\hat{E}$: parabolic envelope of the former hyperbolae, with its focus at S.

3. PARABOLOIDAL SHIELD ASSOCIATED WITH A HOT STAR

    We notice on Fig. 3 that the hyperbolic trajectories tend to cross

the conchoid C all the more that ω becomes large and what we need here
is the envelope E of the family of hyperbolae. E coincides with C at
$\omega = 0$ and bends off, as ω increases, towards S, appearing as the meridian
curve of a surface of revolution representing an insuperable barrier for
the dust grains, insofar as the dust-gas interaction is neglected.

In order to find the equation of envelope E, it proves suitable to
chose $\epsilon = \tan \omega = (e^2-1)^{\frac{1}{2}}$ as parameter rather than ω itself.

The family of hyperbolae being defined by

$$\Phi(r,\theta,\epsilon) \equiv r - \frac{a\epsilon^2}{\cos\theta + \epsilon\sin\theta - 1} \tag{14}$$

the auxiliary relation

$$\partial\Phi/\partial\epsilon = 0$$

yields

$$\epsilon = 2\,\frac{1-\cos\theta}{\sin\theta}$$

and after substituting into (14), we get

$$r = \frac{4a}{1+\cos\theta} \tag{15}$$

representing a paraboloid of revolution of focus S, corresponding to the
inner edge of a dust layer reached by the grains of initial velocity V,
submitted to the radiation pressure generated by the star S. This surface
acts as a paraboloidal shield of S, against the incoming grains, as S
goes through the cloud.

It should be pointed out here that Radzievskii and Dagaev (1969) had
found this same paraboloid (15) using quite another method than the present determination of the envelope; their aim was in fact not to consider
an interstellar cloud passing over a B star, but to derive a density law
for the galactic diffuse medium around a hot field star. This density
law, connected to the paraboloidal shield of the star, was then used to
compute the mean brightness difference of B stars having either positive
or negative radial velocities v_r. The stars with $v_r < 0$, turning the
vertex of their paraboloidal shield towards the observer, were expected
to be fainter than those with $v_r > 0$ and this appeared statistically to
be case, the average magnitude difference being of the order $0^m.7$.

Let us turn back to the envelope described by equation (15) recalling that all the physical characteristics of the grains are contained in
the length a given by (12), via the second term of k in (6). In a thin
shell located just outside the envelope, the grains tend to accumulate
and there are numerous crossings between trajectories of grains approach-

ing and leaving the surface, producing a density enhancement such that
the dust-gas interaction might become important. Anyway, as long as the
dust shell remains optically thin, it will be a locus of maximum light
scattering, and if our line of sight is perpendicular to the velocity V
of the cloud, we shall observe a parabolic fringe at the border of a
reflection nebula.

4. INFLUENCE OF THE GAS DRAG ON THE GRAINS

On account of the enhanced dust density prevailing in the paraboloidal shell, the coupling between the dust grains and gas atoms cannot
be ignored there at least, because the gas streaming through the shell
will presumably be partly stopped by the grains. The drag force on the
dust grains, which is equal and opposite to the damping force on the at-
oms, may be written, per unit volume $\rho_g (\vec{v}_g - \vec{v}_d)/\tau$ in terms of the gas
mass density ρ_g, the respective velocities $\vec{v}_d$, $\vec{v}_g$ of the dust grains and
of the gas atoms (Jancel et Kahan, 1963); further, $1/\tau$ plays the role of
a friction coefficient, directly connected to the mean time τ between
two successive encounters of a dust grain with gas atoms. If σ_d is the
cross-section of the grains with respect to incident atoms and n_d the
dust grain concentration, we have obviously

$$\frac{1}{\tau} = \sigma_d \, n_d \, |\vec{v}_g - \vec{v}_d|$$

so that the drag force on a single grain amounts to

$$f_{drag} = \rho_g \sigma_d \, |\vec{v}_g - \vec{v}_d| \, (\vec{v}_g - \vec{v}_d). \tag{16}$$

This force, in the direction $\vec{v}_g - \vec{v}_d$, has the effect, through its component normal to V, to bring back the grain towards the axis SP_0 of the
paraboloid (Fig. 4). Adding now f_{drag} to the r.h.s. of equation (5) will
complicate the problem, compelling the resort to numerical computation of
the solution. Such a computation has been performed by Arny (1977) with
appropriate choices for the values of ρ_g, v_g and of the grain parameters
σ_d, n_d.

Insofar as the gas-dust interaction was neglected, the star was
shielded, as pointed out in section 3, by the paraboloidal surface described by equation (15), and projecting itself in the parabola of Fig.3
on any plane passing through S and parallel to the velocity V. Therefore,
the ratio of the distances from S to the parabola measured in directions
respectively perpendicular and parallel to V, is equal to 2; we expect
it to be less than 2 when the drag force is taken into account and the
fact that it appears larger than 2 in Fig. 2 of Arny's paper, which illustrates the numerical integration of the equation of motion, should
probably be ascribed to the occurrence of a spurious change of scale in
the y-direction, when using the computer line printer (Arny, 1980).

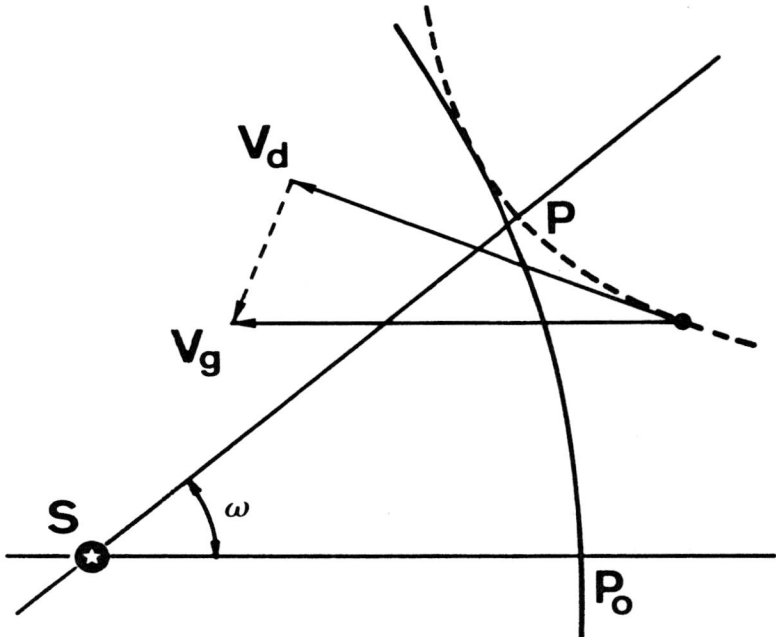

Fig. 4.

5. EXPECTED OBSERVED NEBULOSITY PROFILES

     It is clear that in general, the line of sight from the observer
will not be perpendicular to the velocity V of the passing cloud, but
shall make an angle α with V; consequently the luminous fringe observed
is the projection on the plane of the sky, of the edge of the paraboloidal shield as seen under an angle α with its axis. This fringe appears
again as a parabola, of parameter p = 4a cosecα and of focus S' located
at the projection of star S on the sky along the line of sight. The
demonstration of this result is carried out in the Appendix. So finally
the distance ratio mentioned at the end of section 4 remains equal to 2,
whatever the angle α.

     Let us now turn our attention to the nebulosities surrounding the
bright stars of the Pleiades cluster, as they show up on the Kitt Peak
telescope plate reproduced in Ap.J. vol. 217 in connection with Arny's
paper (1977), or on the Astrophoto Laboratory photograph reproduced in
the Aug. 1979 issue of Sky Publications (Cambridge, Mass.) where, after
enlargement, the location of the illuminating star is determined with
good accuracy.

     In seeking for possible parabolic profiles among these reflection
nebulae, we are faced with a number of circumstances which make it diffi-

cult to specify the exact form of the contour limiting any of the nebulae. First, the presence of several nearby illuminating stars greatly complicates the structure of the radiation field, secondly, the cloud itself has probably an irregular patchy structure, thirdly, most of these nebulae exhibit a finely striated appearance of uncertain origin; in particular the Maia nebula shows a two-fold system of striae, probably due to the influence of neighbouring stars. The striae could be interpreted in terms of a finite number of different species of dust grains present in the cloud; to each species would correspond a slightly different value of a, so that the illuminating star would then be shielded by a superposition of slightly different paraboloids.

Another origin for these striae might reside in some sort of instability (Field, 1971); anyway, the overall aspect of the striated nebulae gets more and more blurred as we pass from Merope to Maia and other bright stars like 25 or 17 Tau and this may well be caused by the latter stars being immersed, deeper than Merope, inside the large interstellar cloud associated with the star cluster.

Therefore, the Merope nebula seems to offer the best clearcut contour and when looking at the south-east side of it, we notice that the ratio of the distances from star 23 Tau to the border of the nebula, along directions respectively parallel and perpendicular to the striae, appears indeed very close to the value 2 typical of a parabolic profile.

This points to a dominant radiation pressure effect of star 23 Tau on the dust component of the cloud, thus indirectly confirming the presence of dust (Andriesse et al., 1977; Jura, 1977) in the immediate vicinity of 23 Tau.

Furthermore, the strong CH^+ lines observed very near to this same star could result from the evaporation of the grains, according to Bates and Spitzer (1951), but if all the grains evaporated before reaching the paraboloidal shield, the parabolic profile of the nebula would simply be washed out, eventually replaced by a circular one. The fact that the Merope nebula does exhibit a distinctly parabolic eastern boundary shows therefore that a large fraction of the dust grains is able to attain the paraboloidal shield, even near the vertex located at 0.1 pc from the star, before being deflected by the repulsive force due to radiation pressure.

APPENDIX

Projection of the Paraboloidal Shield on the Sky.

The geometrical situation is illustrated by Fig. 5 below.
S being the illuminating star surrounded by its paraboloidal shield of vertex P and axis Sx, let L_o be the line of sight from a distant observer, through the Sx axis, to the shield's surface; L_o is therefore tangent at M to the meridian parabola PM lying in the plane Sxy. M shall project itself on the sky at the vertex M' of the contour limiting the

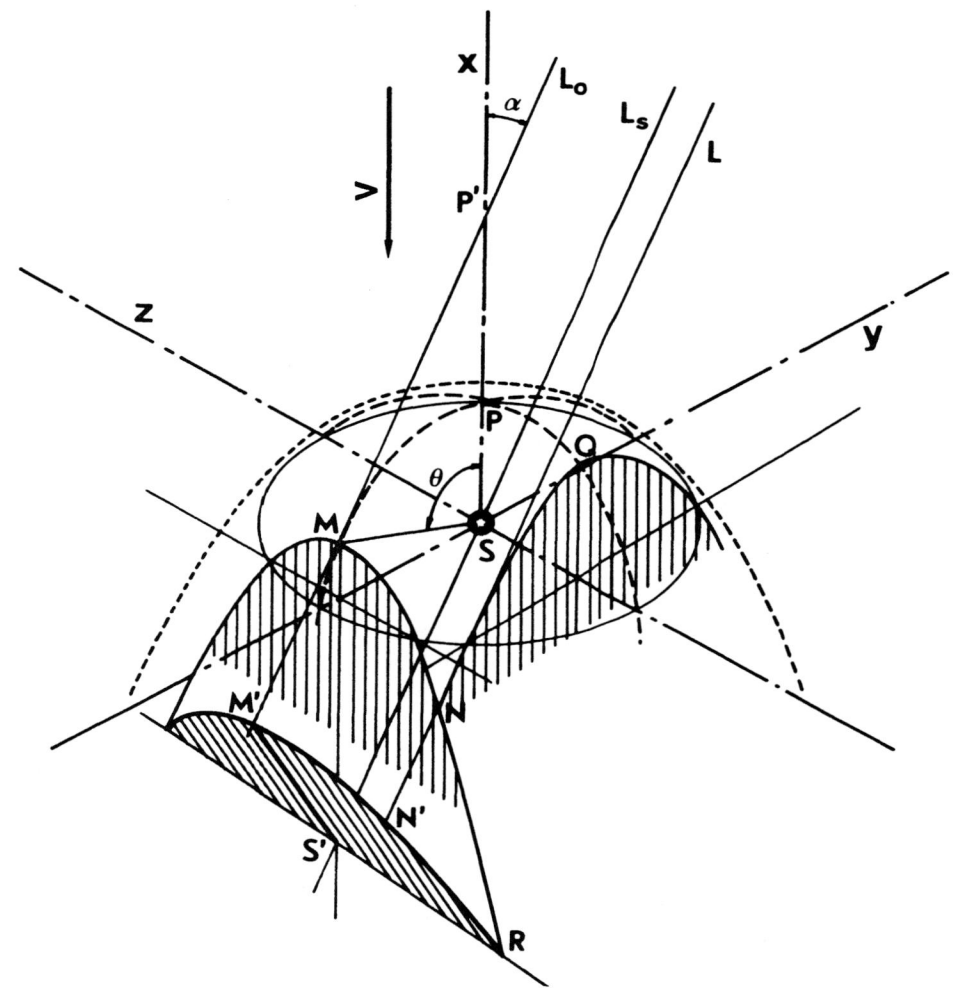

Fig. 5. Projection M'R on the sky of the paraboloidal shield of axis SP, around star S. The distance SS', along the star's line of sight L_S is arbitrary.

reflection nebula associated with S and we denote by L the line of sight to any other point N' of that contour; L is obviously parallel to L_o, both making an angle α with the axis Sx, which is also the direction of the cloud's velocity V. The polar coordinates of M in the xy plane are r = SM and θ, the angle PSM; the paraboloid's surface of revolution is described by equation (15) in section 3, namely

$$r = \frac{4a}{1+\cos\theta} \qquad (A1)$$

where

$$r^2 = x^2+y^2+z^2 \quad, \quad \cos\theta = \frac{x}{r} \quad, \quad 2a = SP$$

The line of sight L lies in a plane z = c parallel to Sxy, and letting z = 0 in (A1) brings us back to the equation of the meridional parabola PM, while if z = c, we find the equation

$$y^2 = -8ax + 16a^2 - c^2 \qquad (A2)$$

describing a parabola of vertex Q, passing through N and contained in the plane z = c; its parameter is 4a, so that its focus lies below the reference plane Syz. The parallel lines L_0, L tangent to the paraboloid at the respective points M and N, have their common slope given by differentiation of (A2), viz.

$$\frac{dx}{dy} = \cot\alpha = - (y/4a) > 0 \qquad \text{if} \quad \alpha < 90°$$

The points M, N have consequently the same abscissae y, equal to

$$y = -4a \cot\alpha$$

and this shows that the locus of the points of the paraboloid grazed by the lines of sight to the edge of the nebula lies on a plane curve, which is the parabola of equation

$$z^2 = -8ax + 16a^2 (1-\cot^2\alpha) \qquad (A3)$$

The vertex of the curve is M and its focus is somewhere below the Syz plane; according to (A3), the ordinate of M is equal to

$$x_M = 2a (1-\cot^2\alpha).$$

On the other hand, the intersection P' of Sx with L_0 has the ordinate

$$x_{P'} = x_M + 4a \cot^2\alpha = 2a \cosec^2\alpha$$

having used equation (A2). Now we see that $x_{p'} = SP' = S'M$, where S' is the projection on the sky, of the star S; in other words, we consider the sky as the plane perpendicular to the lines of sight and passing by S'. Furthermore, the ordinate of S' is equal to

$$x_{S'} = x_M - x_{P'} = -4a \cot^2\alpha$$

and the parabola (A3) intersects the plane $x = x_{S'}$, at a point R of abscissa

$$z_R = 4a \, \mathrm{cosec}\alpha = S'R \qquad\qquad (A4)$$

The distance S'M is projected onto the sky as S'M' where

$$S'M' = S'M \sin\alpha$$

whence the ratio S'R/S'M' equal to 2, whatever α, showing that the parabola (A3) projects itself indeed as a parabola of focus S' (the star as seen on the sky), vertex M' and parameter z_R given by (A4).

REFERENCES

Andriesse, C.D., Piersma, T.R., Witt, A.N.: 1977, Astron. Astrophys. <u>54</u>, 841.

Arny, T.: 1977, Astrophys. J. <u>217</u>, 83 .

Arny, T.: 1980, Private communication.

Bates, D., Spitzer, L.Jr.: 1951, Astrophys. J. <u>113</u>, 441.

Field, G.B.: 1971, Astrophys. J. <u>165</u>, 29.

Jancel, R., Kahan, T.: 1963, Electrodynamique des Plasmas , Dunod, Paris.

Jura, M.: 1977, Astrophys. J. <u>218</u>, 749.

Radzievskii, V.V., Dagaev, M.M.: 1969, Soviet Astron. <u>13</u>, 42.

Wickramasinghe, N.C.: 1967, Interstellar Grains, Chapman & Hall Ltd., London.

TRAPPING OF ORBITS IN BARRED GALAXIES*

G. Contopoulos
University of Athens
Athens, Greece

ABSTRACT

The main factors that affect the trapping of stellar orbits around stable periodic orbits in spiral or barred galaxies are reviewed. Such trapping enhances or destroys the imposed field, therefore it is important in producing self-consistent models of galaxies. Various possible models of self-consistent bars are considered. It seems that most bars end at corotation.

The most important orbits in a dynamical system are the stable periodic orbits. These orbits are followed by sets of non-periodic orbits that are trapped in their neighbourhood and have a similar topological form. In fact these non periodic orbits fill tubes around the periodic orbits.

Among the stable periodic orbits most important are those followed by large sets of non periodic orbits. E.g. let us consider a plane axisymmetric galaxy in a frame rotating with angular velocity Ω_S. The basic family of periodic orbits is composed of circular orbits. However there is also an infinity of resonant periodic orbits. Such orbits appear whenever the ratio $(\Omega - \Omega_S)/\varkappa$ (where Ω is the angular velocity of the circular orbit of radius r, and $\varkappa$ the corresponding "epicyclic frequency"), is equal to a rational number n/m (more accurately when $\vartheta_0/2\pi = n/m$, where ϑ_0 is the angle between two successive apocentra). However these orbits are linearly marginally stable (in fact they are unstable) and no trapping occurs around them. On the other hand if we add a bar perturbation some of these orbits become stable and trap other orbits around them.

In studying the orbits in various plane galactic models we must consider the following main factors that influence the characteristics

---

* Paper presented at the workshop "Orbits in Galaxies", organized at the European Southern Observatory (CERN, Geneva) on 5-6 May 1980.

E. G. Mariolopoulos et al. (eds.), Compendium in Astronomy, 55–65.

of the main types of periodic orbits (Contopoulos, 1979) :

1. The Axisymmetric Model

Of particular importance is the form of the $\Omega - \varkappa/2$ curve. The two most important types of curves are : a) A curve decreasing monotonically, and b) A curve increasing from zero to a maximum value, and then decreasing monotonically (Fig. 1).

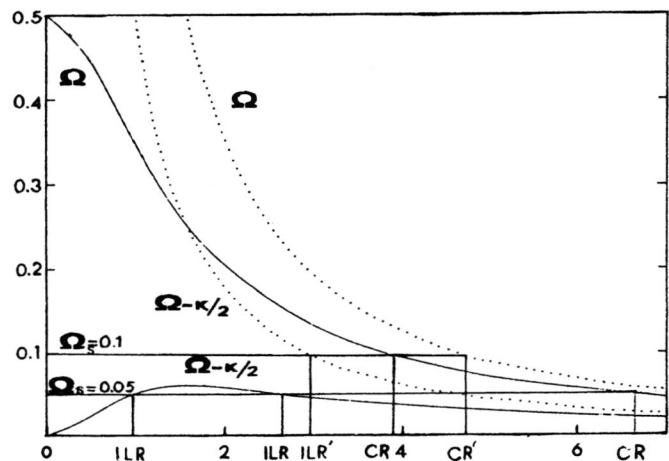

Fig. 1 :The curves Ω and $\Omega - \varkappa/2$ as functions of r in an isochrone model (———) and in a point mass model (....). The straight lines $\Omega_s = 0.05$ and $\Omega_s = 0.1$ intersect the curves $\Omega - \varkappa/2$ and Ω at the inner Lindblad resonances (ILR's) and at corotation (CR). (ILR' and CR' refer to the point mass model).

2. The Angular Velocity of the Spiral, or Bar, Ω_s

In case b) above if $\Omega_s < (\Omega - \varkappa/2)_{max}$ there are two Inner Lindblad Resonances (ILR's), while if $\Omega_s > (\Omega - \varkappa/2)_{max}$ there is no ILR at all. In case a) there is always one ILR (Fig. 1).

3. The Amplitude of the Bar

If the density of the bar is of the order of 1% of the axisymmetric background we call it weak, while if it is of order of 100% we call it strong.

4. The Variation of the Amplitude

The orientation of the orbits depends on the quantity $A' + \frac{4\Omega}{r\varkappa} A$ (where A' is the derivative of the amplitude A with respect to r). The orientation of the periodic orbits changes by 90 % if this quantity changes from positive to negative around the ILR's.

5. Bars vs Tight Spirals

The trapping effects are quite different in bars and in tight spirals (see effect D below). Open spirals are very similar to bars in this respect.

The response density depends mainly on the following :

A. Orientation of the Orbits

The stars stay longer at apocentron inside corotation and at pericentron outside corotation (Fig. 2).

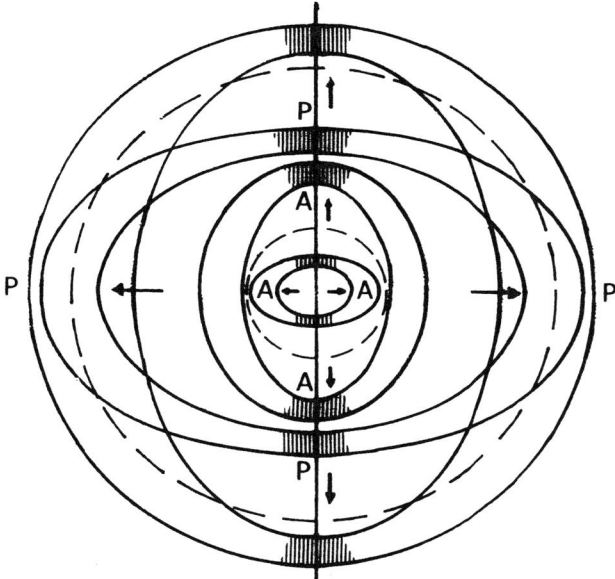

Fig. 2 : The main types of orbits inside and outside the ILR and the OLR (dashed circles) in the case of a bar (heavy straight line). There is only one ILR and the bar is along the strong line. Three main effects influence the response density. (a) The stars stay longer at apocentron (A) inside corotation, and at pericentron (P) outside corotation. (b) The denser parts move outwards along the major axes (arrows). (c) The congenstion of orbits enhances the hatched regions.

B. Axisymmetric Density Variation

The elongation of the orbits brings material from the inner parts of the galaxy outwards along the major axes. In general the density decreases outwards ($\sigma_0' < 0$) thus this effect increases the density along the major axes (Fig. 2).

Both effects A and B are stronger if the elongation of the orbits is larger.

C. Congestion of the Orbits

If the elongation of the orbits changes with the distance r, there is a congestion of the orbits in two diametrically opposite directions (Fig. 2). This effect is strongest near resonances.

D. Differential Rotation of Major Axes

This effect appears in spirals but not in bars, except if the amplitude of the bars changes in time.

In spirals the congestion of orbits produces a spiral density maximum. In tight spirals this is the dominant effect, while in open spirals it is relatively unimportant.

E. Extent of Trapping

If we have two or more stable periodic orbits the question arises what proportion of non-periodic orbits is trapped around each of them. The trapping around higher order resonant orbits is usually very small but the almost circular periodic orbits are followed by large sets of non periodic orbits. The amount of trapped matter depends on the initial distribution function. E.g. in a region near the ILR we have two perpendicular stable periodic orbits. If initially we have a Schwarzschild-type distribution the trapping around the most elongated orbits is small in comparison with the trapping around the less elongated orbits.

F. Stochastic Orbits and Escapes

If the amplitude of the bar is small one can construct a second integral of motion (or an adiabatic invariant) from which one can derive the positions of the periodic orbits.

However if the amplitude of the bar increases the new integral (or adiabatic invariant) is no more valid, and most orbits become stochastic. Inside corotation such orbits in general fill almost circular regions, while outside corotation they usually escape to infinity. If the energy of a star is large enough it can escape even if it starts inside corotation.

We describe now the main families of periodic orbits in barred galaxies.

The Hamiltonian of a barred galaxy is

$$H \equiv \frac{1}{2} \left(\dot{r}^2 + \frac{J_0^2}{r^2} \right) + V - \Omega_s J_0 = h, \tag{1}$$

where h is the numerical value of the hamiltonian, $\dot{r}$ is the radial velocity, J_O the angular momentum, and V the total potential

$$V=V_O(r) + V_1,\tag{2}$$

composed of the axisymmetric background $V_O(r)$ and the bar component, V_1.
A simple form of V_1 is

$$V_1=A(r)\cos 2\vartheta.\tag{3}$$

If we intersect all orbits by the axis $\vartheta=0$, perpendicular to the
bar, we construct a diagram $(r, \dot{r})$ for every value of h. Each orbit in
this diagram is represented by a set of points (Fig. 3). These points
usually form a closed invariant curve surrounding an invariant point which
represents a stable periodic orbit (Contopoulos, 1981). The appearance
of invariant curves indicates the existence of a second integral of motion.
However if the amplitude A is large this integral disappears for most
initial conditions and most orbits are represented by scattered sets of
points.

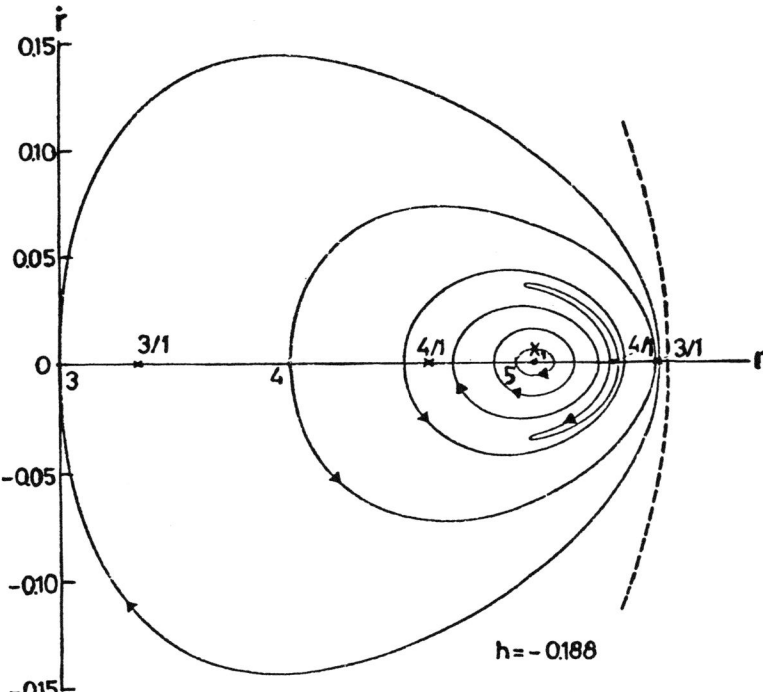

Fig. 3 : Invariant curves in a model galaxy for a given value of the
Hamiltonian h. The invariant curves close around the stable invariant
points x_1 and 4/1 (dots). The unstable points are marked by (x).
(————) is the limiting curve.

It is well known that the orbits cannot cross the so-called curve of zero velocity (where the velocity is zero in an inertial frame). The intersection of the zero velocity curve by the axis $\vartheta=0$ is given by the equation

$$- \frac{1}{2} \Omega_s^2 \, r^2 + V_o + V_1(\vartheta=0) = h. \tag{4}$$

We consider now the main periodic orbits, which cross the ϑ-axis perpendicularly. We find in a diagram (r, h) the characteristic curves of the various families of periodic orbits (Contopoulos and Papayannopoulos, 1980; Contopoulos, 1981).

Equation (4) gives a curve, like a parabola, which we also call a curve of zero-velocity (CZV). Motion is forbidden inside this curve (Figs. 4 and 5).

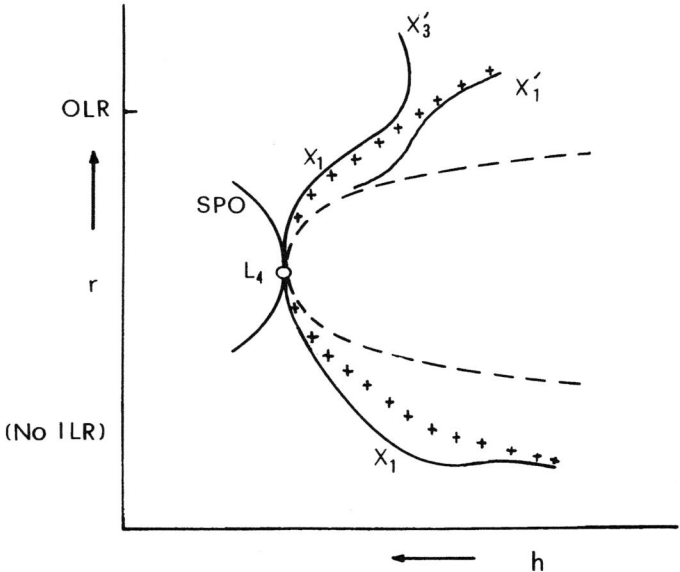

Fig. 4 : Characteristics of the main families of periodic orbits in the case with a single ILR (schematically). The values of r are given as functions of the Hamiltonian h. The unperturbed family of circular orbits (+++) and the CZV (----) are marked. The ILR, OLR and corotation (L_4, representing the stable Lagrangian point) are also given. See description in the text.

As a comparison we start with the axisymmetric case A=0. For each value of h we have a circular orbit of radius r_c. Thus we have the characteristic of the circular orbits (crosses). This curve goes through the Lagrangian point L_4, which lies on the CZV.

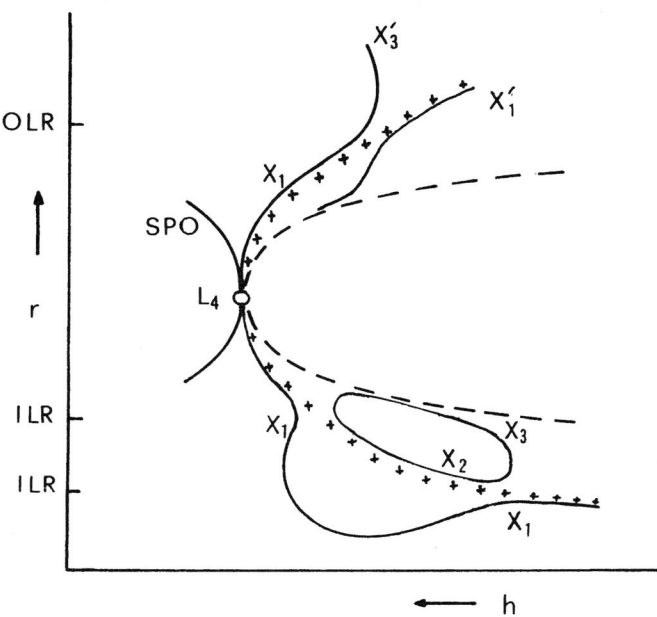

Fig. 5 : Same as in Fig. 4 in the case with two ILR's

If h is larger than the value corresponding to L_4 there is no curve of zero velocity, thus the orbits are not confined by the integral (1). However there is also a second integral of motion, namely the angular momentum, which restricts the motion. In particular there are no escapes unless the energy

$$\frac{1}{2}(\dot{r}^2 + \frac{J_o^2}{r^2}) + V_o(r)$$

is positive (assuming that $V_o(\infty) = 0$).

If we add now a bar perturbation we consider various cases :

(a) Only one ILR

For small enough h (when r_c is inside the ILR) there is a family (x_2) of periodic orbits close to circles but elongated perpendicularly to the bar (Fig. 4).

Beyond the ILR these orbits become very elongated and a new family (x_1) of stable orbits of small elongation appears. This is the main family up to the corotation region. The family x_1 joins the family x_3 of unstable orbits near the ILR. Near corotation this family is composed of elongated long period orbits.

Beyond corotation the continuation of the family x_1 is composed of retrograde orbits elongated perpendicularly to the bar. This family ends by joining the unstable family x_3' near the OLR (Outer Lindblad Resonance). Beyond the OLR there is a new stable family, x_1', of nearly circular orbits elongated along the bar. This family becomes unstable closer to corotation.

To complete the picture we must mention the short period orbits (SPO) that start at L_4 and continue for larger values of h. In a region near L_4 these orbits are stable, and there is some trapping around them. However as the amplitude A of the bar increases the region of stability of these orbits decreases.

There are also many stable orbits of type m/n, which close after n revolutions (n>1). These orbits are unimportant because the set of trapped orbits around them is very small.

(b) Two ILR's

In this case the family x_1 exists all the way from the center to infinity (Fig. 5). Up to the corotation region this family consists of orbits along the bar. Inside the inner ILR and outside the outer ILR the orbits x_1 are close to circles. Between the two ILR's these orbits are very elongated, and a new family of stable periodic orbits, x_2, becomes dominant there, whose orbits are perpendicular to the bar. This happens if the bars are weak and most orbits are initially close to circles (Contopoulos, 1980). If, however, the elongated resonant orbits are populated preferentially then the dominant population is along the bar (Lynden-Bell, 1979).

The family x_2 joins the family x_3 of unstable orbits (Fig. 5). On the other hand close to corotation the orbits x_1 are long period orbits.

Beyond corotation the orbits are similar to the case (a) above. The same is true with the short period orbits.

(c) No ILR

This case is the simplest of all. Inside corotation there is only one family of orbits, x_1, close to circles, elongated along the bar (Fig.6).

As in the cases (a) and (b) above, between the corotation region and the OLR we have stable orbits perpendicular to the bar, while beyond the OLR the stable orbits are along the bar.

(d) Strong Bars

As the bars become stronger we have the following effects :

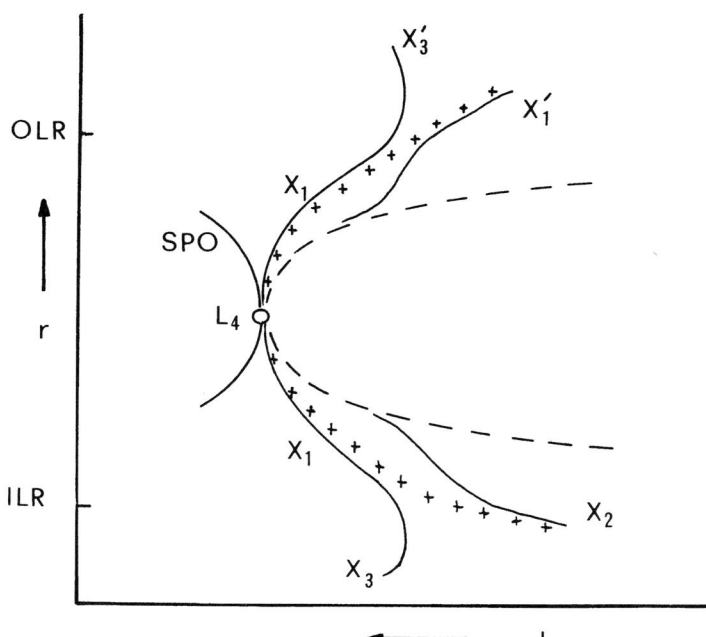

Fig. 6 : Same as in Fig. 4 in the case with no ILR.

A) The deviations of the orbits from circles become larger.

B) In the case of two ILR's the populations x_2, x_3 become less important and disappear for large enough A. If Ω_s is close to the top of the curve $\Omega-\varkappa/2$, even a moderate amplitude A produces the disappearance of these families.

C) Between corotation and the OLR the trapping becomes smaller as the amplitude of the bar increases. A number of new families of simple periodic orbits appear, which restrict the trapping region around the family x_1. In moderately strong bars the family x_1 becomes mostly unstable.

D) Some characteristics inside and outside corotation become tangent to the CZV. The corresponding orbits have an angular point (a cusp). The continuation of such a family consists of orbits with a loop.

When A is large enough the Lagrangian points L_4, L_5 become unstable. In such a case the characteristic of the family x_1 does not reach L_4 but becomes tangent to the CZV and continues with retrograde orbits.

Orbits with cusps or loops are of small importance because they are mostly unstable.

E) In cases of large amplitude we find stochastic orbits inside corotation. Such an orbit fills the whole region inside the corresponding curve of zero velocity, which is an oval elongated along the bar.

Outside corotation such orbits in general escape to infinity. A star may go to large distances, where the effect of the bar is unimportant but then return close to the bar and interact with it (Schwarz, 1979). The escape usually occurs after many interactions with the bar. When the orbits x_1 are unstable most orbits escape to infinity.

(e) Case $A' + \frac{4\Omega}{r\varkappa} A < 0$

In all cases described above we assumed that $A' + \frac{4\Omega}{r\varkappa} A > 0$. If $A' + \frac{4\Omega}{r\varkappa} A < 0$ the orientation of the orbits changes by $90°$.

The above qualitative description is quite general. Similar characteristics were found in different models, used by various authors.

CONCLUSIONS

We can reach the following conclusions as regards possible self-consistent models of bars.

We assume a) that the unperturbed distribution function is peaked around the circular orbits, and b) that $A' + \frac{4\Omega}{r\varkappa} A > 0$.

Then the majority of orbits enhances the bar between the (unique, or outer) ILR and the neighbourhood of corotation. In this region all three major effects in the response density are positive (along the bar).

Between the center and the inner ILR in models with $A' \geqslant 0$ the congestion of the orbits is the strongest effect and we find that the response is negative. However if A' is relatively large negative (even if $A' + \frac{4\Omega}{r\varkappa} A > 0$) the congestion is smaller and the response is positive. Thus we can have a bar from the center to the inner ILR.

If there is no ILR we have positive response all the way between the center and the corotation region.

If the amplitude is sufficiently large we have positive response even between the two ILR's.

Between corotation and the OLR the orientation of the orbits produces a positive response, while the variation of the axisymmetric density produces a negative response. The congestion is negative close to corotation and positive further out. The final result is negative near corotation and positive near the OLR. In any case the effect is small. Furthermore for moderately large amplitudes most orbits between corotation and the OLR escape, thus there is no bar formation in this region (Contopoulos, 1981).

Outside the OLR the orientation of the orbits produces a negative response, while the congestion and the variation of the axisymmetric density produce a positive response. The largest effect is the congestion. Thus it seems that the response outside the OLR is positive in general.

From numerical experiments (e.g. Sellwood, 1980) we have indications that in most models Ω_s is a little above, or a little below the maximum of the curve $\Omega - \varkappa/2$. In such cases the bar should extend all the way from the center to corotation.

Beyond corotation the bar is destroyed in moderate and strong bars (bar densities $\geqslant 10\%$ of the axisymmetric density) by the removal of stars between corotation and the OLR.

Beyond the OLR there may be again an outer bar. This is probably weak.

The case $A' + \frac{4\Omega}{r\varkappa} A < 0$ may also appear in realistic models. In fact in linear self-consistent models one finds that at the ILR's $A' + \frac{4\Omega}{r\varkappa} A$ is exactly zero. In non-linear models this quantity is close to zero and may change sign near the ILR's. Thus we may have a continuation of the bar between the two ILR's. Of course, as A decreases strongly in this region the bar inside the inner ILR is very strong, and it becomes weak further out.

Thus, summarizing, we may have a central bar up to the ILR region, or a bar from the center (or the ILR) to corotation, and also an outer weak bar beyond the OLR.

REFERENCES

Contopoulos, G. : 1979, Astron. Astrophys. <u>71</u>, 221.
Contopoulos, G. : 1980, Astron. Astrophys. <u>81</u>, 198.
Contopoulos, G. : 1981, Astron. Astrophys. (in press).
Contopoulos, G. and Papayannopoulos, Th. : 1980, Astron. Astrophys. <u>92</u> 33.
Lynden-Bell, D. : 1979, Monthly Notices Roy. Astron. Soc. <u>187</u>, 101.
Schwarz, M.P. : 1979, Ph. D. Thesis, Australian National Observatory.
Sellwood, J.A. : 1980, ESO preprint No. 107.

ON THE STABILITY OF THE ASTEROIDS

John D. Hadjidemetriou
University of Thessaloniki
Thessaloniki, Greece

ABSTRACT

It is proved that all the nearly circular orbits of asteroids at
the resonances with Jupiter of the form (2n+1)/(2n-1), i.e. 3/1, 5/3,...
are unstable and gaps in the distribution of the asteroids must be expected there. Since these resonant unstable orbits are very close to
each other and have an accumulation point at the orbit of Jupiter, the
space near Jupiter must be empty of asteroids. The resonant asteroid
orbits, of the form (n+1)/n, i.e. 2/1, 3/2, 4/3, ... are not necessarily
all unstable, and concentrations of asteroids could appear, as is the
case with the 3/2 resonance.

1. INTRODUCTION

It is well known that apart from the main planets of the solar system, many thousands of small planets, or asteroids, revolve around the
Sun mainly in the area between Mars and Jupiter. Most asteroids describe nearly circular orbits but there are also orbits with large eccentricities. If only the effect of the Sun is taken into account then
the orbits are Keplerian. In this approximation one would expect a
smooth distribution of the asteroids. This is not the case however.
The observations have shown that the asteroids avoid the orbits which
are in the 3/1, 5/2, 7/3, 2/1 and 5/3 resonance with the motion of Jupiter, the famous Kirkwood gaps, but they also concentrate at the resonance 3/2 (the Hilda group). Also, very few asteroids are beyond the 3/2
resonance, so practically the area near Jupiter is empty of asteroids
(e.g. Roy, 1979).

The explanation of the above distribution is a longstanding problem
of Celestial Mechanics. This problem could not be solved by the old
methods and techniques. The explanation that the resonance with Jupiter
makes the orbit unstable because of the small divisors which appear, is
not satisfactory because we also have concentration of asteroids at some
resonances. Recent developments in mathematics have shed much light to

67

E. G. Mariolopoulos et al. (eds.), Compendium in Astronomy, 67–78.
Copyright © 1982 by D. Reidel Publishing Company.

this problem and its main features are now becoming clearer. In this paper we shall study the stability of the asteroid orbits by making use of the theory developed in Hadjidemetriou (1981) on the generation of instabilities in periodic planetary-type orbits of the general N-body problem, based on the theory of Hamiltonian perturbations of Krein, Gelfand and Lindskii (see Yakubovich and Starzhinskii, 1975). The results are compared with the theorem of Kolmogorof, Arnold and Moser, known as the KAM theorem, on perturbations of integrable systems. In this way we shall come to conclusions which are consistent with the observations.

We could take into account any number of perturbing planets to study the stability of the asteroid orbits. Since however Jupiter's effect is by far more important than the effect of the other planets of the solar system, and in order to make clearer the presentation of the main ideas, we shall consider Jupiter only as the perturbing body. In the study which follows we shall not assume that the mass of the asteroid is negligible.

As we shall explain in the next section, there is a dense set of periodic orbits of the general 3-body problem, for the Sun, Jupiter and a third body which we shall identify as the asteroid, though most of these orbits are multiple with high multiplicities and large periods. Then the actual motion of an asteroid will be considered as a perturbed motion of some periodic orbit of the above type.

2. FAMILIES OF PERIODIC PLANETARY-TYPE ORBITS OF THE ASTEROIDS

We start by assuming that the masses of Jupiter (denoted by P_1) and asteroid (denoted by P_2) are both zero. Then they describe uncoupled Keplerian orbits around the Sun. We assume that both orbits are circular with frequencies ω_1 and ω_2, respectively, in the same plane and the same direction. We define now a rotating frame of reference Oxy with the Sun at the origin and the x-axis defined by the line joining the Sun with P_1. It can be easily seen that the motion of an asteroid in this rotating frame is periodic with period $T=2\pi/(\omega_2-\omega_1)$. We normalize the units so that

$$m=1, \quad G=1, \quad \dot{\vartheta}_0=1 \tag{1}$$

where m is the total mass of the system, coinciding here with the mass of the Sun, G is the gravitational constant and $\dot{\vartheta}_0$ is the angular velocity of rotation of the rotating frame at t=0, here coinciding with $\omega_1=1$, and also

$$T = \frac{2\pi}{(\omega_2/\omega_1) - 1} . \tag{2}$$

According to the normalization (1), the radius of the orbit of Jupiter is equal to 1, and consequently $x_1=1$, where x_1 is the abscissa of P_1, which is evidently constant. The initial conditions of such a periodic orbit are

$$x_{10}=1, \; \dot{x}_{10}=0, \; x_{20}=R=\text{arbitrary}, \; y_{20}=0, \; \dot{x}_{20}=0, \; \dot{y}_{20}=R^{-1/2}-R, \; \dot{\vartheta}_{o}=1, \quad (3)$$

where we have assumed that at t=0 all bodies lie on the x-axis.

A periodic orbit given by (3) is symmetric with respect to the x-axis of the rotating frame, since at t=0 we have $\dot{x}_1=0$, $y_2=0$, $\dot{x}_2=0$. This orbit belongs to a monoparametric family along which R varies. Equivalently, we can use the ratio of the frequencies of the asteroid and Jupiter, namely ω_2/ω_1, as the parameter of the family. We shall restrict the study to inner asteroid orbits, i.e. R<1, or $\omega_2/\omega_1>1$. A symmetric periodic orbit of the above type can be represented by a point with coordinates $(x_{10}, x_{20}, \dot{y}_{20})$ in the space of (nonzero) initial conditions $x_{10}x_{20}\dot{y}_{20}$. This means that the monoparametric family of asteroid orbits can be represented by a smooth curve in this space. For a qualitative description, we shall use here the projection of this curve on the $x_{10}x_{20}$ plane only. The family (3) is therefore represented by the straight line $x_{10}=1$ (Fig. 1). Each point on this line represents a periodic planetary-type orbit for the system Sun-Jupiter-asteroid, for zero masses of Jupiter and the asteroid. Several resonant orbits are shown on this line.

Let us examine now what happens when the masses of Jupiter and the asteroid are increased. It can be proved (Hadjidemetriou 1975a) that a periodic orbit (3) is continued uniquely, when the masses are increased, to a periodic motion of the general 3-body problem, of the planetary type. This is a symmetric periodic orbit in a rotating frame whose origin is at the center of mass of Jupiter and Sun, and its x-axis contains always these two bodies. The continuation from zero to nonzero masses is not unique when T, given by (2), is equal to $T=2\nu\pi$, $\nu=1,2,3,\ldots$. This corresponds to the resonant orbits $\omega_2/\omega_1=2/1$, 3/2, 4/3, $\ldots$. As a consequence, the degenerate family $x_{10}=1$ (zero masses) breaks down, when the masses are increased, to an infinite number of families of periodic orbits of the planetary type which are "separated" by the resonant orbits 2/1, 3/2, $\ldots$ as it was shown by numerical computations (Hadjidemetriou 1976, Delibaltas 1976). The projections of the first three of these families are shown in Fig. 1.

The first family has a "circular" branch, nearly parallel to the x_{20} axis, which represents the continuation of the circular orbits with zero masses, between the resonances 0 and 2/1. These are the so called periodic orbits of the first kind. There is also an "elliptic" branch which is generated from the resonant orbit 2/1. (We remind that the continuation of the resonant orbit 2/1 is not unique). Along this branch the orbits of Jupiter and the asteroid are nearly elliptic, with eccentricities which may become quite large, but the resonance is almost constant, equal to 2/1. These are the periodic orbits of the second kind, obtained from the continuation of unperturbed elliptic orbits.

A similar situation holds for all other families. We have two

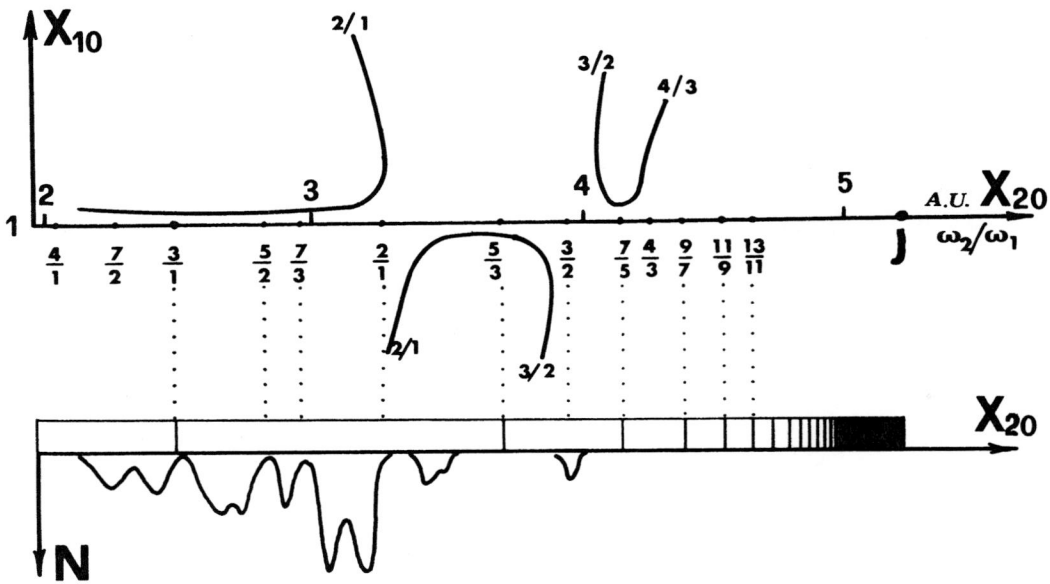

Fig. 1 : Three families of periodic planetary-type orbits for the asteroids (schematically). The resonances on the "elliptic" branches are
indicated. The distribution of asteroids is shown in the lower part,
and the unstable resonant orbits $(2n+1)/(2n-1)$ are shown in the strip.

"elliptic" branches, generated by the resonant unperturbed orbits $(n+1)/n$
and $(n+2)/(n+1)$, respectively, along which the resonance is almost constant, equal to the above values, respectively, and in between a "circular" branch along which the resonance varies between the above two values.

The above description is for the general 3-body problem, i.e., the
mass of the asteroid is not considered as negligible. A particular case
would be that in which the mass of the asteroid is negligible, corresponding to the restricted 3-body problem. This latter case has been studied
by Guillaume (1969) who found the same general behaviour as in the general problem.

All the above described orbits are simple symmetric periodic orbits.
This means that the asteroid starts perpendicularly from the x-axis at
t=0 when $\dot{x}_1=0$ and falls again perpendicularly on the x-axis at the <u>first</u>
intersection ($\dot{x}_1=0$ at that instant also). These orbits will be called
simple periodic orbits. We can assume however that any orbit for zero
masses is a multiple symmetric periodic orbit, i.e. we assume that the
asteroid falls perpendicularly on the x-axis at the nth intersection,
n>1. In particular, if the unperturbed orbit is resonant, $\omega_2/\omega_1=r/s$,
its period as obtained from (3) is $T=2\pi s/(r-s)$ and consequently, if as-

sumed to be described (r-s) times, it has a period which is a multiple
of 2π. Thus, according to the above, it is not continued uniquely as
a multiple periodic orbit of the first kind but, instead, it generates
a resonant "elliptic" branch, when the masses are increased, in exactly
the same way as the simple resonant orbits 2/1, 3/2, ..., shown in Fig.
1. Along this "elliptic" branch we have a constant resonance (rp)/(sp),
where p=r-s. For example, the unperturbed resonant orbit 5/2 is continued as a simple periodic orbit of the first kind, with nearly circular
orbits of Jupiter and asteroid, since its period is $T=10\pi/3$. If however
we consider it as being described 3 times, its period is equal to 10π,
i.e. a multiple of 2π, and consequently it generates resonant branches
when the masses are increased. This is indeed the case, as found by
Hadjidemetriou (1980). There exist two different such resonant families
with resonance 15/6, along which the eccentricities vary.

We note that the set of unperturbed resonant orbits r/s is dense
in the set of all circular orbits. We mentioned before that each unperturbed orbit can be represented by a point on the straight line $x_{10}=1$
in Fig. 1 and consequently the resonant orbits r/s are represented by
points which are dense on this line. When the masses are increased each
of these points generates resonant branches and consequently the whole
space in Fig. 1 is filled by resonant orbits, which however are of high
order. Only the simplest of these are shown in Fig. 1.

The above mentioned periodic orbits of the planetary type can be
considered as closely representing motions of asteroids. If we compare
the results summarized in Fig. 1 with observations (see for example Roy,
1979), we find that most asteroid orbits are confined to the part of the
line $x_{10}=1$ lying in the area between the resonances 2/1 and beyond, up
to the orbit of Mars, and also betwen 2/1 and 5/3 and also at the resonance 3/2. This distribution is shown, for comparison, in the lower part
of Fig. 1. The Kirkwood gaps are evident in this presentation. The
existence of these gaps will be explained by studying the stability characteristics of the described periodic orbits. This will be done in the
next section.

3. THE STABILITY OF THE PERIODIC ORBITS OF THE PLANETARY TYPE

(a) General Remarks

We start by studying first the linear stability of a periodic orbit
of the planetary type. This is a periodic orbit of the general planar
3-body problem and has four degrees of freedom (assuming the center of
mass fixed). As generalized coordinates we can use the abscissa x_1 of
P_1, the coordinates (x_2, y_2) of P_2, in the rotating frame defined in
section 2, and the angle ϑ between the x-axis and a fixed direction in
the inertial frame. It turns out (Hadjidemetriou, 1975a) that ϑ is an
ignorable coordinate and consequently we have the angular momentum integral besides the energy integral. The angular momentum integral can be
used to eliminate the angle ϑ and thus we are left with three degrees

of freedom (coordinates x_1, x_2, y_2) and only one integral, the energy integral.

A symmetric periodic orbit of the general 3-body problem, and in particular of the planetary type, is of the form

$$x_1 = x_1(x_{10}, x_{20}, \dot{y}_{20}, E, L; t), \quad x_2 = x_2(x_{10}, x_{20}, \dot{y}_{20}, E, L; t),$$

$$y_2 = y_2(x_{10}, x_{20}, \dot{y}_{20}, E, L; t) \tag{4}$$

where E, L are the energy and angular momentum constants, respectively. The linear stability of a periodic orbit (4) can be studied by integrating the variational equations and computing the eigenvalues of the monodromy matrix (see for example Whittaker, 1960). We know, by the Hamiltonian nature of the problem, that there exist three pairs of reciprocal eigenvalues λ_1, ... λ_6, for which we have

$$\lambda_1 \lambda_2 = 1, \quad \lambda_3 \lambda_4 = 1, \quad \lambda_5 \lambda_6 = 1. \tag{5}$$

The λ_i's can be also arranged in complex conjugate pairs, and also there always exists a pair of unit eigenvalues, $\lambda_5 = \lambda_6 = 1$, because of the existence of the energy integral. For the other two pairs, we may have the following possibilities, consistent with the above mentioned properties :

(a) $\lambda_{1,2} = e^{\pm i\varphi_1}$, $\lambda_{3,4} = e^{\pm i\varphi_2}$,

(b) $\lambda_{1,2} = e^{\pm i\varphi_1}$, $\lambda_3 = \alpha$, $\lambda_4 = 1/\alpha$, α:real,

(c) $\lambda_1 = \alpha$, $\lambda_2 = 1/\alpha$, $\lambda_3 = \beta$, $\lambda_4 = 1/\beta$, α, β:real   (6)

(d) $\lambda_{1,4} = Re^{\pm i\varphi_1}$, $\lambda_{2,3} = R^{-1} e^{\pm i\varphi_1}$,   R:real

A method to obtain the nonunit eigenvalues λ_1, ... λ_4 is given in Hadjidemetriou (1975b), based on the method of surface of section. The stability properties in terms of the coefficients of the characteristic equation of the monodromy matrix are given in Broucke (1969). Obviously, only the case (6a) is stable, which means that instabilities can be generated if a unit eigenvalue which is originally situated on the unit circle in the complex plane moves out of this circle, as a consequence of a perturbation. In this study we shall consider Hamiltonian perturbations only. Note that a linearly unstable orbit is unstable (Lefschetz, 1977).

Let us now come back to the circular uncoupled Keplerian orbits, represented by the straight line $x_{10} = 1$ in Fig. 1. From Keplerian theory it can be found that the nonunit eigenvalues are given by (e.g. Hadjidemetriou, 1979),

$$\lambda_{1,2}=e^{\pm i\varphi_1}, \qquad \lambda_{3,4}=e^{\pm i\varphi_2},$$

where $\varphi_1=\omega_1 T$, $\varphi_2=\omega_2 T$ and $\varphi_1=\varphi_2 (\mathrm{mod}\,2\pi)$ because of (2). Thus,

$$\lambda_{1,3}=e^{i\varphi}, \qquad \lambda_{2,4}=e^{-i\varphi}, \tag{7}$$

where

$$\varphi=2\pi/(\omega_2/\omega_1 -1). \tag{8}$$

From (7) and (8) we obtain that there exist two double eigenvalues of an unperturbed circular motion which lie on the unit circle and are symmetric with respect to the real axis. These eigenvalues depend on the ratio ω_2/ω_1 and consequently they move on the unit circle as we proceed along the family $x_{10}=1$.

When the masses of Jupiter and the asteroid are increased, we obtain the periodic orbits of the planetary type described in section 2. The study of their stability therefore reduces to the study of the evolution of the eigenvalues of the unperturbed orbit, of the form (7), when the masses are increased. We shall study separately the stability of the "circular" and the "elliptic" branches, as there are qualitative differences between them. The rigorous mathematical treatment is given in Hadjidemetriou (1981). We shall present here the main results.

(b) Linear Stability of the "Circular" Branch

Let us consider an unperturbed periodic orbit for which $\varphi \neq 2n\pi$, or $\varphi \neq (2n+1)\pi$. Then it can be proved that the perturbation due to the increase of the masses makes the eigenvalues to move on the unit circle, thus preserving stability (note that the evolution out of the unit circle, of the form 6d, does not violate the Hamiltonian properties, but it is proved that this never can happen). Let us see now what happens as we approach a periodic orbit with $\varphi=(2n+1)\pi$, moving on the straight line $x_{10}=1$: The double eigenvalues $\lambda_{1,3}$ and $\lambda_{2,4}$ approach to the point -1 on the unit circle, and coincide with it at $\varphi=(2n+1)\pi$, which corresponds to the resonant orbits

$$\omega_1/\omega_2 = (2n+1)/(2n-1), \quad n=1,2,3, \ldots, \tag{9}$$

i.e. 3/1, 5/3, ... At this point it is proved that at the unperturbed resonant orbits (9) there always exists a Hamiltonian perturbation which generates instability. The numerical computations have shown that the increase of the masses is such a perturbation (Hadjidemetriou, 1976). This means that at the resonant asteroid orbits 3/1, 5/3, 7/5, ... we must expect unstable regions.

Note that the third pair of eigenvalues $\lambda_5 = \lambda_6 = 1$ is preserved under a Hamiltonian perturbation, because of the existence of the energy integral.

Note also that there is an accumulation of resonant unstable orbits (9) at Jupiter's orbit. This is shown in the strip in the lower part of Fig. 1.

The case $\varphi = 2n\pi$ corresponds to the unperturbed orbits with resonances 2/1, 3/2, ... which generate elliptic branches, which will be studied next.

(c) Linear Stability of the "Elliptic" Branches

Let us come next to the stability of the "elliptic" branches. As we already mentioned, they are obtained from the continuation of two un-coupled elliptic Keplerian orbits with resonance $(n+1)/n$, and its period is a multiple of 2π, so the eigenvalues are all equal to unity,

$$\lambda_1 = \lambda_2 = 1, \quad \lambda_3 = \lambda_4 = 1, \quad \lambda_5 = \lambda_6 = 1. \tag{10}$$

In this case, only the last pair is conserved when the masses are increased, due to the existence of the energy integral. The other two pairs may move on or out of the unit circle and thus instability may be generated. Whether or not instability is generated depends on the particular branch and also on the magnitude of the masses and the ratio of the masses, as shown by numerical computations (Hadjidemetriou, 1976; Delibaltas, 1976). Thus, contrary to the "circular" resonant orbits 3/1, 5/3, ... , which always become unstable when the masses are increased, "elliptic" resonant orbits 2/1, 3/2, ... may become stable in some cases. This is also true for the resonant orbits 5/2, 7/3, e.t.c. which generate resonant "elliptic" branches of multiple periodic orbits.

(d) Comparison of the Linear Stability Analysis with the KAM Theorem

The Kolmogorof-Arnold-Moser theorem (KAM theorem) deals with slightly perturbed Hamiltonian systems, of the form $H(q,p) = H_0(q,p) + \varepsilon H_1(q,p)$ where the unperturbed Hamiltonian $H_0(q,p)$ is integrable, i.e. it has n independent, analytic, singlevalued first integrals, which we assume to be in involution (the Poisson bracket of any two of them is equal to zero). It can be proved then (see Berry, 1978; Treve, 1978) that for $\varepsilon = 0$ any bounded trajectory in phase space lies on an n-dimensional torus, which we shall call invariant torus. In this case we can introduce the so called action-angle variables and the Hamiltonian takes the form

$$H(q,p) = H_0(p) + \varepsilon H_1(q,p), \tag{11}$$

where the q_i can be considered as the angles on the torus and the p_i as the radii. Evidently, for $\varepsilon = 0$ the motion is stable and the question is what happens when $\varepsilon \neq 0$. The KAM theorem states that most of the above invariant tori are preserved, though distorted, but some of them, forming a set of measure greater than zero, are dissolved and behave in a patho-

logical way, filling the whole phase space. Moreover, this set of dissolved tori is dense in the set of the unperturbed tori (see Moser, 1978).

The dissolved tori correspond to the resonant cases for the unperturbed frequencies, defined by

$$\omega_k = \partial H_0 / \partial p_k, \quad k=1, 2, \ldots n, \tag{12}$$

while the tori which survive correspond to frequencies which are not
commensurable. However, only the low order resonances are of importance,
because only in this case the instability area, corresponding to the dissolved invariant tori, is of practical importance. The instability area
generated by dissolved tori of high resonance is very small. Evidently,
the dissolution of the invariant tori corresponds to the generation of
instability.

It can be proved that the planetary problem can be expressed in the
form (11) by a suitable choice of the variables, where ε is a parameter
which is zero when all the masses of the planets are zero (in our case
the masses of Jupiter and asteroid). The bounded unperturbed orbits
are Keplerian orbits which lie on invariant tori (we consider motion in
the rotating frame where the degrees of freedom have been reduced to
three). This is clear if we take into account that in this case we have
three integrals of motion, two energy integrals for Jupiter and for the
asteroid, respectively, and the angular momentum integral of the asteroid
(the angular momentum integral of Jupiter has been used to reduce to
three the degrees of freedom). The frequencies ω_1 and ω_2, given by (12),
are the frequencies of Keplerian motion of Jupiter and the asteroid, respectively. When $\varepsilon \neq 0$ (nonzero masses) only one integral survives, the
energy integral of the whole system. Now, according to the KAM theorem,
most of the above mentioned tori survive when $\varepsilon \neq 0$, there are however tori which dissolve. These latter tori correspond to rational values of
ω_2/ω_1, i.e. to commensurabilities between the motions of Jupiter and the
asteroid. These commensurable motions are dense, though of measure zero, in the set of all possible unperturbed motions. So the dissolved
tori, when the masses increase, are densely distributed among the tori
which survived. Their measure in phase space is greater than zero but
the instability area is of importance only near the lowest order resonances. This is true especially for the resonances 3/1, 5/3, ...
Note that these resonant orbits become linearly unstable when $\varepsilon \neq 0$, as stated
in section 3b. The generation of linear instability is thus associated
to the dissolution of the invariant tori on which lied the unperturbed
resonant orbits.

Let us consider next an "elliptic" orbit. The invariant torus on
which the unperturbed orbit lies may dissolve if linear instability is
generated when $\varepsilon \neq 0$. Simple "elliptic" orbits are at the resonances 2/1,
3/2, 4/3, ... and the numerical integrations have shown that we do have
instability in some cases, but in other cases the increase of the masses
results in linearly stable perturbed periodic orbits. How is this latter effect reconciled with the KAM theorem ? To study better this we

note that the KAM theorem is applicable when the unperturbed system corresponds to the Hamiltonian (11) for $\varepsilon=0$, i.e. there is no coupling of
the degrees of freedom and there exist n integrals of the motion. Let
ε_0 be the maximum perturbation for which the KAM theorem is applicable.
If the actual perturbation ε is larger than ε_0, $\varepsilon > \varepsilon_0$, then the KAM theorem cannot be applied because if we take the case ε_0 as the unperturbed
problem to increase further the value of ε and thus extend the applicability of the KAM theorem, we immediately see that we no longer have n
integrals of motion at $\varepsilon = \varepsilon_0$. In this case we can study the problem by
a linear stability analysis. It seems that at some "elliptic" resonant
orbits, notably at the resonance 3/2, the actual masses in the solar system are beyond the applicability of the KAM theorem. So the linearly
stable orbits at these resonances imply stability of the system. Thus,
it seems that the increase of the masses stabilizes the system.

In stating that the increase of the masses stabilizes the system
we do not mean that we do not have any more dissolved invariant tori.
These always exist, but by increasing the masses we move away from a low
order resonance, for example 3/2, and go to higher order resonances where
the measure of dissolved invariant tori corresponding to them is negligible. These resonances are between the frequency of a perturbed peri-
odic orbit for $\varepsilon = \varepsilon_0$ and the frequency of the small oscillations near this
periodic orbit, as determined by the purely imaginary characteristic ex-
ponents. Note that the set of all such resonant orbits is dense in the
set of all the periodic orbits for $\varepsilon = \varepsilon_0$ belonging to a monoparametric
family, but their order of resonance is very high.

4. DISCUSSION

We now piece together the previous analysis and we come to the following conclusions on the distribution of the asteroid orbits :

(a) No asteroid orbits can exist near the resonances 3/1, 5/3, ..., as
explained in sections 3b and 3d. At these resonances instability is always generated even for very small masses of Jupiter and the asteroid and
for any ratio of their masses. Also, by the KAM theorem the measure of
the dissolved invariant tori is not negligible since the order of the
resonance is low. This is indeed the case, as obtained by the observations.

(b) No asteroid orbits can exist near Jupiter. Indeed, the unstable resonant "circular" orbits $(2n+1)/(2n-1)$, i.e. 3/1, 5/3, 7/5, ... have an
accumulation point at the orbit of Jupiter, since the limit of the above
sequence is 1/1 (this should not be confused with the Trojan asteroids
which correspond to the same resonance, because these latter are of a
completely different nature and the above results do not apply to them).
A simple calculation shows that the radii of the resonant unstable orbits
$(2n+1)/(2n-1)$ are so close to each other that practically the whole area
near Jupiter is unstable (see Fig. 1). The observations have shown that
the area near the 5/3 resonance is empty and apart from the resonance 3/2,

the whole area beyond this point is empty of asteroids. This has been also tested numerically : Perturbed orbits at the resonances $(2n+1)/(2n-1)$, even for rather small values of n, rapidly escape from this region. This has been also observed by numerical integrations made by Lecar and Franklin (1974). Note that the above result is consistent both with the theory presented in section 3b and with the KAM theorem. The difficulty of the small measure of the dissolved invariant tori in phase space at the high order resonances $(2n+1)/(2n-1)$, in the application of the KAM theorem, is overcome by the fact that these unstable orbits are coming closer and closer to each other as the order of the resonance increases.

(c) It is possible for asteroid orbits to exist at the resonances of the form $(n+1)/n$, i.e. 2/1, 3/2, 4/3, ... These resonant orbits correspond to elliptic orbits and the increase of the masses stabilizes in some cases the system, as explained in section 3c. The eigenvalues of the unperturbed periodic orbits are at the point +1, but the increase of the masses has in some cases the consequence to shift them on the unit circle, away from the dangerous point +1 (we remind that it is from this point that they can move out of the unit circle, thus generating instability). In this way the increase of the masses stabilizes the system. This happens at the resonance 3/2, where the observations have shown a concentration of asteroids, known as the Hilda group. On the contrary, the resonance 2/1 for simple orbits and the resonances 5/2, 7/3 for multiple elliptic orbits are unstable. The above results are consistent with the theory developed in section 6c. The KAM theorem seems not to be applicable in all such cases, as explained in section 6d. There exist many other resonant cases where no gaps have been observed in the asteroids. These resonances are linearly stable, if considered as simple orbits but may become unstable if considered as multiple. This means that a perturbed orbit at the vicinity of the simple orbit can be considered also as a perturbed orbit in the vicinity of the multiple orbit at the same resonance. What will happen then ? For a low order resonance the instability may dominate, as indeed nature shows us to take place at the resonances 7/3 and 5/2. Then the process of the so called Arnold diffusion has depleted the region of asteroids, thus forming the corresponding Kirkwood gaps. For a higher order resonance however, the unstable regions are so small that their effect is negligible.

(d) The nonexistence of asteroids beyond the orbit of Mars is not associated with instabilities, because at that part of space the circular asteroid orbits correspond to resonances which are all linearly stable (the first unstable resonant orbit is 3/1). Their absence must be traced back to the conditions of formation of the asteroids.

Note that the orbits of Mars, Earth, Venus and Mercury are well within the stable region, with respect to perturbations due to Jupiter.

Note also that the 5/2 resonance is unstable for the asteroids, but the same resonance appears for Jupiter and Saturn, which seems to be stable. This means that the KAM theorem is not applicable up to the mass of Saturn, and that the increase in the masses of the perturbing planets stabilizes the resonant system 5/2 by the same mechanism as for the 3/2 resonance, stated in (c) above.

REFERENCES

Berry, M.V. : 1978, in Topics in Nonlinear Mechanics, A Tribute to Sir Edward Bullard (ed. by S. Jorna), American Inst. of Physics, New York, p. 16

Broucke, R. : 1969, J. of AIAA 7, 103

Delibaltas, P. : 1976, Astrophys. Space Sci. 45, 207

Guillaume, P. :1969, Astron. Astrophys. 3, 57

Hadjidemetriou, J.D. : 1975a, Celes. Mech. 12, 155

Hadjidemetriou, J.D. : 1975b, Celes. Mech. 12, 255

Hadjidemetriou, J.D. : 1976, Astrophys. Space Sci. 40, 201

Hadjidemetriou, J.D. : 1979, in Instabilities in Dynamical Systems (ed. by V. Szebehely), D. Reidel, Dordrecht-Holland, p. 135

Hadjidemetriou, J.D. : 1980, Celes. Mech. 21, 63

Hadjidemetriou, J.D. : 1981, Celes. Mech. 23, 277.

Lecar, M. and Franklin, F.A.: 1974, in The Stability of the Solar System and of Small Stellar Systems, IAU Symp. No. 62 (ed. by Y. Kozai), D. Reidel, Dordrecht-Holland p. 37.

Lefschetz, S. : 1977, in Differential Equations : Geometric Theory, Dover, p. 188

Moser, J. : 1978, in Topics in Nonlinear Mechanics, A Tribute to Sir Edward Bullard, (ed. by S. Jorna), American Inst. of Physics, New York, p. 1

Roy, A.E. : 1979, in Orbital Motion, Adam Hilger Ltd., Bristol, p. 233

Treve, Y.M. : 1978, in Topics in Nonlinear Mechanics, A Tribute to Sir Edward Bullard (ed. by S. Jorna), American Inst. of Physics, New York, p. 147

Yakubovich, V.A. and Starzhinskii, V.M. : 1975, in Linear Differential Equations with Periodic Coefficients, 1, Halsted Press, Chic. Ill.

Whittaker, E.T. : 1960, in Analytical Dynamics of Particles and Rigid Bodies, Cambridge Univ. Press, pp. 353, 398.

PART III
SOLAR PHYSICS

SOLAR MAGNETIC FIELDS AND ACTIVITY

V. Bumba
Astronomical Institute Czechoslovak Academy of Sciences
Ondrejov Observatory

ABSTRACT

In the present paper we are trying to summarize the experience of
a solar observer dealing with the dynamics of solar background magnetic
fields formation, including the development of new active regions magnetic fluxes. Single active regions occurrences,complexes of activity and
proton-flare region related large-scale regular magnetic patterns growths
are investigated. The meaning and importance of magnetic active longitudes are indicated. The physical background of magnetic fields generation in connection with convection, differential rotation and before all
with the photospheric surface kinematics of the background fields is discussed. The need for more systematic and complex observations in the
form of physical quantities measurements as well as for their evaluations
is the natural consequence of all presented discussions.

1. INTRODUCTION

An increasing number of astrophysicists realizes that by studying
the regularities of solar activity, investigating the rules of its development and by searching for the physical reasons why it exists, we
investigate at the same time more general physical laws governing the
changes in the photospheres, chromospheres, coronas and in the electromagnetic emission as well as in stellar winds of a substantial part of
stars in the universe. It concerns at least the main-sequence stars of
the Hertzsprung-Russel diagram possessing the convection in their upper
layers, which - according to the numerous solar observational evidence -
is closely related to the production of magnetic fields - the main source
of solar and highly probably also of stellar activities.

We may discuss the physics of these relations, but the observational
facts demonstrating the dependence of magnetic field distribution and
new magnetic flux appearance among other factors also on the hierarchy
of convectional elements in the solar atmosphere go still further : they
line up the Sun through its observations, for example, in integrated

81

E. G. Mariolopoulos et al. (eds.), Compendium in Astronomy, 81–96.
Copyright © 1982 by D. Reidel Publishing Company.

monochromatic light of ionized calcium lines H and K and with the help
of the Wilson-Bappu relationship (Wilson and Bappu, 1957) with other
stars having convectional zones. And what more, all these stars, including the Sun, showing the dependence of their chromospheric activities
produced by their convection and magnetic fields on their luminosities
(Wilson, 1963; Wilson and Skumanich, 1964) suggest that the intensity of
stellar and, of course, of the solar activity may be closely related to
the intensity of processes generating energy in their interiors.

In the present paper we would like to consider more in detail the
possible ways how to obtain from the observed morphological regularities
in the solar magnetic field and activity development better indicators
of physical processes causing the observed phenomena, and how to understand
them more deeply. We would like to do this by investigating in the first
place the new magnetic field appearance and some of the solar activity
consequences of this new magnetic flux evolution in the whole solar atmosphere from its simplest form of one individual active region development to the most complex and complicated form we know until now - the
proton-flare related large-scale magnetic field regular pattern formation.

2. BACKGROUND MAGNETIC FIELDS DURING THE NEW MAGNETIC FLUX APPEARANCE

The development of solar activity, which is practically the same
like the development of a new magnetic flux, is determined by the dynamics of solar background magnetic fields. For example the evolution of
an active region depends not only on the magnetic field state due to the
phase of the eleven- as well as the twenty-two - year cycle and their
mutual relations, but it is influenced by its possible relevance to an
active longitude, by the prehistory of the background field in the given
region or by the fact that it may be a part of a more complex evolutionary process, e.g. of a complex of activity or of a large-scale regular
magnetic pattern. At the same time the kinematics of the background
fields - interacting new local and old fields dissipated from active
regions - is effected not only by the action of individual modes of convective elements, but also by the existence of active longitudes and
latitudes, for example, by the Spörer's law and other regularities in
longitudinal and latitudinal distribution, by the very important action
of differential rotation, by the resulting density, intensity and topology of magnetic fields, by the velocity of their boundaries shift during
their development and area growth and other, up to now not yet known
reasons and forces (Bumba and Howard, 1965b; Bumba et al., 1968). At
the same time we already learned some statistically meaningful regularities governing the development of individual active regions which we
do not yet understand physically (Bumba, 1970; Bumba and Howard, 1965a).
Let us again remember, for example, the close relationship of the new
magnetic field intensification and the related chromospheric emission
extension with the supergranular and granular convectional network (Bumba
and Howard, 1965a), the role of the "center" of a sunspotgroup, which
means the same like the internal magnetic field boundary and the chromospheric as well as flare-activity related to the field singularity and

their time development bound to the individual phases of local magnetic
field and background magnetic field development, as soon as both fields,
the local and the background one, start to interact (Bumba and Tomášek,
1980; Bumba, 1981a). We may also mention the practically constant velocity of about 10 m/sec with which the outer boundary of the local field
shifts during its area increase etc. (Bumba, 1981).

2.1. Individual Active Region Formation

Although the life-time of an active region in the photosphere is
usually counted in days or tens of days, its magnetic field lasts much
longer - in the case of a single active region without a revival of activity and with a constant velocity of area extension of its dissipating
magnetic field it may be often followed for more than five solar rotations
(Bumba and Howard, 1965a). This is also the reason why the "magnetic
active longitudes" are much better recognizable than in other activity
phenomena distribution. Any new appearance of magnetic flux in a given
region prolongs considerably the life-time of the resulting background
field patterns.

The fact that active regions develop rarely far from preceding activity is well known. And they never develop outside the old magnetic
fields. This is the reason why the old and new fields may be considered
as two opposite sides of one background field. Our recent results concerning the details of individual active regions development (Bumba and
Tomášek, 1980; Bumba, 1981a,b) demonstrate that in a relatively narrow
meridional sector of the background magnetic field of one polarity, curved
and stretched out by differentional rotation, influenced by the large-
scale convection, usually several new active regions develop within a
relatively short time interval from one to several days (see Fig. 1) in
the vicinity of its boundary often at different latitudes. At the same
time the whole process of new active region formation does not practically exceed the sector boundary, although other active regions are formed
often in a neighbouring sector of opposite polarity within the same time
interval. During the first days of the new region growths no indications
speaking in favour of their interrelations are observed. The magnetic
field boundary forms the mutually connected gulfs of opposite polarities
in which the new magnetic field grows. Both polarities in each new region do not grow parallel, but alternately; they seem to exchange their
activities and opposite polarity flux growth rates. In one "unipolar"
old field region new fields may develop almost simultaneously on its
eastern as well as western slopes playing the role of the leading as well
as of the following portion and belonging to the southern or northern
hemispheres, to the old or new activity cycle. And all the described
processes take place within the convectional patterns emphasizing in different ways the distribution of each of the two opposite polarities.

As we will see later when studying the development of the solar magnetic background fields on a global scale, we always recognize that the
above described processes concerning the individual active region formation form the basis of a more general magnetic pattern reorganization

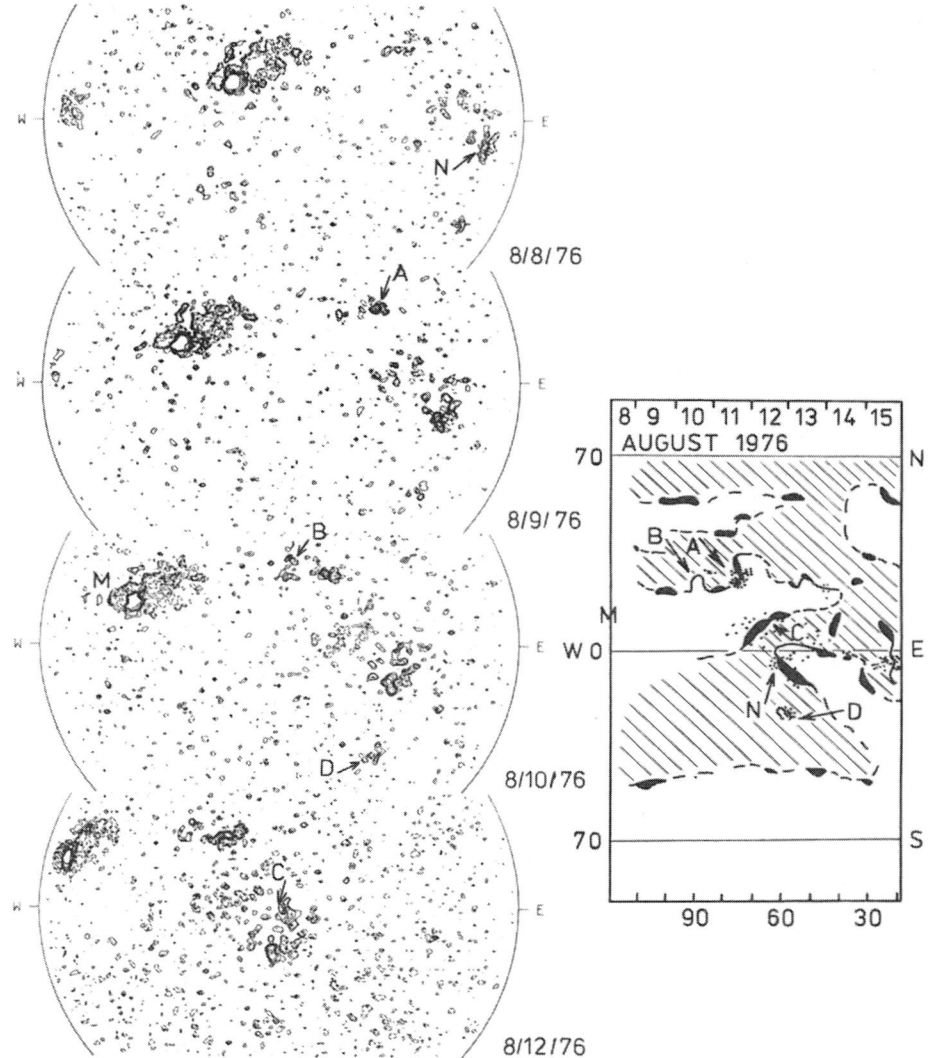

Fig. 1 : On copies of parts of four Mt. Wilson Observatory daily magnetic maps concerning the development of three new active regions it may be seen : On 9 August 1976 a new positive polarity field A (indicated by an arrow) belonging to the new activity cycle is formed in an older negative polarity following field of an older active region of the same cycle, connected with the negative polarity leading field of the old cycle. Its following positive polarity field displays on 10 August a new formation D on the southern hemisphere belonging to the new activity cycle. New secondary formations B and C may be seen on 10 and 12 August, the first one belonging to the new cycle, the second one to the old cycle, both on the northern hemisphere. The gulfs connected with these new active regions formed on the boundary of a positive (hatched) magnetic field sec-

tor are well visible on McIntosh's magnetic map attached on the right side.
and mutual transitions.

2.2. Complex of Activity Formation

We believe that a complex of activity represents a higher organiza-
tional unit of solar activity, if compared with active regions constituting it. For the first time it has been recognized as an "impulse of
solar activity" by Eigenson et al. (1948). The description and main characteristics of a complex were given by Bumba and Howard (1965a) and by
Howard and Švestka (1977). It joins in a time sequence the formation
from several till several dozens of active regions having a normal course
of a solar event: fast - one to three rotations lasting - increase and
slow - many more rotations lasting - decrease. The magnetic field produced by a complex may be not only followed much longer, but it serves
very often as a source of further new activity developments. Its topology is very complex, the individual active regions being interconnected
and form in the chromosphere and before all in the corona a complicated
geometry (Howard and Švestka, 1977).

The characteristic curve of its number of active regions, their
areas, flare activity etc., growth and decrease are accompanied by an
extension of its area with a constant velocity of its boundary shift :
in longitude of the order of about 120 m/sec; the same is true for the
heliographic latitude, but only up to about $\pm 40^{\circ}$. From this latitude
onward the velocity diminishes till about 30 m/sec (Bumba and Howard,
1965a). Very interesting is the process of its area increase. Its outer
boundary shift resembles by its morphological signs the motion of a wave
front. Even in some cases - at a place where two boundaries moving in
opposite directions and belonging to two different complexes meet - next
to the rotation a new secondary complex with only few active regions or
only one or two active regions develops (see on Fig. 2; Bumba et al.,
1968; Bumba, 1980).

During the first phase of a complex formation the greatest sunspot
groups appear. Their sizes diminish during the complex development and
the last types of spot groups visible in the complex are usually the
groups of the Zurich types A and B. The center of gravity of the activity of such a complex may move from one solar hemisphere to the opposite
one and vice versa (Bumba, 1976).

With the time development of a complex known it is relatively easy
to forecast its further progression. Naturally, the identification accuracy of a complex depends on the phase of the activity cycle. It is
very difficult to recognize the formation of such a complex during the
activity maximum. But as the activity uses to be relatively regularly
spaced on the solar surface in such a way that one half of the Sun is
usually more active - often there exist two almost opposite magnetic active longitudes with activities in opposite phases : if, in one longitude,
the activity begins, it just ends in the second one - in such a situation
the identification of activity complexes uses to be more successful.

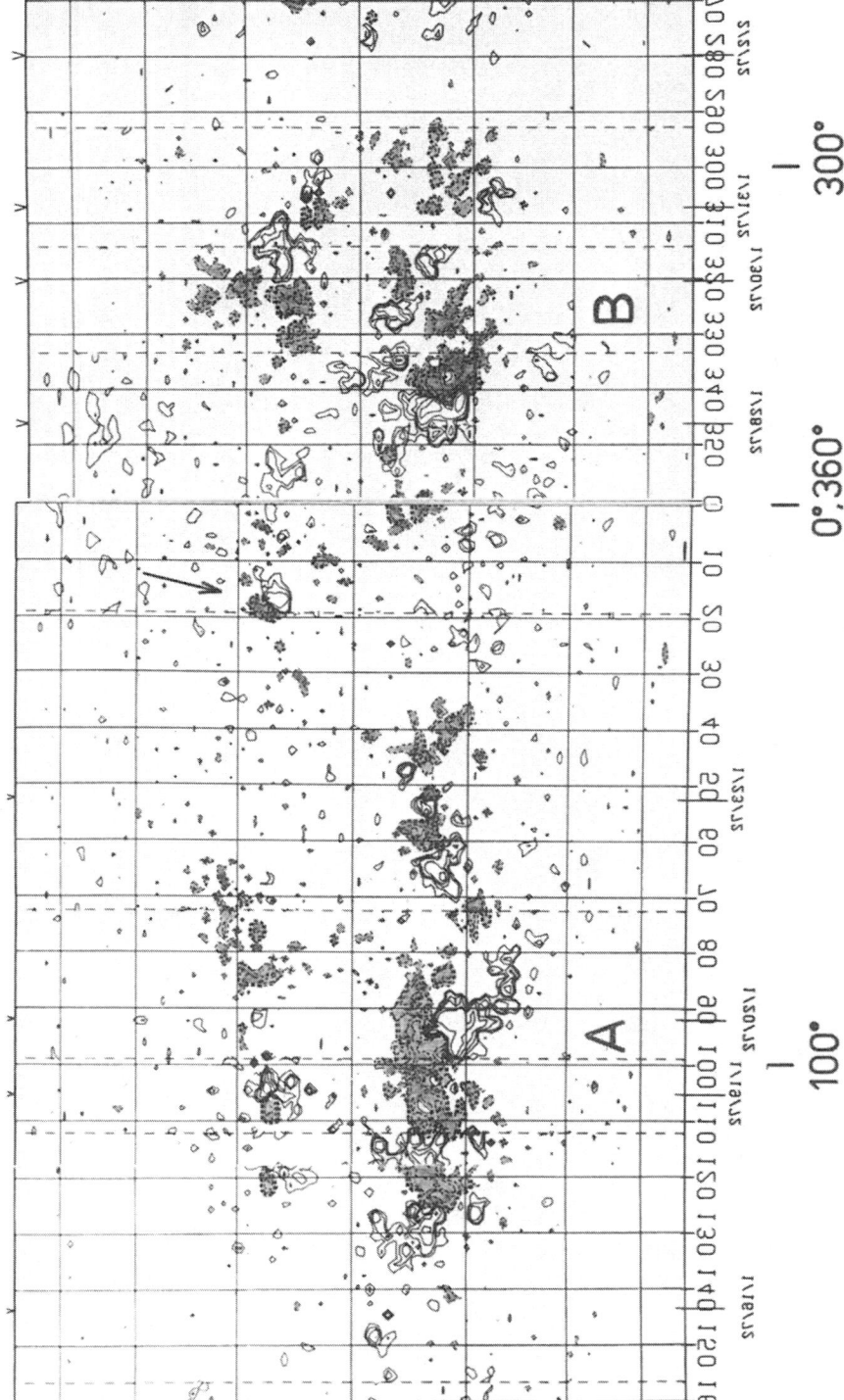

Fig. 2 : A part of the Mt. Wilson Observatory magnetic synoptic chart from 15 January till 2 February 1972, demonstrating two fully developed complexes of activity A and B and a new activity area, indicated by an arrow, which occured at the place where and when the outer boundaries of both complexes met. Both complexes developed during the initial phase of the August 1972 large-scale regular pattern formation. Negative polarity fields are dotted and shadowed.

2.3. Proton-Flare Related Large-Scale Regular Pattern Formation

The process during which usually one to several genetically connected complexes of activity develop as well as many renewals of activity appear is the large-scale regular magnetic pattern formation with a maximum stage of development to which proton-flare region occurrence is usually related (Bumba and Sýkora, 1974; Bumba, 1976). Local magnetic fields of such a proton-flare region possess not only a complex specific magnetic field configuration, but their topology is formed gradually as a part of a slowly changing large-scale distribution of the background magnetic field patterns (Bumba, 1980; Bumba and Hejna, 1981). It seems that the continuous mutual interaction of the new magnetic flux appearance and the old field weakening and dissipation are governed not only by the interference of convection, differential rotation and various magneto-hydrodynamic forces, it seems also that some other agent must cause the regularity of temporal as well as spatial distribution of developed magnetic field structures.

We described many times the complex situation in the background magnetic field distribution connected with a proton-flare region, i.e. the characteristic large-scale configuration and its development (for example : Bumba and Sýkora, 1974; Bumba, 1976; Bumba, 1980). What was said about the magnetic field distribution is true also about its correlation with the green (λ 5303 Å) coronal emission distribution drawn in the form of isophotes on synoptic charts, forming big elliptical features from an enhanced emission, practically identical with the large-scale body of the photospheric magnetic field (again for example : Bumba and Sýkora, 1973, 1974).

The Meudon Observatory synoptic charts of the solar chromosphere as well as McIntosh's magnetic maps derived from H_α solar pictures demonstrate the same visibility of the whole complex large-scale figure. All this means that the large-scale magnetic field characteristic body giving birth to the proton-flare regions occupies in its maximal phase very often not only one half of the solar surface joining the magnetic field products of many active regions, but also practically all visible layers of the solar atmosphere and is probably extended far into the interplanetary space.

The development of the described magnetic as well as activity patterns is a very complex process, since, morphologically, the two magnetic polarities studied separately at a large-scale do not develop simultaneously and in phase with regard to time and heliographic position, and the new activity is closely related to the negative polarity formation (Bumba and Hejna, 1981). The life-time of the whole large structure is of the order of one year, sometimes even more, although its best visibility during the most characteristic form lasts only two-three rotations. Its formation proceeds continuously - the individual magnetic field patterns are transformed successively from one stage into another due to the appearance of new magnetic fluxes in the individual active regions or in complexes of activity, and also because of the weakening,

dissipation and form changes of old fields, their mutual interaction
with new fields resulting in their subsequent weakening or strengthening
etc.. It is possible to decompose this entire very complex process into
individual stages each of them following the rules detected during the
more simple occasions. At the beginning of every secondary phase the
inflow of new magnetic flux into the photosphere may be observed (Bumba,
1980). Also the study of chromospheric and photospheric activity developments characterizes the investigated large and complex magnetic field
body as one physical entity : both activities grow continuously till they
reach their maximum phase coinciding with the proton-flare occurrence,
and then practically stop (Bumba, 1980; Bumba and Hejna, 1981).

Just after the proton-flare occurrence not only the main proton-flare
region but the whole complex magnetic field body disintegrates fast.
Instead of an expansion of the field area and gradual weakening of its
intensity, the field dissipation is much faster and practically in situ,
the field disappearing in various places of the extended region simultaneously (see on Fig. 3). This means that the magnetic field dissipation
process - which in normal active regions lasts for several rotations,
the field being removed off its place of formation and gradually weakening - is accelerated at least twice or three times both in the scale of
proton-flare regions and in the scale of the whole visible solar disk.
Three or four rotations are sufficient for the large-scale background
field - developing often more than one year - to be fully recognized,
desintegrated and nearly totally dissipated. It is surprising that no
photospheric or chromospheric activity with the exception of weak emission and filaments, and no flares are observed during this last period
of field disintegration (Bumba, 1980; Bumba and Hejna, 1981).

When studying the development of large-scale characteristic patterns
using both the Mt. Wilson Observatory magnetic synoptic charts and the
McIntosh maps (1979), where the areas of very weak fields are better indicated, we may get an idea about the formation of the specific magnetic
situation from a global solar scale. And we see once again that the
evolution of such a complex and complicated magnetic field body follows
the rules estimated during the studies of individual active regions :
the first appearance of the new magnetic field is closely related to the
boundary between two old "unipolar" magnetic regions. The field complicates each rotation more and more. Its topology changes from a simple
bipolar case - when its appearance is divided into two gulfs of opposite
polarity - to the more complex one, as for example in the case when one
polarity island separated from one of the two primary gulfs is surrounded
completely by an extended region of opposite polarity etc. (see on Fig. 4).

Through these above described local processes and individual changes
the reorganization of the field distribution on the whole Sun is realized
step by step. At the same time the formation of the regular figure is
a substantial part of the global development of solar surface magnetic
fields (see also on Fig. 4; Bumba and Hejna, 1981; Bumba, 1981). While
the Mt. Wilson Observatory magnetic observational data allow us to estimate the space and time scales of this process as an important part of

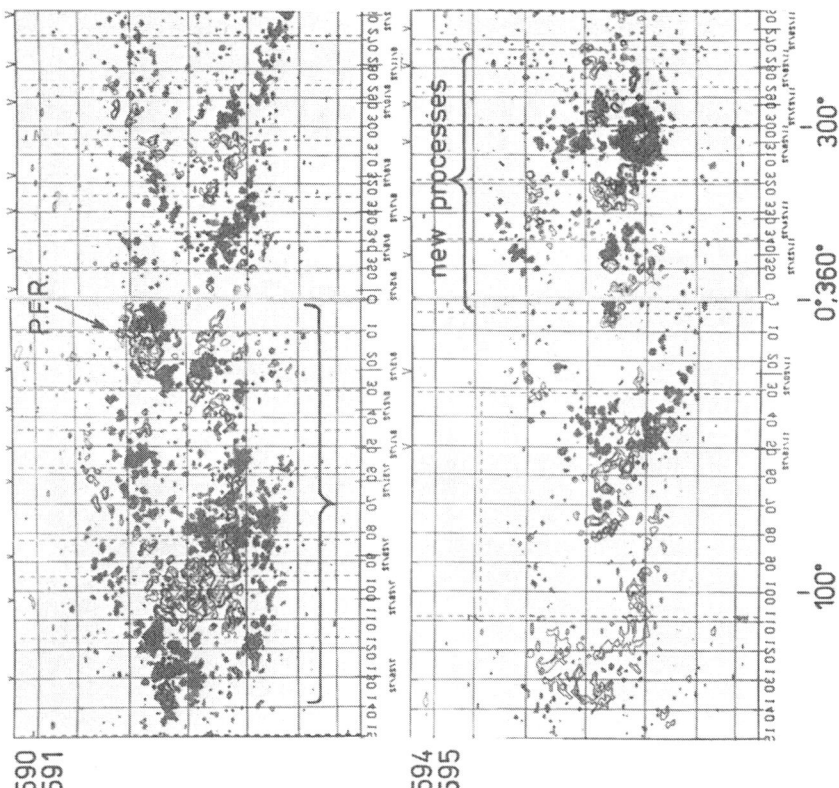

Fig. 3 : On parts of two Mt. Wilson Observatory synoptic charts (Rotations No. 1590+1 and No. 1594+5) the fast disintegration and disappearance of the August 1972 large-scale regular magnetic field body (visible in its maximum stage on the first map) may be seen. The August 1972 proton-flare region (P. F.R.) on the first map as well as new activity processes developed in the eastern half of the second map, not connected with the studied patterns, are indicated. Again, negative polarity fields are dotted and shadowed.

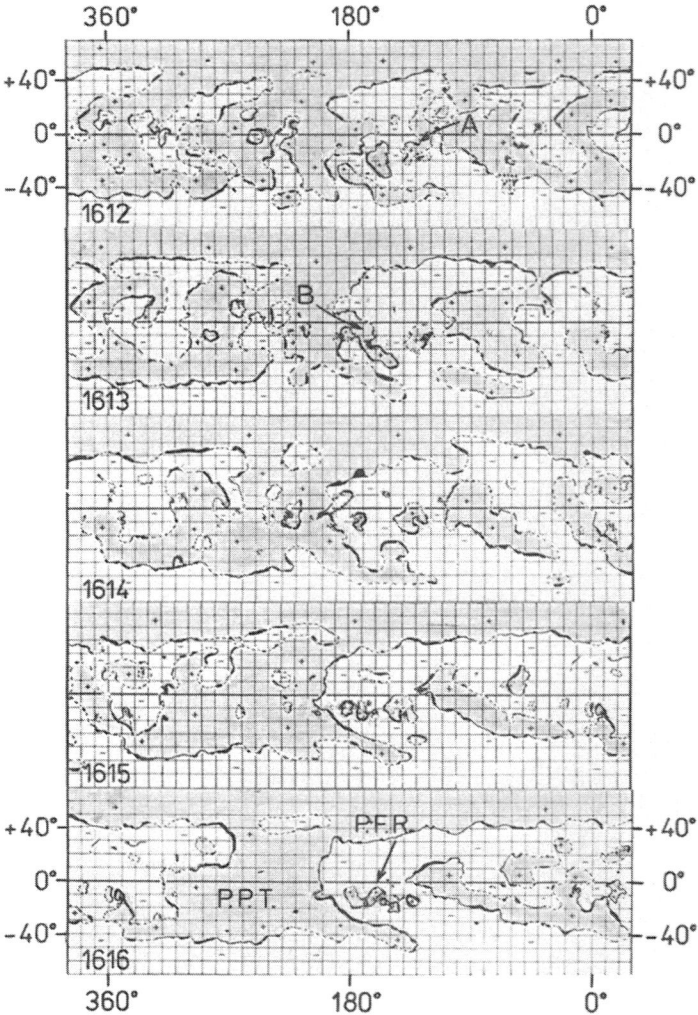

Fig. 4 : Five consecutive McIntosh's synoptic maps (Rotations Nos. 1612
till 1616) demonstrating the development of a complex magnetic situation
connected with the large-scale regular pattern of the June–July 1974
proton-flare region formation are seen. From two large gulfs A and B of
positive polarity fields in the negative polarity head of the large-scale
figure separate positive polarity islands of the complex patterns developed. At the same time the formation of the large positive polarity
tongue (P.P.T.) preceding the large-scale negative field body and reaching
from the northern polar region till the southern latitude of about 50°
is also well visible. The positive polarity fields are shadowed.

an activity cycle, the McIntosh's synoptic charts permit us to investigate its relationship to the "general" field of the Sun.

3. MAGNETIC ACTIVE LONGITUDES AND NEW FLUX AND ACTIVITY FORMATIONS

We have already mentioned the reasons of better visibility of "magnetic active longitudes" : the relatively long life-time of active regions magnetic fields and the fact that the appearance of a new magnetic
field is highly dependent on the presence of residues of old magnetic
fields. Is this the only physical reason for the existence of magnetic
active longitudes ? Certainly not ! There must be a source, a physical
cause, of new field and activity formation rotating with velocities which
differ from those of photospheric visible layers. But as up to now no
serious efforts were made to investigate this problem very important for
solar activity appearance and distribution, we do not yet know the physical background of the magnetic active longitudes formation. But their
existence has been already demonstrated (see for example : Bumba and
Howard, 1969; Ambrož et al., 1971; Bumba, 1970; McIntosh, 1980) together
with the fact of a very long life-time of some of them; it is therefore
possible to follow on magnetic synoptic charts the long living magnetic
active longitudes not only through the periods of the two last activity
minima, but also during the course of the whole solar cycle, although
it is very difficult to estimate during the period of activity maximum
which one of the magnetic active longitudes or streams is the main one
(Bumba, 1976).

During the high level of solar magnetic as well as photospheric, chromospheric and coronal activity the magnetic fields are usually much crowded
in the solar surface layers. But their background changes relatively regularly its patterns in the form of sectors of different polarities spaced
with distances of about 30° in longitude, curved and stretched out by the
differential rotation. But even during this phase of activity cycle two
main active longitudes usually exist; one of them persists practically
through the whole cycle or even further and is separated from the other
one by about 160° or 200°, whose activity develops in an antiphase. This
means that quite often when one of the magnetic active longitudes with
prevailing or more concentrated new negative polarity magnetic fields
reaches a high degree of activity of its new active regions, the opposite
one with prevailing residues of positive polarity fields with possibly
related sources of solar wind is in its decaying phase. Such a situation
in magnetic field distribution may be best observed during the last four-
five years of a solar cycle.

In the case of a proton-flare related large-scale regular patterns
formation we may remember again that their maximum evolutionary phases
are related to the intersection of two magnetic active longitudes with
a different velocity of synoptic rotation : to those of the 27-day and
the 28-29-day longitudes (Švestka, 1968; Bumba, 1976). For example, in
the case of August 1972 and June/July 1974 proton-flare large-scale bodies, the 27-day magnetic active longitude was the same during both pro-

cesses. In those two cases the suspicion concerning the different roles of positive and negative polarities in these two different magnetic active longitudes was deepened (see Fig. 5; Bumba and Hejna, 1981). In the 27-day active longitude it is the negative polarity field which seems to play a more important role. Naturally, especially the more slowly rotating magnetic active longitude - the 28-29-day one - apparently connected with better pronounced positive polarity fields, has to be investigated even more carefully because of its changing velocity of synodic rotation.

The recently published "Annotated Atlas of H_α Synoptic Charts" (McIntosh, 1979) as well as McIntosh's (1980) and Stepanyan's (1981) results demonstrate the usefulness of these maps for the study of magnetic active longitudes existence, regularities in their development, as well as for the research of the physical background of these characteristic morphological features of our synoptic magnetic charts.

4. DISCUSSION

What physical consequences may be drawn from the above demonstrated observational results ? How is it possible to pass over from determined morphological regularities in solar magnetic field and activity development and distribution to the physical laws governing the dynamics of solar background magnetic fields ? Are the dynamomechanisms taking part in the convective zones of stars and of the Sun (Parker, 1979) the only sources of magnetic fields producing the stellar as well as solar activity ? In which way the solar dynamo acts in real conditions of the solar photosphere, on surface of which and above which strong differences between the equatorial activity zone (latitudes of about $\pm 40^\circ$) and the remaining polar regions exist in differential velocity values, velocity of magnetic field propagation, densities of magnetic flux appearance and concentration, as well as in the prevailing orientation of magnetic lines : these are oriented practically parallel to the equator in the equatorial zone and lie practically meridionally at the outer boundaries of this zone; we do not know much about the change of these quantities with the altitude.

We may state that in the distribution of magnetic fields accompanying the development of single active regions mostly the supergranular and granular network plays a very important role. If we pass over to the background magnetic field kinematics, we have still to add the action of giant convective cells (Bumba, 1970) in connection with the influence of convection on magnetic field distribution. Next to the effects of several modes of convective motions we have to take into account the relatively slow process of individual active regions field extension and weakening; in the case of several active regions in a close neighbourhood we must consider the mutual influence of their magnetic fields. As it was demonstrated earlier (Bumba and Howard, 1965a), fields of the same polarity move very often toward each other and merge and strengthen the resulting fields; the fields of opposite polarities avoid one another

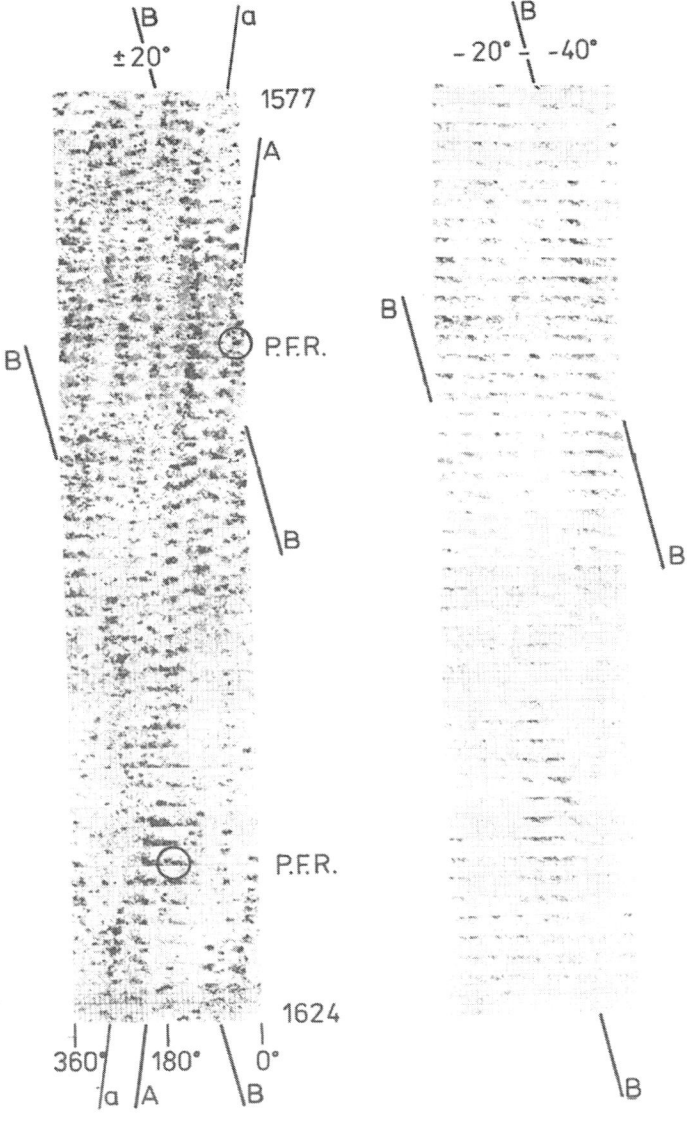

Fig. 5 : Mt. Wilson Observatory magnetic synoptic charts cut into strips representing the equatorial zone (this time only ± 20°) to the left and the southern zone of higher latitudes (-20° till -40°) to the right, arranged in chronological order during the period from 1972 to 1974 (Rotations Nos. 1577 till 1624). The dark regions represent the minus polarity areas. The 27-day ("main") magnetic active longitude (a,A) and the 28-29-day ("secondary") magnetic active longitude (B) are demonstrated on the figure. The August 1972 and June-July 1974 proton-flare regions are also indicated.

and this may lead sometimes to the weakening or disappearance of an area
of one polarity when it is nearly surrounded by the other polarity. We
have also to bear in mind that the interaction of fields of different
origin does not start immediately with their mutual contact. This already complicated process of interacting fields is, moreover, influenced
by the field density and topology and above all by the differential rotation changing the form of resulting sectors of the regularly spaced
background field; in the equatorial zone this change goes to a certain
degree of curvature only. This resulting curved form may be then observed
for many rotations and seen without substantial changes, although in
higher latitudes, outside the equatorial zone, the field is abruptly
stretched out into long tails. We did not yet speak about the propagation of magnetic field boundaries and their mutual interference etc..
We hope that for all the above mentioned morphological processes adequate
physical descriptions and explanations may be found in a near future.

What is the physical reason for magnetic active longitudes existence?
Certainly, the long life-time of produced magnetic fields is very important for renewal of activity and for complexes of activity and large-
scale regular magnetic bodies formation. But why do we observe - in one
heliographic longitude regularly shifted in Carrington's coordinates
long lasting processes of new active regions or of clusters of active
regions appearance with fast increase of such an activity lasting several
rotations and a decrease of new fields production lasting many more rotations, while in the opposite magnetic active longitude the process of
new fields production occurs in an antiphase ? The answer to the problem of the reason for longitudinal organization of new solar magnetic
fields production may serve as a clue to understanding of solar active
regions magnetic field origin. But we have to evaluate far more observational data; in the first place we have to learn more about the velocities of rotation of various magnetic active longitudes and their dependence on heliographic latitude (Stepanyan, 1981).

Very striking and yet hardly comprehensible is the mutual relation
of the old and new fields. It concerns both the first phases of the
new field development and its latest stages and influence on the background patterns formation. I do not think that we are capable of explaining the described facts by a model of buoyant magnetic flux tubes
emerging on the photospheric surface only. For example, the formation
of long tongues of one polarity on the solar surface which reach from
one polar region over the equator into the next solar hemisphere, the
development of waves on the slopes of their boundaries, where the
gulfs representing the bipolar active regions are formed, lead us to the
conclusion that the solar surface magnetic fields are not only passive
objects of various forces and motions operation, but rather that they
themselves contribute very actively to the process of field strengthening
or weakening due to their mutual interactions. The same is true about
the later progress of field development, such as, for example, the formation of islands of one polarity separated from the above mentioned
gulfs, or the still more complicated situations, we observe, before all,
on the Sun in a global scale. In the investigation of the magnetic field

development of individual active regions on McIntosh's magnetic maps, the role of mutual motions of whole "unipolar"sectors of the background field in the vicinity of their boundaries sometimes resembling a shearing motion, sometimes demonstrating various other types of motions, could be very effective, due not only to the differential rotation, but also to the proper propagation of the "unipolar" field tongues or sectors on the surface of the photosphere. All this means that the process of new photospheric flux formation is bound not simply to old magnetic fields, but also to their boundaries and to the dynamics of their formation.

From the development of the large-scale magnetic body we may try to derive one more conclusion : during the beginning of each separate phase of magnetic field growth, in which the new magnetic flux is added into the photosphere, the field seems to be more rooted into the photosphere, at least during the stage of sunspot formation. With its further evolution, however, its lines of force ever more penetrate into the higher layers of the solar atmosphere and the interplanetary space forming ever more complicated geometry. It seems that their depth in the solar atmosphere diminishes with the age of the field, that they are ever more shallow, especially when their concentrations into spots disappear. The powerful chromospheric and coronal processes taking place during the maximum evolutionary stage of the field consume a large part of the accumulated magnetic energy. But the rest of this energy is probably also spent in these upper solar atmospheric layers, since no photospheric or chromospheric activity with the exception of flocculi and prominences is observed during the last phase of the large-scale magnetic field body evolution.

We hope that all that was said about the magnetic field and activity occurrences and distribution, in the first place in the photosphere, does not demonstrate the only possible approach to the determination of physical laws governing the solar field and activity development. We think that a renaissance of solar activity studies consisting in a better knowledge of mutual relations between various solar activity phenomena and magnetic field dynamics may help us to solve far more general problems of astrophysics. To succeed in similar investigations we have to improve our understanding of observations which are not always sufficiently accurate, systematic and complex.

REFERENCES

Ambrož, P., Bumba, V., Howard, R., Sýkora, J.: 1971, in Solar Magnetic Fields, IAU Symp. No. 43 (ed. by R. Howard), D. Reidel, Dordrecht-Holland, p. 696.
Bumba, V.: 1967, Rendiconti della Suola Internazionale di Fisica "E. Fermi" , XXXIX Corso, 77.
Bumba, V.: 1970a, Solar Phys. 14, 80.
Bumba, V.: 1970b, in Solar Terrestrial Physics (ed. by E.R. Dyer), Part 1, D. Reidel, Dordrecht-Holland, p. 21.
Bumba, V.: 1976, in Basic Mechanisms of Solar Activity, IAU Symp. No. 71

(ed. by V. Bumba and J. Kleczek), D. Reidel, Dordrecht-Holland, p. 47.

Bumba, V.: 1980, B.A.C. 31, 351.

Bumba, V.: 1981a, Physica Solari-Terrestris, in print.

Bumba, V.: 1981b, B.A.C. 32, in print.

Bumba, V. and Hejna, L.: 1981, B.A.C. 32, in print.

Bumba, V. and Howard, R.: 1965a, Astrophys. J. 141, 1492.

Bumba, V. and Howard, R.: 1965b, Astrophys. J. 141, 1502.

Bumba, V. and Howard, R.: 1969, Solar Phys. 7, 28.

Bumba, V. and Sýkora, J.: 1973, in COSPAR Space Research XIII, (ed. by M.J. Rycroft, S.K. Runcorn), p. 803.

Bumba, V. and Sýkora, J.: 1974, in Coronal Disturbances, IAU Symp. No. 57 (ed. by G. Newkirk Jr.), D. Reidel, Dordrecht-Holland, p. 73.

Bumba, V. and Tomásěk, P.: 1980, Physica Solari-Terrestris, in print.

Bumba, V., Howard, R., Martres, M.J. and Soru-Iscovici, I.: 1968, in Structure and Development of Solar Active Regions, IAU Symp. No.35 (ed. by K.O. Kiepenheuer), D. Reidel, Dordrecht-Holland, p. 13.

Eigenson, M.S., Gnevyshev, M.N., Ol, A.J. and Rubashev, B.M.: 1948, Solnetshnaja aktivnost i jejo zemnyje projavlenija, Moskva, p. 81.

Howard, R. and Švestka, Zd.: 1977, Solar Phys. 54, 65.

McIntosh, P.S.: 1979, Annotated Atlas of Hα Synoptic Charts for Solar Cycle 20 (1964-1974) Carrington Solar Rotations 1487-1616, World Data Center A for Solar-Terrestrial Physics, NOAA, Boulder.

McIntosh, P.S.: 1980, in Solar and Interplanetary Dynamics, IAU Symp. No. 91 (ed. by M. Dryer and E. Tandberg-Hanssen), D. Reidel, Dordrecht-Holland, p. 25.

Parker, E.N.: 1979, Cosmical Magnetic Fields, Their Origin and Their Activity, Clarendon Press-Oxford, p. 740.

Stepanyan, N.N.: 1981, Izv. Krymsk. Astrofiz. Obs., in print.

Švestka, Zd.: 1968, Solar Phys. 4, 18.

Wilson, O.C.: 1963, Astrophys. J. 138, 832.

Wilson O.C. and Bappu, M.K.: 1957, Astrophys. J. 125, 661.

Wilson, O.C. and Skumanich, A.: 1964, Astrophys. J. 140, 1401.

THE EXPERIMENTAL CURVE OF GROWTH IN FUNCTION OF DIFFERENT SETS OF OSCILLATOR STRENGTHS

N. Gökdoğan, K. Avcıoğlu, D. Koçer
University Observatory
Istanbul, Turkey

ABSTRACT

In order to show the influence of the oscillator strengths measurements on the experimental curve of growth, three curves are constructed for the solar iron, with three different sets of oscillator strengths. The excitation temperatures found in this way are different.

The experimental curve of growth is commonly used in the determination of stellar abundances and in the discussion of local thermodynamical equilibrium in stellar atmospheres.

In order to construct an experimental curve of growth, one needs first of all, as many as possible lines, belonging to different multiplets of the element under consideration. Therefore a big dispersion spectrum and accurate wavelength measurements are needed. Then comes the identification of the lines which is basic for the evaluation of the equivalent widths; and finally the oscillator strengths of the chosen lines must be known.

In this paper we want to show the effects of the different sets of oscillator strengths on the experimental curve of growth. As iron has many lines in the visual spectrum we will take this element for our discussion and we will construct the curve of growth for iron in the solar atmosphere by taking the equivalent widths of the lines from "the solar spectrum 2935 Å to 8770 Å" (Moore et al. 1966).

Many authors have published oscillator strengths values, either experimental, semi-empirical or theoretical. In most recent publications a few number of lines only are given and the number is extended by means of a correlation formula with previous determinations.

Already the laboratory measurements of the oscillator strengths carry some errors and the extension by a functional relation adds some more. In order to avoid this, we will use three large sets of oscillator strengths

97

E. G. Mariolopoulos et al. (eds.), Compendium in Astronomy, 97–103.
Copyright © 1982 by D. Reidel Publishing Company.

Table I

Multiplets	χ	CW θ_{ex}	KP θ_{ex}	MRW θ_{ex}
a^5D	0.00	–	–	–
a^5F	0.86	1.13	0.91	0.91
a^3F	1.48	1.05	1.03	0.69
a^5P	2.17	1.10	1.03	0.97
a^3P	2.27	1.00	0.98	0.86
b^3F	2.55	1.09	0.89	0.94
z^7F^O	2.83	1.08	1.06	0.94
b^3P	2.83	1.06	1.00	0.93
b^3G	2.97	1.06	1.04	0.91
c^3P	3.03	1.04	0.99	0.95
z^5D^O	3.23	1.00	1.02	0.96
b^3H	3.25	1.01	0.95	0.90
a^3D	3.26	1.04	0.98	0.95
z^5F^O	3.37	1.00	0.99	0.90
z^5P^O	3.63	1.00	0.93	0.91
z^3F^O	3.86	0.98	0.90	0.94
z^3D^O	3.90	0.99	0.97	0.91
y^5D^O	4.15	0.98	0.92	0.94
y^5F^O	4.22	1.02	0.90	0.92
z^5G^O	4.30	0.99	0.93	0.92
z^3G^O	4.37	1.06	0.96	0.92

values : C.H Corliss-B. Warner's best values (CM) (1964) R.L. Kurucz-E.
Peytremann (KP) (1975), M. May-J. Richter-J. Wichelmann (MRW) (1974).
Fig. 1 and 2 show the correlation between those sets and Fig. 3, 4 and
5 the experimental curves of growth constructed with those values,
$F = W/\lambda \cdot 10^3$, $\log X_f = \log gf\lambda - \chi\theta$. All iron lines with wavelengths bigger
than 4000 Å, listed in "the solar spectrum 2935 Å to 8770 Å catalogue"
with given oscillator strengths in either one of the above mentioned
lists were taken.

Table I shows the excitation temperatures obtained for different
multiplets with the three different sets of oscillator strengths. In
this calculation the multiplets with fewer than four lines were omitted.
The discrepancy between the temperatures for one set is not big enough
to conclude for a non thermodynamical equilibrium. But the mean excitation temperatures given by the three sets are different, (Fig. 6).

Mean T_{ex} (CW) =   4850 ^{O}K

Mean T_{ex} (KP) =   5200 ^{O}K

Mean T_{ex} (MRW) =   5480 ^{O}K

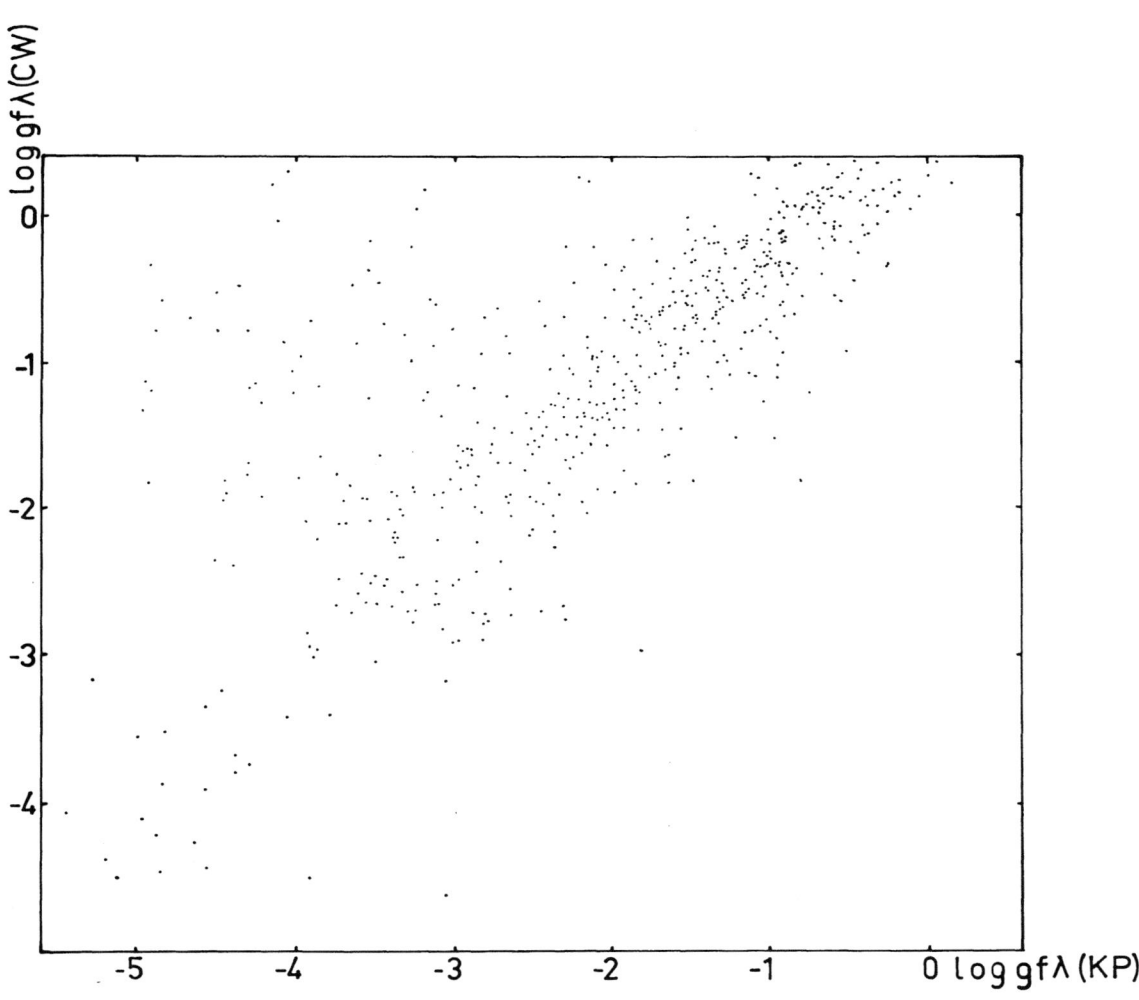

Fig. 1

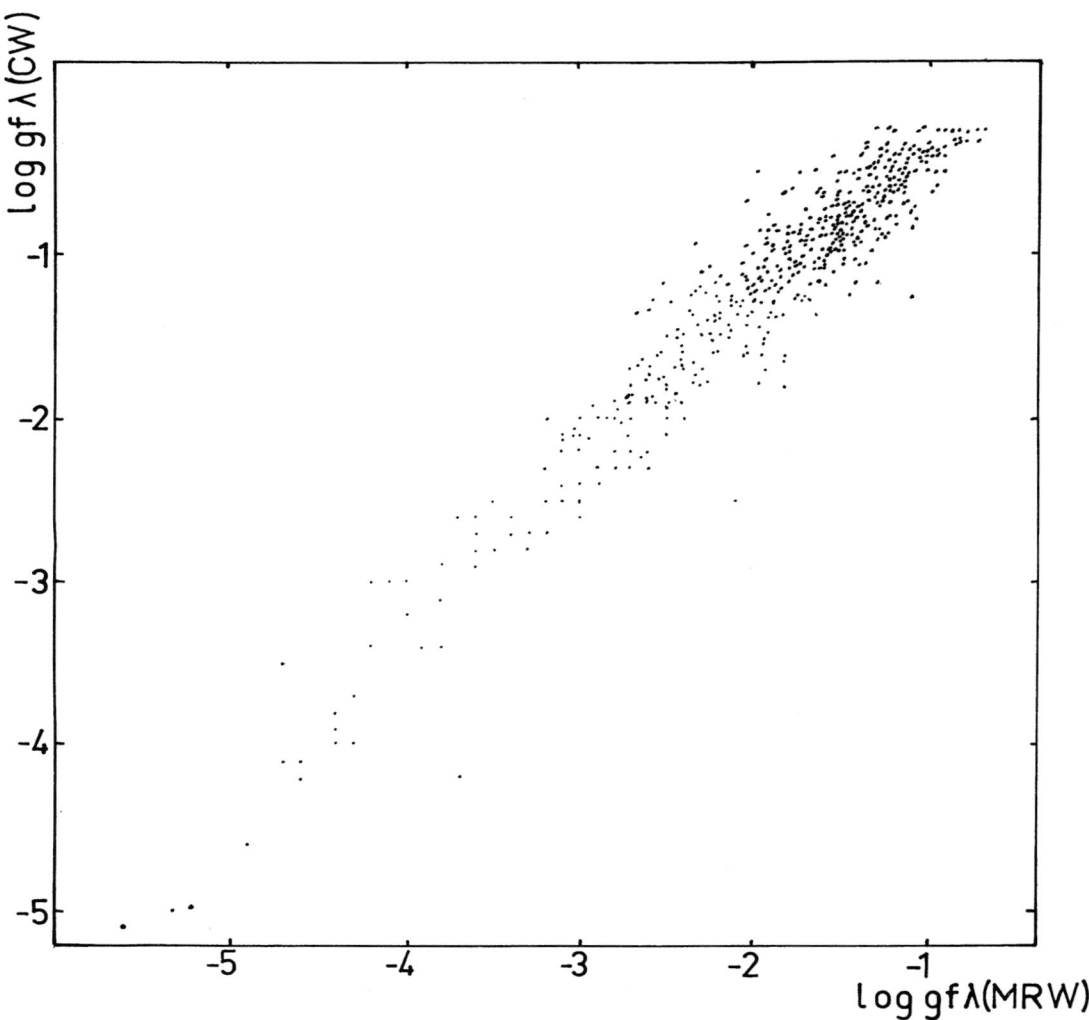

Fig. 2

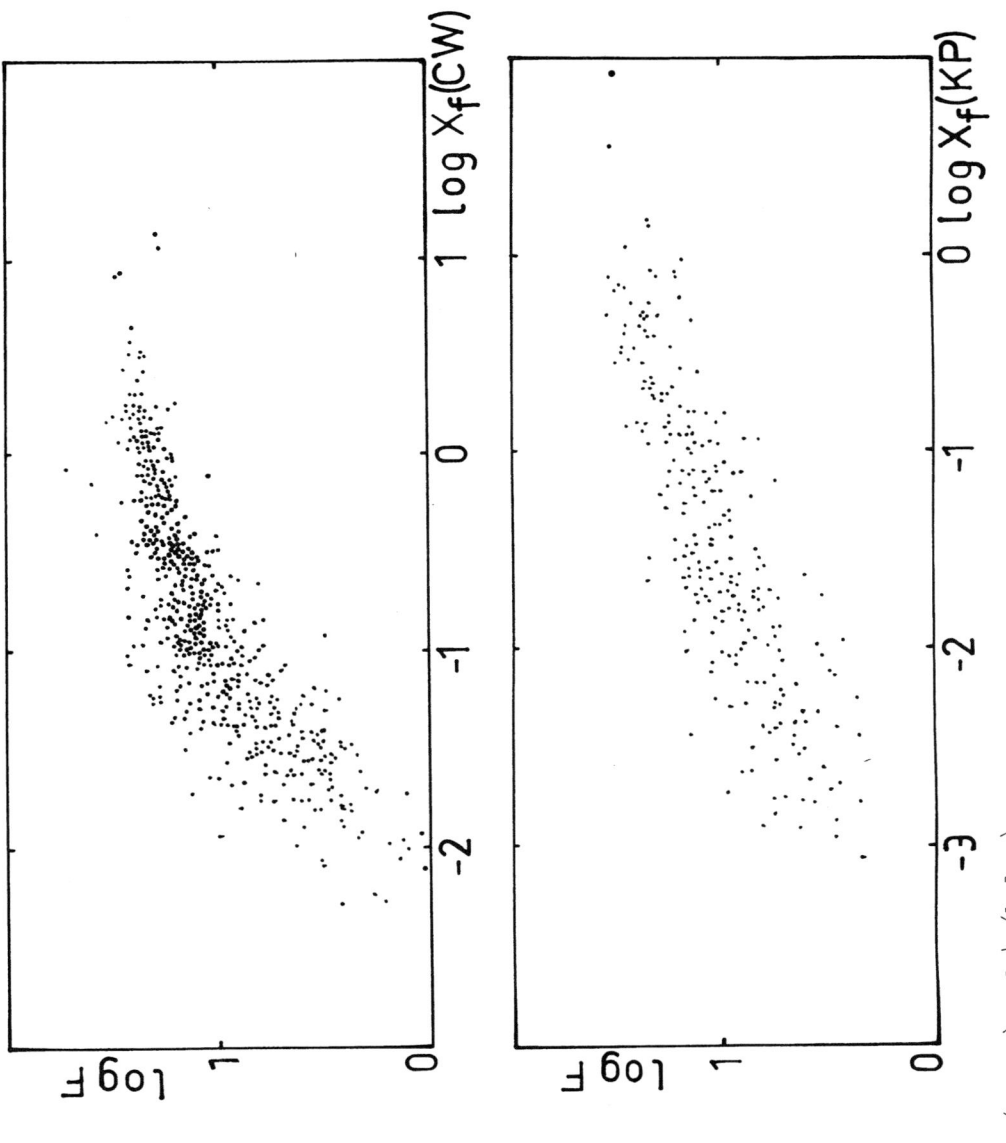

g. 3 (above) and 4 (below).

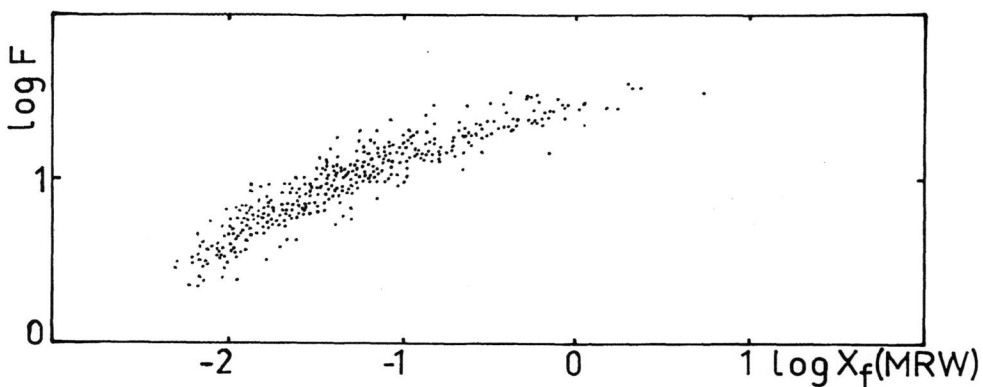

Fig. 5

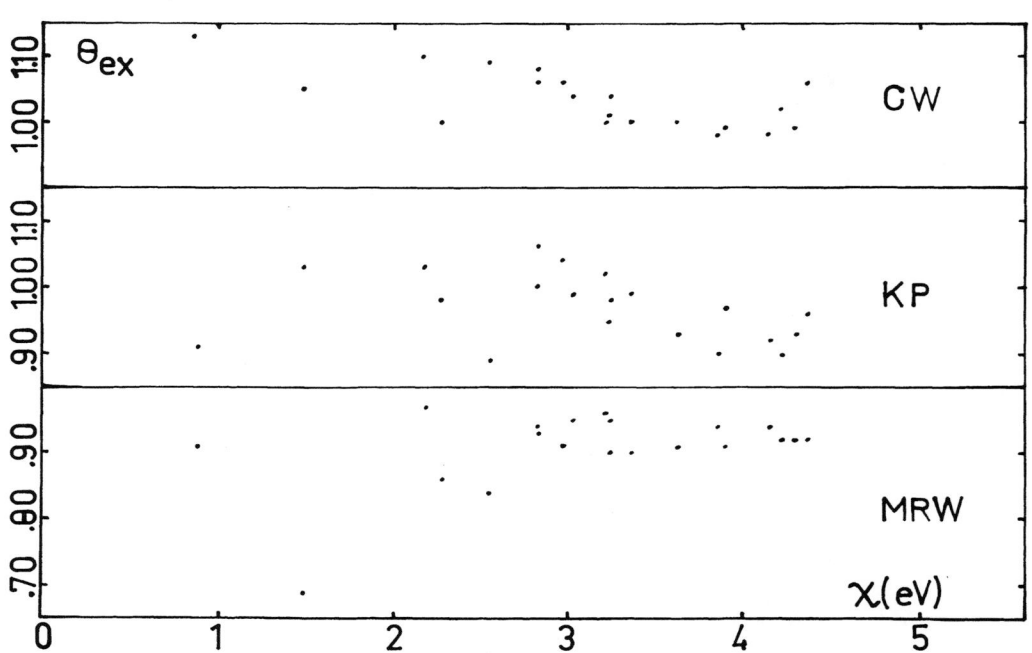

Fig. 6

As it will be seen from Fig. 3,4 and 5, the three curves of growths are clearly different and when used for the determination of the iron abundance, the results will be evidently different.

As a result we want to emphasize once more on the importance of a reliable set of oscillator strengths values for the resolution of problems of this kind.

REFERENCES

Corliss, C.H. and Warner, B. : 1964, Astrophys. J. Suppl. 8, 395.
Kurucz, R.L. and Peytremann, E. : 1975, Smithsonian Astrophys. Obs.,
    Special Rept. 362.
May, M., Richter, J., and Wichelmann, J. : 1974, Astron. Astrophys. Suppl.
    18, 405.
Moore, C.E., Minnaert, M.G.J., and Houtgast, J. : 1966, The Solar Spectrum
    2935 Å to 8770 Å.

RELATIONS OF SOLAR ACTIVITY INDICES

J.Kleczek and J.Olmr
Astronomical Institute, Czechoslovak Academy of Sciences
Ondrejov, Czechoslovakia

ABSTRACT

Currently used and some newly introduced activity indices have been used for individual active regions from the period January 1967 to June 1972. Table 1 shows to what degree the indices are interrelated.

INTRODUCTION

Celebrating the 25th anniversary of Professor John Xanthakis membership of the Academy of Athens we remember, among others, his important contributions to the solar activity research. In particular, he searched for the relation of different indices of solar activity and their dependence upon the time of solar cycle rise, viz. sunspot areas, Wolf numbers, (Xanthakis, 1962a, 1965), spot group numbers (Xanthakis, 1962b), areas of faculae, areas of prominences and sunspot magnetic field strengths (Xanthakis, 1967). It may be, therefore, opportune to include into the prepared volume honouring Professor Xanthakis scientific achievements our contribution from the same field of research.

While the research of Professor Xanthakis concerns mainly the global indices of activity on the whole visible hemisphere, the present contribution concentrates on the solar indices of individual active regions and their interrelations. The latter indices have, therefore, a "local" character, while the currently studied indices have rather a "global" character. It is apparent that the "global indices of solar activity" are fully dependent upon the "local indices of individual active regions".

MATERIAL AND ITS TREATMENT

392 active regions from the five and half years of and around the past maximum of solar activity represent the material for our study. The active regions from the period January 1967 – June 1972 were used the distance of which from the central meridian was less than $60°$, to exclude

E. G. Mariolopoulos et al. (eds.), Compendium in Astronomy, 105–110.
Copyright © 1982 by D. Reidel Publishing Company.

effects of foreshortening. Due to the incompleteness of data for some active regions the actual number of active regions with a solar index may be smaller than 392. For our study, the following published material has been used:

"Solar Geophysical Data" corresponding to the period January 1967 – June 1972. Both sections "Solar Centers of Activity" and "Daily Solar Activity Centers" have been the major source of our data. We complemented them by the "Freiburg Daily Solar Maps", "Solnechnye Dannye" and "Quarterly Bulletin on Solar Activity". Besides the classical indices of solar activity published in the mentioned periodicals we have introduced some new indices. As a result, we tried to describe each active region by the following set of indices:
1) Zurich type of its spot group.
2) s : number of spots in the group.
3) L9(S9) : luminosity of the active region on λ 9.1 cm. A reading y_{ij} in the i-th column and j-th row of the Stanford radioheliograms represents the brightness temperature $T9 = C_9 \cdot y_{ij}$ K, where C_9 is the brightness temperature unit. Its value in the Stanford radio maps is currently 5000 K (Solar Geophysical Data No. 318, p. 20). Each reading y_{ij} corresponds to a solid angle $2.5 \cdot 10^{-7}$ steradian. The total flux from an active region (called here L9 luminosity) is given by the following expression:

$$L9 = 4.25 \cdot 10^{-2} \, \Sigma y_{ij} \quad \text{flux units.} \tag{1}$$

For a detail discussion and derivation of the relation see Kleczek (1981).
4) A9 : size of the 9.1 cm radio plage associated with the respective active region:

$$A9 = n_9 \cdot 3 \text{ square minutes,} \tag{2}$$

where n_9 is the number of y_{ij} reading in the 9.1 cm radio map (Kleczek, 1981).
5) MaxT9 : the highest brightness temperature in the active region on 9.1 cm.
6) PA : area of the ionized calcium plages belonging to the active region.
7) PI : intensity of the ionized calcium plages.
8) Magnetic type according to the Mt. Wilson classification of sunspots.
9) Hmax : the largest magnetic field strength measured in the group.
10) SA : spot area of the group.
11) L21 : luminosity of the active region on λ 21 cm. In analogy to (1) it is given by

$$L21 = 6.8 \cdot 10^{-3} \, \Sigma Y_{ij} \quad \text{flux units,} \tag{3}$$

where Y_{ij} are readings in the 21 cm radio plage (Kleczek, 1981).
12) A21 : size of the 21 cm radio plage:

$$A21 = n_{21} \cdot 6.8 \text{ square minutes,} \tag{4}$$

where n_{21} is the number of Y_{ij} readings in the 21 cm plage.
13) MaxT21 : the highest brightness temperature in the 21 cm plage.
14) PA.PI : the product of calcium plage area and intensity. It is our opinion that the combined index may estimate the magnetic field determining both the area and intensity of the plage.
15) $H^2_{max} \cdot$ SA : we tried this combined index as a rough approximation.

INTERDEPENDENCE OF ACTIVITY INDICES

Let us denote by (X_i, Y_i) two different indices of the same region on the same day. The index i means the serial number in our list of active regions. The interdependence of the two indices is then expressed by the correlation coefficient:

$$r\ (Y,\ X) = \frac{\Sigma(X_i - \bar{X})\ (Y_i - \bar{Y})}{\sqrt{\Sigma(X_i - \bar{X})^2\ (Y_i - \bar{Y})^2}}\ , \tag{5}$$

or

$$r\ (Y,\ X) = \frac{n\Sigma X_i Y_i - \Sigma X_i \Sigma Y_i}{\sqrt{\left[n\Sigma X_i^2 - (\Sigma X_i)^2\right]\left[n\Sigma Y_i^2 - (\Sigma Y_i)^2\right]}}\ . \tag{6}$$

In the expressions (5) and (6) $\bar{X}$ and $\bar{Y}$ are mean values and n is the number of pairs (X_i, Y_i), i.e. of treated active regions. From our set of 15 indices we omitted the Zurich type and magnetic type. They will be treated in another paper (Kleczek and Olmr, 1981). The resulting correlation coefficients for our material are arranged in Table 1.

DISCUSSION

There exists very good correlation between both radio luminosities and maximum brightness temperatures (L9 * MaxT9, L21 * MaxT21).

Both radio plages have a good correlation with calcium plage area (L9 * PA, A9 * PA, MaxT9 * PA, L21 * PA, A21 * PA, MaxT21 * PA). Still better is the relation between the radio plages and the composite index PA.PI for calcium plages.

The interdependence of the sunspot number in an active region with its other indices is good, as may be seen in the second column of Table 1. It is better than for the spot area.

The correlation coefficients in our Table have been deduced from all the active regions, without any differentiation as for size. The correlation becomes better if only active regions with spot area less than 800 millionths of s.h. are considered. On the other hand, there is a poor correlation for spot areas larger than 1000 millionths of s.h. This fact is also illustrated by Fig. 1a and 1b.

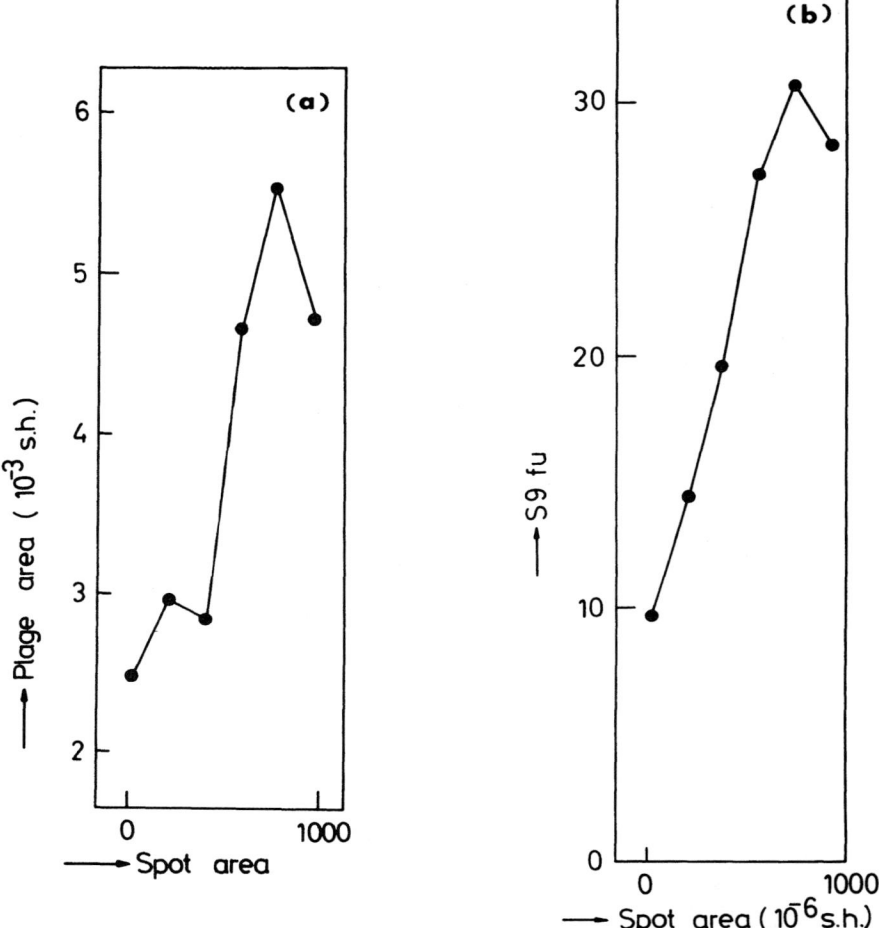

Fig. 1a and 1b. Plage area and 9 cm luminosity dependences upon spot area. Mean values of PA and S9(L9) for 190 millionths of solar hemisphere intervals of spot area (horizontal axis) are plotted.

The dependence of PA and L9 (marked as S9) on smaller spot areas is drawn only. For values larger than 1000 millionths of s.h. the scatter of points is so high that no clear dependence can be found out. Our conclusion is that the interdependence of activity indices may improve, if considered for different sizes of active regions. For that purpose a considerably larger number of active regions should be treated than we could use in our communication.

Table 1

Interdependence of active region activities expressed by coefficients of correlation

X \ Y	s	L9 (S9)	A9	MaxT9	PA	PI	H_{max}	SA	L21	A21	MaxT21	PA.PI
L9 (S9)	0,84	-										
A9	0,52	0,73	-									
MaxT9	0,82	0,92	0,82	-								
PA	0,64	0,73	0,44	0,70	-							
PI	0,38	0,40	0,14	0,62	0,34	-						
H_{max}	0,34	0,38	0,36	0,40	0,28	0,36	-					
SA	0,66	0,39	0,29	0,38	0,30	0,17	0,22	-				
L21	0,74	0,77	-	-	0,68	0,33	0,25	0,32	-			
A21	0,62	-	0,50	-	0,62	0,29	0,21	0,33	0,95	-		
MaxT21	0,68	-	-	0,72	0,56	0,37	0,25	0,34	0,82	0,67	-	
PA.PI	0,68	0,76	0,44	0,79	-	-	0,27	0,30	0,68	0,62	0,56	-
$H^2_{max}\cdot SA$	0,44	0,49	0,31	-	0,37	-	-	-	0,33	0,32	0,39	0,39

Acknowledgement

Our thanks are due to Mrs F. Kopecká for having revised a part of used data. We are obliged to Mrs. O. Jedličková for programming our data treatment.

REFERENCES

Kleczek, J.: 1981, BAC (in print).
Kleczek, J. and Olmr, J.: 1981, BAC (in print).
Solar Geophysical Data; IER - FB 271 to IER - FB 298 and Nos. 299 to 336 (Part I).
Xanthakis, J.: 1962a, Mem. Soc. Astron. Ital. 33, 291.
Xanthakis, J.: 1962b, Ann. Astrophys. 25, 342.
Xanthakis, J.: 1965, Mem. Soc. Astron. Ital. 36, 25.
Xanthakis, J.: 1966, Proc. Acad. Athens 41, 384.
Xanthakis, J.: 1967, in Solar Physics, (ed. by J. Xanthakis), J. Wiley, London, p. 157.

WHY THE TOTAL SOLAR RADIO FLUX AT A WAVE LENGTH OF 10 CM CANNOT FULLY REPLACE THE WOLF RELATIVE SUNSPOT NUMBER

Miloslav Kopecký
Astronomical Institute, Czechoslovak Academy of Sciences
Ondrejov, Czechoslovakia

ABSTRACT

The radio flux at a wave length of 10 cm is an independent, objectively determined solar activity index which, however, cannot have a priori the same time course as the Wolf sunspot number R or as the total sunspot area, and cannot replace them in an equivalent way.

There has been quite a long discussion concerning the possibility of replacing Wolf sunspot number R by the intensity of the total solar radio flux at a wave length of 10 cm. The main reason was the fact that the value of the total radio flux intensity could be, without any doubt, determined more objectively than the value of the Wolf number. It appears, however, that these two solar activity indices do not have always the same course. The fact that these two indices do not have exactly the same course can be verified by the time curves of the daily values of these indices (see e.g. Fig. 6.1 in the book by Kundu (1965) or Fig. 2 in the work by Krüger et al. (1964)) and by the scatter of dependencies of the radio flux daily values on the Wolf number daily values (see e.g. Fig. 6.2d in the book by Kundu (1965)).

I would like to show in this paper that the revealed differences in the time course of these two solar activity indices have their roots in the very substance of these two indices, that they are real and, finally, that there is no reason why these two indices should have the coincident time course - as had been briefly indicated before in connection with other problems (Kopecký et al., 1980).

The basic cause is that the Wolf sunspot number cannot reflect the structure of the sunspot distribution on the solar disk into individual groups, while the intensity of the radio flux is fully dependent on this structure. The Wolf sunspot number can reach the same value even with a different number of groups and sunspots contained therein, while the intensity of the radio flux will be different in these cases. Let us take a concrete example: Let us have two extreme situations on the solar disk.

111

E. G. Mariolopoulos et al. (eds.), Compendium in Astronomy, 111–115.

In case a) let us have 10 small isolated spots; in case b) let us assume
we have one big sunspot group with 100 spots of different sizes. In both
the cases the Wolf sunspot number R will have the value of 110. But the
intensity of the total radio flux at 10 cm will be much larger in case
b), i.e. in case of a large sunspot group, than in the case a), i.e. the
ten small isolated spots.

The same result will be obtained if we compare the time curve of
the total radio flux intensity with the time course of the total sunspot
surface.

This basic idea can be verified by a more general theoretic consideration.

Let us assume there are g spot groups on the solar disk. Out of them
$gN(S)$ has the area from S to S+dS. The total area A of all the sunspot
groups on the solar disk is then given by the equation

$$A = g \int_{0}^{\infty} SN(S)dS, \tag{1}$$

while it holds for $N(S)$ that

$$\int_{0}^{\infty} N(S)dS = 1 . \tag{2}$$

Let the total radio flux at a wave length of 10 cm from the total solar
disk be F_O. Let a radio flux φ from one active region depends on the area
S of the sunspot group; then

$$\varphi = \varphi(S) . \tag{3}$$

Let us assume there are g groups of spots on the solar disk, out of them
$gN(S)$ groups have an area from S to S+dS, so that each of these groups
emits a radio flux of $\varphi(S)$. The total radio flux from groups with an area
from S to S+dS will be $gN(S)\varphi(S)$ and the total radio flux F_O from all the
sunspot groups on the solar disk is then expressed by the equation

$$F_O = g \int_{0}^{\infty} N(S)\varphi(S)dS . \tag{4}$$

Let us assume a case that the number of sunspot groups g changes on
the solar disk as well as their distribution function $N(S)$ according to
area changes in a way that the total area A of the spots remains the
same, i.e.

$$g \int_{0}^{\infty} SN(S)dS = K = const. . \tag{5}$$

If, then, the time courses A and F_O were to be identical, it would have
to hold in this special case that with a change of g and $N(S)$ the total
radio flux also remains constant, i.e.

$$g \int_{0}^{\infty} N(S)\varphi(S)dS = aK , \tag{6}$$

where a is a constant. Equation (6) can be satisfied only, if

$$\varphi(S) = aS ,\qquad\qquad(7)$$

i.e. if the radio flux from one group is directly proportionate to the area of sunspot group. Equation (7) cannot however, be regarded as valid and in case of any form of the function $\varphi(S)$ other than the expression (7) equation (6) will not be satisfied with the equation (5) holding. In other words, if the radio flux from one group of spots is not directly proportionate to the area of this group, than the total radio flux F_O cannot be directly proportionate to the total area of spots A.

Let us now consider the Wolf sunspot number R. Let us assume that n is the number of spots in one group. Let $g\alpha(n)$ groups out of g spot groups on the solar disk have from n to n+dn spots. The total number f of spots on the disk is then given by the equation

$$f = g \int_0^\infty n\alpha(n)dn ,\qquad\qquad(8)$$

while

$$\int_0^\infty \alpha(n)dn = 1 .\qquad\qquad(9)$$

In that case the Wolf sunspot number R is determined by the relation

$$R = g \{10+ \int_0^\infty n\alpha(n)dn \}.\qquad\qquad(10)$$

Let us assume that the radio flux φ from one sunspot group is the function of the number of spots n in this group, so that

$$\varphi = \varphi(n) .\qquad\qquad(11)$$

The total radio flux F_O is then determined by the equation

$$F_O = g \int_0^\infty \alpha(n)\varphi(n)dn .\qquad\qquad(12)$$

Analogically as in case of the total sunspot area let us assume that g and $\alpha(n)$ change in such a way as not to make the Wolf sunspot number changed so that

$$g \{10 + \int_0^\infty n\alpha(n)dn\} = K = const.\qquad\qquad(13)$$

Further on let us assume that the radio flux from one group of spots is directly proportionate to the number of spots in the group, i.e.

$$\varphi(n) = bn,\qquad\qquad(14)$$

where b is the constant of proportionality.
Then

$$F_O = gb \int_0^\infty n\alpha(n)dn.\qquad\qquad(15)$$

From the validity of the condition (13) it follows that

$$_0\!\int^{\infty} n\alpha(n)dn = \frac{K}{g} - 10 \; . \qquad\qquad (16)$$

By substitution of (16) for (15) we obtain

$$F_0 = b(K-10g) \; . \qquad\qquad (17)$$

It follows from equation (17) that if the number of spot groups g and their distribution function $\alpha(n)$ change so that the Wolf sunspot number R remains the same, then even in the case of the radio flux from one spotgroup proportionate to the number of spots in the group, the total radio flux from the total solar disk is not constant but changes depending on the number of g spot groups.

We can see from the above considerations that there is, in fact, no reason why the time curve of the relative sunspot number and of the total radio flux at 10 cm wave length should have to be necessarily identical.

In reality, the situation is far more complex, because it cannot be assumed that the radio flux from one spot group is a clear function of the mere area of the spot groups or of the mere number of spots in the group. The physical conditions in the chromospheric layers above the sunspot group responsible for the intensity of the radio flux are not determined by the number of spots in a group or by the total area of the spot, but rather by the entire configuration of the magnetic fields in these layers. The latter, for its part, is determined mainly by the intensity of the individual local magnetic fields, their mutual distribution including the polarity of the magnetic fields. These realities are not reflected by the Wolf sunspot number R or by the total area of the spots. Let us bear in mind that the mere number of spots entering the value of the relative number R says nothing about their area and mutual distribution of spots of different area inside a group. A group with about 30 sunspots can have easily 2 big spots and 28 very small ones, or, in another case, 5 big spots, 10 medium size and 15 very small spots; at the same time the spots in these two groups can have different magnetic field intensities and quite a different distribution of the magnetic field polarities. It could be hardly expected that the intensity of the radio flux can be the same from these two groups of spots, although the number of spots in them is identical and the two groups, consequently, contribute with the same value 40 to the total Wolf sunspot number R.

It may therefore be concluded:

The radio flux at a wave length of 10 cm is an independent, objectively determined solar activity index which, however, cannot have a priori the same time course as the Wolf sunspot number R or as the total sunspot area, and cannot replace them in an equivalent way.

REFERENCES

Kopecký, M., Kuklin, G.V., Růžičková-Topolová, B.: 1980, B.A.C. <u>31</u>, 267.
Krüger, A., Krüger, W., Wallis, G.: 1964, Z. Astrophys. <u>59</u>, 37.
Kundu, M.R.: 1965, Solar Radio Astronomy. Interscience Publishers,
      New York.

INCREASING TREND OF FLARE ACTIVITY BEFORE THE PROTON FLARE APRIL 30, 1976

L. Křivský
Astronomical Institute, Czechoslovak Academy of Sciences
Ondrejov, Czechoslovakia

ABSTRACT

A number of examples have been given earlier to document (first treatment by Křivský, 1969) that the identification of the increasing trend on the summation curves of the flare index from an active region, where a flare with an emission of cosmic or sub-cosmic radiation is later generated, can be used for forecasting these energetic flares.

Using the active region McMath 14149, S 09°, W 47°, CMP Apr. 27, 1976, we are able to demonstrate another similar case, in which the onset of a steep flare trend was associated with the generation of a small satellite group of C and D type sunspots adjacent to an old H-type group which outlived a live active region from the preceding solar rotations.

ACTIVITY OF THE REGION ON THE DAYS AROUND THE PROTON FLARE

For the active region McMath 14149, in which a proton flare was generated "unexpectedly" at 20 47 UT on Apr. 30, 1976 in the year of the cycle minimum (Bucknam and Coffey, 1977), we conctructed a summation curve of flare activity according to the system used earlier, in which the product I x D (importance I times duration D) was determined for the individual flares (Krivsky, 1975). It was found that between 22 Apr. and 29 Apr., when the sunspot group was of type H or J, the trend of the flare activity F was very gradual (F =3 over 24 hours), but when the C-type satellite sunspot group occurred on 29 Apr., it changed to a steep trend (F = 160 over 24 hours) and this prevailed until the occurrence of the proton flare in the evening of 30 Apr., terminating after the flare. Graphically the whole process is shown in Fig. 1. The vertical axis represents the flare index F for the summation curve, the horizontal axis the time in days with the appropriate types of sunspots. We can see the distinct role played by the generation of the young satellite C- and D-type group in the immediate vicinity of the old H-type group in the simultaneously onset of the steep trend of 29 Apr., which simultaneously conditioned the generation of the proton flare after 50 hours.

117

E. G. Mariolopoulos et al. (eds.), Compendium in Astronomy, 117–118.
Copyright © 1982 by D. Reidel Publishing Company.

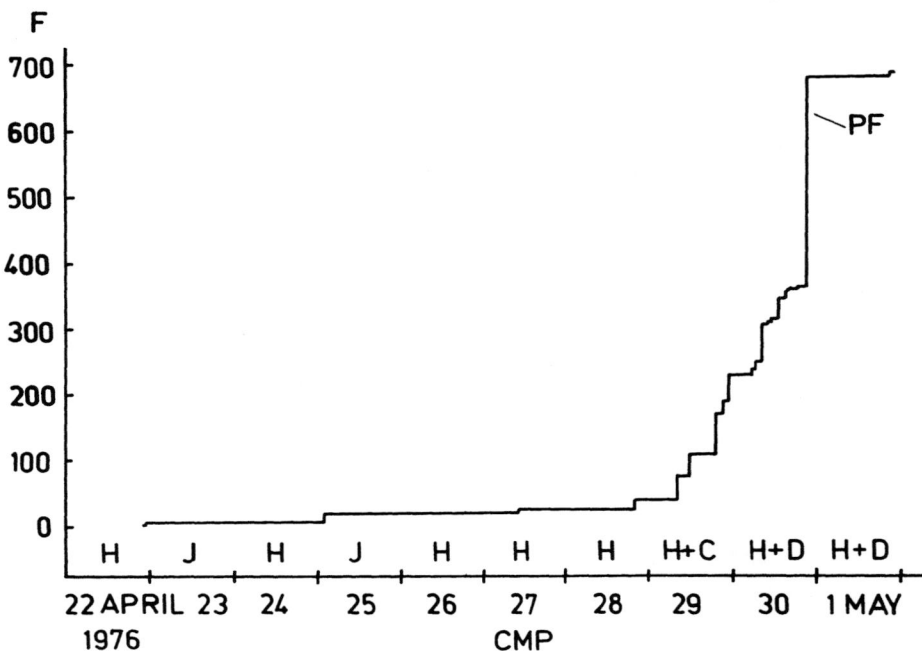

Fig. 1. Summation curve of the flare index F for the time interval before the proton flare April 30, 1976.

It is evident that this relation expresses the interaction of magnetic systems of the two adjacent sunspot groups within the same complex. The paper of Křivský and Obridko (1968) drew attention to the importance of satellite sunspot groups in association with the occurrence of mighty energetic flares.

The authors wish to thank Mrs. F. Kopecká for preparing the data and carrying out the computations.

REFERENCES

Bucknam, D.B. and Coffey, H.E.: 1977, in Study of Travelling Interplanetary Phenomena (ed. by M.A. Shea, D.F. Smart, and S.T. Wu), D. Reidel, Dordrecht-Holland, p. 395.
Křivský, L.: 1969, in The Proton Flare Project Ann. IQSY 3 (ed. by A.C. Stickland), MIT Press, p. 135.
Křivský, L.: 1975, B.A.C. 26, 203.
Křivský, L. and Obridko, V.: 1968, Solar Phys. 6, 418.

CENTIMETER WAVELENGTH OBSERVATIONS OF ACTIVE REGIONS AND FLARES WITH A FEW ARC SECOND RESOLUTION

M. R. KUNDU
Astronomy Program, University of Maryland,
College Park, Maryland, ~~20742~~, U.S.A.

In recent years large aperture synthesis instruments such as the Westerbork Synthesis Radio Telescope (WSRT) and the Very Large Array (VLA) with resolution of a few arc seconds at cm wavelengths have been used for solar observations. These observations have provided valuable information on the structures of radio emissive regions in active regions, and flares on scales of a few arc seconds. As a result we have much improved understanding of the origin of radio emission in these regions, and consequently of the processes of electron acceleration in the flaring regions and of the build-up of energy in the preflare active regions.

The synthesis observations give the total intensity (I) and circular polarization (V) maps for active regions. The parameter which has a direct physical significance is the brightness temperature in the I maps, which provides an estimate of the electron temperature in the region of formation of the radiation. Similarly, the circular polarization maps give the temperature difference between the levels of formation of the extraordinary and ordinary radiation and provide information about the steepness of the transition zone. In the absence of temperature inversions in the atmosphere, the sense of circular polarization is determined by the sign of the longitudinal component of the magnetic field. Information about the magnitude of the field can be obtained from the requirement that, for gyroresonance radiation, the observing frequency must be a harmonic of the local gyrofrequency. Finally, since the structure of the cm-λ emission is largely determined by the local magnetic field, the maps can be used for comparison of the magnetic field structure in the transition zone and the low corona with the photospheric magnetic field. The I and V maps for an active region are shown in Fig. 1 (Kundu and Alissandrakis, 1975). The source brightness temperature in maps like these is between 0.5×10^6 and 2.7×10^6 K. The sunspot. associated sources are usually circularly polarized in the extraordinary sense. The polarization maps such as those obtained with the WSRT can possibly be used as coronal magnetograms, provided there is no radio-wave propagation effect in the corona or superthermal loops embedded in the ambient

E. G. Mariolopoulos et al. (eds.), Compendium in Astronomy, 119–137.
Copyright © 1982 by D. Reidel Publishing Company.

atmosphere; the absence of such effects can be verified from the correspondence of the radio polarities with those of the photospheric magnetic field.

Centimeter Wavelength Observations of Active Regions with the VLA

The Very Large Array (VLA) of the National Radio Astronomy Observatory (NRAO) has been used to produce synthesized maps of active regions with spatial resolution as good as 2" arc at 2, 6 and 20 cm wavelengths (Lang and Willson, 1980; Velusamy and Kundu, 1981; Kundu and Velusamy, 1980; Kundu et al. 1981a). In agreement with the earlier results, the VLA observations have confirmed that the radiation from an active region is dominated by a few intense cores with angular sizes $\leq$ 10" arc, brightness temperatures $\sim 10^6$ K and polarization $\sim$ 30-80%. Further, as in the case of the WSRT observations, 6 cm-λ sources observed in active regions are associated with magnetic neutral lines and the overlying filaments or arch filament systems seen in Hα (Kundu et al., 1981a). It has been established from x-ray photographs and EUV spectroheliograms of ATM/Skylab that an association exists between x-ray, EUV loops and neighboring regions of opposite magnetic polarity in active regions. In such data, x-ray and EUV arcades are observed over filaments (Vaiana et al., 1973; Schmahl, 1980) and bright "persistent" loops are seen over arch filament systems in areas of emerging magnetic flux (Webb and Zirin, 1981). Therefore, it is to be expected that bright sources should appear above magnetic neutral lines, given sufficient opacity in the arcades or loops.

Kundu et al. (1981a) pointed out that the hottest 6 cm-λ components in their data were displaced from sunspots, but still within the regions of strong photospheric magnetic fields. This is consistent with EUV and x-ray observations of Foukal et al. (1974) and Webb and Zirin, (1981) which show that the hot (T $\geq 2 \times 10^6$ K) portions of the slowly varying component in active regions tend to be displaced from directly above sunspots to outside the umbra. This near absence of hot corona above sunspots would cause difficulties for gyroresonance interpretation of the slowly varying component if the hot microwave sources were always found to overlie spots (see e.g. Pallavicini, et al. 1979; Schmahl, 1980). Therefore, the frequently observed displacement of the 6 cm-λ sources from the spot positions is consistent with both the gyroresonance model (which requires strong magnetic fields) as well as with the EUV and x-ray observations.

The large dynamic range, high sensitivity and, most importantly, high resolution (3".5 arc) of the VLA permitted Kundu and Velusamy (1980) to demonstrate, for the first time, the existence of a loop-like structure connecting sunspots of opposite polarity in an active region at 6 cm wavelength (Fig. 2). This loop structure is reminiscent of the x-ray loops as observed from Skylab. The "loop" is delineated by contours of lower brightness temperature ($\sim 10^6$ K) than those at the foot points of the "loop" where the brightness temperatures are as high as 5×10^6 K. The intense 6 cm emission at the

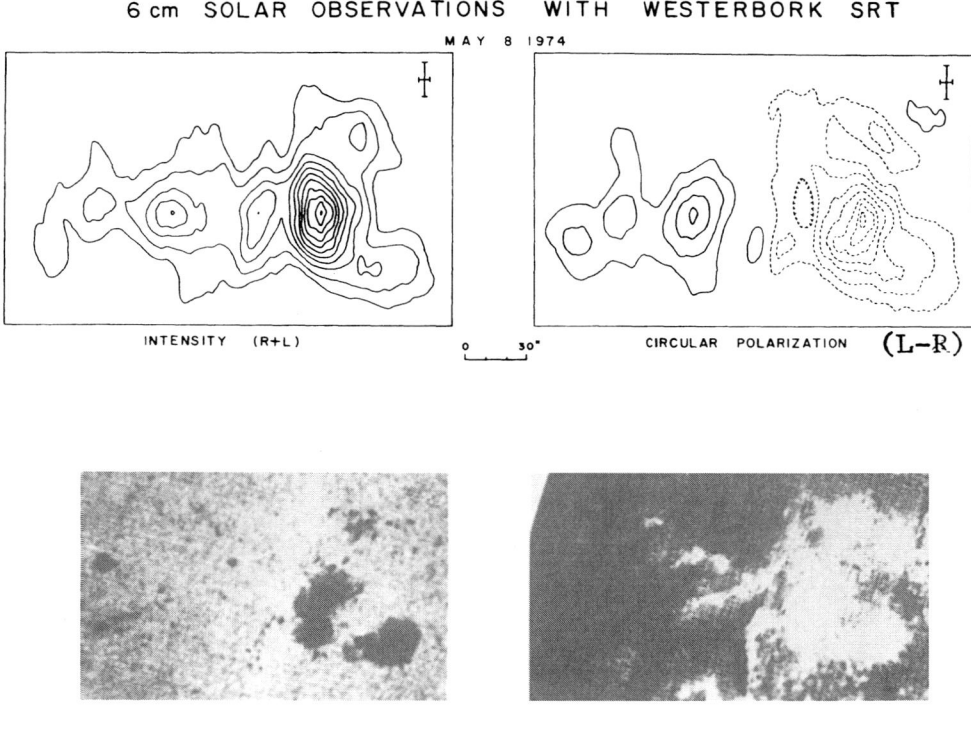

Fig. 1 Total intensity (I) and circular polarization (V) maps of an active region near S12W05 on May 8, 1974. The beam size is indicated on the right hand corner. The peak T_b on the I map is 2.7 x 10^6 K, and on V-map it is 0.7 x 10^6 K (From Kundu and Alissandrakis, 1975).

foot points located near the sunspot penumbra must be due to gyro-resonance process. It is possible that the emission in the "loop" (outside of the foot points) is entirely due to thermal bremsstrah-lung; the lower T_b of $\sim 10^6$ K is probably due to the fact that it is optically thin at 6 cm. This will also explain its lower temperature compared to the frequently observed x-ray temperatures of $\sim 2 - 5$ x 10^6 K. The many structural details that are seen in the loop (Fig. 2) are possibly due to the superposition on the loop emission of radio emitting regions which are not necessarily associated with the loop. This is certainly true of the many high brightness regions seen along the Hα filaments and/or magnetic neutral regions; these must be generated by the gyroresonance absorption process, which greatly increases the optical depth when the angle between the line of sight and magnetic field approaches 90° (see e.g. Kundu et al., 1977). There are also weaker sources associated with bright plage regions and their emission may be due to bremsstrahlung or gyroresonance process, depending upon their magnetic field strengths.

Kundu and Velusamy (1980) also pointed out that several highly polarized compact high brightness emission peaks existed over emerging flux regions near one of the spots. Some of these sources appeared to be associated with arch filament systems (AFS) in this region (Fig. 3). These associations suggest that the 6 cm sources ($T_b \sim 3 - 6$ x 10^6 K) may be related to the neutral sheets that may be produced at higher levels by the emergence of flux at AFS.

Preflare Changes in Active Regions

Observations of an active region prior to the onset of a flare are of obvious importance to our understanding of the flare build-up process. At short centimeter wavelengths (2-6 cm), one observes this flare build-up in the form of increased intensity and increased polarization of the active region, both of which suggest that the magnetic field increases or becomes more ordered prior to the occurrence of the flare. The flare-associated bursts originate in these intense sources, and the probability of occurrence of bursts increases with the increasing intensity of these narrow bright regions (Kundu 1959). This behavior of the active region prior to the onset of flares has also been observed on spatial scales of a few arc seconds (Kundu et al., 1974a; Lang, 1974; Kundu, 1980).

With the VLA, it was possible to produce two-dimensional syn-thesized maps over short periods before the start of a flare. In the case of a hard x-ray associated impulsive centimeter burst observed on June 25, 1980, we had good active region data for about one hour prior to the flare start. We produced several 15-minute synthesized maps in total intensity (I) and polarization (V). There was a definite trend for the active region undergoing brightness and polarization changes. The region consisting of several components became more and more compressed and the maximum degree of polarization increased upto 80-90%. Most importantly, the sense of polarization of one of the

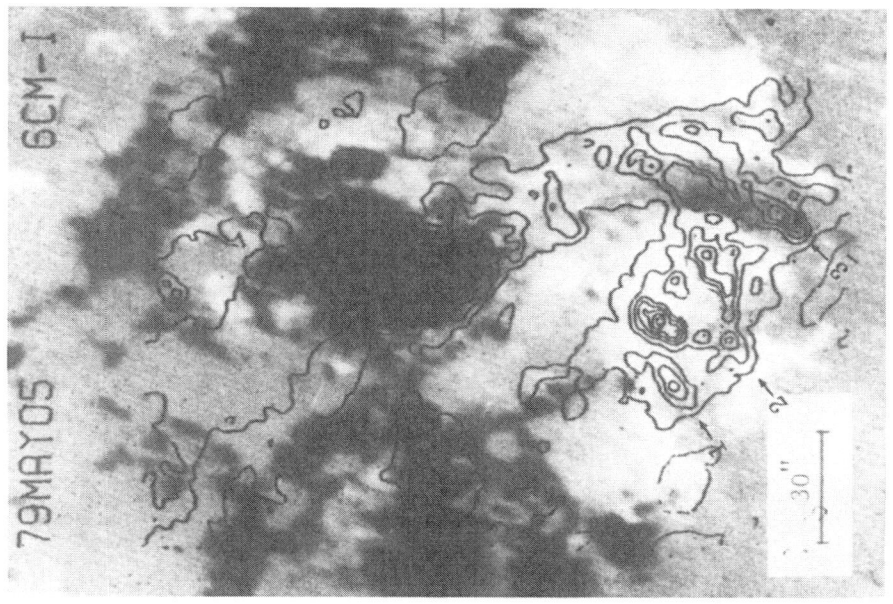

Fig. 3 Overlays of left circularly polarized intensity map on the Hα filtergram obtained at the Big Bear Solar Observatory at 1900 UT. The arrows indicate the radio sources associated with arch filament systems (AFS) (From Kundu and Velusamy, 1980).

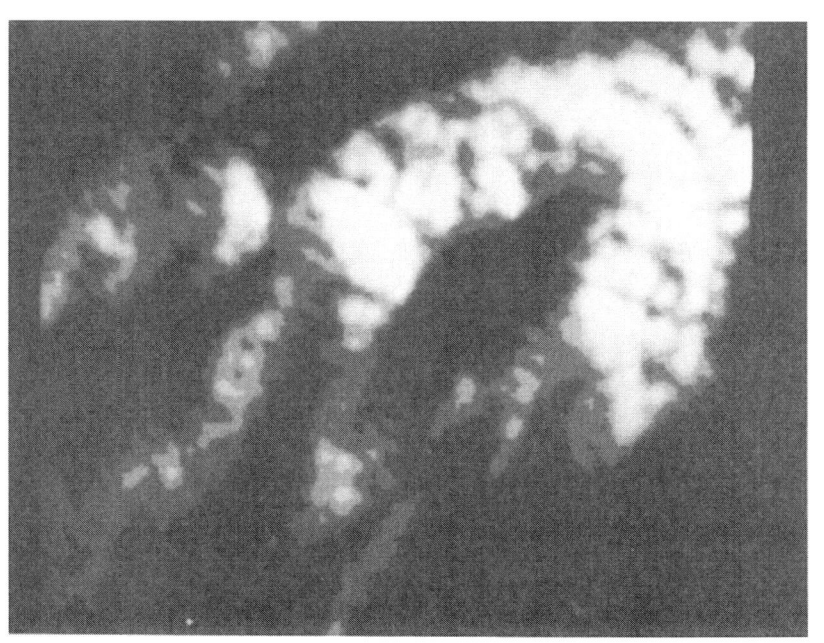

Fig. 2 Shaded contour photographic represent-ation indicating the loop structure in the high resolution (3".5 x 3".5) map of the right-circularly polarized intensity (From Kundu and Velusamy, 1980).

components changed (see Fig. 4) and another component greatly increased in polarized intensity. This might imply the emergence of a flux of reverse polarity at coronal levels. Consequently, we feel that we may be dealing with a situation considered in the model of Heyvaerts, Priest and Rust (1977). The basic concept is that current sheets develop between a newly emerging flux and a pre-existing flux and mechanisms occur in the current sheet that produce various aspects of flare activity. Heyvaerts et al.'s analysis led to the determination of an equilibrium temperature of the plasma in the current sheet as a function of height; if the sheet rises to a certain height, there will be no neighboring equilibrium situation, implying that there will be a sudden increase of temperature. This analysis indicates that the configuration is subject to a thermal instability, and that its onset is explosive, meaning the start of a flare.

Observations of Centimeter Bursts with a Few Arc-second Resolution

High resolution observations of Kundu (1959) have shown that the centimeter burst source starts with a size of ~ 1', condenses into a smaller region (< 1' arc) and then expands to a size of $\geq$ 3' during the post-maximum decay phase. The burst is polarized mainly during the impulsive phase, the post-burst phase being essentially unpolarized. This source size evolution appears to be true even on scales of a few arc seconds (Hobbs et al.,1974; and Kundu et al.,1974b).

One-dimensional fan-beam observations with a resolution of 6" arc at 6 cm by Alissandrakis and Kundu (1978) have confirmed the polarization and spatial characteristics of microwave bursts on a scale of several arc seconds and have revealed several other interesting features. Bursts of intensity 1 to 10 sfu (10^{-22} Wm^{-2} Hz^{-1}) occur quite often near the neutral line of the magnetic field, as determined by the polarization maps of 6 cm active regions (Kundu et al.,1977). At the time of maximum (impulsive phase), the source is most compact (size $\leq$ 10"), it is strongly polarized and the brightness temperature generally exceeds 10^8 K. After the maximum (post-burst phase), the burst core expands often with a velocity < 30 km/s up to a size of > 1' arc; it is unpolarized and the brightness temperature is generally ~ 10^6 K. The impulsive phase is polarized in only one sense over the entire extent of the burst source; the post-burst phase is essentially unpolarized. This suggests that, if the burst is associated with loop structures, the emission must be associated with one leg of the loop. On the other hand, Enomé et al. (1969) observed bursts of stronger intensity ($\gtrsim$ 100 sfu) above active regions with angular resolution of 24". They observed both right and left circular polarization presumably coming from opposite magnetic polarities of the loop. Qualitatively one can explain this polarization characteristic if one assumes that the magnetic fields in the two radiating sources are of unequal strengths, and that the number and distribution of nonthermal electrons are equal (Kundu and Vlahos,1979).

Marsh and Hurford (1980) used the VLA at 2 cm and 1.3 cm to

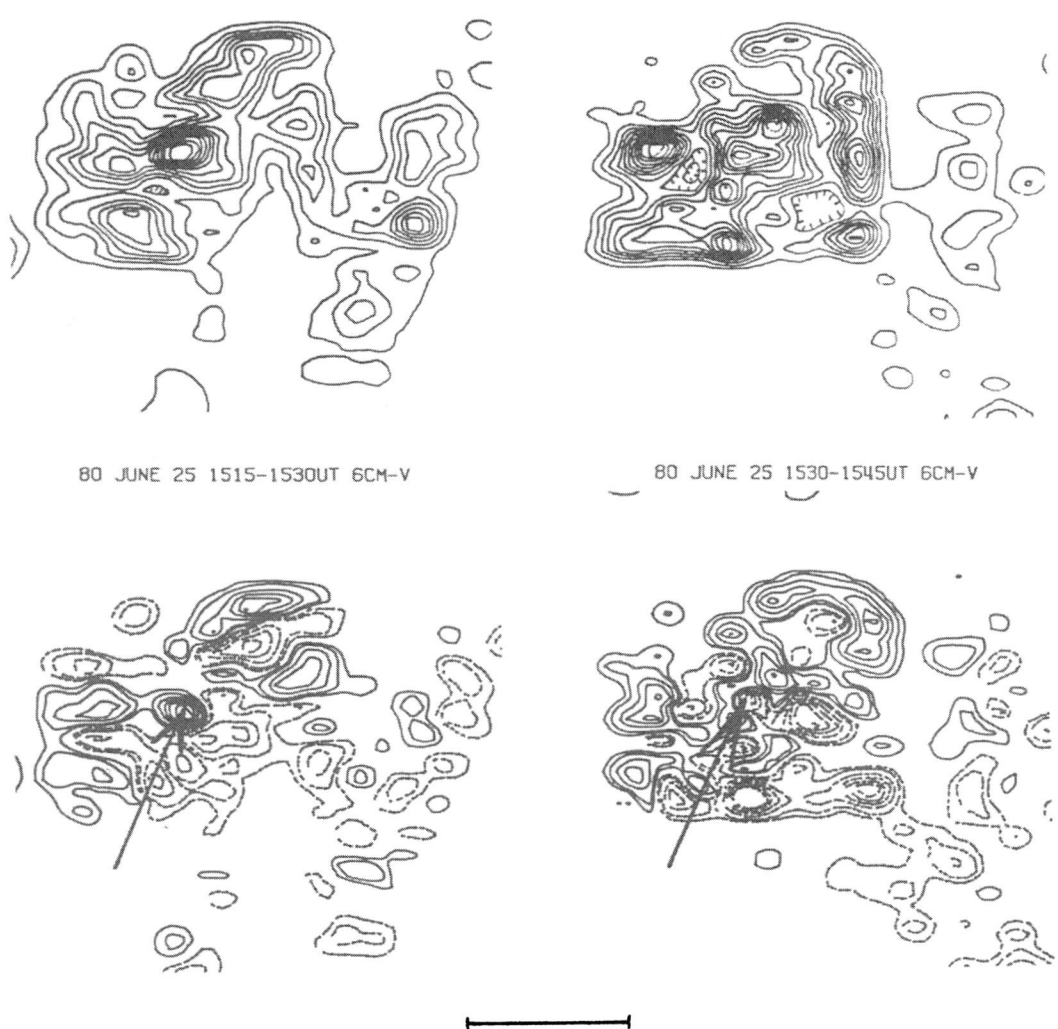

80 JUNE 25 1515-1530UT 6CM-I

80 JUNE 25 1530-1545UT 6CM-I

80 JUNE 25 1515-1530UT 6CM-V

80 JUNE 25 1530-1545UT 6CM-V

30 arc sec

Fig. 4 Pre-flare 6 cm maps in total intensity (I) and circular polarization (V) synthesized over 15 minute intervals before the burst of June 25, 1980. Several bipolar sources appear, the strongest of which reversed in polarization immediately before the flare (see arrows).

produce two-dimensional images with angular resolution as high as 1" x 0.75". Comparison with optical data showed that in the impulsive phase, the microwave emission was dominated by a compact source (~ 2") located between the Hα kernels. In the post-impulsive phase, the microwave source became larger and elongated in a direction consistent with the orientation of the magnetic field lines joining the Hα kernels. In one case, the microwave structure exhibited some correlation with the Hα kernels, and formed a connecting bridge between them. Marsh and Hurford interpreted these results to imply that the initial energy release occurred near the top of the magnetic arch joining the Hα kernels, and set an upper limit of 2" for the size of the energy release region. A size of 2" was obtained for a 3.7 cm burst observed by Kundu et al. (1974b) with corresponding brightness temperature of 8×10^9 K.

Using the VLA, Kundu and his collaborators (Kundu, Schmahl, Velusamy, 1981b; Kundu, Velusamy and Schmahl, 1981c) studied several impulsive 6 cm bursts and tried to relate their observations to the observations made in hard x-rays by Hard X-ray Burst Spectrometer (HXRBS) experiment aboard the SMM and to interpret their results in terms of existing flare models.

Magnetic Structure of a Flaring Region Producing Impulsive Microwave and Hard X-rays

As we know the microwave bursts are very closely associated with hard x-ray bursts. In order to interpret this result Takakura and Kai (1966) first proposed the electron trap model in which the energetic electrons are magnetically trapped in the corona. Over the years, this simplistic trap model has undergone changes. In particular it has become clear that precipitation of electrons must occur in trap models, resulting from Coulomb scattering of electrons into the loss cone (i.e. diffusion in pitch angle) at a rate of about 2-3 times the collisional energy loss rate. The trap models generally include a population of energetic electrons contained within a magnetic loop, whose foot points are rooted in the photosphere. The impulsive microwave emission is produced by electrons of energy > 100 Kev, depending on the spectral index of the electron energy distribution, while the associated hard x-ray emission is produced by electrons in the energy range 20-100 Kev. This difference in electron energy plays an important role in trap models. The primary energy release that occurs at the loop top causes bulk energization of the local plasma (Brown, Melrose and Spicer, 1979). Electrons with energies greater than a critical value, which depends on the energy distribution in the trap, will precipitate out of the energy release region.

It is clear that the accurate determination of the relative positions of microwave and hard x-ray burst sources is crucial in resolving the question of thermal versus nonthermal origin of microwave and hard x-ray bursts. Kundu, Schmahl and Velusamy (1981b) presented a detailed description of a complex burst observed with the

VLA with 1" x 2" arc resolution at 6 cm wavelength during its
impulsive phase, with special reference to its brightness, polarization and positional characteristics. The hard x-ray time profile
in the energy range 20-300 Kev for this burst was available from the
SMM-HXRBS experiment (Figure 5).

The 6 cm burst region is located near the neutral line of the
oppositely polarized regions near the center of the preflare active
region. As one can see from samples of the 1 minute synthesized maps
in total intensity (I) and polarization (V), shown in Figures 6a and
6b the burst source during the impulsive phase is very complex; it
consists of several component sources of which the most intense one
(A) is located near the center. This strong component undergoes
intensity changes, having peaks at 1551-52, 1554-55 and 1556-57 IAT,
the peak at 1554-55 IAT being the strongest. This is also the time
when the Hα flare area is largest and the two ribbons (#1, #2) most
developed. The polarization (V) maps (Fig. 6b) show that most of the
component regions have bipolar structure. Several components, including the strong central source have almost the same finite extent in
total intensity and polarization (~ 5" x 10"); the neutral line (line
of zero polarization at 6 cm) passes through the peak of total intensity and divides it into two regions of opposite polarity across the
shorter extent of the region. Figure 6a also shows the Hα flaring
region at the time of each snapshot map as a grey overlay on the contours. Essentially there are two main ribbons. This may be seen in
Fig. 6a at 1555-56, where two large parallel ribbons straddle the
dashed line, which is the neutral line of the photospheric
magnetogram.

The V maps (Fig. 6b) show that the bipolar structures A, B, C, D
have left-hand circular polarization over ribbon #1 and right-hand
polarization over ribbon #2. Comparison with photospheric magnetograms shows that the left (right) circular polarization overlies
positive (negative) magnetic flux. This indicates that these sources
are radiating in the o-mode. According to the theory of gyrosynchro-
tron emission, the sources then must be optically thick.

The main conclusions that emerge from these high resolution (~ 1"
arc) observations at 6 cm wavelength are that the emission from this
complex microwave burst originates from a very complex region,
consisting of several component sources, although there is a strong
dominant component located near the center of the complex. These
component sources are usually in the form of arcades of loops, and the
6 cm emitting region occupies a substantial portion of the flaring
loop including its top. Clearly the emission occurs on both sides of
bipolar loops, as evidenced by the bipolar structure and finite extent
(5"-10" arc) of the burst source. The component burst sources clearly
appear in the form of a series of arcades of loops with bipolar
structure. This is best represented in the radio photo of Figure 7,
where each image in I and V corresponds to a 10-second snapshot map.

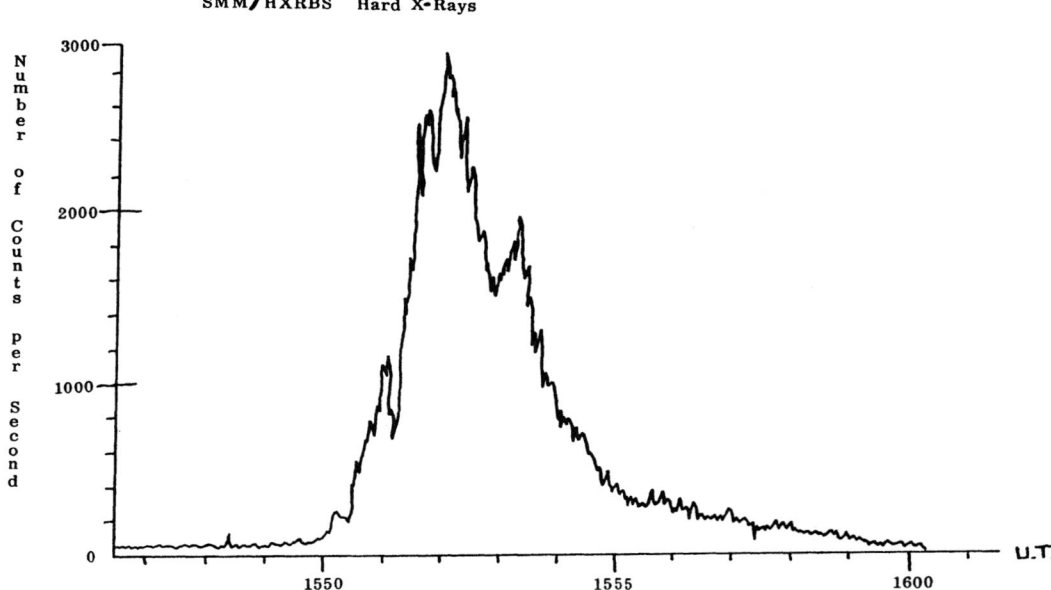

Fig. 5 Time profile of the 25 June 1980 burst in hard x-rays (SMM-HXRBS data, courtesy of K. Frost).

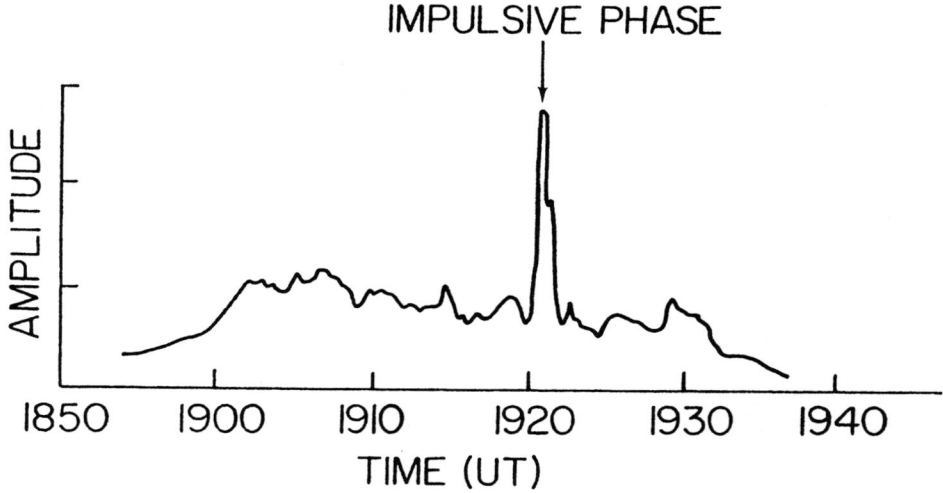

Fig. 8 Plot of the visibility amplitude at a spacing of 5000 λ at 6 cm during the burst of May 14, 1980.

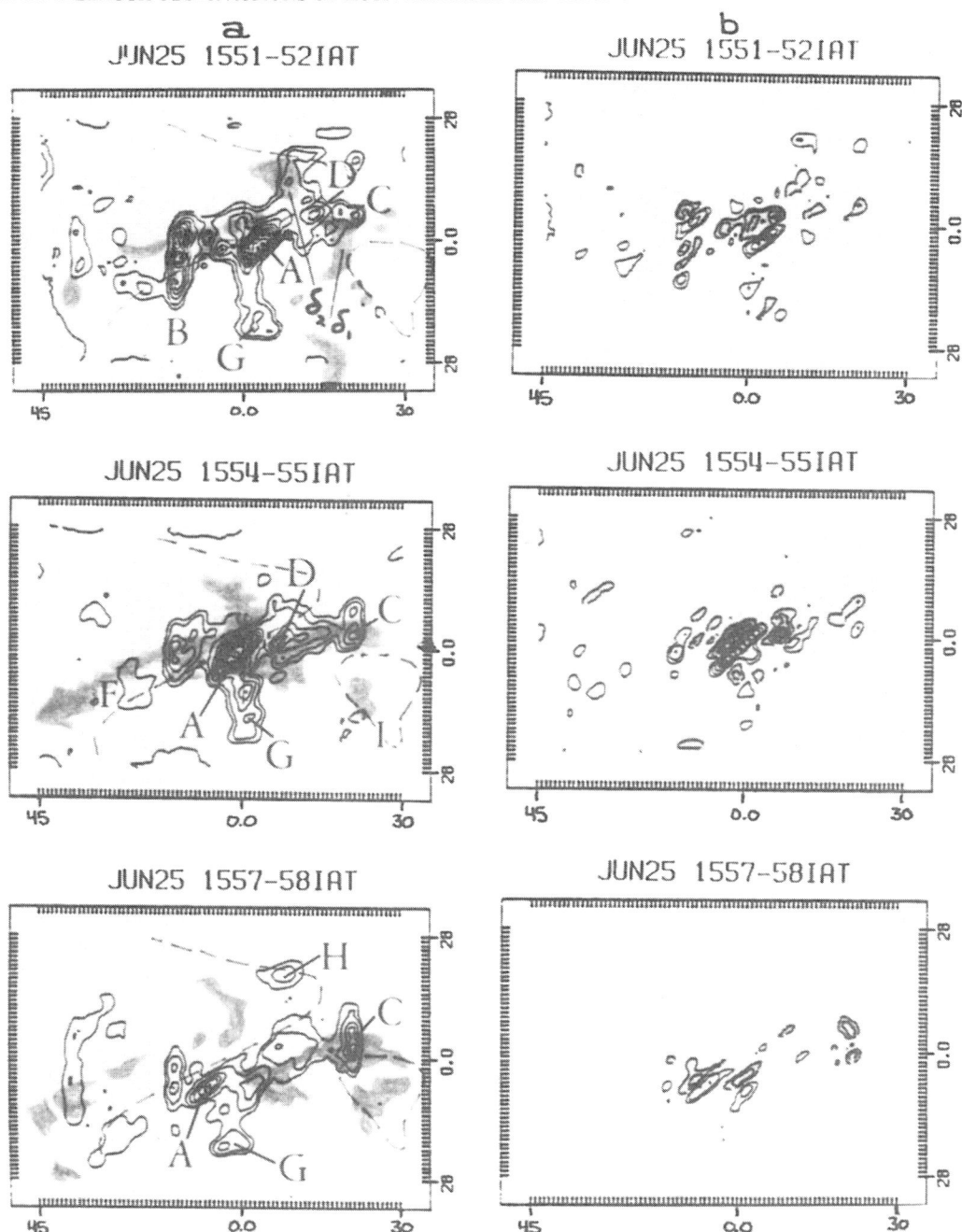

Fig. 6 Samples of VLA one-minute "snapshot" maps of the 6 cm burst sources. Maps (a) show total intensity (I) and maps (b) show circular polarization (V). The grey regions represent Hα flaring footprints. East is to the left (From Kundu et al., 1981b).

I   V   I   V

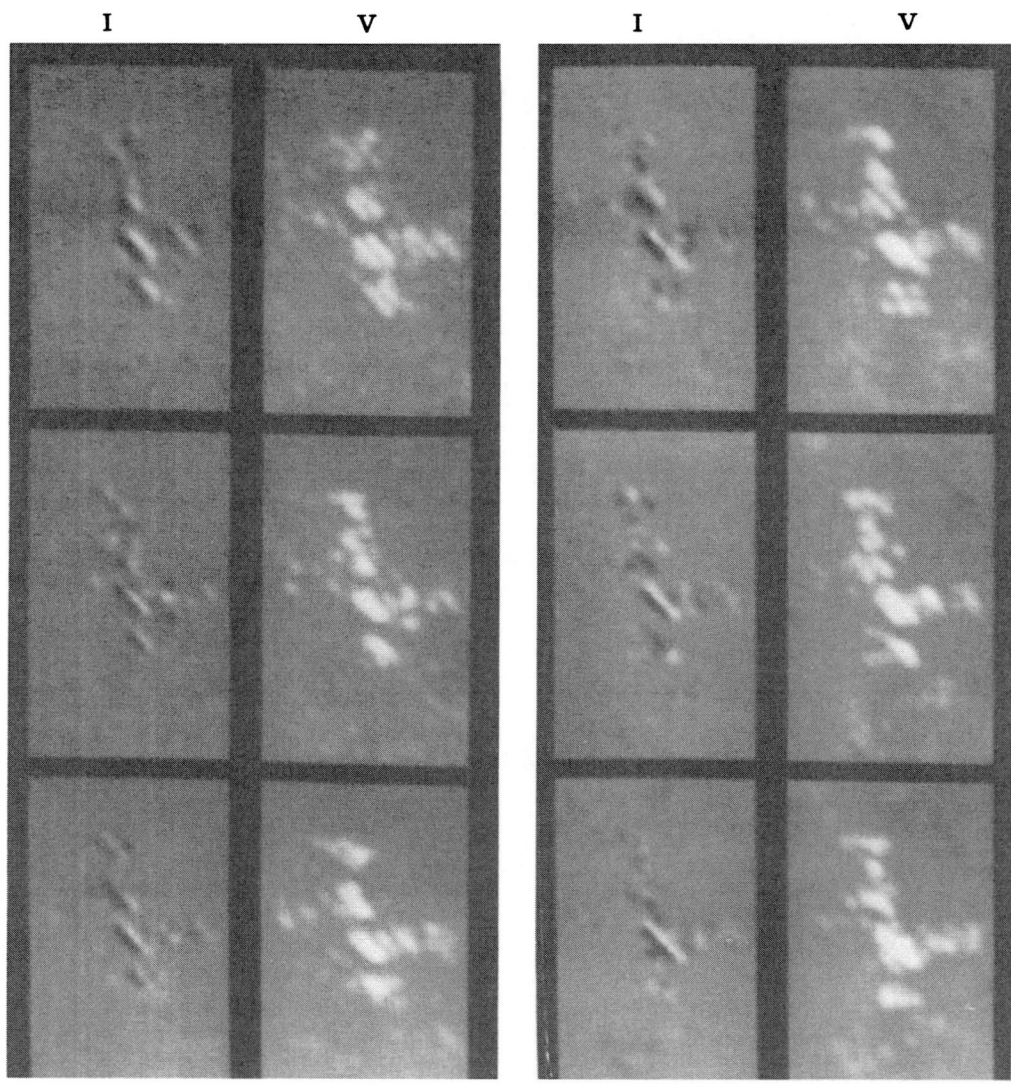

Fig. 7 A sample of consecutive 10-second VLA snapshot radio photo-graphs in intensity (I) and polarization (V) during a period of the impulsive phase of the 6 cm burst. North is to the right (From Kundu et al., 1981b).

Current Sheet, Magnetic Reconnection and Impulsive Microwave Burst

An impulsive burst superimposed on a gradual burst was observed at 6 cm-λ, using the VLA with a resolution of ~ 2" arc. This centimeter burst (Fig. 8) occurred in association with a two ribbon Hα flare of importance 2B and a filament eruption. A soft x-ray burst (class M1) started at 1900 UT, coincident with the onset of the 6 cm burst at 1859 UT. A hard x-ray burst of duration ~ 2 minutes was observed by the SMM-HXRBS experiment, in coincidence with the 6 cm impulsive phase at 1920 UT. The 6 cm burst source shows large structural changes in both total intensity and polarization over the entire duration of the burst. These changes were studied with a time resolution of 5 minutes during the gradual phase and 10 seconds during the impulsive phase. The evolution of the burst source can be summarized as follows:

The map of the region made over the period 1806 – 1848 UT (Fig. 9a), showed intense emission (with peak T_b ~ 10^7 K). As can be seen from the polarization map, this emission was extended along a neutral line situated approximately in the east west direction. However, there was a kink (at an angle of ~ 120°) in the neutral line near its center.

At the onset of the gradual phase (1854-1859 UT, Fig. 9b), the east-west structure in intensity and polarization of the burst source was quite similar to the pre-flare structure. The most conspicuous feature of the burst source evolution was that from 1904-1919 UT (Fig. 9c) a gradual brightening consisting of several blobs (mostly bipolar) developed in the north south direction. The polarization map at 1914-19 UT, just before the impulsive burst clearly showed a north-south neutral line and also showed two juxtaposed bipolar regions, S, N, N, S; this magnetic configuration implies two intersecting bipolar loops (Fig. 10b) and it is possible that it was somehow related to the impulsive energy release.

In the beginning of the impulsive phase, there was a small relatively faint bipolar source which grew in intensity very gradually; there was also a fainter elongated feature with longer axis at a position angle -45° with T_b ~ 100×10^6 K, apparently originating from an arcade of loops. Definite changes in the polarization structure took place in this region before the peak of the impulsive phase. Starting at 191935 UT, this arcade of loops gradually changed and ultimately developed into two strong bipolar regions or a quadrupole structure (at 191955 UT) whose orientations were such that near the loop tops the field lines were opposed to each other (see Figs. 9 and 10c). It may be significant that this particular field configuration occurred at just the impulsive peak, and so must be related to the impulsive energy release at its peak. At the time of maximum (191955 UT) the central compact source brightened to a high brightness temperature of ~ 1100×10^6 K, with little change in its size. The bright compact source is obviously related to the region of energy release by

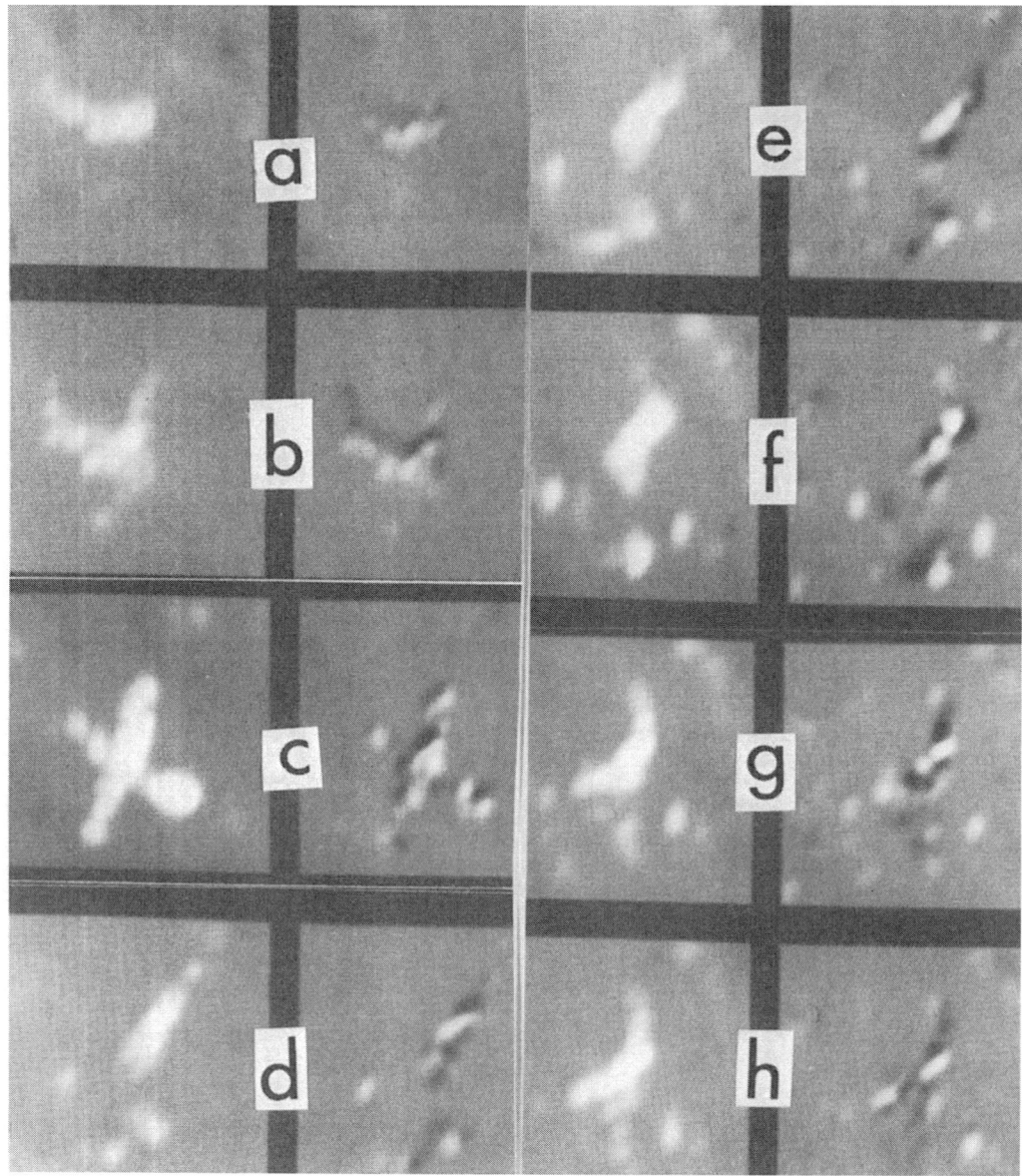

Fig. 9   Photographic representation of 6 cm maps in total intensity I
(left) and circularly polarized intensity V (right) during the gradual
and impulsive phases.   (a) preflare map at 1806-1845 UT; (b) and (c)
are 5 minute maps during the gradual phase at 1856-1901 and 1914-
1919 UT.   (d)-(h) are 10 second maps from 191935 to 192015 UT.   Field
of view is 48" x 48" and synthesized beam is 3" x 2".

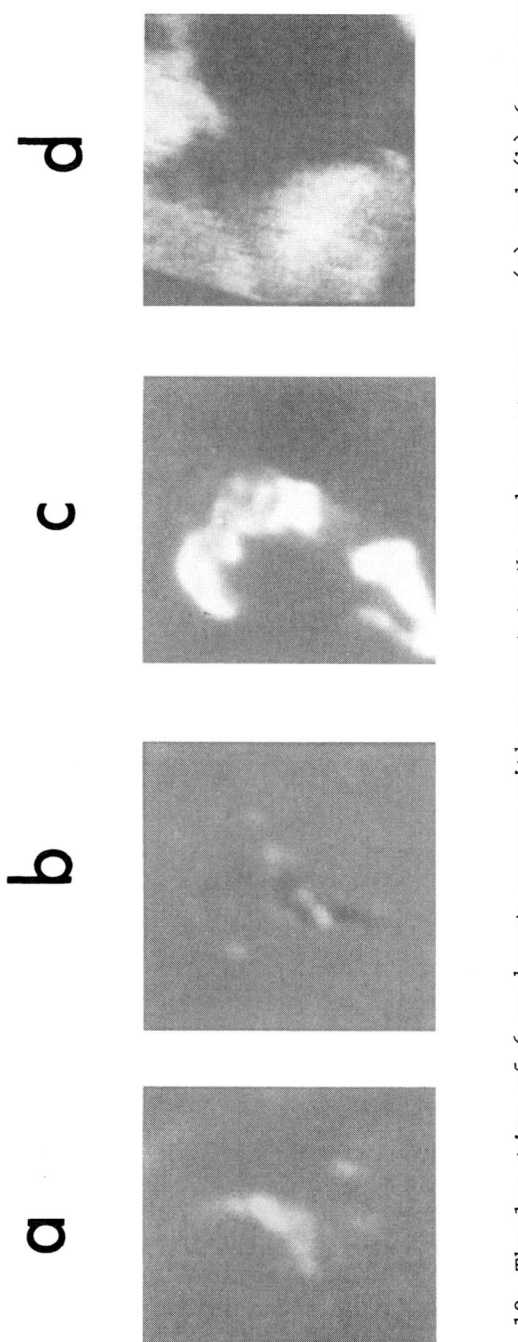

Fig. 10 The location of 6 cm burst source with respect to Hα and magnetogram. (a) and (b) 6 cm maps of total and polarized intensity at 192005 UT; (c) Hα photograph at 192013 UT (SOON data, courtesy D. M. Rust); and (d) magnetogram at 1844 UT (KPNO data, courtesy J. Harvey).

Fig. 11 A schematic sketch of the geometry of magnetic field lines at the times indicated. The shaded areas represent location of strongest 6 cm emission.

some kind of magnetic reconnection of the field lines originating from the two bipolar regions between which this compact region is located. Soon after the maximum, a loop-like structure with a lower brightness temperature ($<300 \times 10^6$ K) developed around the compact source (Fig. 9g). The footpoints of the loop had predominantly opposite polarization and the two ribbon Hα flare occurred near the footpoints of this loop of microwave emission (Fig. 11). The intense compact source is located approximately near the top of this loop.

The appearance of the north-south neutral line at the start of the impulsive phase must be indicative of the appearance of a new system of loops, possibly due to the development of new discontinuities between two loop systems at coronal levels or due to reconnections. The pressing of two loop systems with opposite polarities must ultimately be responsible for the acceleration of electrons responsible for the impulsive phase. The polarization map of Figure 9f suggests such a possiblity.

As can be seen from the polarization maps shown in Figure 9, the magnetic field structure appears to be rather complex. A simple picture of two loops of opposite polarity seems to fit well the extended loop-like emission. These loops are seen clearly soon after the impulsive maximum, partly due to the fact they brightened up as a result of the impulsive burst. Clearly, the magnetic field configurations inferred from our polarization data is similar to that proposed by Gold and Hoyle (1960) although it is not possible for us to observe in the radio data the "twists" required for the explosive release of energy. In any case, the quadrupole magnetic field configuration observed just prior to the impulsive energy release at its peak appears to be related to the explosive reconnection of magnetic fields necessary for the flare phenomenon.

Post-Flare Loops

Using the VLA, Velusamy and Kundu (1981) observed burst emission at 20 cm wavelength from post-flare loops. On May 19, 1979, a 20 cm burst occurred near the west limb and was associated with an Hα flare and a soft x-ray burst. Most of the intense radio emission occurred at the end of the Hα flare and in the decay phase of the x-ray burst. This post-flare radio emission was clearly to be identified with an emission similar to the Hα and x-ray emissions originating in post-flare loops. Indeed "snapshot maps" of the burst source obtained with a resolution of 12" x 24" showed that the radio emission occurred in loop-like structures (Fig. 12), similar in size and shape to those observed in x-rays from P78-1. Associated with the post-flare loops above (Fig. 12), several Hα eruptive prominences or surges and green line loops extending far outside the limb were also seen. The 20 cm emission in the form of loop-like structures with long extensions in the southwest direction developed and persisted for at least 2.5 hours. The structure showed a gradual ascension into the corona. The increasing height of 20 cm emission as a function of time at 20-40

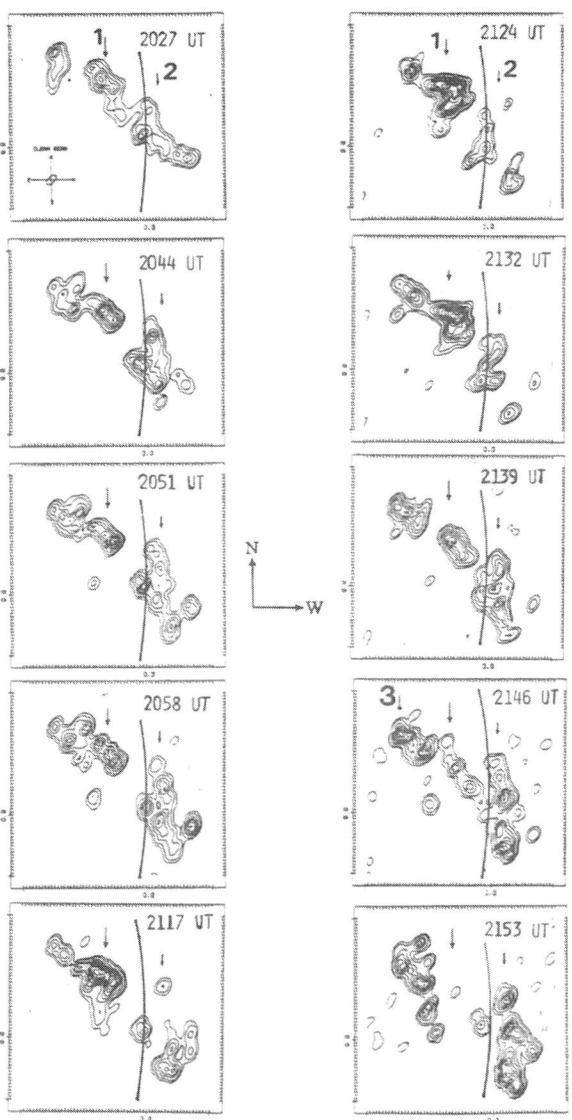

Fig. 12 Sequential snapshot synthesis maps of the burst source at 20 cm
wavelength. Each map represents a time average over 6-8 minutes at the
times indicated. The first two contours are 250,000 and 500,000 K, and
the contour interval is 500,000 K for the higher contours. The field
of view is 6' x 6' and the synthesized beam (size 24" x 12") is shown
at the left corner of the first map. The arrows 1 and 2 refer to the
emissions associated with regions close to McMath 16014 and 15999,
respectively. The arrow 3 refers to a third loop system that formed
later (From Velusamy and Kundu, 1981).

km/s, appears to be the same as the apparent motion observed in loop
prominence systems. Intepreting this post-flare loop radio emission
as thermal gyroradiation Velusamy and Kundu (1981) obtained an
estimate of magnetic field in the loop of ~ 120-170 Gauss. If one
interprets it as nonthermal synchrotron radiation, the lower limit of
the burst duration (2.5 hours) results in an upper limit of 160 gauss
for the strength of the magnetic field in the loops. Observation of
coronal radio emission in post-flare loops clearly offers an important
means of determining magnetic fields in loops.

This research was supported by NSF grant ATM 7821762 and NASA
grant NGR 21-002-199.

REFERENCES

Alissandrakis, C.E. and Kundu, M.R.: 1975, Solar Phys., 41, 119.
Alissandrakis, C.E. and Kundu, M.R.: 1978, Astrophys. J., 222, 342.
Brown, J.C., Melrose, D.B., and Spicer, D.S.: 1979, Astrophys. J., 228,
    592.
Enomé, S., Kakinuma, T., and Tanaka, H.: 1969, Solar Phys. 6, 428.
Foukal, P.V., Huber, M.C., Noyes, R.W., Reeves, E.M., Schmahl, E.J.,
    Timothy, J.G., Vernazza, J.E., and Withbroe, G.L.: 1974, Astrophys.
    J., 193, L143.
Gold, T. and Hoyle, F.: 1960, Monthly Notices Roy. Astron. Soc., 120, 89.
Heyvaerts, J., Priest, E.R., and Rust, D.M.: 1977, Astrophys. J., 216, 123.
Hobbs, R.W., Jordan, S.D., Webster, W.J., Jr., Maran, S.P., and Caulk,
    H.M.: 1974, Solar Phys., 36, 369.
Kundu, M.R.: 1959, Ann. Astrophys., 22, 1.
Kundu, M.R., Becker, R., and Velusamy, T.: 1974a, Solar Phys., 34, 185.
Kundu, M.R., Velusamy, T., and Becker, R.: 1974b, Solar Phys., 34, 217.
Kundu, M.R. and Alissandrakis, C.E.: 1975, Nature, 257, 465.
Kundu, M.R., Alissandrakis, C.E., Bregman, J.D., and Hin, A.C.: 1977,
    Astrophys. J., 213, 278.
Kundu, M.R. and Vlahos, L.: 1979, Astrophys. J., 232, 595.
Kundu, M.R.: 1980, in Radio Physics of the Sun (ed. by M.R. Kundu and
    T.E. Gergely), D. Reidel Publ. Co., Dordrecht, Holland, p. 157.
Kundu, M.R. and Velusamy, T.: 1980, Astrophys. J., 240, L63.
Kundu, M.R., Schmahl, E.J., and Rao, A.P.: 1981a, Astron. Astrophys., 94,
    72.
Kundu, M.R., Schmahl, E.J., and Velusamy, T.: 1981b, Astrophys. J.,
    (submitted).
Kundu, M.R., Velusamy, T., and Schmahl, E.J.: 1981c, Astrophys. J., in
    preparation.
Lang, K.R.: 1974, Solar Phys., 36, 351.
Lang, K.R. and Willson, R.F.: 1980, in Radio Physics of the Sun (ed. by
    M.R. Kundu and T.E. Gergely), D. Reidel Publ. Co., Dordrecht,
    Holland, p. 109.
Marsh, K.A. and Hurford, G.: 1980, Astrophys. J., 240, L111.
Pallavicini, R., Vaiana, G.S., Tofani, G., and Felli, M.: 1979, Astrophys.
    J., 229, 375.

Schmahl, E.J.: 1980, in Radio Physics of the Sun (ed. by M.R. Kundu and
     T.E. Gergely), D. Reidel Publ. Co., Dordrecht, Holland, p. 71.
Takakura, T. and Kai, K.: 1966, Publ. Astron. Soc. Japan, 18, 57.
Vaiana, G.S., Davis, J.M., Giacconi, R., Krieger, A.S., Silk, J.K.,
     Timothy, A.F., and Zombeck, M.: 1973, Astrophys. J., 185, L47.
Velusamy, T. and Kundu, M.R.: 1981, Astrophys. J., 243, L103.
Webb, D. and Zirin, H.: 1981, Solar Phys., 69, 99.

A CORRELATION BETWEEN VARIOUS INDICES OF SOLAR ACTIVITY AND SOLAR CYCLES

Yu.I. Vitinsky
Pulkovo Observatory
Leningrad, USSR

ABSTRACT

A variation with time of various parameters of relationship between relative numbers and total areas of sunspots is considered on the basis of Zürich and Greenwich data for 1879-1964. A significant difference of these indices on the descending and ascending branch of the 11-year cycle is found. Their variation is essentially different in the even and odd 11-year cycles. Secular variations in parameters of relationship W-S were found for the ascending branch of 11-year cycles. An interpretation is given of the revealed time variations in the relationship between W and S.

It is generally thought that there is a stable correlation between annual relative numbers and total areas of sunspots as Waldmeier (1955) stated. This permits to singly determine the total sunspot area S using the Wolf number W in the formula

$$S = 16.7 \ W. \tag{1}$$

However, Xanthakis (1960) showed that the annual ratio S/W is greater during the epoch of maximum of the 11-year solar cycle than during its minimum. Moreover, the value of the ratio varies in a different way than the Wolf numbers W during the cycle (Kopecký and Růžičková-Topolová, 1978) and from cycle to cycle (Mergentaler, 1959). All these made us doubt the stable correlation S-W.

The present paper reports on the investigation of the variation in parameters of relationship between monthly Wolf numbers and total sunspot areas (correlation coefficient r, the ratio S/W, and their standard deviation ratio σ_S/σ_W) with the 11-year and 22-year and secular solar cycle. The results of observations made in Zürich and at the Greenwich Observatory in the years 1879-1964, i.e. 8 eleven-year solar cycles (cycles No. 12-19 according to Zürich) were used as the basic data.

E. G. Mariolopoulos et al. (eds.), Compendium in Astronomy, 139–148.
Copyright © 1982 by D. Reidel Publishing Company.

1. THE ELEVEN-YEAR CYCLE

First of all the variation in the above parameters of relationship W-S with the phase of an 11-year solar cycle was studied. The detailed results are published in Vitinsky (1980). Here we shall dwell upon only main points which will be useful for our present study of the stability with time in the relationship W-S. The year of maximum of Wolf numbers was used as the zero-point, because the year of minimum is more difficult to determine due to the overlapping of cycles (Vitinsky, 1973). The parameters were averaged for the corresponding years of an 11-year solar cycle in relation to this zero-point (M-3, M-2, M-1, M, M+1, M+2, M+3, M+4, M+5). Table 1 gives values r (W,S), $\overline{S/W}$ and σ_S/σ_W. It also contains the values of a comparison of Wolf numbers with the density of solar radio flux at 2800 MHz (Ottawa data) during the years 1948-1976 : r (W,F), $\overline{F/W}$, σ_F/σ_W.

Table 1

Relationship between Various Indices of Solar Activity in an
11-Year Solar Cycle

Phase of the cycle	r (W, S)	r (W, F)	$\overline{S/W}$	$\overline{F/W}$	σ_S/σ_W	σ_F/σ_W
M-3	0.88	0.95	15.5	7.30	22.7	0.57
M-2	0.84	0.96	15.4	2.33	23.0	0.76
M-1	0.87	0.88	15.9	1.28	21.4	0.89
M	0.89	0.97	17.1	1.28	20.0	0.96
M+1	0.90	0.95	16.2	1.35	18.8	0.95
M+2	0.94	0.93	16.5	1.35	19.2	0.94
M+3	0.94	0.95	14.7	1.55	17.6	1.00
M+4	0.86	0.90	14.7	1.80	19.8	0.94
M+5	0.86	0.72	13.5	2.54	16.8	0.83

As is seen from Table 1 the closest correlation between W and S takes place not during the epoch of maximum, but 2-3 years later. While

on the ascending branch of the 11-year cycle the value r varies but in-
significantly, the correlation coefficients r (W, S) are significantly
different in the (M-2) and (M+2) or (M+3) and (M+4) or (M+5)-th years
of the cycle. It should be noted that it is on the descending branch
that the largest sunspot groups are usually observed.

The variation $\overline{S/W}$ and σ_S/σ_W is regular on the ascending branch of
the 11-year cycle but on the descending branch the regularity (particularly for σ_S/σ_W) is broken.

Thus, the correlation of Wolf numbers and total sunspot areas is
different on the ascending and descending branch of the 11-year solar
cycle. That is why they should be considered separately in our study as
well. It should be noted that a similar approach to the relationship
between Wolf numbers and the density of solar radio flux at 2800 MHz
did not give so definite results partly because the used material was
too scanty. The most typical peculiarity of the variation in parameters
of this relationship is maximum values of r and minimum values of $\overline{F/W}$
in the maximum of the cycle. However, the maximum value of the ratio
σ_F/σ_W shows up only three years after the maximum. The largest difference of the correlation coefficients r (F, W) exists between the (M-1)
and M year for the ascending branch and the (M+5) year and the other
years for the descending branch of the cycle. All this shows that even
within one 11-year solar cycle the density of the solar radio flux at
2800 MHz behave otherwise than Wolf numbers, and hence can hardly be
used as their equivalents regardless of a very close correlation between these solar indices.

2. THE 22-YEAR SOLAR CYCLE

The 22-year solar cycle is characterized apart from the inversion
of the magnetic field polarity both in near-equatorial and polar zones
by the Gnevyshev-Ohl rule, i.e. each odd 11-year cycle is usually higher
than the even one (Gnevyshev and Ohl, 1948). There is evidence that the
number of solar flares correlates differently with the Wolf number in
different 11-year cycles (Smith and Smith, 1963). That is why it seemed
interesting to see if Wolf numbers and total sunspot areas are connected
similarly in even and odd 11-year solar cycles.

Using the same method as for 11-year cycles we obtained the values
of r (W, S), $\overline{S/W}$ and σ_S/σ_W separately for odd and even cycles (Table 2).

As is seen from Tables 1 and 2, the mean variation in the parameters
of relationship W-S for all 11-year cycles is in general similar to the
variations in even cycles. As to the odd cycles, the variation has a more
complicated form.

A detailed consideration of the correlation coefficient r variation
with the phase of an odd and even 11-year cycle using the Kholmogorov
z-criterion (Mitropolsky, 1961) enables to make the following conclu-

sions. While in even cycles values of r are significantly different in

Table 2

A Relationship between Wolf Numbers and Total Sunspot Areas
Within Odd and Even 11-Year Solar Cycles

Phase of the cycle	r (W, S)		$\overline{S}/\overline{W}$		σ_S/σ_W	
	even	odd	even	odd	even	odd
M−3	0.89	0.82	15.6	12.6	24.5	22.4
M−2	0.72	0.86	15.2	15.6	24.6	21.7
M−1	0.83	0.76	16.6	15.7	27.6	20.2
M	0.84	0.91	18.1	16.5	21.1	20.4
M+1	0.86	0.83	15.8	13.8	18.9	18.1
M+2	0.94	0.95	16.0	17.0	18.7	20.5
M+3	0.92	0.82	14.0	15.0	17.6	16.4
M+4	0.93	0.77	15.4	14.5	21.4	17.4
M+5	0.91	0.84	13.2	13.5	15.5	19.7

the (M−2) and (M+2) year, as on the average for all the 11-year cycles, in odd cycles apart from the primary maximum in the (M+2) year there is a secondary maximum in the year of the Wolf number maximum. This shows up in significantly different values of r in the (M−1) and M year and in the (M−1) and (M+2) year. If during even cycles the correlation coefficient r is practically invariable beginning with the (M+2) year of the descending branch during odd cycles it oscillates greatly manifesting a significant difference between its values in the (M+1) and (M+2), and (M+2), (M+4) and (M+2), (M+5) and (M+2) years.

The variation of $\overline{S}/\overline{W}$ in even 11-year cycles on the descending branch is not so prominent and of a more monodirected character than in odd cycles. Moreover, in even cycles the maximum value of $\overline{S}/\overline{W}$ takes place in the year of maximum Wolf number, in odd cycles two years after. It is of interest to note, that these years have secondary maxima of $\overline{S}/\overline{W}$ in odd and even cycles respectively (see Table 2). As a comparison of Table 1 and 2 shows, the variation of $\overline{S}/\overline{W}$ for all 11-year cycles is generally determined by odd cycles, while the maximum value of this parameter takes place in the year of maximum, as in even cycles.

The variation of the ratio σ_S/σ_W in even and odd cycles is of the inverse character with the increase (decrease) to the (M-1) year and decrease (increase) with some fluctuations, which are particularly well detected in odd cycles. The difference of odd and even cycles shows up in the regression equation, determining the connection between W and S (Formulas (2) and (3)). As is seen from the formulas the coefficient W_{M+i} is on the average larger in even cycles than in odd cycles, and all the free terms are, while in odd cycles the terms are positive in three cases out of nine.

$$S_{M-3} = 21.8 \ W_{M-3} - 251 \qquad S_{M+1} = 16.2 \ W_{M+1} - 34$$

$$S_{M-2} = 17.7 \ W_{M-2} - 112 \qquad S_{M+2} = 17.6 \ W_{M+2} - 112$$

$$S_{M-1} = 22.9 \ W_{M-1} - 418 \qquad S_{M+3} = 16.2 \ W_{M+3} - 97 \qquad (2)$$

$$S_{M} = 17.7 \ W_{M} - 22 \qquad S_{M+4} = 19.9 \ W_{M+4} - 186$$

$$S_{M+5} = 14.1 \ W_{M+5} - 14$$

$$S_{M-3} = 18.4 \ W_{M-3} - 46 \qquad S_{M+1} = 15.0 \ W_{M+1} - 154$$

$$S_{M-2} = 18.7 \ W_{M-2} - 119 \qquad S_{M+2} = 19.5 \ W_{M+2} - 132$$

$$S_{M-1} = 15.4 \ W_{M-1} + 23 \qquad S_{M+3} = 13.4 \ W_{M+3} + 105 \qquad (3)$$

$$S_{M} = 18.6 \ W_{M} - 262 \qquad S_{M+4} = 13.4 \ W_{M+4} + 29$$

$$S_{M+5} = 16.5 \ W_{M+5} - 83 \ .$$

Thus, it is clearly seen that the variation in parameters of relationship between relative Wolf numbers and total sunspot areas in odd and even cycles is distinctly different especially on the descending branch.

3. THE SECULAR SOLAR CYCLE

The available material does not allow to get reliable conclusions on secular variations of the parameters of relationship of W-S. However, we had a chance to obtain some preliminary results on the subject, because the material includes the descending branch of the previous cycle and maybe the whole ascending branch of the current secular solar cycle.

Taking into account the different character of the correlation indices of W and S on the ascending and descending branch of the 11-year solar cycle, we examined them separately for each cycle from No. 12 to

No. 19. The year of maximum of Wolf numbers was assumed as part of the ascending branch and only the first four years after the maximum were considered for compatibility.

The numerical values of r (W, S), $\overline{S}/\overline{W}$ and σ_S/σ_W for the ascending and descending branch of the 11-year cycle separately are given in Table 3. It also presents their smoothed values, computed from the formula

$$(X_i)_{sm} \quad \frac{X_{i-1} + 2X_i + X_{i+1}}{4} . \tag{4}$$

Smoothed bracketed values for the 12th and 19th solar cycle were calculated without taking into consideration the term W_{i-1} or W_{i+1} respectively and should be considered as estimations only.

Table 3

The Relationship between Relative Numbers and Total Sunspot Areas for the Ascending Branch of an 11-year Cycle in the Secular Solar Cycle

Number of cycle	r	$\overline{S}/\overline{W}$	σ_S/σ_W	r_{sm}	$(\overline{S}/\overline{W})_{sm}$	$(\sigma_S/\sigma_W)_{sm}$
12	0.87	15.7	25.6	(0.89)	(16.0)	(23.3)
13	0.94	16.6	18.8	0.92	16.2	21.3
14	0.91	15.8	21.9	0.92	15.6	19.7
15	0.94	14.4	16.1	0.86	15.6	19.6
16	0.65	17.7	24.3	0.78	16.3	21.2
17	0.85	16.5	19.9	0.81	17.2	21.2
18	0.89	17.9	20.6	0.90	17.5	19.8
19	0.97	16.0	17.7	(0.94)	(16.6)	(18.7)

As is seen from Table 3 differences between numerical values of r are insignificant on the descending branch of the secular cycle and are significantly different on the descending branches (cycles No. 16–19) in cycles No. 16 and No. 17, No. 18 and No. 19 for ascending branches of the 11-year solar cycle using the Kholmogorov z-criterion. For the descending branch of the 11-year solar cycle the picture of the secular variation of r is more complicated (Table 4) : a significant difference between the correlation coefficient r (W, S) shows up on the descending branch of the secular cycle in cycles No. 12 and No. 13 and on its ascending branch in No. 18 and No. 19. It should be noted that the 22-year cycle manifests itself rather distinctly : values of r in even cycles are smaller than in odd ones.

Table 4

A Relationship between Relative Numbers and Total Sunspot Areas for the Descending Branch of an 11-Year Solar Cycle in the Secular Cycle

Number of cycle	r	$\overline{S}/\overline{W}$	σ_S/σ_W	r_{sm}	$(\overline{S}/\overline{W})_{sm}$	$(\sigma_S/\sigma_W)_{sm}$
12	0.94	15.6	19.1	(0.91)	(15.7)	(19.3)
13	0.85	15.8	19.8	0.91	15.6	20.2
14	0.93	15.5	22.2	0.89	15.6	20.3
15	0.86	15.6	17.1	0.90	15.8	20.6
16	0.94	16.7	25.9	0.91	16.5	23.3
17	0.89	17.0	24.1	0.88	16.4	23.6
18	0.81	14.7	20.3	0.86	16.8	21.4
19	0.95	15.9	20.7	(0.90)	(15.5)	(20.8)

A more distinct long-period variation of r may be found out from its smoothed values. For the ascending branch of the 11-year solar cycle the secular variation r is detected distinctly with its minimum value in the 16th solar cycle (Table 3). This is an additional evidence for a special role of the ascending branch in an 11-year solar cycle for solar cycles as a whole (Waldmeier, 1935; Xanthakis, 1962). For the descending branch of the 11-year solar cycle the long-period variation of r is practically unnoticable, although there is a tendency of a decrease of r towards the 18th solar cycle, whose amplitude appears to be within the errors (Table 4).

The secular variation of the $\overline{S}/\overline{W}$ ratio is most distinctly detected for the ascending branch of the 11-year solar cycle with the minimum in the 15th solar cycle, although the amplitude of the variation is only 0.6 on the descending branch. As to the variation of the parameter for the descending branch of the 11-year solar cycle, it' has an increasing character from the 13th to the 18th solar cycle with small oscillations, but within the errors. It is of interest that the 22-year solar cycle in the $\overline{S}/\overline{W}$ value is detected most distinctly for the descending branch of the 11-year cycle, while for the ascending branch its most typical peculiarity is violated ($\overline{S}/\overline{W}$ in the even cycle is higher than in the odd one), except cycles 12-13 (Tables 3 and 4). It is probable that the 22-year solar cycle shows up differently in the variation of the $\overline{S}/\overline{W}$ value for the ascending and descending branch of the 11-year cycle.

A long-period variation of σ_S/σ_W (Tables 3 and 4) does not show any secular variation pattern and must be characterized by some longer cycle. In the case of the ascending branch of the 11-year cycle a regular decrease of the σ_S/σ_W is noted with a slight increase between the 15th and 16th solar cycle. It is of interest that this fluctuation took place near the epoch of minimum of the secular cycle. In the case of the descending branch of the 11-year solar cycle, on the contrary, an increase of σ_S/σ_W from the 12th to the 17th cycle is noted and afterwards a decrease as in the previous case. It should be noted that the 22-year solar cycle shows up definitely only for the ascending branch of the 11-year solar cycle as far as σ_S/σ_W concerns, while for the descending branch 44-year variations of this characteristic of relation are better traced.

A likely variation of the secular or longer-period type is noted in the equation of linear regression W-S for the ascending and descending branch respectively in various 11-year solar cycles, (5) and (6).

$$
\begin{aligned}
S_{12} &= 22.3 \; W_{12} - 514 & S_{16} &= 15.8 \; W_{16} - 420 \\
S_{13} &= 17.7 \; W_{13} - 108 & S_{17} &= 16.9 \; W_{17} - 206 \\
S_{14} &= 19.9 \; W_{14} - 197 & S_{18} &= 18.3 \; W_{18} - 189 \\
S_{15} &= 15.1 \; W_{15} - 92 & S_{19} &= 17.2 \; W_{19} - 163
\end{aligned}
\tag{5}
$$

$$
\begin{aligned}
S_{12} &= 18.0 \; W_{12} - 138 & S_{16} &= 24.3 \; W_{16} - 303 \\
S_{13} &= 16.8 \; W_{13} - 210 & S_{17} &= 21.4 \; W_{17} - 554 \\
S_{14} &= 20.6 \; W_{14} - 347 & S_{18} &= 16.4 \; W_{18} - 594 \\
S_{15} &= 14.7 \; W_{15} - 76 & S_{19} &= 19.7 \; W_{19} - 605.
\end{aligned}
\tag{6}
$$

It is of interest to note that it is in the 15th solar cycle, which is practically in the minimum of the secular cycle, that the minimum values of the coefficient of W_i and the free term in both systems of equations are seen.

4. CONCLUSIONS AND FINAL REMARKS

As it follows from the present paper the relation between relative values and total area of sunspots cannot be considered as permanent in time. Its characteristics show variations, which are in all likelihood due to peculiarities of solar cycles. There is a significant difference between the ascending and descending branch of the 11-year solar cycle, odd and even 11-year solar cycle forming a 22-year solar cycle, in the variation of the characteristics of relationship W-S (r, $\bar{S}/\bar{W}$, σ_S/σ_W). A secular and a longer-period variation of these characteristics shows up.

It is known that relative numbers and total areas of sunspots reflect different aspects of solar activity, although they are very closely connected, and they behave differently in the 11-year and secular solar cycle (Vitinsky, 1973). The found regularities show most and foremost that larger sunspot groups occur not in the epoch of maximum, but on the beginning of the descending branch of the 11-year cycle. This affirm the earlier conclusion that the most powerful active regions appear after the maximum of the 11-year solar cycle (Gnevyshev, 1977). Moreover, while the ratio $\bar{S}/\bar{W}$, particularly on the ascending branch, in the average is greater in even cycles than in odd ones, the difference of the regressions connecting the number of flares with the relative number of sunspots becomes understandable.

It is of great importance that the secular variation of the characteristics of relationship between relative numbers and total sunspot areas occurs only for the ascending branch of the 11-year solar cycle. This means that the ascending branch is not a determinant within the 11-year solar cycle but in the secular one as well. But it should be noted, however, that the latter conclusion needs an additional verification using much more data.

The problem of stability of the relationship between various indices of solar activity, as is seen from the present paper, in which only indices of sunspots and density of the radio flux at the frequency 2800 MHz, has a great importance for the problem of solar cycles as a whole. The resolution of it would permit to approach the study of main peculiarities of solar cycles in greater detail, taking into consideration the correlation of various solar phenomena. The main difficulty in studying the problem lies in a very small number of homogeneous series of various indices, which cover at least several cycles. But even a study of the variation of their relation only in the 11-year solar cycle could be of great value and importance.

REFERENCES

Gnevyshev, M.N. : 1977, Solar Phys. 51, 175.
Gnevyshev, M.N. and Ohl, A.I. : 1948, Astron. Zh. 38, 18.
Kopecký, M. and Růžičková-Topolová, B. : 1978, BAC, 29, 65.
Mergentaler, J. : 1959, Acta Astron. 9, 107.
Mitropolsky, A.K. : 1961, Tekhnika Statisticheskikh Vychislenii, Moscow.
Smith, H.J. and Smith, E. v.P. : 1963, Solar Flares, Macmillan Co., N.York.
Vitinsky, Yu.I. : 1973, Tsiklichnost i Prognozy Solnechnoi Aktivnosti,
     Leningrad.
Waldmeier, M.: 1935, Astron. Mitt. Zürich, Nr. 133.
Waldmeier, M.: 1955, Ergebnisse und Probleme der Sonnenforschung, 2.
     Aufgabe, Leipzig.
Xanthakis, J. : 1960, Geofis. Pura e Appl. Milano, 46, 11.
Xanthakis, J. : 1962, Ann. Astrophys. 25, 342..

PART IV

SOLAR-TERRESTRIAL RELATIONS

CARACTÈRES MÉTÉOROLOGIQUES DU CLIMAT, ET ACTIVITÉ SOLAIRE : QUELQUES REMARQUES

Jean-Claude Pecker
Collège de France
Paris, France

Que le Soleil ait sur les couches élevées de l'atmosphère terrestre, celles où se forme la composante principale du géomagnétisme une influence importante, - cela est avéré et depuis longtemps. Que ces couches, en revanche, transmettent aux manifestations climatiques, voire météorologiques, les effets, même atténués, étalés sur le temps, distribués sur la planète, de ces phénomènes solaires, cela est fortement contesté : les influences solaires sont en général noyées dans des influences locales plus manifestes.

Nous ne tenterons pas ici de faire une large bibliographie de ces questions, et voudrions seulement apporter à leur étude une contribution statistique sans doute nouvelle.

(Je voudrais préciser que ces calculs ont été menés à bien par Claire LEFEVRE, ancienne élève de l'Ecole Polytechnique, à l'occasion de son stage au Laboratoire d'Astrophysique Théorique du Collège de France, en 1975-76. Malheureusement, Melle LEFEVRE a quitté depuis lors la voie de la recherche; il n'a pas été possible de lui demander son accord sur ce texte, dont elle aurait dû bien évidemment partager la paternité avec moi; je demande donc à Claire LEFEVRE, et aux lecteurs, de vouloir bien excuser l'absence de Claire LEFEVRE du titre de l'article. Jean-Claude PECKER).

LES RESULTATS DE J. XANTHAKIS

C'est dans un article de 1973 que le Professeur XANTHAKIS, à qui il me plait, par cette note, de rendre un amical et juste hommage, à l'occasion de son vingt-cinquième anniversaire comme membre de l' Académie d' Athènes, a posé le problème en termes clairs.

Travaillant sur des données statistiques portant sur 80 années (8 cycles solaires), il analyse les corrélations entre la pluviosité mesurée P dans différentes zones de latitude, avec l'activité solaire moyen-

151

E. G. Mariolopoulos et al. (eds.), Compendium in Astronomy, 151–160.
Copyright © 1982 by D. Reidel Publishing Company.

ne, définie par l'indice I_a (P et I_a sont définis dans la légende de la figure 1).

Le résultat de XANTHAKIS est remarquable, et déroutant. Il est représenté sur la figure 1; XANTHAKIS montre clairement que :
(a) dans la zone polaire (12 stations localisées entre 70° et 80° de latitude Nord) une bonne corrélation existe entre l'activité solaire et la moyenne des précipitations. Le coefficient de corrélation est, entre P et I_a, égal à 0.77, donc très élevé.
(b) dans la zone nordique (22 stations localisées entre 60° et 70° une anticorrélation existe, de - 0.64, -très significative également).
(c) dans la zone semi-nordique (71 stations, entre 50° et 60°) on trouve deux périodes, l'une d'anticorrélation (1890-1913); l'autre de corrélation (1914-1960); pour chacune, le coefficient de corrélation est de -0.76 et 0.81, respectivement.
(d) dans la zone moyenne (80 stations de 40° à 50°) un comportement assez complexe est trouvé; les corrélations, pendant de courtes périodes, sont assez marquées; -mais moins que dans les cas précédents, et ces corrélations sont ici peu significatives.

D'autres auteurs trouvent des résultats intéressants, mais, dans une certaine mesure, contradictoires :
A Adélaïde (Cornish, 1954) une corrélation existe -mais avec le cycle de 22 ans. Dans une station brésilienne, et trois stations sud-africaines, King retrouve un phénomène analogue, -à 22 ans. Certains auteurs ont passé les données au crible de l'analyse de Fourier, et ont retrouvé aussi les périodicités à 22 ans. Lamb seul fait aussi mention de périodicités de 170 et 200 ans; mais il est clair que les données sont insuffisantes pour affirmer avec confiance l'existence de si grandes périodicités.

ETUDE DE LA PLUVIOSITE DANS HUIT STATIONS EQUATORIALES

Il nous a semblé que l'utilisation d'un grand nombre de stations risque de fausser les données, au lieu d'améliorer les corrélations : en éliminant, par des moyennes peut-être mal équilibrées, les effets locaux, elles tentent de ne conserver que les influences planétaires; mais, précisément, elles manquent alors l'un de leurs buts essentiels, ne permettant pas, éventuellement, les prévisions locales. D'autre part, les régions équatoriales devraient donner des résultats meilleurs, peut-être moins affectés que les latitudes plus élevées par les effets saisonniers. Nous avons donc étudié un certain nombre de stations équatoriales, -et l'unique station tempérée de Paris (48°49 N, 02°20 E). Toutes les données sont tirées des WORLD WEATHER RECORDS, sauf dans le cas de Paris, où nous les devons à l'obligeance des responsables de la Météorologie Nationale. Plusieurs techniques d'analyse ont été employées. Mais c'est l'analyse de Fourier qui donne les résultats les plus clairs.

La carte de la figure 2 montre la localisation des huit stations étudiées.

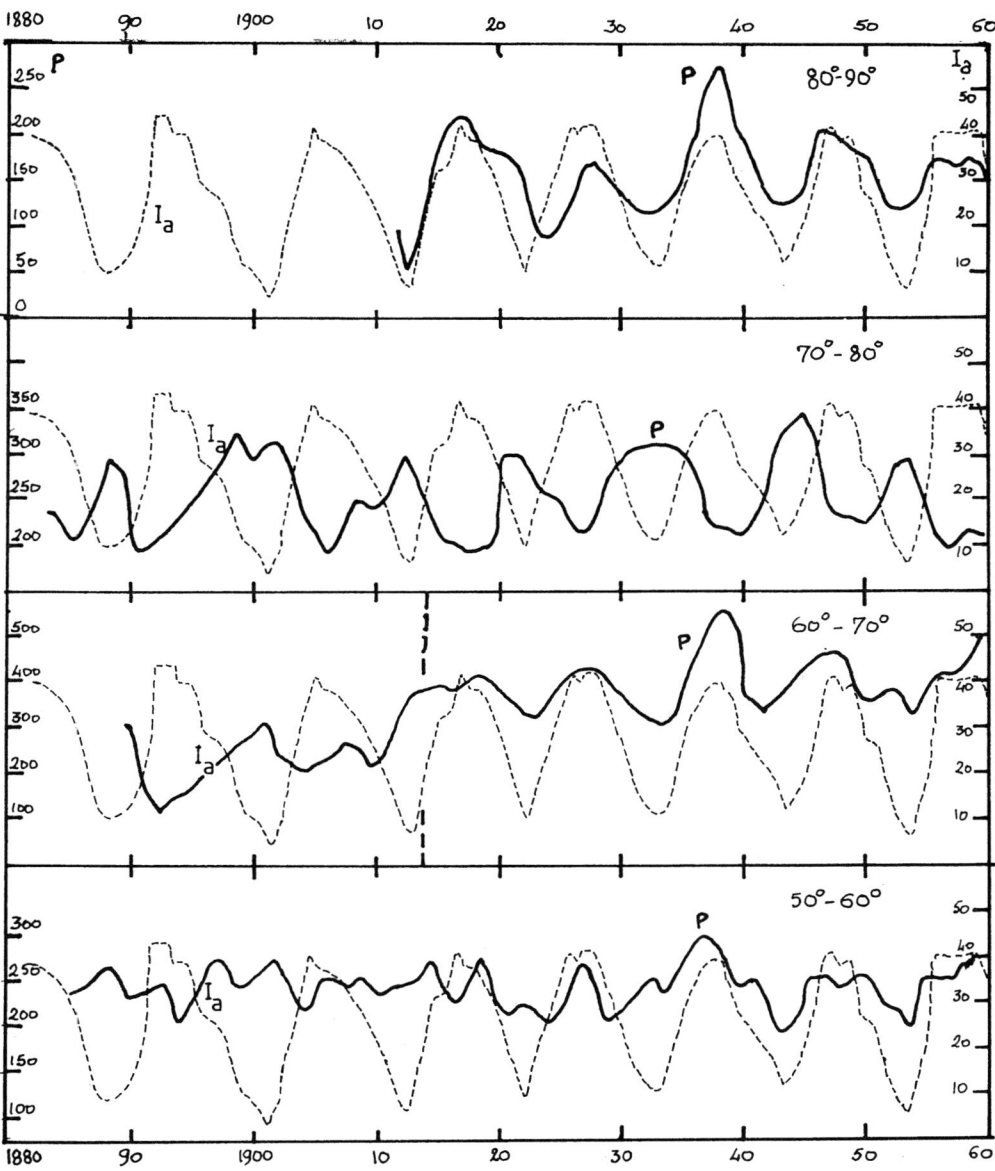

Fig. 1.: L' analyse de Xanthakis. Pour 4 groupes de stations définies par l' intervalle de latitude où elles se trouvent, en fonction du temps, deux fonctions sont représentées: en traits pointillés, l' indice I_a d'activité solaire, défini par Xanthakis; en traits pleins épais, indice de pluviosité P (identique à la quantité $R-R_o$, définie par Xanthakis).

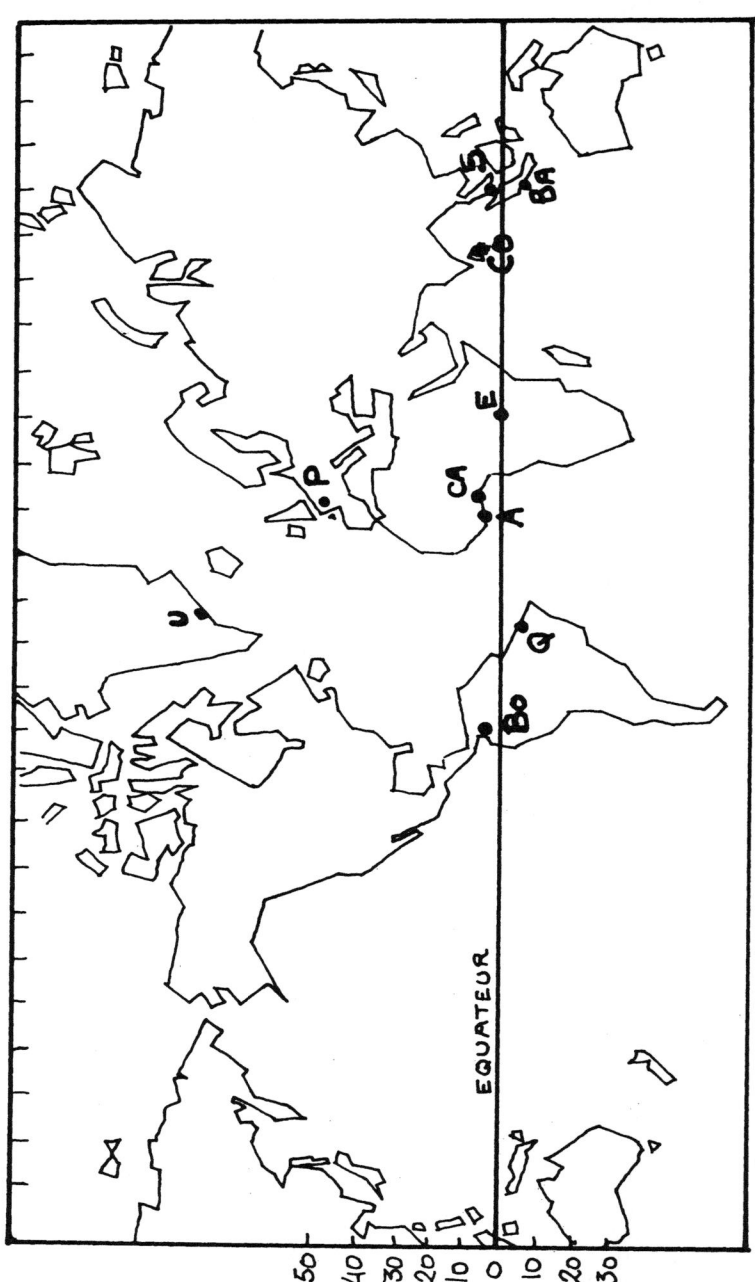

Fig. 2.: Localisation des stations équatoriales étudiées (et des deux autres stations – Paris, Upernivik – étudiées). Voir aussi Tableau I.

Les figures 3a à 3h, et le tableau I donnent les résultats de cette analyse pour les diverses stations équatoriales etudiées.

Ces différents résultats peuvent être commentés en plus de détails, comme suit.

(a) La première remarque évidente, est la <u>grande diversité</u> des courbes obtenues. De façon assez arbitraire, nous les avons représentées, avec la même échelle des abscisses bien sûr, mais avec des échelles d'ordonnées différentes, et sans avoir calculé (tel n'était pas notre propos!) le degré de signification statistique des "pics" des spectres de puissance obtenus.

Seules les trois stations BO, Q et BA ont un maximum correspondant à la période de l'activité solaire. Mais on notera, sur toutes les courbes, des périodes plus faibles, allant de 3 à 9 jours, et plus longues, atteignant 18.5 jours. Dans certains cas (stations africaines A et E), les maximums sont très marqués; dans d'autres stations, ils sont au contraire très peu marqués, bien que significatifs (BA ou Q). Pour BO, ils ne sont sans doute même pas significatifs; peut-être ce fait est-il lié à l'altitude de Bogota, qui lui confère une situation géographique bien particulière.

Aucune corrélation avec l'activité solaire n'est donc marquée. Mais dans certains cas, d'évidentes périodicités, liées à des conditions géographiques, sont mises en évidence. Ainsi, la courbe correspondant à Entebbe devrait aider à y prévoir les années sèches et humides.

(b) La seconde remarque concerne, par comparaison, la station unique, très <u>nordique</u>, d'Upernivik, qui est incluse dans les données de Xanthakis. Elle confirme la corrélation avec le cycle solaire, que l'ensemble des données de Xanthakis de la figure 1 montre aussi, un peu moins nettement (figure 4).

(c) Notre troisième remarque concerne la corrélation avec l'activité solaire, qui n'est pas mise en évidence par les spectres de puissance (ils confondent d'ailleurs corrélation et anticorrélation ...). On a cherché, pour chacune des huit stations, la corrélation entre les indices d'activité et la pluviosité; la corrélation, pour les périodes étudiées ne depasse jamais 0.3; -ce qui est faible ! Cependant, on peut légèrement améliorer cette corrélation en notant que les effets météorologiques peuvent être corrélés avec les phénomènes solaires, à condition de tenir compte d'un décalage temporel ΔT, -constante de temps de l'interaction possible-. En faisant varier ce décalage ΔT, nous avons trouvé que la corrélation augmente régulièrement pour ΔT allant de 0 à 3 jours, valeur correspondant à un maximum de la corrélation, -qui reste néanmoins faible.

(d) Une dernière remarque concerne le comportement des <u>températures</u> dans les huit stations étudiées, étude que nous avons abordée par les mêmes méthodes. En général, peu de corrélation existe entre pluies et

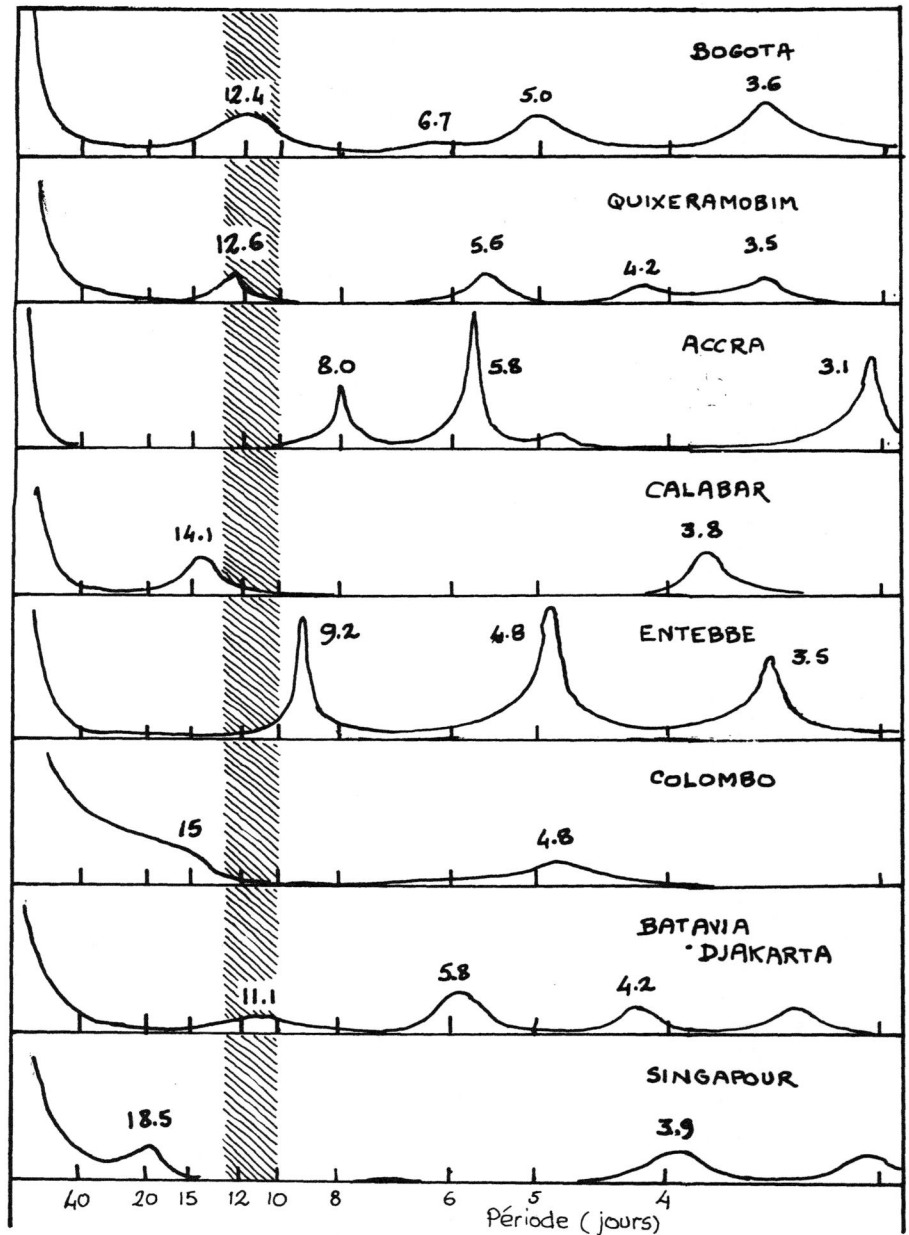

Fig. 3.: Spectres de puissance pour les huit stations équatoriales étudiées. Les hachures correspondent à la périodicité du cycle d' activité solaire.

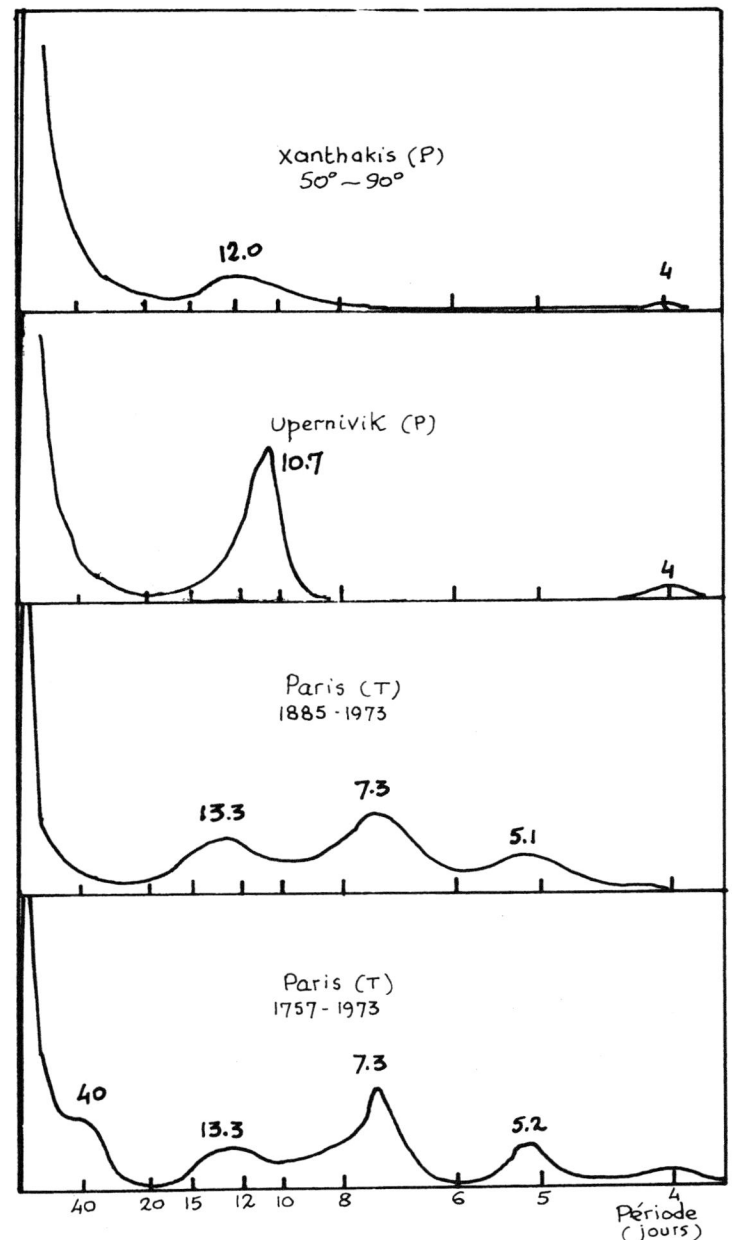

Fig. 4. : Spectre de puissance correspondant aux stations de latitude élevée. De haut en bas: (a) stations de Xanthakis (b) station d' Upernivik (c) Paris, températures, période 1885-1975 (d) Paris, températures, période 1757-1975.

TABLEAU I

HUIT STATIONS EQUATORIALES

STATIONS	BOGOTA	QUIXERAMOBIN	ACCRA	CALABAR	ENTEBBE	COLOMBO	BATAVIA (DJAKARTA)	SINGAPOOR
	BO	Q	A	CA	E	CO	B	S
Période d'observation	1866-1960	1896-1960	1888-1960	1866-1960	1905-1960	1881-1960	1866-1960	1911-1960
Latitude	4°36' N	5°16' S	5°36' N	4°58' N	0°05' N	6°54' N	6°11' S	1°18' N
Longitude	74°5' W	38°50' W	0°10' W	8°19' E	32°29' E	79°53' E	106°53' E	103°53' E
Altitude	2651 m	199 m	65 m	12.2 m	1.17 m	7.3 m	8 m	2.7 m
Corrélations (pluies)								
$T=0$	0.039	0.068	-0.109	-0.034	-0.190	-0.108	0.127	-0.074
1	0.002	-0.031	-0.054	-0.110	-0.225	-0.018	0.067	-0.064
2	-0.041	-0.137	0.071	-0.146	-0.121	-0.055	0.037	-0.054
3	0.005	-0.141	0.157	-0.219	-0.072	-0.027	0.048	0.080
4	0.132	-0.146	0.092	-0.262	-0.278	-0.058	-0.106	-0.005
5	0.164	-0.050	0.069	-0.263	-0.272	-0.237	-0.119 ?	0.061
Périodes (ans)/ Pluies	12.4 6.7 5.0	12.6 5.6 4.1	8 5.3	14.1 3.8	9.2 4.8 3.6	15 4.8	11.1 5.8 4.2	18.5 3.9
Températures	- - -	11.4 5.6 3.9	15 8.3	- - -	21.8	néant	10.6 5.6	10.4 5.6

températures; la température est corrélée plutôt négativement avec l'activité solaire; la corrélation reste faible. Une station unique montre une corrélation presque parfaite entre la pluviosité et la température : c'est la station Q.

LE CAS DE PARIS

A Paris, nous avons pu disposer de 217 valeurs annuelles, que nous avons étudiées de façon plus détaillée. Elles concernent malheureusement seulement les températures.

Tout d'abord quelques remarques qualitatives sont intéressantes. On note que les années 1782, 1784, 1799, 1829, 1879 et 1917 ont une température moyenne inférieure à 10 C; 1879 a été particulièrement froide (8.7 C). Les années chaudes (T > 12 C) sont 1934, 1945, 1947, 1959, 1961,... Un réchauffement moyen est donc assez clair. Peut-on lier ce réchauffement à une eventuelle période de 170 ans? Cela est évidemment bien hasardeux!

Plus qualitativement, les corrélations entre température et activité restent faibles, - de l'ordre de 0.1 et 0.2, -valeurs de toute évidence non significatives.

L'analyse de Fourier (figure 4, courbes du bas) met en évidence des périodes assez nettes, de 5.1 à 5.2 ans, 7.3 ans, 15.3 ans, -et peut-être 40 ans. Il est difficile d'associer ces périodes à l'activité solaire, en tous cas jusqu'en 1885. A partir de 1885, les corrélations sont bien meilleures, et vont de 0.35 à 0.47 pour des décalages ΔT de 0 à 3 ans. Il semble bien qu'alors les températures soient minimales au voisinage du minimum solaire, et qu'elles augmentent assez régulièrement lors qu'on progresse vers le maximum. Encore l'analyse de Fourier doit-elle nous inciter à beaucoup de réserves dans l'affirmation de ce résultat ! Mais comment comprendre que cette corrélation soit soudainement apparue vers 1885 ? Et à quoi donc peut correspondre la période de 40 ans, qui était nettement marquée avant cette date ?

CONCLUSION .... PROVISOIRE

On sait que le débat que nous avons ouvert à nouveau ici est un débat difficile, qui a entraîné des prises de positions passionnelles, et souvent très exagérées. Les uns, au mépris des règles élémentaires de la statistique, trouvent un peu trop souvent ce qu'ils cherchent; d'autres, justement critiques des mauvaises analyses statistiques, s'autorisent trop facilement de ces critiques pour rejeter en bloc l'idée d'une influence du Soleil sur les conditions météorologiques.

A l'issue de cette recherche, nous pensons, avec le Professeur XANTHAKIS, que le dossier ne saurait être refermé, et que si le climat est, si peu soit-il, sensible à l'activité solaire, il importe de le chercher avec soin.

Nous pensons aussi que le climat est, en certains lieux géographiques, plus nettement influencé par l'activité solaire qu'en d'autres. Localement, divers effets (influences des reliefs, des sols, des océans) conjuguent leurs effets. Il est remarquable de noter que cette conjugaison conduit à la mise en évidence de périodicités significatives, quoique non nécessairement (et même généralement non) liées à la périodicité de l'activité solaire. Cette remarque devrait au moins, indépendament de toute conclusion sur l'effet de l'activité solaire, conclure à des prévisions moyennes; un exemple remarquable devrait pouvoir être celui d'Entebbe. Mais on notera aussi que les influences solaires semblent plus importantes aux latitudes élevées.

Enfin, on notera l'amélioration des corrélations (même faibles) lorsqu'on introduit un retard ΔT entre l'activité solaire et les phénomènes météorologiques. Il semblerait possible de considérer que du Soleil à la Terre, il faut deux ou trois ans pour que les influences se propagent ... Il est bien clair que ce résultat est hautement spéculatif, à ce stade, - et que nous ne sommes pas en mesure d'en proposer la moindre explication ...

Quoiqu'il en soit, force nous est de constater la richesse de la question posée, les insuffisances statistiques des analyses poursuivies jusqu'à présent, et par suite, le caractère encore très actuel de ce domaine de recherches.

BIBLIOGRAPHIE

Jagannathan, P., Bhalme, H.N.: 1973, Monthly Weather Rev., 101, 691.
King, J.W.: 1975, Astronaut. Aeronaut., 13, 10.
Lefèvre, C.: 1976, Activité Solaire et Météorologique, Ecole Polytechnique promotion 1973, Mémoire de Stage, Option Astrophysique.
Pittock, A.B.: 1979, in Solar-Terrestrial Influences on Weather and Climate, (ed. by B.M. McCormac and T.A. Seliga) D. Reidel, Dordrecht-Holland, p. 181.
Xanthakis, J.: 1973, in Solar Activity and Related Interplanetary and Terrestrial Phenomena (ed. by J. Xanthakis) Springer-Verlag, N,Y.

ABSTRACT

The variation of pluviosity and air temperature in 8 stations of the equatorial zone and in Paris was studied using various techniques including Fourier analysis. The correlation with solar activity is rather weak. This correlation might however become stronger if we compare the solar activity at a given time with the pluviosity and the air temperature at a later time.

A SUGGESTED APPROACH TO RESEARCH ON THE SUN-WEATHER PROBLEM

Walter Orr Roberts and Roger H. Olson
Aspen Institute for Humanistic Studies
Boulder, Colorado, U.S.A.

ABSTRACT

We have tried to show that there are both theoretical and empirical reasons to suppose that some sort of mechanism involving atmospheric electricity is likely to be a key to understanding sun-weather relationships. We propose that international, as well as national, efforts to improve the theoretical and observational basis for understanding atmospheric electricity be pushed vigorously by scientists interested in this problem. Perhaps the one most important project at the moment would be to initiate the suggestion by Markson for continuous tethered balloon or kite flights to monitor the electric potential of the ionosphere at two or more widely separated stations.

INTRODUCTION

Despite the large amount of work that has been done on sun-weather connections over more than one hundred years, and more specifically over the past decade or so, the research has not advanced significantly beyond the empirical-statistical state. We have a large number of empirical results, some of which appear to pass rigid tests of significance. However, progress will remain slow until we have established at least a few viable physical models to explain what the key processes are.

In this brief paper, we offer arguments that atmospheric electricity is a particularly promosing field in which to seek a defensible physical mechanism to explain a wide array of the empirical results. This does not mean that other mechanisms should be neglected - but rather that we believe this one shows unusual potential for progress.

HISTORY

Herman and Goldberg (1978) summarize the history of research on connections between solar activity and electrical parameters in the atmos-

161

E. G. Mariolopoulos et al. (eds.), Compendium in Astronomy, 161–167.
Copyright © 1982 by D. Reidel Publishing Company.

phere. McCormac and Seliga (1979) give additional background. From
these two references, and others, we summarize some of the evidence for
short-term relationships that appear to us to merit special research effort at this time. We will not discuss the longer-term relationships,
such as 11 year and 22 year effects, even though these occupied a great
deal of the creative talent of John Xanthakis, to whom this volume is
most fittingly dedicated. We concentrate, instead, on short-term relationships in the expectation that faster progress in discovering mechanisms
will probably come in this way.

One of the early workers in the field was H. Flohn. He found (Flohn,
1950, 1951) that, as large sunspot groups approached central meridian,
there was a period of several days over which thunderstorm activity in
central Europe increased. Following this, Bossolasco et al.(1972, 1973)
found that thunderstorms over the Mediterranean area increased a few
days after large solar flares, particularly after flares near central
meridian.

In our own studies, shortly to be published, we confirm the work
of both Flohn and Bossolasco, using thunderstorm data gathered by Lethbridge over the eastern two-thirds of the United States. Markson (1971)
has added a new dimension to the analysis. Using the Lethbridge thunderstorm data mentioned above, he finds that the position of the Earth in
respect to the solar sector boundaries of the interplanetary magnetic
field is important in modulating thunderstorms. He and others, e.g.
Cobb (1967) have found changes in the vertical atmospheric electric potential gradient associated with actively flaring solar regions, and
Markson speculated that the change in atmospheric electricity is the
cause of the thunderstorm activity enhancement over the United States and
Europe. One of Markson's findings is that the frequency of thunderstorms
is significantly increased when the Earth is in a negative solar sector
of the interplanetary magnetic field, as opposed to a positive magnetic
sector. (Negative sectors are characterized by an interplanetary magnetic field vector pointing toward the Sun; in positive sectors it points
the opposite way). Combining the results of Flohn and others with those
of Markson, one would expect a cohesive picture to emerge if large active
regions can be shown to occur more often in negative sectors than in positive. This indeed appears to be the case, as shown by Dittmer (1975).

Dittmer, following earlier results of Dodson and Hedeman (1972),
showed that there is a large maximum in the frequency of central meridian
passages of strongly flaring active regions on the Sun about five days
before minus to plus sector boundaries at Earth. Five days is the transit time of the solar wind from Sun to Earth, and solar sectors are usually 10-12 days of solar rotation in width. Because of these facts, the
flaring region is at the solar central meridian on the average, at the
time that the Earth is in the center of the negative solar sector.

Thus, when the Earth is near the center of a negative sector, and
there is a large plage and flare region near the center of the Sun, there
also appears to be heightened thunderstorm activity in North America and

Europe. We have no way to know, at this stage, whether the plage or the sector magnetic field is the causative agent in the thunderstorm relationship.

However, in what follows we will work mainly on the assumption that the active plage region is primarily responsible, and that the sector effect is an interesting corollary.

Important research over many years by Reiter (1979) in the Bavarian Alps generally confirms the association between thunderstorms and plages described above.

THE SOLAR ACTIVE REGION AND ITS TERRESTRIAL EFFECTS

It has been known for more than 100 years that large active regions on the Sun cause various terrestrial effects such as magnetic storms and auroras, particularly when the solar activity takes place near central meridian. Many investigators, including the present authors, have studied the apparent influences of magnetic storms on terrestrial weather. We suggested some years ago a mechanism to explain short-term sun-weather effects observed on the Gulf of Alaska area.

We postulated aurora-induced cirrus clouds forming over the relatively warm Gulf of Alaska in winter, and surmized that they affected weather by modulating the atmospheric radiation balance (Roberts and Olson, 1973). However, this mechanism required several days to make itself felt. This time scale is appropriate to geomagnetic and auroral effects, in which the auroral-geomagnetic event lags the solar event by a few days and the weather response in turn lags the auroral-geomagnetic effect. But it acts too slowly to explain one important body of empirical results that suggest nearly immediate weather responses. The cirrus mechanism may be valid in some circumstances, but we wish, in the present paper, to focus on mechanisms that can act in the much more rapid manner semingly demanded by recent statistical studies.

Analysis by Schuurmans (1979) suggests that weather responses to large flares occur in less than 24 hours, making our cirrus mechanism implausible as the cause of this particular relationship. Moreover various weather effects from active solar regions seem to begin a few days before the active region passes central meridian, so that the mechanism must act very rapidly. This suggests that solar electromagnetic radiation or relativistic solar particles are involved. For instance, increased ultraviolet emission from the active regions, increased x-rays or solar cosmic rays from flares, all of which reach the Earth within some minutes, merit added consideration as the initiating cause of sun-weather effects. It seems plausible that all of these fast-arriving emissions are associated with strongly flaring solar plages.

However, there is a serious problem, as has been frequently stated in the literature, in transferring the effects of these perturbing en-

ergy sources from the high tenuous atmosphere down into the troposphere
in so short a time. There is also a problem in understanding how the
very small gross energy involved in these solar emissions can provide
sufficient impact to affect materially the far denser air of the low troposphere. Before examining these problems, however, we wish to mention
some of the correlations between the passage of active regions over central meridian and the resulting radiation flux at Earth.

Heath and Wilcox (1975) showed that a measurable increase in ultra-
violet emission from the Sun occurs characteristically about 5 days before the passage of sector boundaries at Earth. Taking into account the
work of Dittmer referred to earlier, one concludes that this implies an
increase in ultra-violet radiation at Earth at the time of central meridian passage of a major active region. It also implies that the effect
seen by Wilcox and Heath should be stronger when one looks specifically
at the ultra-violet associated with minus to plus sector boundaries.

However, the picture is far from simple. An active region contains
not only bright regions such as plages and faculae, but dark, cool sunspot regions. It is not obvious whether the result should be a net increase or a net decrease in ultra-violet radiation, as seen at Earth.
Recent work by Willson et al.(1981) and by Hickey et al.(1981) strongly
suggest that the total electromagnetic irradiance, or "solar constant",
goes through a distinct minimum of a few tenths of one percent over a
period of several days at the time of central meridian passage of a large
sunspot group. We need to evaluate the question of whether it is reasonable to have solar ultra-violet and total flux behave in opposite directions as a large active region transits central meridian.

A DIRECTION OF SEARCH FOR CAUSAL MECHANISMS

If we accept the fact that thunderstorms and other electrical parameters are modulated by the passage of flaring regions across the solar
disc, there are several ways of looking for a causal mechanism. One way
is to seek to discover a synoptic weather effect produced by the solar
activity that, in turn, is favorable to the development of thunderstorms.
Indeed various authors, including ourselves, have found (Olson et al.,
1978) that the central meridian passage of the most active plages is accompanied by an increase in the area of the Northern Hemisphere covered
by strong cyclonic vorticity, as measured by the hemispheric vorticity
area index. Increased vorticity should favor thunderstorm development,
and should do so with the appropriate time relationship. However, studies
by MacDonald and Ward (1974) and MacDonald (1976) suggest that it is not
only the size of troughs that affects their thunderstorm-producing capa-
bility. Rather it seems that troughs whose principal axis is aligned in
a northwest-southeast direction are the ones most likely to produce thunderstorms in the United States. Thus a careful synoptic study may be
merited in order to examine this possibility.

Another complicating factor is that there is a relationship between

the position of the Earth in the solar sectors and the size of troughs
that sweep towards North America from the Pacific Ocean. Wilcox et al.
(1979) found that if the Earth is in a positive sector at the time a typical trough starts its motion from the Gulf of Alaska into the adjoining
continental areas of North America, the trough is destined to become larger
than normal. Again a careful synoptic study should be carried out to
determine if this result is consistent with the result of Markson quoted
earlier, namely that thunderstorms are maximized in Eastern North America
when the Earth is in a negative sector. Since the time required for the
Earth to pass from a positive to a negative sector is approximately the
time required for a trough to move from the central Pacific to Eastern
North America, the timing appears to be no problem, but a detailed study
would be needed to confirm this.

   The second way to search for a causal relationship is to assume that
the thunderstorms are initially caused by some sort of solar effect on
the Earth's electrical environment, as earlier suggested by Markson. If
thunderstorms can be created or enhanced in some such way, then there are
good possibilities for the thunderstorm energy released into the atmosphere to alter the atmospheric circulation patterns, and in some circumstances to invoke positive feedback loops to pyramid the effects. In the
last part of this paper we look at this possibility.

   At least two schemes have been suggested involving mechanisms where-
by solar activity might induce thunderstorms directly, rather than through
a change in atmospheric circulation. One scheme was advanced by Herman
and Goldberg (1978). They reason as follows : During a solar event, in
which relativistic solar particles are emitted, the ionization of the
upper atmosphere is increased, and the atmospheric conductivity likewise
increases. Shortly after the particle event, a disturbance of the Earth's
magnetic field begins, and as a consequence there is a reduction in galactic cosmic rays falling on Earth. This decreases the ionization and
conductivity of the lower atmosphere, and in turn increases thunderstorm
activity. The authors do not specify an exact mechanism by which these
electrical changes cause an increase in thunderstorms. However it appears
to us that thunderstorm researchers should be able to assess the plausi-
bility of this suggestion. It also appears worthwhile to initiate a study
of plausible meteorological responses to large relativistic solar proton
events that are followed by the so-called Forbush decreases in cosmic ray
flux.

   A second suggested model is that of Markson (1979). His proposed
chain of events is as follows : (1) Solar wind modulation of galactic
cosmic rays and occasional solar particle events control the ionization
and conductivity in the upper atmosphere. (2) Increased conductivity
increases the intensity of the global electric circuit. This circuit is
driven by currents flowing upward to the ionosphere from tropical or mid-
latitude thunderstorms that are always present. The circuit is completed
by the down currents spread out through the fair weather part of the atmosphere. When the electrical resistance above the tropical thunderstorms
is reduced, the thunderstorms drive the global circuit more strongly, and

produce a higher voltage in the ionosphere, thus also increasing the re-
turning fair weather electric field gradient. (3) The heightened electric field gradient in the global circuit then enhances cloud drop formation in mid latitude convective clouds thus increasing cloud and thunderstorm formation in middle latitudes.

In a more recent study, Markson and Muir (1980) provide still another
set of suggestions on the electrical connection. They find that the speed
of the solar wind correlates positively with the electrical potential of
the ionosphere. Since the ionospheric potential is a measure of the total thunderstorm activity of the globe, this means a correlation between
solar wind speed and thunderstorm activity, according to this theory.
They propose a low cost experiment to test this theory. It involves
tethered balloon or kite flights to measure the ionospheric potential.
Since the ionosphere is to a rough approximation, an equipotential surface, one location could conceivably be enough to monitor global thunderstorm activity, though this assumption should probably be checked by
having measurements done at a minimum of two widely separated locations.

In our view, the electrical connection offers a very good possibility of explaining at least some of the known sun-weather correlations,
for the following reasons : (1) It provides the possibility of a physical link between the upper and lower atmospheres, something most other
hypotheses have stumbled over. (2) It provides the possibility of a
very rapid terrestrial response to a solar signal, whereas other mechanisms, such as the cirrus mechanism, require several days to be effective. (3) The energetics look qualitatively reasonable, although detailed
energy studies have not yet been done. The enhancement of droplet formation and coalescence by increased electric field gradients amounts to
a powerful "trigger" mechanism, because latent heat of condensation is
released in the process that far exceeds the electric energy involved.
If the rain falls to the ground rather than being re-evaporated, heat
is made available, perhaps sufficiently to alter the general circulation.

For these reasons, we believe that improved monitoring of electrical
phenomena should be pursued. Some observing programs, such as the balloon
or kite program proposed by Markson, would be very inexpensive. Another
inexpensive and highly worthwhile program would be the improvement of
vertical electric field gradient monitoring from a modest network of
widely dispensed, standardized electric field mills.

Other methods would involve much more expensive satellite programs,
but are worth considering as adjuncts to planned space observing plat-
forms. One very promising space technique is the use of the VHF inter-
ferometer as a lightning detector. Such interferometers in space have
an obvious advantage over the visual detectors commonly used in that
they would be able to see lightning flashes through the heavy clouds
that normally accompany lightning. The technique has been well proven
by Warwick et al. (1979) and appears to be feasible for a satellite
observing program.

REFERENCES

Bossolasco, M., Dagnino, I., Elena, A., and Flocchini, G., : 1972, in Studi in Onore di G. Aliverti, Instituto Universitario, Navale Di Napoli, p. 213.

Bossolasco, M., Dagnino, I., Elena, A., and Flocchini, G., : 1973, Rivista Italiani di Geofisica (cont. of Geofisice E. Meteorologia), 12, 293.

Cobb, W., : 1967, Monthly Weather Rev., 95. 905.

Dittmer, P.H., : 1975, Solar Phys., 41, 227.

Dodson, H.W. and Hedeman, E.R., : 1972, in Solar Activity Observations and Predictions,(ed. by M. Dryer and P.S. McIntosh), Prog. Astronaut. Aeronaut., Vol. 30, p. 19.

Flohn, H., : 1950, in Das Gewitter, (ed. by H. Israel), Vol. 25, Akad. Verlag Ges., Leipzig, p. 143.

Flohn, H., : 1951, Arch. Meteor. Geophys. Bioklimatol, Vol. 3, No. 314, p. 303.

Heath, D.F. and Wilcox, J.M., : 1975, NASA SP - 366, p. 79.

Herman, J.R. and Goldberg, R.A., : 1978, NASA SP-426.

Herman, J.R. and Goldberg, R.A., : 1978, J. Atmos. Terr. Phys., 40, 121.

Hickey, S., Griffin, F., Alton, B., and Maschheff, R., : 1980 or 1981, Solar Parameter Measurements from Nimbus Satellites and Other Solar Constant Experiments, to be published.

McCormac, B.M. and Seliga, T.A., : 1979, Solar-Terrestrial Influences on Weather and Climate, D. Reidel, Dordrecht-Holland.

McDonald, N.J. and Ward, F.W., : 1974, Div. Ital. Geofis., 23, 79.

McDonald, N.J., : 1976, Monthly Weather Rev., 104, 1618.

Markson, R., : 1971, Pure Appl. Geophys., 84, 161.

Markson, R.,: 1979, in Solar-Terrestrial Influences on Weather and Climate (ed. by B.M. McCormac and T.A. Seliga), D. Reidel, Dordrecht-Holland, p. 161.

Markson, R. and Muir, M., : 1980, Science, 208, 979.

Olson, R.H., Roberts, W.O., Prince, H.D., and Hedeman, E.R., : 1978, Nature, 274, 140.

Reiter, R., : 1979, in Solar-Terrestrial Influences on Weather and Climate (ed. by B.M. McCormac and T.A. Seliga), D. Reidel, Dordrecht-Holland, p. 289.

Roberts, W.O. and Olson, R.H., : 1973, Rev. Geophys. Space Phys., 11, 731.

Schuurmans, C.J.E., : 1979, in Solar-Terrestrial Influences on Weather and Climate (ed. by B.M. McCormac and T.A. Seliga), D. Reidel, Dordrecht-Holland, p. 105.

Warwick, J.W., Hayenga, C. and Brosncehan, J.W., : 1979, J. Geophys. Res., 84, 2457.

Wilcox, J.M., Duffy, P.B., Schatten, K.H., Svalgaard, L., Scherrer, P.H., Roberts, W.O., and Olson, R.H., : 1979, Science, 204, 60.

Willson, R.C., Gulkis, S., Janssen, M., Hudson, H.S., Chapman, G.A., : 1980 or 1981, Observations of Solar Irradiance Variability, to be published.

PART V
PLANETARY SYSTEM

OBSERVATIONS OF ZODIACAL LIGHT IN EGYPT AND ITS VARIATIONS

A.S. Asaad
Helwan Institute of Astronomy and Geophysics
Helwan, Egypt

ABSTRACT

Observations of brightness and colour of zodiacal light in blue, yellow and red regions at Abu Simbel during the period 1975 - 1979 are given. Dependency of observations on geographical latitude and elevation above sea level, besides lunar and solar variations are discussed. Possible explanation for short and long period variations are mentioned.

INTRODUCTION

Eye estimates of the brightness of the zodiacal light (ZL) started three centuries ago while several photographic and photoelectric observations have been carried out during the 20th century. Earlier observers noticed that the ZL is a variable phenomenon. Thom (1939) has shown from eye estimates that the brightness is periodically changing with two solar cycles. Elvey and Roach (1937) noticed seasonal variations while Huruhata (1951) showed irregular variations. Blackwell and Ingham (1961) noticed that the brightness of the ZL was enhanced by a solar flare, while Peterson (1961), Vande Noord (1970) and Misconi (1976) found no effect of flares on ZL.

VARIATION OF ZL WITH SOLAR CYCLE

Weinberg (1964) showed graphically large discrepancies between reliable photometric measurements of brightness and polarization of ZL in the period 1953 - 1964. Such discrepancies were partly explained by Asaad (1967) as due to solar activity in which the brightness decreases and the polarization increases with the solar activity. Weill (1966) and Dufay (1966) from their own observations extended over many years showed that the brightness of the ZL is changing in opposite phase with the solar cycle but with two years phase shift. In 1964, during the quiet solar activity Asaad et al. (1979a) carried out observations for the brightness of the ZL at Daraw which is a small town

171

near Aswan where the observations of Divari and Asaad (1959) were carried out in 1957. The observations showed clearly a decrease of the ZL brightness with solar activity. Re-analysis of observations by Weinberg (1970) showed apparent relation with solar cycle.

LUNAR VARIATION

Lunar variation of the ZL has been reported by Divari (1963). This has been confirmed by Asaad et al.(1979b) from their own observations, see Fig. 1 where the ZL brightness expressed in number of stars of 10th magnitude per square degree is plotted as ordinate against the lunar age as abscissa. The curves are given for blue, yellow and red regions and for different elongations, ϵ, from the sun. From the figure it is clear that the brightness of the ZL has a peak at new moon after its separation from the sun by 0.5 day in the blue and 1.5 days in the yellow and red regions. The curves show a tendency for the brightness of the ZL to increase at full moon. The mean increase of brightness at blue, yellow and red wave length ranges are respectively 1.32, 1.42 and 2.00 times. The increase in the red region is more than in other regions. Space measurements however, did not confirm such effect on ZL brightness, Regener and Vande Noord (1967), Vande Noord (1970) and Burnett (1976).

VARIATIONS OF COLOUR INDICES OF ZL

Colour indices (C.I), (B-V) of the ZL have been determined by many investigators. Behr and Siedentopf (1953),Lillie (1972) and Leinert et al. (1974) found that the ZL is slightly redder than the sun by about 0.2 mag. while Peterson (1961, 1967), Tanabe and Huruhata (1967), Hayakawa et al. (1970) and Hofmann et al. (1973) stated that the ZL colour is identical with that of the sun. Divari and Asaad (1959), Divari and Krylova (1963) and Wolstencroft and Rose (1967) found it slightly bluer than the sun by not more than 0.2 mag. Large disagreement in the variation of the CI (B-V) of the ZL with elongations have been found by many investigators. No study from ground stations has been done for CI (B-R).

VARIATIONS WITH THE GEOGRAPHICAL LATITUDES AND ELEVATIONS

Comparison by Asaad et al. (1979c) for observations obtained by different investigators indicated that the brightness of the ZL observed at lower geographical latitudes, φ, are higher than that at higher latitudes, see Fig. 2 where the brightness is plotted against the elongation for different observers. Curves (a) and (b) are for observations at high and low solar activity respectively and for the spectral regions indicated on the diagrams. The brightness at lower latitudes is higher by a factor of 1.3 for about 10° change in φ. This may be due to the fact that the ZL in the northern latitudes is observed nearer to the horizon where it is much affected by the scattering of the earth's

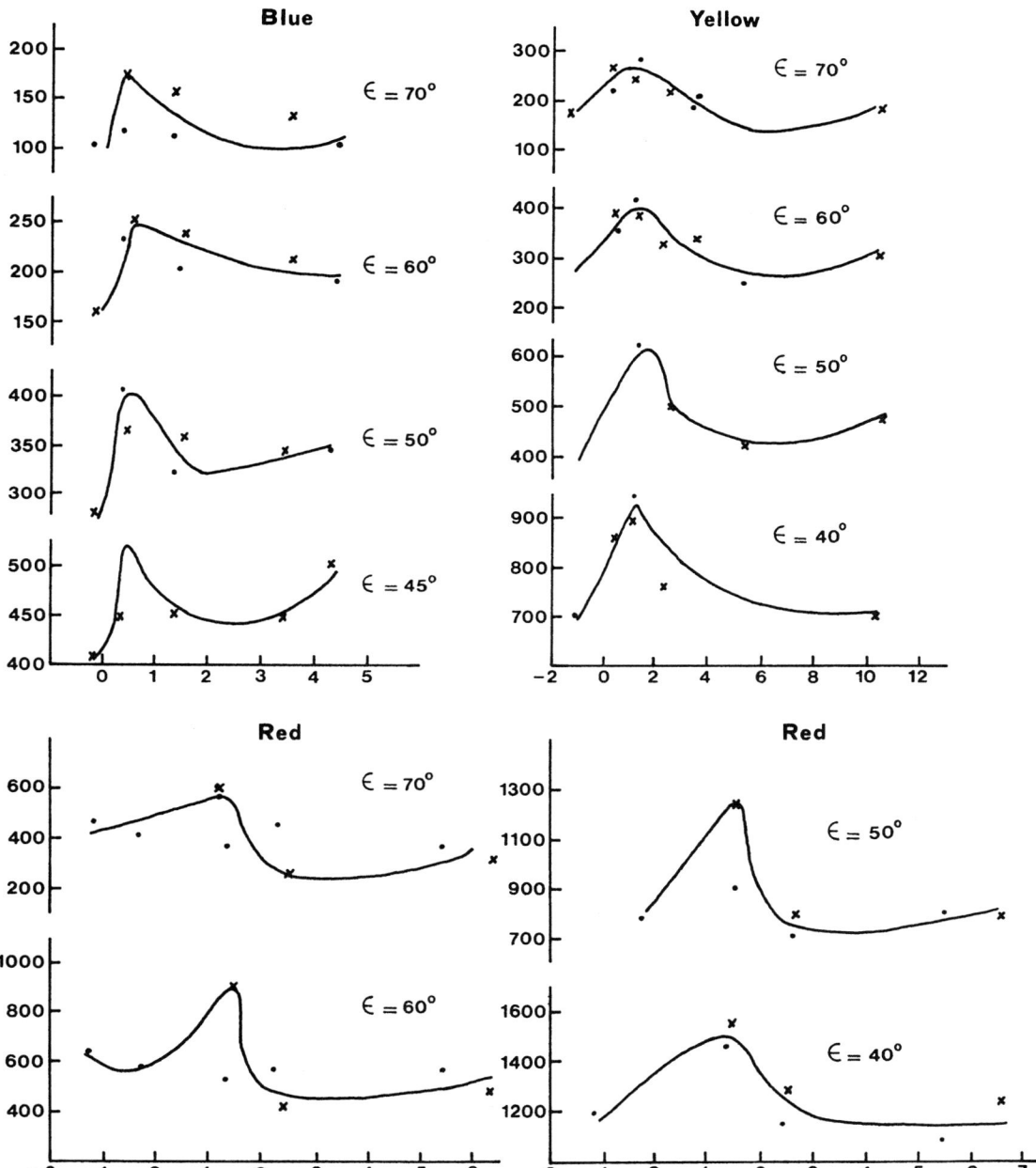

Fig. 1. Variations of ZL brightness with lunar age in days for elongations, ϵ, for yellow, blue, and red regions. Ordinates: ZL brightness; Abscissae: lunar age.

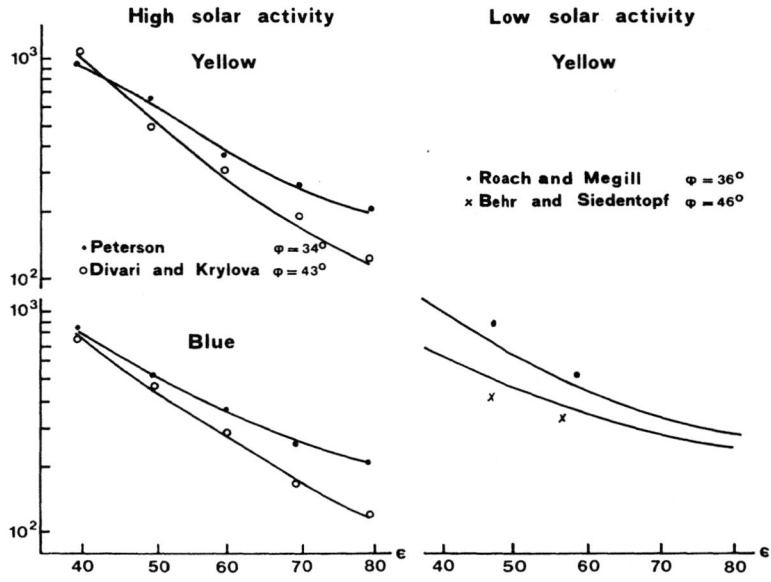

Fig. 2. Variation of ZL brightness in yellow and blue regions, at different elongations, ϵ, observed at different geographical latitudes φ.

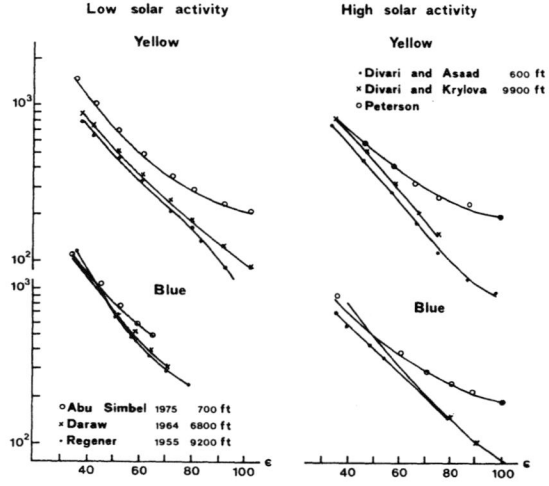

Fig. 3. Variation of ZL brightness in the yellow and blue regions at different elongations, ϵ, observed at different heights.

atmosphere.

        Comparison indicates also that the brightness of the ZL observed
at higher altitudes above sea level exceeds that obtained at lower
altitudes by a factor of about 1.5 and 2 for small and large elongations,
see Fig. 3. This may be due to the improving observing conditions at
sites of higher altitudes above sea level which are almost free from
dust and haze.

OBSERVATIONS AT ABU SIMBEL

        To study the variations of the ZL with the solar cycle, in view of
the above analysis,one cannot consider the results of different spectral
bands obtained at different sites for short periods. It is much useful
to have consistent observations over a long period, besides extending
the observation to the red region. For such reasons, it has been decided
in 1970 to establish a site at Abu Simbel ($\lambda=31^{\circ}$ 38$'$E, $\varphi=24^{\circ}$ 22$'$N) where
the latitude is favourable for observing ZL. The observations are being
carried out by A.S. Asaad, J.S. Mikhail and S. Nawar for two periods of
one month each in the years 1975 to 1980. A semi automatic photoelectric
photometer with three band filters, blue, yellow and red centered at
4140, 5500, and 7900 Å are used.

        To obtain the brightness of the ZL it is necessary to allow for
absorption, scattering and attenuation of light by the earth's atmosphere.
The observed value includes components for ZL, airglow and integrated
star light (ISL). To correct for atmospheric effect, the transparency
is determined each night from observations of stars. Table I gives the
mean values of the transparencies in the different regions with the p.e.

Table I

Month	Transparency		
	Blue	Yellow	Red
Nov. 1975	0.78 ± 0.02	0.86 ± 0.01	0.95 ± 0.01
Nov. 1976	0.74 ± 0.02	0.83 ± 0.02	0.92 ± 0.02
Apr. 1977	0.69 ± 0.05	0.76 ± 0.05	0.80 ± 0.05
Dec. 1977	0.76 ± 0.02	0.85 ± 0.01	0.93 ± 0.01
Oct. 1979	0.70 ± 0.05	0.87 ± 0.04	0.82 ± 0.04
Apr. 1980	0.71 ± 0.01	0.77 ± 0.01	0.85 ± 0.02

The values indicate that the transparency at Abu Simbel is high as
compared with other sites in north of Egypt which is a characteristic
of the country. The transparency improves in the winter than in the

summer and tends to be lower at sunspot maximum.

INTEGRATED STAR LIGHT OBSERVATIONS

       As regarding the correction for ISL it is usually taken from charts
and tables given by Roach and Megill (1961) but it is preferred to use
ISL measured at the same spectral regions used for ZL observations. The
distribution of ISL with galactic latitude b^{II} in the yellow, blue and
red are measured at Abu Simbel and given in Tables 2 and 3. For comparison the tables give the corresponding values by Roach and Megill (1961)
and by Tanabe and Mori (1976). The values of Pioneer 10 are taken from
Tanabe and Mori from values supplied by Wienberg. Table 2 shows that
for $b^{II} > 25^{\circ}$ the ISL values of Abu Simbel are in good agreement with the
values of Pioneer 10 and slightly lower than that of Roach and Megill,
while for lower b^{II} the values are slightly higher than of Pioneer 10

Table 2

Yellow

b^{II}	10°	15°	20°	25°	30°	40°	50°
Abu Simbel	113	93	77	68	60	46	38
Roach & Megill	123	98	83	74	68	54	40
Pioneer 10	100	80	65	63	60	45	38

Table 3

Blue and red

Region	Blue 4100	Blue (3500-5000)	Red 7900	Red (6200-6700)
Investigator	Abu Simbel	Tanabe & Mori	Abu Simbel	Tanabe & Mori
b^{II}				
21.9°	49	35.7	207	200
27.0°	38	32.2	162	102.6
31.8°	30	25.0	133	77.6
44.9°	22	22.1	90	74.0
49.3°	19	23.6	80	89.2

and slightly lower than that of Roach and Megill. Table 3 gives the values of the ISL in units of 10th magnitude star per square degree for

the blue and red regions besides that of Tanabe and Megill which are obtained photographically for the regions 3500-5000 Å (blue) and 6200-7000 Å (red). Although the wave length regions are slightly different, yet the values of Tanabe and Megill are lower than the values of Abu Simbel by a factor of 1.13 in the blue and 1.23 in the red. The values determined for the ISL are used in the reduction of observations at Abu Simbel. The correction for ISL, however, is small as the observations of ZL are carried away from the Milky Way.

The correction for the airglow is accounted for in the usual manner by considering points away from the ZL and assuming that the airglow component does not change with azimuth and is determined for each scan of ZL observation.

BRIGHTNESS OF THE ZL AT ABU SIMBEL

The observations of the ZL are corrected for ISL, airglow, scattered light and tropospheric scattering. The results obtained at Abu Simbel expressed in number of stars of 10th magnitude per square degree are given in Tables 4, 5 and 6 for the blue, yellow and red regions for the years 1975, 1976, 1977 and 1979. The values are given for different elongations, ϵ , from the sun and along the ecliptic. The tables include values of 1957 at Aswan by Divari and Asaad (1959) and at Daraw for 1964.

Table 4

Brightness of the ZL in the blue

Year	30°	35°	40°	45°	50°	55°	60°	70°	80°	Wave length
1975	--	1063	603	456	360	293	240	162	113	4410
1976	1016	836	637	483	338	256	196	109	76	4410
1977	--	686	550	430	325	230	180	110	80	4410
1979	710	560	445	340	250	200	160	110	80	4410
1957	--	509	386	297	230	----	168	123	90	4140
1964	--	950	685	508	586	302	244	161	105	4140

VARIATION OF ZL BRIGHTNESS WITH SOLAR CYCLE

The brightness of the ZL brightness given in Tables 4, 5 and 6 are plotted against the relative sunspot numbers for elongations between 40° and 70° for blue and between 40° and 80° for yellow and red respectively in Fig. 4. The scatter in the figure is very small except perhaps at small elongations where the ZL is observed near the horizon. The figure indicates that the brightness of the ZL is decreasing with number of sunspots by a factor of 1.38 in the mean during the four years observations at

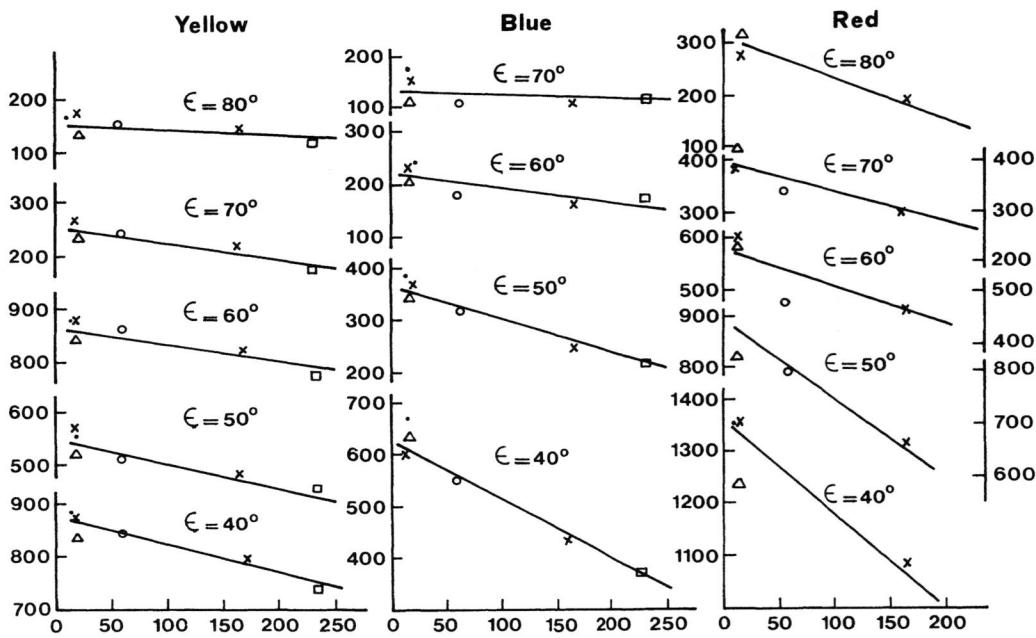

Fig. 4. Variation of ZL brightness at different elongations, ε, for yellow, blue and red regions. Ordinates:ZL brightness; Abscissae, Relative number of suspots.

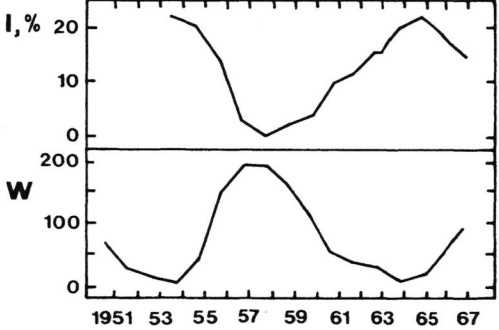

Fig. 5. Variation of Cosmic Rays intensity I, %, and the Wolf number, W, from 1951-1965 (S. Kudo and M. Wada).

Abu Simbel and it is somewhat higher in the blue and slightly less in the yellow.

Table 5

Brightness of the ZL in the yellow

Year	30°	35°	40°	45°	50°	55°	60°	70°	80°	Wave length
1975	--	1302	866	694	570	465	376	256	173	5500
1976	1733	1080	818	638	513	428	343	225	158	5500
1977	--	--	840	660	520	430	360	245	155	5500
1979	1440	1060	785	590	480	390	315	210	140	5500
1957	--	--	740	570	450	---	280	180	120	5410
1964	--	1325	878	692	553	452	372	252	172	5410

Table 6

Brightness of the ZL in the red

Year	30°	35°	40°	45°	50°	55°	60°	70°	80°	Wave length
1975	1910	1610	1370	1153	953	767	607	397	280	7900
1976	2318	1628	1244	996	826	684	578	430	330	7900
1977	--	1680	1400	1100	800	610	480	350	--	7900
1979	1600	1270	980	780	660	565	460	310	190	7900

COLOUR INDICES OF THE ZL

The colour indices (CI) of the ZL deduced from observations at Abu Simbel during four years are given in Table 7 for (B-V). The last column of the table gives the difference between the CI of the ZL and that of the sun which is taken as 0.63 from Allen (1976). For comparison the values obtained by other investigators are given. The table shows that the CI of the ZL does not differ from that of the sun by no more than ± 0.20 mag. and is slightly bluer. It is noted from observations that there is a tendency for the ZL colour to become redder as one moves away from the sun as well as with increasing sunspot number.

The results of CI (B-R) is given in Table 8 for observations at Abu Simbel. The value of 1.15 for CI (B-R) of the sun is taken from Allen (1976). Here again the ZL is always bluer than the sun by not more than 0.20 mag. There is also a tendency for the CI to become redder with the elongation from the sun.

Table 7

Site or Investigators		CI (B-V)	Difference
Abu Simbel	(1975)	0.46	-0.16
	(1976)	0.53	-0.11
	(1977)	0.62	-0.01
	(1979)	0.66	+0.004
Daraw (Egypt)	(1964)	0.44	-0.018
Divari & Krylova	(1963)	0.47	-0.16
Peterson	(1961)	0.48	+0.03
Divari & Asaad (Egypt)	(1959)	0.35	-0.10
Behr & Siedentopf	(1953)	0.63	+0.20

Table 8

Year	CI (B-R)	Difference
1975	0.98	-0.17
1976	1.12	-0.12
1977	1.06	-0.09
1979	1.00	-0.15

CONCLUSION

The observations at Abu Simbel are consistent among themselves as they are carried out with the same equipment and reduced in the same way. It will be no doubt useful to correct the observed brightness for multiple scattering near the horizon by the method developed by Staude (1975). However applying the correction to observations of Dumont and of Weinberg by Staude indicated that both have accounted to a certain extent for the Mie-scattering.

It seems that the multiple scattering will affect the brightness of the ZL at large zenith distances but it will not remove the solar activity effect. This is due to the fact that the parts of the ZL at the same elongation are usually observed at the same zenith distance. The correlation of the ZL brightness with sunspots number are plotted separately for each elongation.

The observations at Abu Simbel indicate that the ZL seems to undergo short period variations with the lunar phase and long period variations with the solar cycle. The lunar phase variations are detected from ground based observations and not from outer space as indicated previously. In such case the variations are less explained by lunar tides on the dust cloud surrounding the earth but may be due to tides of the moon on the earth's atmosphere. The atmosphere in this case will have an ellipsoidal shape with its bulge axis in the direction of the moon. Such case is similar to the earth's magnetic field lunar variations found in 1854, Chapman and Bartels (1940). Lunar variations of the earth's magnetic field has a regular cycle of unequal amplitudes during the lunation and with one or two days shift from new moon. During the lunation the magnetic field is minimum near the beginning and near the middle. It will be interesting to determine the polarization of the ZL during several lunations, as it is expected to behave in a manner similar to the earth's magnetic field.

As for the solar cycle changes, the process suggested by Asaad (1967) seems to hold. During maximum solar activity the strong solar wind may destroy, evaporate or push away some of the fine particles in the interplanetary medium. It is evident from the curves given by Kudo and Wada (1968) that the cosmic ray intensity is changing with the 11 years solar cycle but with opposite phase. The curves are given in Fig. 5 for the variations of both cosmic rays, I% , and Wolf number of sunspots, W. If such is the case, the cosmic rays are stronger during minimum solar activity and may increase the ZL brightness by fluorescence of the dust particles by collision.

To investigate the problem further, it is required to carry out observations from ground based sites at the time when observations from outer space are taken. Results from outside the Earth are needed for at least five years.

REFERENCES

Allen, C.W.: 1976, Astrophysical Quantities,The Athlone Press, London,
    3rd ed.
Asaad, A.S.: 1967a, Observatory 87, 83.
Asaad, A.S.: 1967b, Nature 214, 259.
Asaad, A.S., Mikhail, J.S., and Nawar, S.: 1979a, Helwan Obs. Bull.
    No. 197.
Asaad, A.S., Mikhail, J.S., and Nawar, S.: 1979b, Helwan Obs. Bull.
    No. 203.
Asaad, A.S., Mikhail, J.S., and Nawar, S.: 1979c, Helwan Obs. Bull.
    No. 204.
Behr, A. and Siedentopf, H.: 1953, Z. Astrophys. 32, 19.
Blackwell, D.E. and Ingham, M.F.: 1961, Monthly Notices Roy. Astron.
    Soc. 122, 113.
Burnett, G.B.: 1976, in Interplanetary Dust and ZL, Lecture Notes in
    Phys. No. 48 (ed. by H. Elsässer and H. Fechtig), p. 53.
Chapman, S. and Bartels, J.: 1940, Geomagnetism,The Clarendon Press,
    Oxford.
Divari, N.B. and Asaad, A.S.: 1959, Astron. Zh. 36, 856; Soviet
    Astron. 3, 832, 1960.

Divari, N.B.: 1963, Soviet Astron. 7, 547.

Divari, N.B. and Krylova, S.N.: 1963, Soviet Astron. 7, 391.

Dufay, J.: 1966, Compt. Rend. Acad. Sci. Paris 263, 947.

Elvey, C.T. and Roach, F.E.: 1937, Astrophys. J., 85, 213.

Hayakawa, S., Matsumoto, T., and Nishimura, T.: 1970, Space Res. X, 248.

Hofmann, W., Lemke, D., Thum, C., and Fahrbach,U.: 1973, Nature,Phys.
    Sci., 243, 140.

Huruhata, M.: 1951, Publ. Astron. Soc. Japan, 2, 156.

Kudo, S. and Wada, M.: 1968, Eleven Year Variations of Cosmic Ray Inten-
    sity, Report of Ionosphere and Space Res., Japan, 22, No. 3, 137.

Leinert, C., Link, H., and Pitz, E.: 1974, Astron. Astrophys., 30, 411.

Lillie,C.F.: 1972, NASA SP - 310, 95.

Misconi, N.Y.: 1976, Astron. Astrophys. 51, 357.

Peterson, A.W.: 1961, Astrophys. J., 133, 668.

Peterson, A.W.: 1967, in The ZL and Interplanetary Medium, (ed. by J.L.
    Weinberg) NASA SP-150, p. 23.

Regener, V.H. and Vande Noord, E.L.: 1967, in The ZL and Interplanetary
    Medium, (ed. by J.L. Weinberg) NASA SP-150, p. 45.

Roach, F.E. and Megill, L.R.: 1961, Astrophys. J., 133, 228.

Staude, H.J.: 1975, Astron. Astrophys. 39, 325.

Tanabe, H. and Mori, K.: 1976, Interplanetary Dust and ZL, Lecture Notes
    in Phys. No. 48 (ed. by H. Elsässer and H. Fechtig), p. 36.

Tanabe, H. and Huruhata, M.: 1967, in The ZL and Interplanetary Medium,
    (ed. by J.L. Weinberg) NASA SP-150, p. 37.

Vande Noord, E.L.: 1970, Astrophys. J. 161, 309.

Weill, M.G.: 1966, Compt. Rend. Acad. Sci. Paris, 263, 943.

Weinberg, J.L.: 1964, Summary Report on ZL, Hawaii Inst. of Geophys.

Weinberg, J.L.: 1970, Space Res. X, 233.

Wolstencroft, R.D. and Rose, L.J.: 1967, Astrophys. J., 147, 271.

ON THE GLOBAL SIZE OF OORT'S CLOUD OF COMETARY NUCLEI AND THEIR TOTAL
NUMBER

L. Biermann
Max-Planck-Institut für Physik und Astrophysik
Garching bei München

ABSTRACT

The size of Oort's cloud and the total number of the cometary nuclei
it contains are rediscussed on the basis of the new data presented in
Marsden, Sekanina and Everhart's paper of 1978 and of Oort's working model of the "cloud". It is shown that a larger part of the 15 comets for
which reliable values of the original 1/a-values are available seem to
come from a rather narrow range of aphelion distances $Q \simeq 2a$, between approximately 50.000 and 55.000 a.u., and 60.000 a.u. are proposed as a
conservative estimate for the radius (R) of Oort's cloud. The rather
drastically smaller number of $(1/a)_{orig}$-values below 34 and above 42
units (of 10^{-6} $(a.u.)^{-1}$) - this interval contains 7 of the 15 values,
the others ranging from 11 to 27 and from 49 to 82 units - is ascribed
to the tidal forces arising during traversals of extended dense interstellar clouds and to lack of uniformity in velocity space for small ve-
locity components $v_{\perp}$ ($\perp \underline{r}_{\odot}$ the comet-sun vector). The total number of
nuclei is evaluated using the new value of R; it is found that the total
number changes only by $\approx 20\%$ inspite of much larger changes in the volume
(by a factor 1/37) and of the average number per unit volume (by a factor
29), such that the balance between losses from this cause (which are now
known to be < 2/3 of those assumed by Oort) and the number of nuclei
present remains essentially unchanged.

The concept of the "Oort Cloud" of cometary nuclei surrounding the
solar system at distances of some 10^4 a.u. (astronomical units) or more
was proposed already 30 years ago (Oort, 1950). Early doubts in its existence have been dispersed largely by the work of Marsden and Sekanina and
of Everhart, by which orbital elements of very high precision are now
available for 200 long-period comets (Marsden et al., 1978) and by considerations on the chemical and physical constitution of cometary nuclei
on one hand and that of dense interstellar clouds on the other*; as to
their origin, the gravitational instability of a dust layer in the equa-

---

* cf. the last tri - annual reports of IAU commission 15.

E. G. Mariolopoulos et al. (eds.), Compendium in Astronomy, 183–190.

torial plane of a rotating fragment of a dense interstellar cloud appears
to be a plausible mechanism by which cometary nuclei with the observed
proportion of volatile and of solid material and the observed size can
originate (Biermann, 1980; further references there).

It is the object of this note to discuss the present state of knowledge concerning the size of Oort's cloud and the number of cometary nuclei it contains, and to show that the model of the "cloud" proposed by
Oort in 1950 is still consistent with the much richer body of observational data available now, if the downward revision of the value of its
radius, which is under discussion since 1973, is taken into account.
Marsden & Sekanina (1973) had already pointed out that the size of Oort's
cloud, on the basis of the data gained by them at that time, could be
only of the order of 10^5 a.u. instead of the several 10^5 a.u. assumed by
Oort. The data published 1978 by the same authors and Everhart (Marsden
et al., 1978) permit the conclusion that the true radius is larger than
50.000 a.u., but probably by not more than 10 or 20%. Stellar perturbations seem to establish approximate statistical equilibrium in (six dimensional) phase space for nuclei of very low angular momentum of revolution, which is characteristic for those approaching the sun to within
the orbit of Jupiter or Saturn, only for comets with aphelia near the
outer boundary of Oort's cloud, whereas for the inner parts the number
of such nuclei should be a fraction only of those demanded by statistical
equilibrium. The total number of nuclei in Oort's cloud is evaluated
taking into account this new situation; the resulting figure turns out
to be not very different from the one obtained by Oort 30 years ago, such
that the relative rate of loss due to perturbations by the major planets
remains about the same. The proportion of comets coming from the inner
parts - from distances less than about 50.000 a.u. - must however be
somewhat larger than discussed by Oort, particularly during encounters
with extended relatively dense interstellar clouds.

The list of Marsden, Sekanina and Everhart (1978) contains 17 orbits
of comets for which $1/a_{orig} < 100$ $(10^6$ a.u.$)^{-1}$ and the perihelion distance
q is larger than 2.5 a.u.; (Table I). The condition first mentioned is
commonly used to distinguish a comet approaching the inner solar system
for the first time from one on its 2nd or a subsequent return, as has
been discussed amply in the literature on the subject. The value of the
semi-major axis a_{orig} is the one which a comet had before entering from
the outside the region of the major planets, the effect of which on the
orbit of a comet being eliminated by computation (which actually takes
into account also the terrestrial planets, although their effect on the
orbit is much smaller). The gravitational attraction of the major planets
changes the $1/a_{orig}$ -values on the average by (positive or negative) a-
mounts of the order of 500 $(10^6$ a.u.$)^{-1}$ for comets with values of q up
to the semi-major axis of Jupiter's orbit; after having left again the
region of the planets, only around 10% of these comets are still in the
range $0 < 1/a_{orig} < 100$ units (Weissman, 1978) (roughly half of the comets
in question get into hyperbolic orbits and leave the solar system for
interstellar space).

Table I

comets	q(a.u.)	$(1/a)_{orig}$	$(m.e.)_{osc}$	$\Delta(38)$
1974 XII	6.019	11	15	-27
1936 I	4.043	19	6	-19
1903 II	2.774	26	9	-12
1914 III	3.747	27	20	-11
1947 VIII	3.261	34	10	-4
1925 VI	4.181	35	9	-3
1975 VIII	3.011	36	3	-2
1951 I	2.572	37	4	-1
1956 I	4.077	39	3	+1
1959 X	4.267	40	11	+2
1955 VI	3.870	42	3	+4
1972 VIII	2.511	49	13	+11
1975 II	6.881	68	16	+30
1972 IX	4.276	69	7	+31
1954 V	4.496	82	16	+44

$(1/a)_{orig}$ values (unit 10^{-6} $(a.u.)^{-1}$) and the mean errors of the osculating orbits of the "new" comets in the list of Marsden et al. (1978) with q > 2.5, which did not split; Δ is the difference between $(1/a)_{orig}$ and 38 units, the adopted average for the seven comets with $|\Delta| < 4$ (see text).

For comets with q-values between 5 and 10 a.u. the amount of the change of 1/a is on the average only about 85 such units (Everhart, 1968; private communication, 1980), that is to say slightly more than twice the $1/a_{orig}$-value of a typical "new" comet (see below). The condition mentioned above concerning the perihelion distance, q > 2.5 a.u., is a consequence of the fact that the orbits of comets with smaller q-values are often rather severely disturbed by non-gravitational effects resulting from the reaction on the comet of the gases streaming away from the surface anisotropically with a speed of the order of 1 km/sec; this was a central point in Whipple's work of 1950 (Whipple, 1950). The quantitative formulation currently used has been given by Marsden (1974). The discussion (particularly Fig. 1 and 2) in the aforementioned work of 1978 shows that for the comets discussed there the limit of 2.5 a.u. appears to guarantee the absence of any non-gravitational effect to speak of, with the exception possibly of two comets which split into two parts during the time they were observed and which indeed might have been somewhat affected in their orbital motion by such forces (cf. Marsden & Sekanina,1973). If these are removed we are left with 15 comets, collected in Table I, 7 of which have $1/a_{orig}$-values between 34 and 42 units, whereas of the remaining values we have 4 with smaller and 4 with larger

values of $(1/a)_{orig}$, the smallest and the largest being 11 and 82 units.
One notices that roughly half of the comets..are in less than 1/10 of the
total interval. Looking at the mean errors of the osculating orbits near
perihelion, it is seen that part of this distribution is due to the mean
errors: those belonging to the inner interval (between 34 and 42 units)
have a (root mean square) average error of 7 units, whereas the remaining
8 have an average mean error twice as large. Five of these 8 are indeed
separated from the middle of the interval 34 to 42 units, 38 units, which
is also the average of the 7 values between 34 and 42 units (see below),
only by amounts of the order of one or at most 1.8 respectively 1.9 times
the mean error of the osculating orbit (2 comets). There remain only 3
cases with $1/a_{orig}$ values separated from the middle of the aforementioned
interval by amounts, called Δ in Table I, so large that a simple error
appears to be improbable. One of these could well be a comet on his second or subsequent return, of which the sample would in any case be expected to contain around 10% of those outside the interval 34-42 units;
the probability that any .of the comets in this interval of only 9 units
is on his second or subsequent return is of course smaller in the ratio
10:1 or ≈1%.

Another cause for a value of $1/a_{orig}$ outside that of most others
could be that with the smaller radius of Oort's cloud as known now the
relative influence of an encounter with an above average massive star or
stellar system (or one with an unusual small relative velocity with respect to the sun) should become apparent, which was not the case in 1950
(see Oort l.c. pag. 98/99). We come back to this point below.

This distribution of the $1/a_{orig}$-values (Table I) suggests that a
genuine average value is best formed using only the 7 comets in the interval 34-42 units. This average is then found to be 37,6 units, or 38
units, if higher weight is given to the 4 comets with mean errors of the
osculating orbits only of 3 or 4 units*. The fact that the average mean
deviation Δ of the 7 comets in this interval is smaller than the average
mean error of their osculating orbits (7 units) suggest then that the
true distribution might be still more strongly peaked around 38 units,
say between 36 and 40 units or so or Q between 50.000 and 55.000 a.u.
The question arises what is the origin of the sharp drop towards smaller
values, that is towards larger values of a_{orig} and second the likewise
pronounced drop inward.

The dynamical state of Oort's cloud differs from that of a star
cluster in that there is effectively no interaction whatsoever between

---

* The 5 comets outside the interval 34-42 units with $\Delta \leq 1.9$ times the
mean error of the osculating orbit have an average $1/a_{orig}$-value of 36.2,
close to the average of the 3 comets of the first 7 with mean errors between 9 and 11 units (36.3); it is clear that the adopted value of 38
units is uncertain by at least one unit. It seems however suggestive how
close the three averages 38.5 (for the 4 best determined osculating orbits), 36.3 and 36.2 are together.

one nucleus and any other, their energy and their angular momentum of revolution with respect to the sun being changed (as long as they are outside 35 a.u. from the sun) only by stellar perturbations and by encounters with dense interstellar clouds. Over very long periods of time, upwards from some 100 million years, the latter ones should be more important (Biermann and Lüst, 1978; Biermann,1978), and a value of the radius of Oort's cloud substantially larger than 50.000 a.u. would be inconsistent with the effect of the tidal forces arising during an encounter with an interstellar cloud of some 100 mass units per cm^{-3} or more, whereas the limit of the radius set by the tidal forces of the central parts of our galaxy is considerably larger.

The model proposed by Oort in 1950 is characterized by a uniform density in velocity space up to a limit $L(r)$ (cm/sec), where L depends on r (r =solar distance, R = radius of Oort's cloud) by the relation

$$L^2 = \frac{2G \, M_\odot}{R} \; (\frac{R}{r} - 1) = L^2 \, (\frac{R}{2}) \, \frac{R-r}{r}$$

Oort then shows that the density in ordinary space is under the given circumstances $\sim L^3$, such that the density in phase space is constant along a trajectory (Liouville's theorem) (Biermann, 1977). The angular momentum of revolution $[\{\underline{r} \; \underline{v}\}]$ is, for orbits with $q \ll a$, related to the perihelion distance q by

$$[\{\underline{r} \; \underline{v}\}] = a_+ \, v_+ \, \sqrt{q_{a.u.}} = 0.63 \cdot 10^{20} \, \frac{cm^2}{sec} \, \sqrt{q_{a.u.}}$$

Nuclei with $q \lesssim 5$ a.u.* are removed from Oort's cloud by the change due to Jupiter and Saturn of the total energy $(GM_\odot/2) \, (1/a)$, as described above, but stellar perturbations should, at 50.000 a.u. $((3/4) \cdot 10^{18} cm)$ from the sun and beyond, suffice to keep the velocity space approximately filled also for values of $v_\perp$ (component $\perp$ r) $\lesssim 10^2$ cm/sec (Oort,1950; Rickman,1976) (Rickman, 1976). The total number of cometary nuclei with a value of $q < q_o$ passing inwards per century (say) through a sphere around the sun with radius r_o (suitably chosen) is then

$$\sim v_1^2 \, L^2 \, r_o^2 \, v_{ph} = (v_1^2 \, r_o^2) \, L^2 v(r_o)/(4\pi/3) L^3 \quad (century)^{-1}$$

(v = number of nuclei per $(a.u.)^3$, v_1 limiting value of $v_\perp$ as given by q_o and r_o) with $v_{ph} = v(r_o)/(4\pi/3) L^{3**}$. The large Q-comets (Q aphelion distance) spend most of their time at large distances from the sun, and at a solar distance r only comets with $a > r/2$ can be present and suffer, by a star or stellar system passing by or an interstellar cloud traversed, a change of $\underline{v}$ and $[\{\underline{r} \; \underline{v}\}]$, which results in their becoming a visible comet (q < some a.u., depending on the size of a cometesimal).

---

* Jupiter's orbit has a = 5.2 a.u.
** Oort's expression (22) with $u_2^2 = L^2(r_o)$, $u_1^2 = 0$.

The observed sharp drop towards smaller values, of the frequency of occurrence of $1/a_{orig}$ -values below 37 or 36 units (cf. Table I upper part) may, using Oort's model, be related to the value of

$$\frac{d\ln L^2}{d\ln r} = \frac{-R/r}{(R-r)/r} = -\frac{R}{R-r}$$

Physically, the related drop of the density toward r = R should have been caused by the tidal forces which have arisen during the last encounters with dense interstellar clouds; this process should be particularly effective during an encounter with an interstellar cloud of such dimensions that the time needed for a traversal is no longer small compared to the period of revolution of comets with Q in the vicinity of R. From Table I it appears that the sharp drop in the number of comets per unit interval (equal to 7/9 in the interval 33.5 to 42.5 units) occurs over an interval of a few units only; from Table II we conclude that the inverse of $1/2a \simeq 35/2$ or $34/2$ units should be very close to the outer boundary R, for which we take the round value 60.000 a.u. (r/R=33.3 units).

Table II

r/R	$L^2/L^2(\frac{1}{2})$	$\dfrac{d\log L^2}{d\log r}$	$L/L(\frac{1}{2})$	$L^3/L^3(\frac{1}{2})$	$r^2/R^2 \cdot L^3/L^3$
0.25	3	1.3	$\sqrt{3} = 1.73$	5.2	$(3/4)\sqrt{3}$
0.5	1	2	1	1	1
.75	$\frac{1}{3}$	4	$1/\sqrt{3}$	.19	$\frac{q}{4} \cdot \frac{1}{3\sqrt{3}} = .43$
.80	$\frac{1}{4}$	5	$\frac{1}{2}$	.125	$8/25 = 0.32$
.85		6.7			
.90	$\frac{1}{9}$	10	$\frac{1}{3}$	.037	0.120
.95	$\frac{1}{19}$	20	$\frac{1}{4.4}$	.012	≈ 0.05

The drop towards larger values of 1/a beyond 42 units, that is inwards, may be understood in the following way. Oort noticed already that statistically stellar perturbations are hardly sufficient to maintain isotropy in velocity space for small values of $v_{\perp}$ ($\lesssim 10^2$ cm/sec) inside of r = 50.000 a.u. Although Oort's quantitative treatment needs revision what regards details, the main result came out again in recent work of Rickman (1976), who fixed attention to comets with q = 5 a.u. and estimated the perturbation of $v_{\perp}$ by stars passing by during one revolution as a function of the aphelion distance Q. His result (cf. fig. 6 l.c.,

p. 101) was that the total relative transverse impulse per revolution is a very steep function of Q, depending on Q approximately as $Q^{7/2}$ or Q^4 (particularly between Q $\simeq$ 20.000 a.u. and 50.000 a.u.). At Q = 50.000 a.u. it is found to be just slightly larger than $v_\perp$ itself, but, as recognized by the author, the combination of the number of stellar systems per pc^3 (0.08 pc^{-3}, which excludes only stars of very low mass) and average mass ($1M_\odot$) is likely to result in slight overestimate of the computed perturbation. Weighing these and one or two other points of detail, it seems rather well established that for orbits with Q $\lessgtr$ 50.000 a.u., i.e. $(1/a)_{orig.} \gtrless$ 40 units, a rather sharp transition occurs in the sense that the perturbations by stars or stellar system penetrating Oort's cloud are no longer sufficient, statistically, to lead to observable comets. That individual relatively massive stars may do so also at smaller r - by diminishing $[r\,v_\perp]$ to values $\lessgtr 10^{20} cm^2/sec$ - in limited regions of space, is a different matter which was raised already by Oort (1950) and will, using the results obtained over the last few years, be discussed elsewhere (L. Biermann and W. Huebner, 1981, to appear), where it is shown that one or two of the comets outside the range 34-42 units may indeed be due to such an event which would in particular favour comets with larger values of $(1/a)_{orig.}$.

It thus appears that Oort's cloud has now to be regarded as smaller in the ratio 10:3 (approximately) than appeared in 1950, and that stellar perturbation leading to observable comets ($[\{r\,v\}] \lessgtr 10^{20}$ cm^2sec) are adequate for doing so only for a rather limited outer region, roughly between r $\simeq$ 50.000 a.u. and the outer boundary (R $\simeq$ 60.000 a.u., see above). This makes it necessary to make a new estimate of the total number of nuclei in the cloud, in order to verify that it is still large enough to meet the losses. In doing so, and in order to make a comparison with Oort's estimate easier, we retain the figure of 97 new comets per century used by Oort (the list in Marsden, Sekanina and Everhart's 1978 work contains 24 new comets for the 3 1/2 decades beginning 1941, which for various reasons should be on the low side by several 10%).

Oort's formulae and figures have to be rewritten as follows. $L^2 (\frac{R}{2})$ for R = 60.000 amounts to $(80/27)\cdot10^8 (cm/sec)^{2*}$ instead of $(8/9)\cdot10^8$ $(cm/sec)^2$ for R = 200.000 a.u. (Oort 1950). L^2 (50.000 a.u.) now becomes $(16/27)\cdot10^8 (cm/sec)^2$ against $(8/3)10^8 (cm/sec)^2$ for R = 200.000 a.u., such that L^3 is smaller by almost a factor of 10 $\{(9/2)^{3/2} = 9.55\}$ and the coefficient in Oort's formula (22) becomes $58\cdot10^{-3}$, which, when multiplied with the new $L^2(50.000$ a.u.), is $34.4\cdot10^5$ instead of $16.3\cdot10^5$ ($\sim L^{-1}$, if the small contribution from r < 50.000 a.u. in Oort's table 4 is disregarded). This figure, multiplied with $\nu(50.000)$, the space density of nuclei per $(a.u.)^3$, should be equal to the number of new comets, with q < 1.5 a.u. per century (97 according to Oort which figure is retained here as stated above). $\nu(50.000)$ is thus found to be $(93/34.4)\cdot10^{-5} =$ = $2.7\cdot10^{-5}(a.u.)^{-3}$ instead of $5.7\cdot10^{-5}$. The ratio of $\nu(R/2)$ to $\nu(5/6\,R)$ is $5^{3/2} = 11.2$ such that $\nu(R/2)$ becomes $31.6\cdot10^{-5}$ against $(5.7/3\sqrt{3})\cdot10^{-5}=$

---

* For this estimate $GM_\odot$ is taken to be $(4/3)\cdot10^{26} cm^3/sec^2$.

$= 1.096 \cdot 10^{-5}$, but the volume of Oort's cloud is now smaller in the ratio 1000:27. The total number is therefore still smaller in the ratio 1096: :(27x30.2) or $\simeq 4{:}3$, and the total number of comets in the "cloud" becomes $1.41 \cdot 10^{11}$ which have still to be reduced somewhat (by $\gtrsim 10\%$) in order to take into account the events inside the reference surface of 50.000 a.u. radius (with the large radius R assumed in 1950 by only 4%). Encounters with interstellar clouds do not seem to have occurred over the last 5 million years and are therefore unlikely to have contributed to comets seen now as "new" objects. On the other hand, the cross section of the "loss cylinder" in velocity space produced by planetary perturbations (Oort, 1950) is now known (Everhart, 1968; Weissman, 1978) to be smaller than in 1950 by about 40%, such that the balance between loss by planetary perturbations and the total number present remains essentially unchanged.

Acknowledgements

The author is indebted to E. Everhart, W.F. Huebner and Rhea Lüst for discussions and valuable comments.

REFERENCES

Biermann, L.: 1977, Sitzber. Bayer. Akad. Wiss., Math. Naturwiss. Klasse.
Biermann, L.: 1978, Astronomical Papers dedicated to Bengt Strömgren, presented at a Symp. held in Copenhagen, May 30-June 1, p. 327.
Biermann, L. and Lüst, Rh.: 1978, Sitzber. Bayer. Akad. Wiss., Math. Naturwiss. Klasse.
Biermann, L.: 1980, Contribution to a meeting for discussion "Planetary Exploration" organized by the RAS, London Nov. 4/5, 1980.
Everhart, E.: 1968, Astron. J. 73, 1039.
Everhart, E.: 1980, Private communication.
Marsden, B.G., Sekanina, Z.: 1973, Astron. J. 78, 1118.
Marsden, B.G.: 1974, Ann. Rev. Astron. Astrophys. 12, 1.
Marsden, B.G., Sekanina, Z. and Everhart, E.: 1978, Astron. J. 83, 64.
Rickman, H.: 1976, B.A.C. 27, 92.
Weissman, P.R.: 1978, Thesis (UCLA).
Whipple, F.L.: 1950, Astron. J. 111, 375.

VOYAGER ENCOUNTERS WITH JUPITER'S MAGNETOSPHERE: RESULTS OF THE LOW ENERGY CHARGED PARTICLE (LECP) EXPERIMENT

Stamatios M. Krimigis
Applied Physics Laboratory
The Johns Hopkins University
Laurel, Maryland 20810, U.S.A.

ABSTRACT

The two Voyager spacecraft encountered the planet Jupiter and its magnetosphere during March and July 1979. Included in the payload was an instrument designed to measure hot plasmas and high energy charged particles ($E_{ion} \gtrsim 30$ keV, $E_e \gtrsim 15$ keV). These measurements have revealed that the Jovian magnetosphere is filled with a hot (20-40 keV) multicomponent plasma consisting primarily of hydrogen, helium, oxygen, and sulfur ions which is moving in the corotation direction out to the magnetopause on the dayside (50-80 R_J) and to a distance of 100-150 R_J on the nightside. Beyond ~ 150 R_J the plasma flow changes abruptly to a generally anti-solar direction and continues in that direction to $\gtrsim 360$ R_J at the 3 AM meridian. Hot plasma velocities in this region range from ~ 300 to $\gtrsim 1000$ km/sec, and the composition is similar to that in the inner part of the magnetosphere. This phenomenon has the characteristics of a wind and has been named "magnetospheric wind". An overall phenomenological model incorporating these features is presented and discussed.

INTRODUCTION

Like the Earth, Jupiter has a magnetic and charged particle environment -- a mangetosphere -- which presents an obstacle to solar wind plasma flowing outward from the sun. Using a gas dynamic analogy, one envisions a viscous interaction in which solar wind flow molds the magnetospheric obstacle into an essentially comet-like shape. The magnetosphere itself is subject to internal forces which may alter this picture somewhat, although the basic idea of an obstacle-in-a-flow is a useful analogy. Filled with intense fluxes of high energy particles, the Jovian magnetosphere extends over 10 million kilometers away from the planet and thus represents an object of truly astrophysical scale.

E. G. Mariolopoulos et al. (eds.), Compendium in Astronomy, 191–200.
Copyright © 1982 by D. Reidel Publishing Company.

MISSION AND INSTRUMENT DESCRIPTION

     Launched in the summer of 1977, the twin Voyager spacecraft flew
past the planet Jupiter in March and July, 1979. The Voyager trajectories
enabled the spacecraft to encounter the Galilean satellites (the four
largest moons of Jupiter) and to make in situ measurements of the Jovian
magnetosphere. As seen in Figure 1, both spacecraft approached the planet
in prenoon local time meridians. Voyager-1 left Jupiter in the meridian
near 5 AM and Voyager-2 left further downstream in the meridian of about
3 AM. In effect, the outbound Voyagers sampled different regions of the
Jovian magnetosphere. Both spacecraft remained close to the ecliptic
plane at low Jovigraphic latitudes.

     Each Voyager probe carried 11 scientific experiments designed to
investigate Jupiter, its major satellites, its particle and field environment, and the interaction of that environment with the solar wind.
Included in these were the Low Energy Charged Particle experiment (LECP).
Designed and built by the Applied Physics Laboratory of The Johns Hopkins
University in collaboration with several other institutions. The LECP
instrumentation is capable of measuring ions of energies $\gtrsim$ 28 keV and
electrons of energies $\gtrsim$ 15 keV (Krimigis et al., 1977). The LECP can also
provide compositional information about the particle environment it samples. The LECP detector array is mounted on a scan platform which rotates
the instrument in 8 discrete 45° steps so as to provide 360° coverage.
The instrument has an effective dynamic range of 11 orders of magnitude
and a fast time resolution of 400 milliseconds in data readout. The LECP
has three basic modes of operation: cruise (low time resolution, low intensity), far encounter (high time resolution, low intensity), and near
encounter (high time resolution, high intensity). Each mode is appropriate
for a particular environment the spacecraft encounters. A comprehensive
description of the instrument is given by Krimigis et al., 1977.

OBSERVATIONS

     Several weeks before the Voyagers entered the Jovian magnetosphere,
the LECP instrument observed intense particle bursts coming from the
planet (Krimigis et al., 1979a). The Jovian nature of these events was
recognized by their strong anisotropies away from Jupiter and by their
non-solar composition. These interplanetary "upstream" events lasted
from a few minutes to a few hours and were seen over $\sim$ 0.4 Astronomical
Units (AU) (1 AU = 150×10^6 km) from Jupiter.

     Inbound near the noon meridian, both Voyagers crossed the magneto-
sphere boundary (or magnetopause) at distances over 50 R_J (1 R_J = 1 Jupiter radius = 71,400 km). Owing to fluctuations in the solar wind-mag-
netosphere pressure balance, this boundary moved back and forth several
times across the spacecraft (see Figure 2). Each encounter with the magnetopause was marked by abrupt changes of an order of magnitude or more
in the low energy particle fluxes. Furthermore, the ion flows changed
direction at the magnetopause. Within the magnetosphere, the ion flows

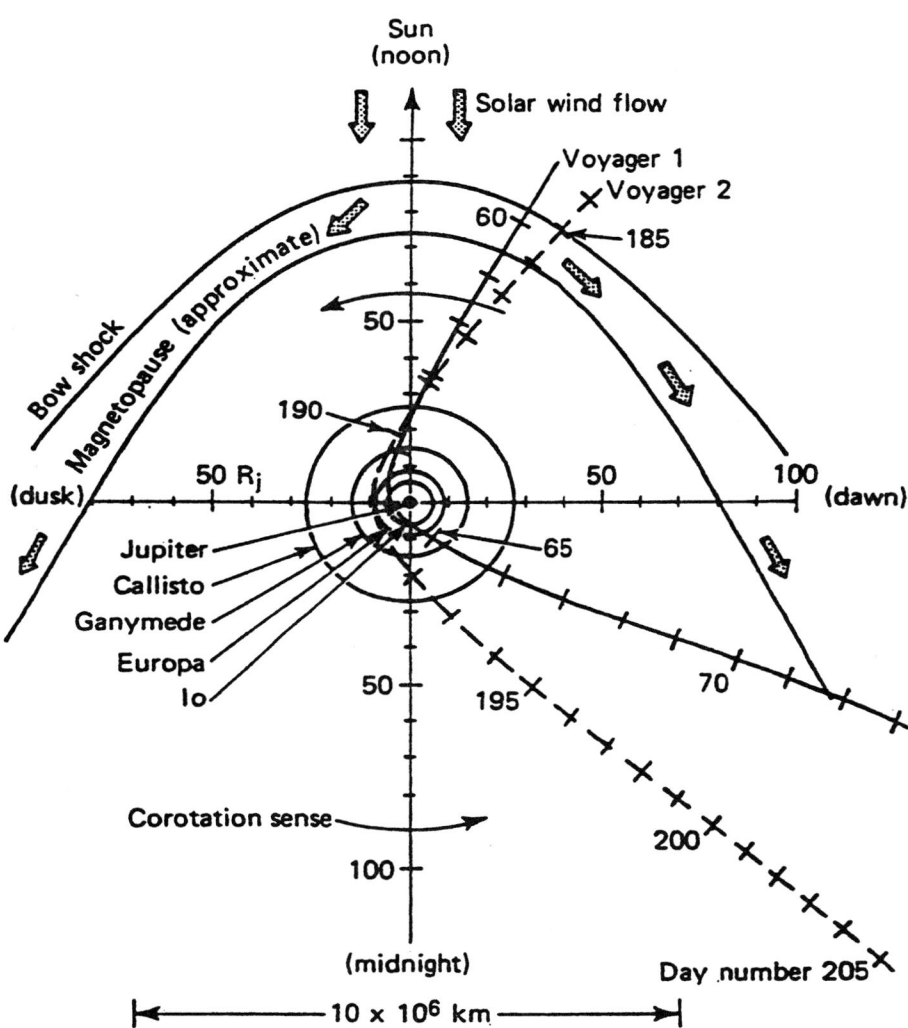

Fig. 1. Polar plot of Voyager 1 and 2 trajectories for the Jupiter encounters. The "magnetopause" curve represents the approximate boundary of the Jovian magnetosphere while the "bow shock" curve represents the detached shock generated by supersonic solar wind flow past the magnetosphere (black arrows). The orbits of the four Galilean satellites are indicated.

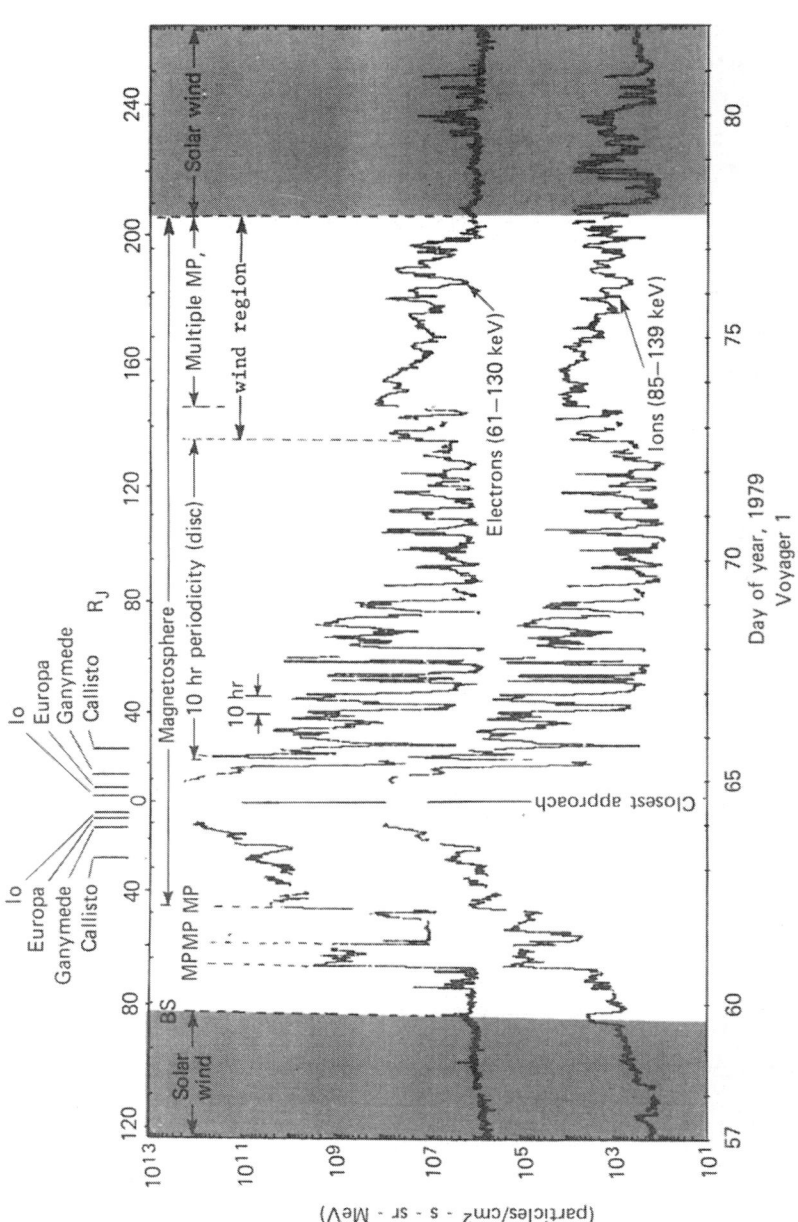

Fig. 2a. Overview of representative particle data for the Voyager 1 encounter. The plot is log-linear, with flux intensity on the ordinate and time on the abscissa. The data at closest approach (day 64) from Voyager-1 are not fully corrected. Multiple bow shock (BS) and magnetopause (MP) crossings were observed inbound. Crossings of satellite orbits are marked. Strong 10-hour periodicities in the fluxes were seen outbound. Beyond ∼140–160 R_J, the periodicities break down as the spacecraft enters the wind region.

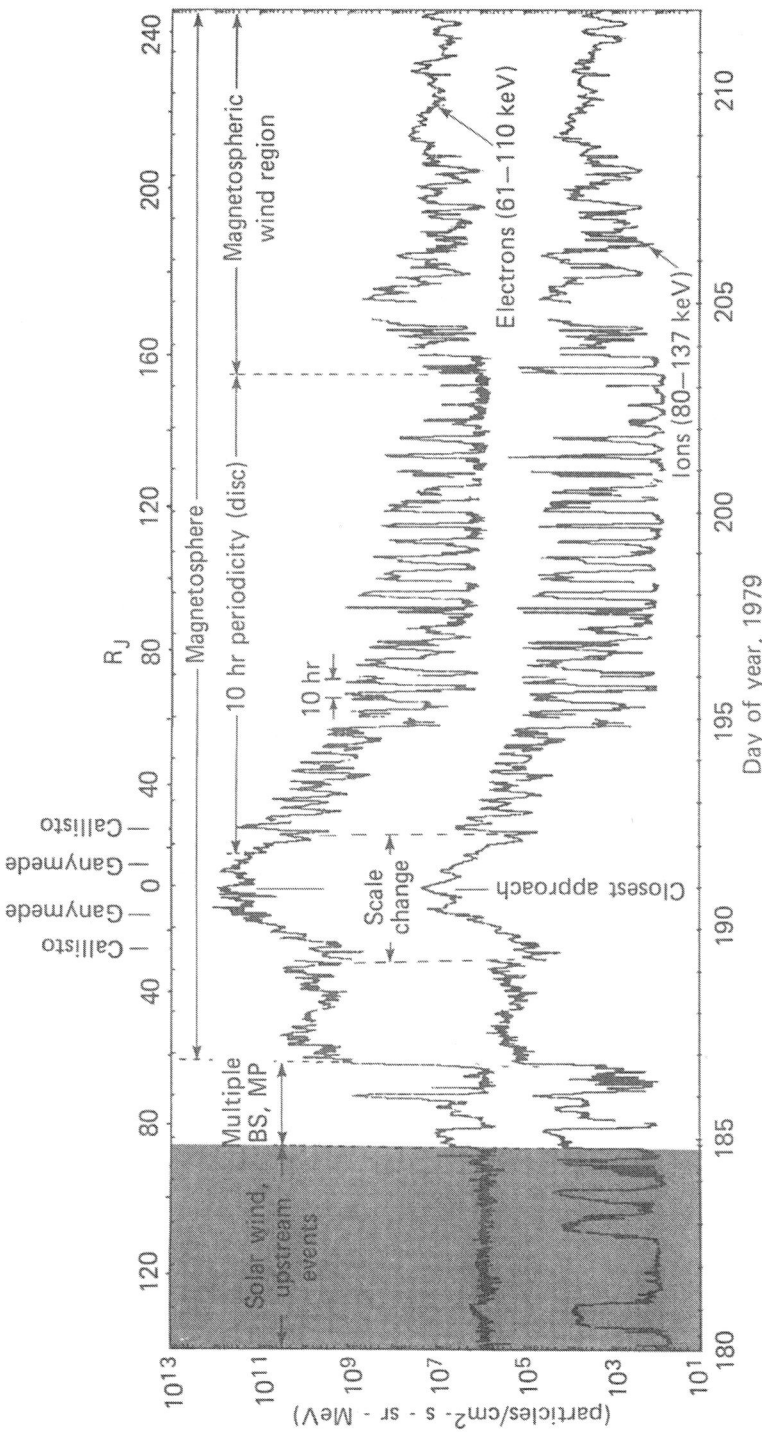

Fig. 2b. Overview of representative particle data for the Voyager 2 encounter. The plot is log-linear, with flux intensity on the ordinate and time on the abscissa. Multiple bow shock (BS) and magnetopause (MP) crossings were observed inbound. Crossings of satellite orbits are marked. Strong 10-hour period-icities in the fluxes were seen outbound. Beyond ~140-160 R_J, the periodicities break down as the spacecraft enters the wind region.

were strongly corotational (i.e., in the direction of Jupiter's rotation).

In the outer magnetosphere, the LECP experiment discovered extremely hot, corotating plasma. Maxwellian fits to the observed spectra revealed temperatures of 20-30 keV (about 250 million degrees K) and flow speeds of up to $\sim$ 1000 km/sec. Though much hotter than the solar corona (at $\sim$ 10^6 °K), this Jovian plasma is exceedingly tenuous and has densities of $10^{-3}-10^{-1}$ ions cm^{-3}. The hot plasma consists primarily of a non-solar mix of hydrogen, helium, oxygen and sulfur.

The four Galilean satellites (Callisto, Ganymede, Europa and Io) orbit within the inner Jovian magnetosphere (i.e., inside about 30 R_J) and were primary objects for study by the Voyager mission. The innermost satellite, Io, was especially interesting since it was known to exert an influence over Jupiter's decameter radio emissions. The Voyager-1 spacecraft flew within 20,570 km of Io to make close-range observations. In conjunction with the Io encounter, the LECP observed an abrupt decrease in higher energy particle fluxes while fluxes at lower energies showed no clear Io effect; the Io data are still under intensive study. General decreases in LECP fluxes were associated with crossings of the orbit of Io. Similar though less pronounced features were observed in connection with Europa and Ganymede, indicating that the Jovian satellites act as absorbers of particles measured by the LECP.

However, the satellite Io is undoubtedly a major source of low energy particles for the Jovian magnetosphere. Other experiments observed a toroidal plasma "cloud" coincident with Io's orbit (Broadfoot et al., 1979). Composition measurements by the LECP detectors revealed that the proportions of oxygen and sulfur increased as the spacecraft approached this torus (Krimigis et al., 1979b). This composition reflects volcanic origin, since a primary volcano emission is SO_2 gas. Indeed, the entire Jovian magnetosphere is filled with sulfur and oxygen from the Ioan volcanoes.

One of LECP's most dramatic observations was the 10-hour periodicities in the particle fluxes. The fluxes may vary by several orders of magnitude with the 10-hour rotation period of the planet. As indicated in Figure 2, these modulations are especially evident on the outbound, or "tailside", parts of the Voyager trajectories (Carbary, 1980).

The outbound periodicities break down at distances of 140-160 R_J. Beyond this distance, intense, variable fluxes are still observed but they no longer occur periodically. Also, the particles flow in directions approximately away from Jupiter. These flows are so extremely anisotropic that they may at times closely resemble the mono-energetic beams of laboratory accelerators (Krimigis et al., 1980). Composition measurements in this region reveal abundant oxygen and sulfur, a mix quite similar to that found in the inner magnetosphere. Thus, the spacecraft was probably still within the Jovian magnetosphere. This outflow region has been christened a "magnetospheric wind". The wind region may extend many hundreds of R_J down the Jovian magnetotail.

DISCUSSION

These recent observations of the Jovian magnetosphere have prompted a revision of ideas held previous to the Voyager encounters. One new version of the Jovian magnetosphere is illustrated in Figure 3 (Krimigis et al., 1979b; 1981). As with older models, particles are confined to a thin "magnetodisc" centered on Jupiter. The disc wobbles in response to the precession of the magnetic axis $\vec{M}$ about the rotation axis $\vec{R}$. This periodic wobbling motion causes the magnetodisc to wave up and down with an approximate 10-hour period (see Figure 3a), thus giving rise to the periodicities observed by the Voyagers. On the frontside of the magnetosphere, solar wind pressure acts to blunt the disc configuration, so strong periodicities are not seen in this region. Furthermore, the flow of particles is in the corotational sense throughout the disc region. A principal feature of this model is the hot plasma within the disc. Preliminary calculations suggest that the pressure of this gas is sufficient to stand off the solar wind pressure as well as balance the magnetic forces outside the disc. A second important feature of the new model is the disruption of the disc at large radial distances and the generation of the magnetospheric wind. Similar disruptions may cause the upstream bursts. At any rate, substantial portions of the magnetodisc seem to be ejected outward from the magnetosphere.

It is possible to make a rough estimate of the energy loss through the magnetospheric wind. Let us assume that the wind originates over the nighttime hemisphere of Jupiter at a mean distance of $\sim$ 200 R_J. Then the rate of energy loss $\dot{E}$ is related to the density of the plasma n, the wind velocity V, and the thickness of the plasma sheet t as follows:

$$\dot{E} = nkT \cdot V \cdot \pi R \cdot t$$

where k is the Boltzman constant. From the observations we obtain

$nkT \sim 10^{-11} erg/cm^3$

$V \sim 500 - 1000 \text{ km/sec}$

$t \sim 4 R_J$

$R \sim 200 R_J$.

Then, $\dot{E} \simeq (6.5 - 13) \times 10^{19}$ erg/sec or $\sim 10^{13}$ watts.

This is a large amount of energy which must be supplied by the interaction of Jupiter with the solar wind or by Jupiter itself. If it is assumed that a small fraction ($\sim$ 10%) of the rotational energy of Jupiter is lost in $\sim 10^{10}$ years (the age of the solar system), then energy becomes available at a rate of $\sim 10^{23}$ ergs/sec. Jupiter would have slowed down by only $\sim$ 1% in that time interval (Carbary et al., 1976). Thus, energy loss through the magnetospheric wind is only $\sim$ 0.1% of probable total available energy in this model. Similar numbers are obtained if it

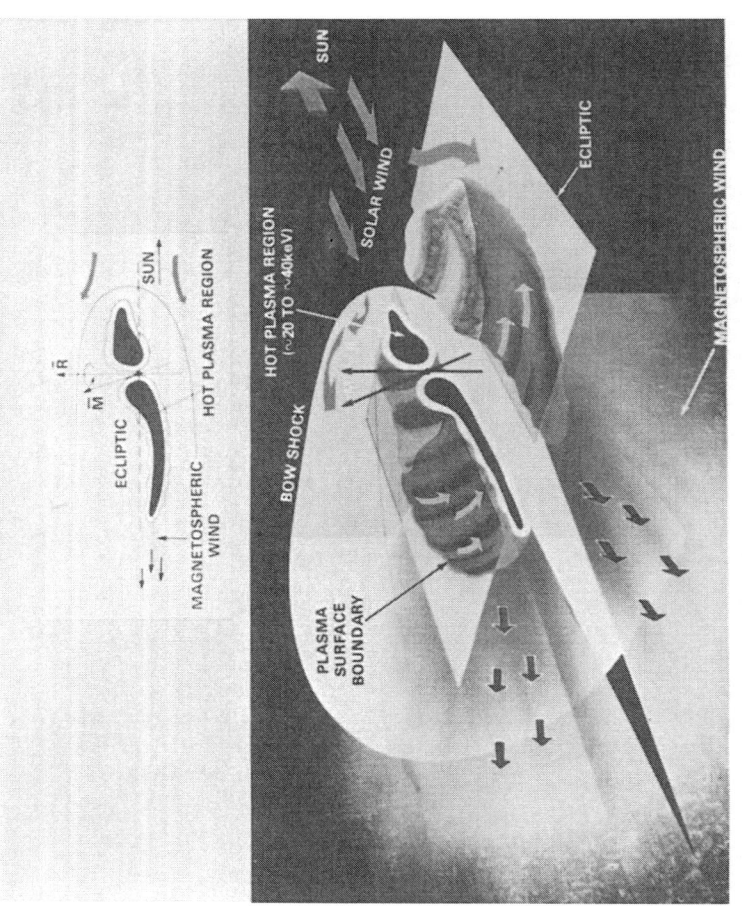

Fig. 3. The hot plasma model of the Jovian magnetosphere which has resulted from analysis of the LECP data. The hot plasma is confined to a disc-like region which waves latitudinally with the precession of the magnetic axis $\vec{M}$ about the rotation axis $\vec{R}$. The upper part shows a noon-midnight meridional cross section. The lower part is a 3-dimensional view. The pink or red regions denote the hot plasma of the corotating magnetodisc. Solar wind pressure (blue arrows) causes the disc to be blunt on the dayside and extended on the nightside. In the far magnetotail (yellow), the disc is disrupted and particles are expelled away from the planet in a magnetospheric wind.

is assumed that the energy is supplied by the solar wind through magnetic reconnection (Kennel and Coroniti, 1977).

The total mass lost through the wind can be calculated in a similar fashion, by assuming that the mean mass in the ion mixture (protons, helium, sulfur, oxygen) is oxygen.

$$\dot{N} = n \cdot V \cdot \pi R \cdot t$$

From the data

$$n \sim 2 \times 10^{-4} \text{ cm}^{-3}$$

$$V \sim (500 - 1000) \text{ km/sec}$$

$$R \sim 200 \text{ R}_J$$

$$t \sim 4 \text{ R}_J$$

resulting in $\dot{N} \sim 2 \times 10^{27}$ ions/sec. The volcanic output of Io is estimated from 4×10^{27} to 4×10^{29} ions/sec (Johnson et al., 1979). Thus mass loss through the wind can be easily maintained by volcanic activity on Io.

A fuller analysis of the LECP encounter data is published elsewhere (_Journal of Geophysical Research_, Special issue on Voyager, 1981), with papers dealing separately with plasma dynamics, density and temperature, composition, satellite interactions, and upstream particle emissions into the interplanetary medium. The LECP instrument has now performed similar measurements in the magnetosphere of Saturn with Voyager-1, with additional measurements expected during the Voyager-2 encounter in August 1981. In spite of intense particle radiation at Jupiter, the instrument operated flawlessly through both encounters.

Acknowledgments

This work has been supported by NASA. I am grateful to my Co-Investigators for many useful discussions, and to J.F. Carbary for preparation of some of the figures. The LECP investigator team consists of T.P. Armstrong, W.I. Axford, C.O. Bostrom, C.Y. Fan, G. Gloeckler, E.P. Keath, S.M. Krimigis (Principal Investigator) and L.J. Lanzerotti.

REFERENCES

Broadfoot, A.L., Belton, M.J.S., Takacs, P.Z., Sandel, B.R., Shemansky, D.E., Holberg, J.B., Ajello, J.M., Atreya, S.K., Donahue, T.M., Moss, H.W., Bertaux, J.L., Blamont, J.E., Strobel, D.F., McConnell, J.C., Dalgarno, A., Goody, R., and McElroy, M.B.: 1979, Science, 204, 978.
Carbary, J.F.: 1980, Geophys. Res. Letters, 7, 29.

Carbary, J.F., Hill, T.W., and Dessler, A.J.: 1976, J. Geophys. Res.,
    81, 5189.
Johnson, T.V., Cook, A.F., II, Sagan, C., and Soderblom, L.A.: 1979,
    Nature, 280, 746.
Kennel, C.F. and Coroniti, F.V.: 1977, Ann. Rev. Astron. Astrophys.,
    15, 389.
Krimigis, S.M., Armstrong, T.P., Axford, W.I., Bostrom, C.O., Fan, C.Y.,
    Gloeckler, G., and Lanzerotti, L.J.: 1977, Space Sci. Rev., 21, 329.
Krimigis, S.M., Armstrong, T.P., Axford, W.I., Bostrom, C.O., Fan, C.Y.,
    Gloeckler, G., Lanzerotti, L.J., Keath, E.P., Zwickl, R.D., Carbary,
    J.F., and Hamilton, D.C.: 1979a, Science, 204, 998.
Krimigis, S.M., Armstrong, T.P., Axford, W.I., Bostrom, C.O., Fan, C.Y.,
    Gloeckler, G., Lanzerotti, L.J., Keath, E.P., Zwickl, R.D., Carbary,
    J.F., and Hamilton, D.C.: 1979b, Science, 206, 977.
Krimigis, S.M., Armstrong, T.P., Axford, W.I., Bostrom, C.O., Fan, C.Y.,
    Gloeckler, G., Lanzerotti, L.J., Hamilton, D.C., and Zwickl, R.D.:
    1980, Geophys. Res. Letters, 7, 13.
Krimigis, S.M., Carbary, J.F., Keath, E.P., Bostrom, C.O., Axford, W.I.,
    Gloeckler, G., Lanzerotti, L.J., and Armstrong, T.P.: 1981, J.
    Geophys. Res., 86, in press.

THE WEATHER ON MARS ON THE BASIS OF THE MEASUREMENTS CARRIED OUT BY THE VIKINGS MISSION

C.J. Macris and B. Ch. Petropoulos
Research Center for Astronomy and Applied Mathematics
Academy of Athens
Athens, Greece

ABSTRACT

   This paper summarizes some new results concerning the Mars atmosphere obtained after the Vikings mission. On the basis of the results of the measurements made by the Viking 2 lander and Viking orbiter, the values of pressure and density corresponding to the altitudes from 28 to 100 km and different molecular weights have been computed. The computed values have been compared with the ones measured by Viking 2.

1. INTRODUCTION

   In a previous paper (Macris and Petropoulos, 1973) a model of the Mars atmosphere based on the measurements made by Mariner 6 and 7 was given. Four years later, the seasonal variation of the weather conditions on Mars (i.e. pressure, density and winds), was studied, taking into account the Mariner's 9 measurements (Macris and Petropoulos, 1977). In the present paper the measurements made by Viking 1 and 2 (lander) and Viking orbiter, which cover one Martian year have been used, in order to compute the values of pressure and density corresponding to the altitude from 0-100 km and different molecular weights. The results thus obtained have been then compared with the climatic variations observed by the Viking orbiter and their influence to pressure and density.

2. THE ATMOSPHERE OF MARS AFTER THE VIKING MISSION

   The weather has been monitored at the two landing sites for Lander 1 and Lander 2, for more than one Martian year. The variation of the mean daily surface pressure, in these two sites, as determined by Tillman is given in Fig. 1 (Snyder, 1979). In this Figure the values of the meteorological elements on Mars have been plotted as a function of the areocentric solar longitude L_S used as a measure of time. $L_S = 0^\circ$ corresponds to the normal equinox of the northern hemisphere. The Vikings observations began at $L_S = 205^\circ$ (February 10, 1977) i.e. before perihelion which corre-

E. G. Mariolopoulos et al. (eds.), Compendium in Astronomy, 201–208.

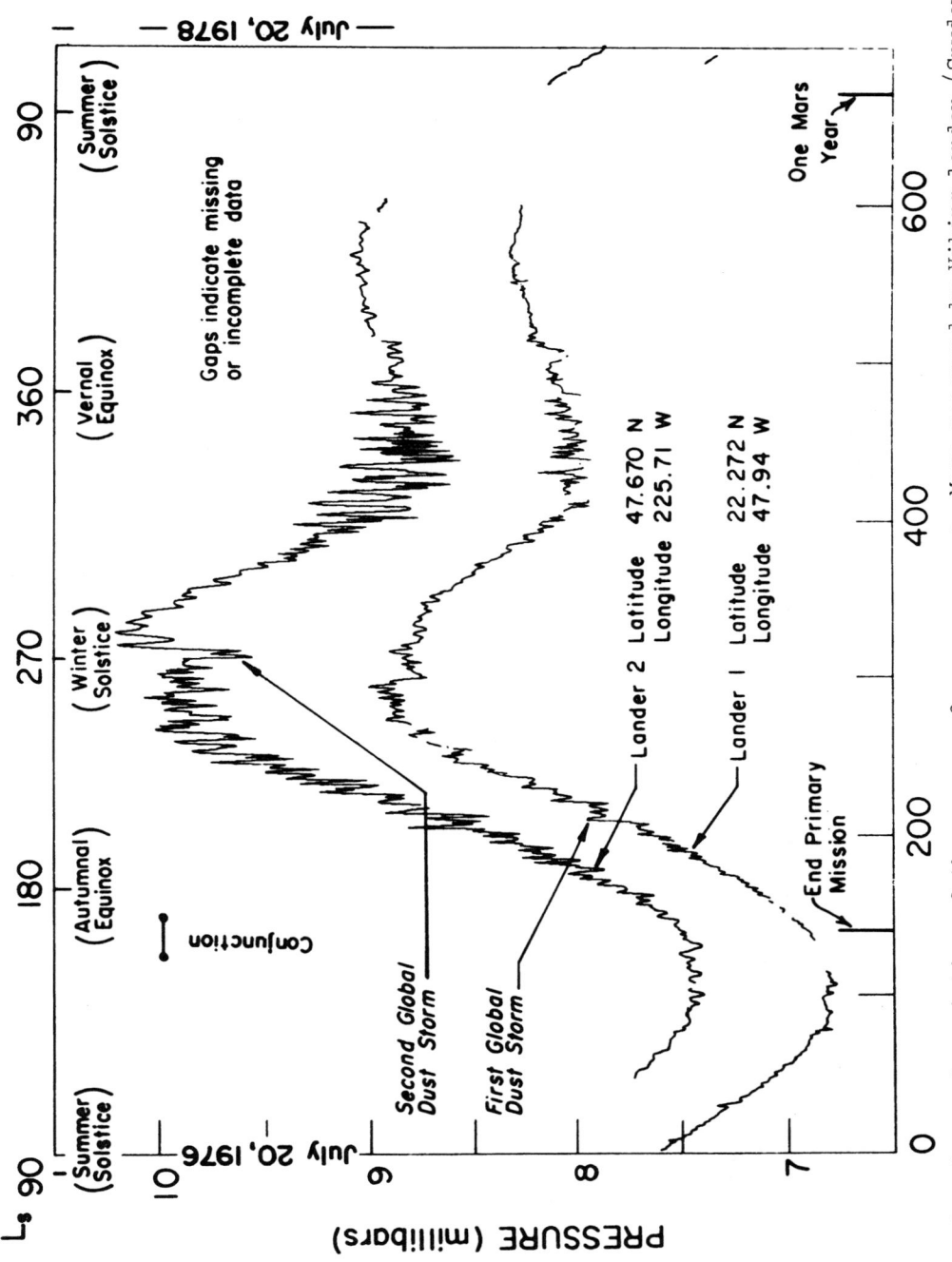

Fig. 1 : Seasonal variation of the mean surface pressure on Mars measured by Viking lander (Snyder, 1979).

sponds to $L_s=250^o$ and continued for about one Martian year. The main features of this pressure cycle were predicted by the radiative model of Leighton and Murray (1966) and the correspondance between this model and the observations indicates that the atmospheric mass is controlled by the sublimation and condensation of the polar caps. The amplitudes of surface pressure variations predicted by this model, are slightly less than 1 mb. around the mean. The observed amplitudes of the surface pressure variation, however are about 2 mb. (Fig. 1), and have been attributed to solar tidal phenomena, predicted by the model of Leovy and Mintz (1969).

Seasonal variations of the near surface atmospheric temperature T_s, have been also measured by Viking 1 and 2 landers (T_s varied from 209^oK to 218^oK in summer at the Viking 1 site, and from 201^oK to 208^oK at the Viking 2 site). In winter $T_s=200^o$K for both Viking 1 and 2 sites (Seiff, 1978).

Besides the variation of surface pressure and temperature, the following meteorological phenomena have been also observed by Vikings :

1) A first dust storm which began at $L_s=205^o$(February 10, 1977) i.e. before the Mars perihelion. The atmospheric opacity, had returned to its prestorm value, when at June 5, 1977 at $L_s=274^o$a second dust storm began and its effect had dissipated by about $L_s=350^o$(Fig. 1).

Major dust storms were observed on Mars by telescopes or by spacecrafts about the following years : 1922, 1924, 1943, 1956, 1958, 1971, 1973 (Mariner 9), 1977 (two storms the same year observed by Vikings), 1979 (Snyder, 1979). The dust in the storm of 1977 was uniformly mixed in the atmosphere up to at least 30 km (Pollack et al.,1977), and resulted in a decrease of the maximum of T_s by more than 10^oK and its minimum by about 15^oK in the Viking 1 site. In the Viking 2 site the effect of the storms was hardly noticeable in the temperature.

2) The day to day variation in the mean pressure began to increase more noticable at the lander 2 than at the lander 1 site (Fig. 1). Just before the equinox the temperature extremes, showed a similar behavior. Viking orbiter observed a regular sequence of cyclones and anticyclones, moving eastward to the north of lander 2, which can be correlated with the above pressure oscillations.

3) The Viking orbiter has photographed spiral clouds at 65°-81° northern latitude during the summer at $L_s=105^o$ to 140°. Hunt and James (1979) concluded that they are water ice clouds at an altitude of about 6 to 7 km above the ground. Clouds have been also observed near the poles at 80 km altitude (Snyder, 1979). These clouds might have been produced from CO_2 condensations, or water condensations (Briggs et al., 1977).

Near the region of Tharsis clouds of type "W", have been observed by the Viking orbiter; some of them were at 10 km above the 27 km peaks of the mountains (Briggs et al., 1977). They appear in the morning over the top of the volcanos. Mariner 9 has also observed them about the afternoon (Leovy et al., 1973; Pickersgill et al., 1979).

4) Fog has been also observed by landers and was sometimes visible in Viking's orbiter pictures over the canyons. It was near the ground for both landers sites (Pollack et al., 1977).

We can also note that the water vapor in the Martian atmosphere is not near the ground, but at an altitude of about 10 km above it, as shown by the Viking orbiter observations (Snyder, 1979). It is well mixed with the other molecules and its quantity shows seasonal variations. This quantity depends also on the dust contained in the atmosphere. The maximum of the water vapor is near the North pole.

From the above considerations we can conclude that the chemical composition of the Mars atmosphere depends on the quantity of the water and dust contained in lower altitudes and it seems probable that the molecular weight changes with the above meteorological phenomena. For this reason in the present paper we have computed the pressure and the density for the Mars atmosphere for the altitudes from 28 to 100 km and three different molecular weights measured by the Viking landers.

3. THE COMPUTED VALUES OF PRESSURE AND DENSITY

The molecular weight of the Mars atmosphere for the altitudes from 0 to 5 km has been measured by Viking 1 and 2 using different methods. The results are given in Table 1. It should be noted that from spectrometric data the value 43.486 (Seiff et al., 1978; Nier and McElroy, 1977) was found.

Table 1

H (Altitude)	m	Spacecraft	References
1.5-4 km	43.36±0.4%	Viking 2	Seiff et al., 1977
2.5 km	43.82	Viking 2	Seiff et al., 1977
0-5 km	44.36±1.0%	Viking 1	Seiff et al., 1977

In the present paper the pressure and density of the Mars atmosphere have been computed using the molecular weights given at Table 1, which were assumed to remain constant for the altitudes from 0 to 100 km.

The computations have been carried out with the help of a programme given by Pitts (1968), which is based on the hydrostatical assumption. Furthermore, the following data obtained by Viking 2 were used :

1) The pressure near the surface P_s=7.82 mb, and the temperature near the surface T_s=225$.^{\circ}$6 K (Seiff et al., 1977).

2) The distribution of the temperature from 0 to 100 km (Seiff et al., 1977).

3) The chemical composition of the Mars atmosphere near the surface (Owen, 1977).

4) The radius of the planet (3389 km) (Seiff, 1978).

The values of the physical parameters thus obtained will be given in a furthcoming paper. The values obtained for the pressure and density corresponding to the altitudes from 28 to 100 km and the three different assumptions concerning the weights are given in Table 2 together with the values measured by Viking 2.

It is known that the pressure can be expressed as a function of the molecular weight with the help of an exponential function of the form

$$P = P_o \exp \left| \int - \frac{mg}{KT} \, dz \right| ,$$

where z is the altitude, K Boltzmann's constant, T the absolute temperature, m the molecular weight and g the acceleration of gravity. With the help of this formula the variation of logP for the given values of the molecular weight and z and T constant has been computed. Finally the values of the molecular weight for the altitudes from 28 to 100 km have been computed using the method of linear regression. These values are given in Table 2 and plotted in Fig. 2 as a function of the altitude.

4. CONCLUSIONS

From Table 2 and Fig. 2 we see that the molecular weight assumes a constant value m=44.8 for the altitudes from 44 to 76 km. For the other altitudes the variations of the molecular weight are not very noticable. At the altitude of 28 km a serious discrepancy appears between the measured and computed values of pressure and density. As a matter of fact, if we used the measured values we obtain m=65.60. This discrepancy is perhaps due to an error in the Vikings measurements. It should be noted, however, that the values measured by Viking 2 (Table 2) are similar to the ones measured by Mars 6 (P=$3.52.10^{-1}$ mb and d=$1.17.10^{3}$ kg.m^{-3} at the same altitude (Marov et al., 1976).

It is interesting to note that Seiff (1978) has also computed the pressure and density for the altitudes from 0 to 100 km, using data similar to those measured by Viking 2 which were used in the present paper (P_s=7.3 mb, and T_s=214°K). He found for the altitude of 28 km the values m=44.36, P=$4.68.10^{-1}$ mb and d=1.38 kg.m^{-3}. These values are very closed to the ones obtained in the present paper (Table 2) and different from the measured ones (Seiff, 1977).

We note also that the temperature distribution for the altitudes

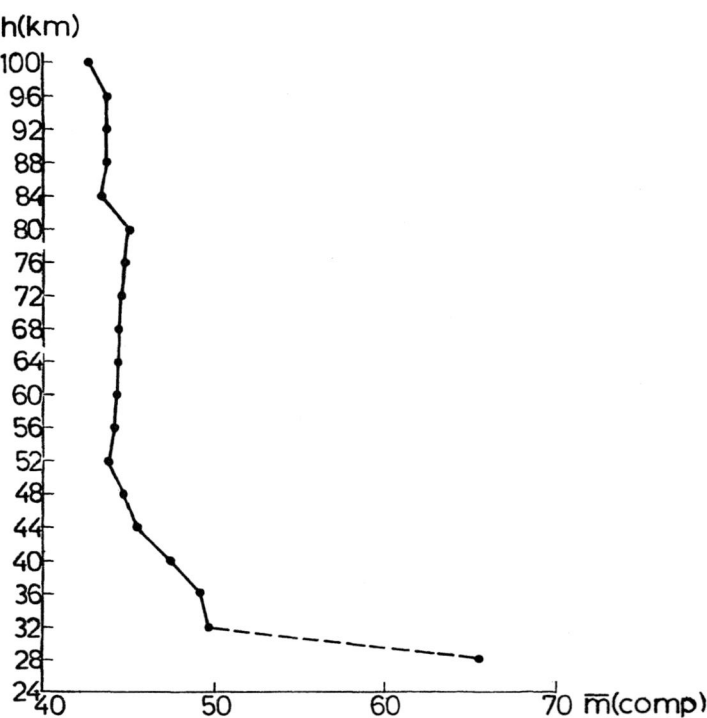

Fig. 2.: The computed molecular weight distribution for the altitudes from 28 to 100 km.

from 5 to 28 km adopted by Seiff (1978) is also very close to the one used in the present paper which was taken from Seiff (1977).

We can conclude that the discrepancies between the computed and measured values of pressure and density noted above can be caused either by errors in the measurements or by various meteorological phenomena,* which change the chemical composition at the altitude of 28 km and 80-100 km above the Viking 2 lander site.

In fact Viking orbiter observed in August 1978 a cyclonic storm at the altitude of 28 km and above this region exist clouds of CO_2 and H_2O (West et al., 1977; Herr et al., 1970; Rossow, 1978). Viking 2 lander in August 1978 measured at the Martian latitude $48°$N, very strong winds.

---

* This second point of view agrees with the recent observations of Vikings. In fact four large clouds floating at an altitude of 28 km are visible in a mosaic of 102 frames taken by Viking, 22 February 1980. (COSPAR, Information Bulletin No. 89, December 1980, page 62).

Table 2

Values of the pressure and density for the altitudes from 28 to 100 km in the Mars atmosphere

h	$m_1=43.36$ P_1(mb)	$m_2=43.82$ P_2(mb)	$m_3=44.36$ P_3(mb)	P_{mes}(mb)	$m_1=43.36$ d_1(gr/cm^3)	$m_2=43.82$ d_2(gr/cm^3)	$m_3=44.35$ d_3(gr/cm^3)	d_{mes}(gr/cm^3)	m(comp)
28	$4,83 \cdot 10^{-1}$	$4,81 \cdot 10^{-1}$	$4,79 \cdot 10^{-1}$	$4,04 \cdot 10^{-1}$	$1,35 \cdot 10^{-6}$	$1,36 \cdot 10^{-6}$	$1,39 \cdot 10^{-6}$	$1,22 \cdot 10^{-6}$	65,60
32	$3,20 \cdot 10^{-1}$	$3,17 \cdot 10^{-1}$	$3,09 \cdot 10^{-1}$	$2,54 \cdot 10^{-1}$	$9,28 \cdot 10^{-7}$	$9,30 \cdot 10^{-7}$	$9,41 \cdot 10^{-7}$	$7,92 \cdot 10^{-7}$	49,73
36	$2,03 \cdot 10^{-1}$	$2,01 \cdot 10^{-1}$	$1,95 \cdot 10^{-1}$	$1,58 \cdot 10^{-1}$	$6,47 \cdot 10^{-7}$	$6,45 \cdot 10^{-7}$	$6,28 \cdot 10^{-7}$	$5,04 \cdot 10^{-7}$	49,23
40	$1,28 \cdot 10^{-1}$	$1,25 \cdot 10^{-1}$	$1,20 \cdot 10^{-1}$	$9,87 \cdot 10^{-2}$	$4,06 \cdot 10^{-7}$	$4,03 \cdot 10^{-7}$	$4,10 \cdot 10^{-7}$	$3,14 \cdot 10^{-7}$	47,39
44	$8,05 \cdot 10^{-2}$	$7,87 \cdot 10^{-2}$	$7,17 \cdot 10^{-2}$	$6,19 \cdot 10^{-2}$	$2,51 \cdot 10^{-7}$	$2,48 \cdot 10^{-7}$	$2,58 \cdot 10^{-7}$	$1,94 \cdot 10^{-7}$	45,54
48	$5,11 \cdot 10^{-2}$	$4,97 \cdot 10^{-2}$	$4,19 \cdot 10^{-2}$	$3,92 \cdot 10^{-2}$	$1,60 \cdot 10^{-7}$	$1,57 \cdot 10^{-7}$	$1,57 \cdot 10^{-8}$	$1,20 \cdot 10^{-8}$	44,67
52	$3,22 \cdot 10^{-2}$	$3,11 \cdot 10^{-2}$	$2,41 \cdot 10^{-2}$	$2,46 \cdot 10^{-2}$	$1,06 \cdot 10^{-7}$	$1,04 \cdot 10^{-7}$	$9,22 \cdot 10^{-8}$	$8,23 \cdot 10^{-8}$	43,82
56	$1,95 \cdot 10^{-2}$	$1,87 \cdot 10^{-2}$	$1,38 \cdot 10^{-2}$	$1,47 \cdot 10^{-2}$	$6,99 \cdot 10^{-8}$	$6,80 \cdot 10^{-8}$	$5,23 \cdot 10^{-8}$	$5,27 \cdot 10^{-8}$	44,22
60	$1,14 \cdot 10^{-2}$	$1,09 \cdot 10^{-2}$	$8,08 \cdot 10^{-3}$	$8,54 \cdot 10^{-3}$	$4,31 \cdot 10^{-8}$	$4,17 \cdot 10^{-8}$	$2,87 \cdot 10^{-8}$	$3,26 \cdot 10^{-8}$	44,26
64	$6,67 \cdot 10^{-3}$	$6,35 \cdot 10^{-3}$	$4,88 \cdot 10^{-3}$	$4,94 \cdot 10^{-3}$	$2,43 \cdot 10^{-8}$	$2,34 \cdot 10^{-8}$	$1,66 \cdot 10^{-8}$	$1,82 \cdot 10^{-8}$	44,34
68	$3,92 \cdot 10^{-3}$	$3,72 \cdot 10^{-3}$	$2,97 \cdot 10^{-3}$	$2,91 \cdot 10^{-3}$	$1,47 \cdot 10^{-8}$	$1,41 \cdot 10^{-8}$	$1,05 \cdot 10^{-8}$	$1,07 \cdot 10^{-8}$	44,41
72	$2,24 \cdot 10^{-3}$	$2,11 \cdot 10^{-3}$	$1,75 \cdot 10^{-3}$	$1,68 \cdot 10^{-3}$	$9,41 \cdot 10^{-9}$	$8,95 \cdot 10^{-9}$	$6,70 \cdot 10^{-9}$	$6,71 \cdot 10^{-9}$	44,53
76	$1,15 \cdot 10^{-3}$	$1,08 \cdot 10^{-3}$	$9,83 \cdot 10^{-4}$	$9,22 \cdot 10^{-4}$	$5,47 \cdot 10^{-9}$	$5,16 \cdot 10^{-9}$	$4,00 \cdot 10^{-9}$	$3,83 \cdot 10^{-9}$	44,77
80	$6,11 \cdot 10^{-4}$	$5,54 \cdot 10^{-4}$	$5,67 \cdot 10^{-4}$	$5,08 \cdot 10^{-4}$	$2,53 \cdot 10^{-9}$	$2,37 \cdot 10^{-9}$	$2,18 \cdot 10^{-9}$	$2,11 \cdot 10^{-9}$	45,04
84	$3,41 \cdot 10^{-4}$	$3,23 \cdot 10^{-4}$	$3,14 \cdot 10^{-4}$	$2,88 \cdot 10^{-4}$	$1,38 \cdot 10^{-9}$	$1,28 \cdot 10^{-9}$	$1,17 \cdot 10^{-9}$	$1,06 \cdot 10^{-9}$	43,33
88	$1,97 \cdot 10^{-4}$	$1,93 \cdot 10^{-4}$	$1,80 \cdot 10^{-4}$	$1,74 \cdot 10^{-4}$	$7,15 \cdot 10^{-10}$	$7,05 \cdot 10^{-10}$	$6,53 \cdot 10^{-10}$	$5,79 \cdot 10^{-10}$	43,65
92	$1,20 \cdot 10^{-4}$	$1,13 \cdot 10^{-4}$	$1,10 \cdot 10^{-4}$	$1,08 \cdot 10^{-4}$	$4,29 \cdot 10^{-10}$	$4,27 \cdot 10^{-10}$	$3,95 \cdot 10^{-10}$	$3,63 \cdot 10^{-10}$	43,72
96	$7,03 \cdot 10^{-5}$	$6,61 \cdot 10^{-5}$	$6,37 \cdot 10^{-5}$	$6,60 \cdot 10^{-5}$	$2,72 \cdot 10^{-10}$	$2,42 \cdot 10^{-10}$	$2,50 \cdot 10^{-10}$	$2,30 \cdot 10^{-10}$	43,66
100	$1,14 \cdot 10^{-5}$	$4,03 \cdot 10^{-5}$	$3,73 \cdot 10^{-5}$	$4,01 \cdot 10^{-5}$	$1,49 \cdot 10^{-10}$	$1,34 \cdot 10^{-10}$	$1,36 \cdot 10^{-10}$	$1,42 \cdot 10^{-10}$	42,66

The variation of the molecular weight at the altitudes from 80 to 100 km is probably caused by the observed clouds or by cluster molecules (Aikin, 1972).

In conclusion we have found that the molecular weight is equal to m=43-50 and can be considerated as constant from 32-100 km.

REFERENCES

Aikin, A.C.: 1972, Nature 235, 10.
Briggs, G.A., Klaasen, K., Thorpe, T., Wellman, J., Baum, W.: 1977, J. Geophys. Res. 82, 4121.
Herr, K.C., Pimentel, G.: 1970, Science 167, 47.
Hunt, G.E., James, P.B.: 1979, Nature 278, 531.
Leighton, R.B., Murray, B.C.: 1966, Science 153, 136.
Leovy, C., Mintz, Y.: 1969, J. Atmospheric Sci. 26, 1167.
Leovy, C., Zureck, R.W., Pollack, J.B.: 1973, J. Atmospheric Sci. 30, 749.
Macris, C.J., Petropoulos, B. Ch.: 1973, in Solar Activity and Related Interplanetary and Terrestrial Phenomena, Proc. of the First Europ. Astron. Meeting, Vol. 1, (ed. by J. Xanthakis), Springer-Verlag, Berlin, p. 140.
Macris, C.J., Petropoulos, B. Ch.: 1977, Compt. Rend. Acad. Sci. Paris 285, 239.
Marov, M. Ya., Kerzhanovich, V.V., Avduesky, V.S., Rozhdestvensky, M.K.: 1976, Proc. of the XVIII COSPAR Meeting, Space Res. XVI, Springer-Verlag, Berlin, p. 1020.
Nier, A.O., McElroy, M.B.: 1977, J. Geophys. Res. 82, 4341.
Owen, T., Biermann, K., Runshneck, D.R., Biller, J.E., Howarth, D.W., Lafleur, A.L.: 1977, J. Geophys. Res. 82, 4635.
Pickersgill, A.O., Hunt, G.E.: 1979, J. Geophys. Res. 84, 8317.
Pitts, D.E.: 1968, NASA TN. D-4292.
Pollack, J.B., Colburn, D., Kahn, R., Hunter, J., Camp, W. van, Carlston, C.E., and Wolf, M.R.: 1977, J. Geophys. Res. 82, 4479.
Pollack, J.B., Colburn, D.S., Flasar, F.M., Kahn, R., Carlstron, C.E., and Pidek, D.: 1979, J. Geophys. Res. 84, 2929.
Rossow, W.: 1978, Icarus, 36, 1.
Seiff, A., Kirk, D.B.: 1977, J. Geophys. Res. 82, 4634.
Seiff, A.: 1978, 21st COSPAR Meeting, Innsbrukh, Austria (in press).
Snyder, C.W.: 1979, J. Geophys. Res. 84, 8487.
West, G.S., Wright, J.J., Euler, H.C.: 1977, NASA TM - 78119.

PART VI
STELLAR ASTRONOMY

ON THE ORIGIN OF NEBULAE

V.A. Ambartsumian
President, Academy of Sciences Armenian S.S.R.
Yerevan, Armenian S.S.R., U.S.S.R.

With great pleasure and warm sympathy the author dedicates this paper to Academician J. Xanthakis.

1. Introductory Remarks

The origin and the evolution of stars has always attracted the attention of astrophysicists. The solution of this problem is considered by many of them as one of the main aims of their science. The theoreticians devoted to this problem their great efforts.

Much less attention has been paid to the problem of the origin and evolution of nebulae as individual objects. In the textbooks the nebulae are often considered in the chapters dedicated to "the interstellar matter". This indirectly makes an impression that a nebula is something deprived of individuality like a fluctuation of "internebular matter". Actually the nebulae are discrete objects and their mutual distances are as a rule much larger than their diameters. Therefore, one must assume that they are to some degree mutually independent.

It is true that some of the nebulae are dispersing with time in the surrounding space (for example the planetary nebulae). But this fact is related rather to their final fate than to their origin.

Even the superficial study of known facts concerning the galactic nebulae is sufficient to conclude that :

a) The observations give us much more direct and rich information on the dynamical changes and physical processes in them than in the case of the stars, where our hopes to obtain a modest amount of direct information on the internal structure from observations of neutrino fluxes at least for one star have not yet materialized.

b) The changes occurring in nebulae are in many cases closely connected with some turning points in the life of some stars. Therefore, any

E. G. Mariolopoulos et al. (eds.), Compendium in Astronomy, 211–218.
Copyright © 1982 by D. Reidel Publishing Company.

conclusion on the origin and evolution of nebulae can serve as valuable
information on the evolution and perhaps even on the origin of stars.

With this connection between the two problems in mind we try to give
here   a short review of ideas on the origin of nebulae. Our aim is to
attract the attention of readers to this more accessible side of the complex evolutionary processes taking place in the Galaxy.

2. The Case when the Nebula Is Connected with Only One Star

There are several classes of nebulae of more or less regular shape,
where some kind of connection with a star is almost obvious. From the
point of view of contemporary Astrophysics the solution of the problem
of the origin for some of such classes of nebulae is almost trivial. Let
us consider them :

a) During the outbursts of Novae we observe the formation of small
expanding nebulae around them. The velocity of expansion is of the order of 1000 km s^{-1} and after some decades the expanding nebula disappears
in the space surrounding the Nova. There is no doubt that the nebula is
ejected from the star and consists of the material formerly belonging to
the outer layers of the star. The mass ejected is usually of the order
of 10^{-5} $M_\odot$.

b) The formation of a planetary nebula is the result of the ejection
of external layers of its nucleus. This time the nebular object formed
is much more massive having the mass between 0.01 $M_\odot$ and 0.1 $M_\odot$. The
planetary nebulae also expand into surrounding space, but they are accessible to observers during the time of the order of 10^5 years. Apparently during the life of our Galaxy hundreds of millions planetaries
have formed and dispersed in it.

c) The supernova remnants are nebulae originated from giant stellar
explosions. Their initial masses are believed to be of the order of one
solar mass. However, during their expansion the original mass often draws
in the surrounding interstellar matter. Thus, the mass of the expanding
shell can increase enormously. Thus, sometime the nebulae of large mass
are formed.

d) It seems now quite certain that the cometary nebulae are formed
from the matter ejected by variable stars which we observe in their "heads".

e) Some WR stars in our Galaxy are surrounded by nebulae of circular
form like NGC 6888. Similar cases have been observed in LMC. The observational data related both to these stars and the surrounding nebulae
suggest the formation of such nebulae from the matter and impulse ejected by WR stars in the same way as SNR are formed as consequence of SN
outbursts.

Five cases considered above cover all known classes of nebulae
of more or less regular shape. We see that in all five cases the evolu-

tionary transitions of matter between the dense stellar bodies and the
rarified nebular state are going on in one way :

   dense matter →   diffuse matter.

3. Processes in Diffuse Nebulae

About 35 years ago, when we started the study of stellar associations
we were much impressed by the fact that almost every OB association contains one or more large diffuse nebulae. From this it was concluded that
the formation of groups of young stars must proceed simultaneously with
the formation of nebulae since the very forms of nebulae were suggesting
their instability and youth.

However, proceeding from the ideas of classical Cosmogony many theoreticians have hurried to conclude from the coexistence of young stars
with nebulae in stellar associations that we witness in them the immediate
transformation of nebular masses into young stars.

Various mechanisms of the so-called collapse have been proposed but
at that time we needed more the observational data rather than elaborated
models of condensation of matter.

The accumulation of necessary data has accelerated in the subsequent
period as a result of the application of new observational methods (21
cm observations of HI, radio observations of molecules, infra-red observations, VLBI observations of fine details). A number of qualitatively
new phenomena has been discovered. Among them :the compact HII regions
deep in the cold and dark parts of nebulae, the OH and H_2O masers, hot
regions of infra-red emission. It was shown that many optically bright
emission nebulae which surround the groups of young OB-stars are expanding with considerable velocity. For example, in the Rosette nebulae
around the cluster NGC 2244 the expansion velocity is of the order of
20 km s^{-1}. It was quite clear that such cases almost directly contradict to the idea of condensation.

However, later, when it was shown that nebulae in OB associations
contain large cold clouds of H_2 and other molecules and that the velocity
gradients in them are as a rule very small, the trend of scientific opinion turned again in favour of the process on condensation. And the discovery of compact HII regions inside the molecular clouds has been considered as the direct proof of the process of collapse within the molecular clouds.

The truth is that the discovery of compact HII regions in such clouds
is the direct evidence only of star formation going on in them but not
the direct evidence of collapse processes.

4. Infra-Red Sources in Diffuse Nebulae

According to the classification of Rowan-Robinson (1979) all diffuse nebulae belong to one of two following classes :

a) Cold nebulae without appreciable radiation in the infra-red region 1-20 μm and,

b) the clouds which have infra-red source (IRS) (or sources) in them. The average masses of clouds of this second type are much higher than the average mass of a cold cloud. The clouds of the second type as a rule are situated in OB-associations.

Very often the presence of IRS coincides with the presence of a compact HII region. In these cases there is quite natural explanation of the origin of IR radiation. The dust in the cloud absorbs completely the radiation of the OB star (stars) and is heated to the temperatures of several hundred degrees.

However, there are cases when the cloud contains an IRS without the radio-continuum. As is known such a continuum is an inevitable consequence of the presence of the HII-region.

These cases were considered by adherents of the collapse-hypotheses just as places where the collapse of the surrounding molecular cloud has produced a very young star which is still accreting the material infalling from the cloud. The best example of such IRS is the Kleinman-Low source with its infra-red maximum IRc$_2$. It was assumed that at such an early stage of the formation of the star the absorption of the Ly-continuum by infalling dust is sufficiently strong to prevent the formation of a HII region.

The observations have shown that in both cases (the presence or absence of compact HII region) the source is accompanied by a maser or by a group of masers (in the molecular lines of OH or H$_2$0). They have been explained as consequence of pumping of gases by infra-red radiation of the source.

At the end of the seventies many theoreticians were convinced that the further detailed study of the objects like Kleinman-Low-region in Orion will result in the clear picture of collapse-processes in molecular clouds and of the formation of "cocoon stars".

5. New Observational Data

During the last three years new observational data have been obtained which completely changed the situation described above.

a) The measurements with sufficient angular resolution of the profiles of the CO radio-lines in the IRS of the type described above (ten of arcseconds) have shown the Doppler-broadening of the order of 80 km s^{-1}. This means that the velocities relative to the centre of mass of the object are of the order of 40 km s^{-1}. In the case of the infall of material to the condensation of the mass of the order of 10^3 M$_\odot$ the velocities expected can hardly exceed 10 km s^{-1}. This alone is sufficient to deny the picture of the gravitational contraction.

In the case of the Kleinman-Low nebulae the profiles look like the superposition of two profiles : one of high velocity flow (plateau about 100 km s^{-1} wide) and another of low velocity flow (40 km s^{-1} wide).

b) In each case dispersion of radial velocities of H_2O masers in and around of such IRS is in good general agreement with the wide of CO profiles of the infrared region. This apparently is an evidence of close connection between the system of H_2O masers and molecular flow producing the line broadening. The masers are immersed in the flow.

c) As a result of a series of very accurate VLBI observations of the positions and the determination of the proper motions of H_2O masers it was established that the system of masers in and around Kleinman-Low infrared region in Orion nebula is expanding (Genzel et al., 1981). The center of expansion was determined with considerable accuracy and coincides within the errors of determination with the infra-red source IRc$_2$.

There are two groups of masers in expansion. One with expansion velocity of 18 km s^{-1} and the second with velocities higher than 40 km s^{-1}. This is in fairly good agreement both with the dispersion of radial velocities of masers and with the profiles of CO lines.

A more complicated picture of tangential motions has been obtained from the similar study of proper motions of H_2O masers in W51-Main. Here the number of H_2O masers is larger and the general pattern deviates from the picture of radial expansion. However, the whole region is in vigorous motions and there is no doubt that the large velocities again are caused by processes of outflow. In any case the collapse-model seems quite impossible.

6. The Estimate of the Ejected Mass

In the case of the outflow in Orion Molecular Cloud it was estimated by Genzel et al. (1981) that the intensity of ejection of mass is of the order of 10^{-3} M$_\odot$ per year. At the same time the duration of outflow as is inferred from the size of KL nebula is not shorter than 2.10^3 years. Therefore, the total mass ejected from the central body during the outflow must be larger than 2 M$_\odot$.

The first important conclusion from these observations is that the Orion Molecular Cloud-1 is gaining the mass from the body situated at IRc$_2$.

We don't know the exact value of the total mass of Orion Molecular Cloud-1 (OMC-1). The value of 10^3 M$_\odot$ seems to be of the right order of magnitude. Comparing the gain from the outflow observed in the region of KL nebulae with this mass of OMC-1 we see that the relative increase of the mass of the nebula is relatively small - about 10^{-3} of the present mass of OMC-1.

If however the phenomenon of outflow is recurrent such gains can play an important role in the formation of the mass of OMC-1. Let us consider the evidences in favour of recurrence.

7. The Recurrence of Ejections

According to Downes et al. (1981) some of the properties observed in OMC-1 and W51-Main (wide molecular lines, the presence of masers) are typical for molecular clouds with infrared sources within them. Therefore, it is very probable that the observations of the molecular lines in such clouds with great angular resulution as well as the determination of the proper motions of their H_2O masers will reveal in them the similar kinematical pattern. And, since the majority of OB associations contain the diffuse nebulae with infrared core, this means that <u>during most part of the life of OB-associations the similar phenomena of ejections persist in them</u>.

Since the life-time of an OB-association is of the order of 10^7 years we can assume with some confidence that during 5.10^6 years the enrichment of the clouds belonging to the association is continuing. Of course the centre of the outflow can change its place and the ejection processes will happen from different bodies. But the <u>total gain</u> of mass by the clouds in the association during its lifetime can reach 10^4 $M_\odot$.

Thus, we can assume that the ejections from some unknown sources similar to that we observe in OMC-1 can play <u>essential</u> if not decisive part in the buildup of the clouds under consideration.

Of course we don't know what kind of stars or other dense bodies are the sources of ejections of such large masses. But we know that side by side with outflow phenomena in KL-like regions the ejection occurs from the WR and O stars we observe in associations. The quantity of matter ejected per year by an O star is at least two orders of magnitude lower than in the case of KL-region but the duration of ejection is longer. It is necessary to take into account that usually there are several O-stars simultaneously in an O-association. Nevertheless, it is quite possible that the total input of O and B stars into the mass of nebula is smaller than from outflows of KL-type. Another source of nebular mass is the large quantity of T Tauri variables. It may happen that their integral input is larger than the total input of OB stars. It is known that observations made with IUE have shown that the large part of T Tauri stars show P Cygni type absorption components and never the redshifted absorption. And it may well happen that their input can be larger than expected.

Our conclusion is that solely from observational grounds we can assume that <u>large nebulae in OB-associations are in the process of growth. They are fed by masses ejected from dense bodies present there</u>.

Is there any need for other agents producing the nebulae? We cannot give a definitive answer to this question. However, the unified picture of the origin of <u>all nebulae</u> in the Galaxy from masses ejected by dense bodies seems now more attractive than ever.

8. The Common Origin of Stars and Nebulae

The purpose of this article was to show that we can try to follow the origin of diffuse nebulae without the speculative assumptions and remaining on purely observational grounds. The possibility of such approach is connected with the fact that each nebula is transparent for some of the frequencies we are able to use for observations. In the case of some nebulae of regular shape the optical observations alone provide the necessary data. Now the radio observations in frequencies of molecular lines give the solution for diffuse nebulae. The VLBI measurements of masers provide us with delicate data on the internal kinematics of nebulae in the regions of most recent star formation. But we cannot obtain similar data on the internal structure and dynamical processes in stars which are in the process of formation.

However, even the partial solution of the problem of the origin of nebulae contains very important information on the origin of the stars. First and the most important information is the total absence of collapse phenomena in nebulae. Instead, there are phenomena of the ejection and vigorous motions of large quantities of material taking place in the regions of star formation. Therefore it is clear that both the processes of formation of stars and nebulae are going on together. The idea of their common origin now seems very probable.

During our studies of OB associations we have expressed the opinion that the process of star formation in associations proceeds in small groups. The Trapezium of Orion is one of such groups. Perhaps θ^2 Orion is an example of an older group. Since the Kleinman-Low nebula is very near to these multiple stars it seems that we observe here the simultaneous process of the formation of a new stellar group and of ejection of nebular mass from some very massive body. It may happen also that first the nebular matter is ejected and then the stars are formed.

In both cases some body should exist from which the material of stars and of the ejected diffuse material is produced. Thus, we return to the idea of <u>protostars</u> (Ambartsumian, 1948).

In the middle of this century the idea of massive protostars (inaccessible as yet for observers) has found little sympathy among the theoreticians who prefer to continue to produce the models of gravitational collapse. The whole generation has been nourished by construction of these models. Though the idea of collapse has produced large numbers of PhD-s working on the models on condensation it was almost fruitless in explaining how the stars have been formed.

REFERENCES

Ambartsumian, V.A. : 1960, Scientific Papers 2, 179.
Downes, D., Genzel, R., Becklin E.E., and Wynn-Williams, C.G, : 1981,
    Astrophys. J. 244, 869.
Genzel, R., Raid, M.J., Moran J.M., and Downes, D. : 1981, Astrophys. J.
    244, 884.
Rowan-Robinson, M. : 1979, Astrophys. J. 234, 111.

ON MASS LOSS FROM STARS

N. Dallaporta
Instituto di Astronomia, Universita di Padova
Padova, Italy

ABSTRACT

A general overview is drawn concerning the present day status on the problem of mass loss from stars. After an introductory part summarizing its classical interpretations, a general comparison is sketched between mass conserving and mass losing evolutions for supergiant stars, for which mass loss effects are expected to have the largest importance; and at first sight, a general agreement seemed to have been reached between interpretation and observation. More recent and detailed data have now completely upset this apparently settled situation : a general discussion of the main inconsistencies is presented, whose principal outcome may be focalized as a polarization into opposite opinions among workers in the field, one of which still assumes that parametrized formulae for mass loss are adequate to foresee the main happenings for evolving stars, while the opposite view denies any dependence of mass loss from general parameters of stars and claims its attribution only to surface phenomena.

As is generally well known, from its origin some thirty years ago up to very present times, researches on stellar evolution have been conducted by assuming as a leading principle the constancy of the star mass in all phases it is going through during the course of its evolutionary path; and all results up to now have been strictly dependent on this main assumption. It is only very recently, not more than about five years ago, that mass loss from stars, which had incidentally been experimentally detected already in single cases about twenty years before, has begun to be systematically introduced in the evolutionary calculations, as an important factor able in some cases to strongly modify the preceding results. And the fact that the stars more affected by the largest loss rate and whose evolution is therefore the most strongly modified are those of largest mass, that is the supergiants, brings as a consequence that the understanding of mass loss phenomena has acquired an impact which goes well beyond the purely local interest of stellar evolution by itself, for some reasons which I like to present as an introduction to the subject.

E. G. Mariolopoulos et al. (eds.), Compendium in Astronomy, 219–231.

One of the chief developments not only of astrophysics but also of cosmology is nowadays centered on the chemical evolution of galaxies. Such an evolution is essentially due to the yield of material given back, by dying stars, to the clouds from which new generations of stars will be borne, which will therefore be different and enriched in heavy elements in respect to the original material from which the first generation was initially formed. The chemical evolution will therefore be so more intense as larger are both the yield of metals from the dying stars and the number of generations of stars succeeding each other. Now it is well known from general properties of stellar evolution that, from one side, the lifetime of a star strongly decreases with increasing mass, so that the number of generations of supergiant stars since the beginning of the universe is by far the largest in respect to that of any other star type; and from the other side, that, as supergiants end their lives as supernovae, they eject in space not only their outer unevolved envelopes as the smaller mass stars do, but also great part of their chemically evolved core, thus yielding to the outer medium the largest amount of metals.

From both these reasons it appears that supergiants are likely to be the main contributors to the chemical evolution and metal enrichment of the galactic medium; and therefore that the problem of evolution of galaxies is strictly dependent on the detailed assumptions concerning the evolution of such large mass stars, and in second instance on those concerning mass loss. Thus the impact of these assumptions on the chemical evolution of the universe has become a subject of great importance, located at the center of the interests of present day astrophysics.

Now, up to a short time ago, it did appear that on the whole the phenomenon could be understood according to some relatively simple physical models, and on this line much new evolutive work on large mass stars has been done. However, lately, some increasing doubts have arisen about the rightness of these interpretations and their capacity of really explaining what is observed; and quite recently, these doubts have grown up to the point of throwing the whole field in a somewhat confused and mistrustful state which is threatening to prevent much further advance on its issues. For this main reason, it may not be perhaps useless to try in the present moment to draw a general picture of the situation reducing it to its main essential lines, not so much in order to interpret the real meaning of the different kinds of discrepancies between observations and explanations which are at the source of the present day doubts, but rather in the intent to focus these discrepancies themselves, as the necessary starting point of any possible future clarification. The present paper, intended to present a very condensed summary of the different phases which the problem of mass loss from stars has been passing through since its first origin up to its present state, may be considered as a small step towards this aim.

The first evidence concerning stellar winds from evolved K and M giants and supergiants has been obtained by Deutsch (1960) as late as 1956, and soon after by Weymann (1962, 1963), by the detection of narrow

P Cygni profiles in the cores of strong resonance lines, interpreted as evidence of mass loss from the red giants, and by showing, through the analysis of the same distortions on the lines of the spectra of fainter companions, (in cases such as α Her and α Sco) that the flowing mass extends to several hundred stellar radii, where the rate of expansion, of the order of 10 to 30 km/sec, exceeds the local escape speed. A general review concerning evidence of stellar winds from red giants and supergiants through optical, infrared and radio studies, has been given more recently by Reimers (1975). The identified rates of mass loss range from 10^{-8} to 10^{-5} $M_\odot$/year with a marked tendency of increasing towards the upper right end corner of the H-R diagram. Through a careful analysis of the observational data, Reimers summarizes the main general properties of the phenomenon in an empirical formula :

$$M \sim \eta \, \frac{L}{gR} \sim \eta \, \frac{LR}{GM}$$

with a single value constant η equal about to 10^{-13} when all quantities, mass, radius, luminosity and gravitational acceleration are written in solar units. Moreover, it is generally surmised, owing to the large convective envelopes of the red giants and the moderate velocity of the outflowing mass, that the observed phenomenon could be interpreted according to the solar wind mechanism.

Most interest in stellar winds however in the past decade was stimulated by the discovery, by Morton (1969) and coworkers (Morton, Jenkins and Brooks, 1969), through ultraviolet observation in rocket flights, of outflows of high velocity (up to 3000 km/sec) from O and B supergiants, of the order of 10^{-6} to 10^{-5} $M_\odot$/year, revealed in this case by broad P Cyg shaped profiles on lines of a moderate state of ionization such as C_{III}. Much more extended data on the same subject have been later collected and discussed by different teams of workers (Barlow and Cohen, 1977; Morton and Wright, 1978; Lamers and Rogerson, 1978; Bruhweiler, Morgan and van der Hucht, 1978; Conti and Frost, 1977), from the whole of which it was early concluded that, owing to the high flow velocities and to the lack of superficial convective zones in hot stars, contrarily to the case of red giants, the solar wind mechanism could not be held responsible for the observed data. Therefore, since the beginning, a duality of interpretation for the facts concerning early and late type stars was implicitly assumed to exist, and up to a very recent time has been always implicitly maintained.

The main reason for the insistence on this duality has to be looked in the fact that, while for red stars the assumption of the similarity of the stellar to the solar wind did represent the least effort to interpret experimental evidence, a rather simple mechanism was devised to explain the already much more extensive data concerning early type stars, which did appear, according to an accurate analysis due to Barlow and Cohen (1977), to well fit the following formula :

$$\dot{M} = a \, L^{1,1\pm0,6},$$

(1)

where a is a constant. Now such a rough proportionality of $\dot{M}$ to the luminosity results from the line driven wind theory (Lucy and Salomon, 1970), and may be visualized in its main lines as follows.

One assumes a stationary state for the envelope of the star above the photosphere, consisting in the different layers expanding with a fixed velocity increasing outwards, and function only of distance from the photosphere. Then, any layer moving with the velocity $v(r)$ away from the photosphere, and containing some atoms with an absorption frequency ν_0, will absorb a blue shifted frequency from the continuum of the photosphere according to the Doppler effect relation

$$\frac{\nu - \nu_0}{\nu_0} = \frac{v(r)}{c} \, ,$$

so that the different layers moving with different velocities will absorb a whole band whose upper limit ν_∞ will correspond to the asymptotic value v_∞ of the velocity of the outermost border of the star. Then, the maximum momentum loss transmitted from the absorbed radiation to the outer gas layers will be of the order

$$\dot{M}v_\infty = \frac{L\nu_0(\nu-\nu_0)}{c} = \frac{L\nu_0 \, \nu_0 \, v_\infty}{c^2} \, ,$$

and for a single strong absorption line situated about at the maximum of the blackbody emission curve of the photosphere,

$$\dot{M} \sim L/c^2,$$

(2)

which for a B-0 main sequence star yields a value of $\dot{M} \sim 10^{-8}$ which is somewhat too low. But if instead of a single strong line one assumes several lines such that the ensemble of the bands $(\nu_\infty - \nu_0)$ corresponding to each of them covers all together the whole blackbody spectrum, then

$$\dot{M}v_\infty = \sum \frac{(\nu_\infty - \nu_0)L\nu_0}{\nu_0 \quad c} \sim \frac{L}{c} \, ,$$

(3)

and hence

$$\dot{M} = \frac{L}{v_\infty c} \, ,$$

which is about a hundred time larger, that is of the right order of magnitude observed.

A detailed theory has been elaborated by Castor, Abbott and Klein (1975) to handle quantitatively the phenomenon, on the following lines: a) assumptions of mass conservation and momentum conservation for the gas outflow in a stationary configuration ($\dot{M}$ constant, $v(r)$ independent of time); b) determination of the velocity field as solution of a differential equation to which the behaviour of increasing velocity with distance from the photosphere and some other boundary conditions are imposed. These conditions lead to a prescription for the mass loss rate, given by the formula :

$$\dot{M} = \frac{L}{C \ v_{th}} \ \frac{\alpha}{\Gamma} \ [\frac{1-\alpha}{1-\Gamma}]^{(1-\alpha)/\alpha} \ (kT)^{1/\alpha} \ , \tag{4}$$

where v_{th} is the thermal velocity, C and α are two constants, and

$$\Gamma = \frac{\sigma L}{4\pi GcM} \quad \text{(with } \sigma = \text{Thomson scattering cross section)},$$

is also roughly constant on the surface layers. It is seen that $\dot{M}$ turns out to be directly proportional to the luminosity, although the absolute value is dependent on the constants whose order of magnitude can only be guessed or postulated by the theory and are therefore expected to be determined by comparison with observation. Roughly speaking, both the agreement with data of the order of magnitude of the phenomenon given by (3), and the good fitting of the theoretical prediction (1) with (4) are the main factors responsible for the general adoption of the line driven wind theory as a general explanation of mass loss from early type stars. One therefore expects that in this case, mass loss rate should depend on luminosity only, with no regard for other stellar parameters.

For the late type stars instead, as already mentioned, the presence of extended convective envelopes was a priori supporting the solar wind model, in the sense that convection should generate mechanical acoustic waves in the plasma, which, propagating outwards should become shock waves heating the outer layers and the corona, with a subsequent escape of high velocity particles whose bulk will constitute the stellar wind observed.

An application of this line of thought to the evolution in the red phase of low mass stars has been tentatively undertaken by Fusi-Pecci and Renzini (1975), with the adoption of a mass loss rate given by a Reimers type formula :

$$\dot{M} = \eta \ \frac{R \ L_{ac}}{GM} \ , \tag{5}$$

where however L_{ac} is taken to be the acoustic luminosity generated by turbulent convective motions and given by the formulae of Proudman (1952) and Lighthill (1952) theory as :

$$L_{ac} = \frac{1}{2} \int P_{ac} \ dv \qquad P_{ac} = 40 \ \frac{\rho}{\lambda} \ \frac{v_c^8}{v_s^5} \ ,$$

where ρ is the density, λ the mixing length of the convection, v_c the average convective velocity and v_s the sound velocity. By using these formulae for determining the evolutive tracks during the first and the

second ascent of population II stars to the red giant tip, Fusi-Pecci
and Renzini were able to show that a single value for the constant η in
(5) did yield the right value for four important characteristics of the
evolution of low mass stars. This success could of course be considered
as an indirect proof of the validity of the solar wind model; so that, on
the whole, all data collected up to that point concerning mass loss could
be considered as sufficiently well accounted for by the pair of theories
proposed, each one valid in its own range of stellar effective temperatures.

As stressed however since the beginning of this paper, and as directly deduced from the first interpretative successes of formulae (1)
and (4), the most characteristic field where mass loss phenomena were
expected to play an important role, whose impact on the general field of
astrophysics should be most determining, was the evolution of high luminosity stars; and in fact since about five years ago, a number of papers
devoted to this increasingly developing subject was to appear (Chiosi
and Nasi, 1974; de Loore et al., 1977; Chiosi, Nasi and Sreenivasan,
1978; Czerny, 1979; Dearborn et al., 1978).

At first sight, it could seem that the results obtained by the different authors could hardly be compared, as different formulae for the
mass loss rate for the O and B evolving stars have been assumed. If
however one considers that, whichever formula one employs, the exclusive
dependence on the luminosity is always the same, and that in both cases
the formula contains normalization parameters, it is easily understood
that the only difference between different approaches consists in the
normalization value, which, not being deducible a priori from the theory,
may be determined in order to best fit the observed H-R diagram for supergiants. The main line of thought is then to calculate evolutionary supergiant tracks with different mass loss constants and to compare the results
with the experimental data.

Let us first consider the changes introduced by mass loss on the
standard conservative evolution of large mass stars. As is generally
well known, there has been for several years in the case of conservative
evolution for supergiants a double uncertainty concerning the convection
criterion to be applied and the relative importance of neutrino losses.
Nowadays, the existence of the electron neutrino weak interaction is almost universally assumed, and for convection, the Schwarzschild criterion
is generally preferred. In such a frame of assumptions, conservative
evolution of large mass stars can be characterized by the following facts:

i) Above about 12 $M_{\odot}$, central core helium ignition starts in the
blue region of the HR diagram, just after the star has left the main sequence (central H-burning phase region).

ii) The crossing to the red region of the diagram occurs slowly,
according to a nuclear time scale.

iii) The further phases of evolution (shell He-burning and more advanced burnings) are considered to occur in the red region up to the presupernova stage.

It is also well known that this pattern of evolution did present discrepancies in respect to the experimental composite diagram (of Humphreys, 1970), for supergiant stars, not easily understandable in the conservative frame of evolution.

The main results concerning mass losing evolution - following more closely the Chiosi et al. (1978) results, which for some time have been the most extended ones, as the evolution was pursued beyond the red phase, using for it the Fusi-Pecci and Renzini type formula - may be condensed, in agreement also with the finding of other authors, in the following main propositions :

a) The introduction of mass loss changes the internal structure of the star during the core H-burning phase, in the sense that the semiconvection and full convection intermediate zones practically disappear; a kind of evolution, mimicking the one obtained for lower mass stars when convection is treated according to the Ledoux criterion, may then be expected a priori, causing the star to make a loop in the HR plane which reconducts it to the blue region after having reached the red one when core H-burning has been completed.

b) This fact mostly occurs as a consequence of the simultaneous behaviour of two competing factors acting in an opposite sense in the H-R diagram. From one side the normal trend towards right (from blue to red), which for the lower range of supergiants (20 to 40 $M_\odot$) is the dominant one, and even displaces towards right the evolutionary track calculated with mass loss in respect to the conservative one; from the other, the losing of mass, peeling off the outer layers, tends to uncover the evolved helium core, and the neighbouring of helium layers to the surface has the immediate effect of pushing backwards the evolving star to the left (from red to blue). The importance of this effect, in part due to the very heavy amount of mass loss experienced by the star in the red region of the diagram, is strongly increasing with the mass, so that in a larger supergiant range (60 to 80 $M_\odot$) the turning back due to the peeling off of the outer layers occurs before the star reaches the red region, the earlier of course the larger is the mass loss rate; while for the largest mass stars, they turn back in their evolution even before the end of the core H-burning phase, and may go as far as to reach the left of the main sequence.

c) The core helium burning is generally ignited on the return track phase after its outmost tip towards the right has been reached.

Now when comparing the preceding results with the experimental composite diagram of all supergiants (Humphreys, 1970), the following remarks may be made :

a) Experimental points on the whole fit much better the theoretical helium burning zone calculated with a rather large rate for mass loss, than those calculated with smaller rates or no loss at all.

b) One can draw, in the diagram, a strip inclined as the main sequence and to its right, which represents the locus of the utmost right extremity reached by the evolutionary tracks. It turns thus out that large mass supergiants are confined in the blue region and never become red, and in a similar way, no Cepheids with very large periods are expected to be found. These predictions are in fact in agreement with observed facts.

c) The strong peeling off of the outer hydrogen layers results in the possibility of the formation of helium stars with very small or no hydrogen content : these could be identified with the Wolf-Rayet stars, - it is believed nowadays that single Wolf-Rayet stars do exist - whose origin therefore should quite naturally be justified in this scheme, and which happen to lay in the HR diagram just where the strongly hydrogen peeled off models are expected to lie. WN stars would then be those for which the peeling off has discovered the CNO cycle processed layers, while WC stars would correspond to more advanced states when the 3α cycle processed layers have already been uncovered.

Without entering into a number of finer details which on the whole should contribute to strengthen and make appear more plausible the present scenario, it appeared at this point almost safe to conclude that the main characteristics of the mass loss phenomenon had been rightly understood and that one was on the right way concerning the interpretation of evolution for large mass stars.

It is only about two years ago that a new set of data concerning mass loss from early type stars, collected by different authors (Conti and Garmany, 1980; de Loore et al., 1977; Lamers et al., 1980) has quite unexpectedly upset this apparently settled situation. The main reason is that while up to that moment most data considered were referred only to early type stars evolving beyond the main sequence, as are OI and Of stars, the new data did contain also several main sequence ones, and even at first sight, for equal total luminosity, the rate of mass loss for main sequence stars turned out to be between one or two orders of magnitude lower than for evolved ones; generally speaking, $\dot{M}$ did increase with the amount of evolution already undergone by the star, from OV to OI, then to Of and finally, largest of all, for WR stars. This unexpected discovery has been the starting point for the development of two main lines of approach, which have arisen in order to try to get over the incipient contradiction between data and theory :

a) From one side, a widespread belief has been gradually growing that the line driven wind theory, even when accepting that radiation forces do in fact modulate the flow, is unsufficient in itself to define its value; and in order to attain this aim, first some older theories, intended both to complete or to substitute the line driven wind mechanism, and which up to that time had been somewhat overlooked, have begun to be actively reconsidered: in particular, the basic approach of Cannon and Thomas (1977), surmising that any perturbation of the velocity field must lead to a general instability of the star surface, and locating the origin of the matter flux and of the formation of the chromosphere, the corona and the emission lines in some subphotospheric layers acting as a reservoir of non thermal kinetic energy; however the complication of the considered phenomena is such that no simple applicable result for calculating the mass loss rate could be derived. Then some further elaborate models have been proposed such as the warm corona model of Lamers and Rogerson (1978), or the small hot corona model proposed by Hearn (1975) and Cassinelli et al. (1978); finally, a completely new approach, the statistical model of Andriesse (1979), which assumes that mass loss, whichever its real origin, occurs in a discontinuous way according to some stochastic process, and as a consequence of general laws of thermodynamics, succeeds in deriving a rather simple formula for the rate :

$$\dot{M} = \{\frac{T_D}{T_K}\}^{1/2} \frac{LR}{GM} \sim G^{7/4} L^{3/2} \{\frac{R}{M}\}^{9/4} \, , \tag{6}$$

T_D and T_K being respectively the free fall time and the Kelvin time, justifying the observed fact of different rates for the same luminosity. Generally speaking, all these different approaches share the common tendency of rejecting the up to now assumed dichotomy of interpretation for the cause responsible of the shedding of mass in early and late type stars, and of considering this shedding as a single phenomenon acting in all cases.

b) From the other side, independently from any purely theoretical stimulus, the new data have provoked the elaboration of some phenomenological analysis, leading to the derivation of a star parameter dependent expression for the mass loss rates, such as to give due account of the changes caused by evolution. And one can say that an indication of success for both lines of approach could be found in the fact that the first deduced phenomenological formula by Chiosi (1980a) could be shown to possess about the same dependence on stellar parameters as reached in (6) by the statistical approach of Andriesse (1979).

However, the optimism due to this coincidence has soon been damped by the early recognition that such a success was mainly a consequence of an inconsistency. The deduction of the phenomenological Chiosi formula was obtained by attributing, to the stars employed for deriving it, mass values calculated with evolutionary tracks computed using a mass loss rate as given by (1) or (4). Now as the true rate was found to be given by (6), it is necessary as a first task to recalculate supergiant evolutionary tracks with the help of (6), and recalibrate the mass of the test stars in respect to these new tracks. This work in fact has recently been done*, and as now mass loss instead of being practically constant -

_____

* In course of publication on the IAU Colloquium No. 59 on "Effects of Mass Loss on Stellar Evolution", Trieste 1980.

because evolution occurs at about constant luminosity – during the whole of the blue phase, as was the case with (1), was now much lower for the early post-main sequence evolution than later, these new tracks have turned out to be quite different from the preceding ones, and much more akin to those obtained with the conservative evolution. Although it could be shown by Chiosi and Greggio (1980) that even this new aspect could be reconciled with the observational one of the composite HR diagram of super-giants, still it turned out that mass values of stars calculated with the new tracks did result to be rather different from those obtained pre-viously, and as a consequence the new phenomenological formula derived using these new mass values (Lamers, 1980) did not fit any more the An-driesse theoretical result.

It may well turn out that a converging series of successive approx-imations will be needed in order to arrive at a self consistent system of evolutionary tracks such that the rate of mass loss employed to con-struct them should coincide with the one deduced from the analysis of star data calibrated with mass values as deduced from these same tracks. But for the moment this refined self consistency has not been reached.

The present disagreement between data and any of the existing the-ories has mainly contributed to increment a growing feeling of mistrust concerning the whole interpretation of mass loss phenomena which has, paradoxically enough, essentially started with the increasing improvement, both in number and quality, of experimental data. I think that a brief summary of the main points implied in this present crisis does constitute the most up to date status concerning the general field of mass loss from stars.

Three main types of disagreements may first of all be recognized among the most recent data presented by different authors and particularly by Conti (1980) and Lamers (1980).

a) Both authors put into evidence a dependence of the mass loss rate not only from the luminosity, but also from the luminosity class. How-ever, they disagree among each other as to the dispersion of this depen-dence; the variation of it for stars with the same luminosity being of the order of a factor 100 according to Conti, and only of a factor 10 ac-cording to Lamers.

b) The second difficulty connected with the improved quality and quantity of data refers to the as yet unexplained occasional discrepancies obtained for the same star by different methods of observation; the spreads are frequently so large as to prevent any physical sense to a mean value; therefore one may nowadays be compelled to foresee the necessity of a general revision of all the main assumptions which underline the different methods. However these occasional changes could be due not to the dif-ferent technique, but to an intrinsic variability of the mass loss rate of the star.

c) A third fundamental difficulty appears to be connected to the fact that sometimes rather similar stars seem to behave quite differently in respect to the detailed profiles of their spectral lines and therefore presumably concerning their regime of mass loss. Such apparently random discrepancies, which of course, also in this case, could be due to intrinsic variability of the source, may further be caused by a list of other possible unresolved issues : non-shpericity and non-homogeneity in the emission of the wind, role of the source of the X rays sometimes observed, instability of the flow, rotation, turbulence, convection, and magnetic fields; the consideration of each of which bringing automatically drastic changes in the whole aspect of stellar winds.

In front of these apparent inconsistencies, upsetting the general asset reached for mass loss phenomena, a number of increasing reactions concerning their real significance has been gradually developing and acquiring impact, in such a way as to ultimately converge towards two main extreme attitudes which, roughly speaking, are marking the outmost borders of the spectrum of opinions presently shared by the different scientists involved in that field of research. At one of these extremities, we have the view, mostly supported by Thomas, that the whole scenario of mass loss, as shown by the observed inconsistencies with current theories, is almost entirely grounded on surface phenomena, that is on subphotospheric energizing mechanisms possibly generated by rotation, vorticity, convection or magnetic fields, unknown up to now in their effective interplay but, in any case, with not much, or even not at all, relation to basic stellar quantities; while the other extreme point of view, following mostly Lamers, stresses that although some irregularities, due to possible surface phenomena, cannot be denied, still the main influence in determining the rate of mass loss has to be reconducted as before to the dependence on luminosity, gravity, radius and mass. Of course both views rely on some important experimental evidence which they both tend to focus at exclusion of all others; for the Lamers view, variability which influences so much line profiles, does not necessarily affect as much the value of the mass loss rate, which should change much less than the structure of the envelope; while for the Thomas view, stars with quite similar main stellar parameters, but different line behaviour have completely different rates of mass loss.

Now are these two positions really incompatible ? and must one be really compelled to choose between a theory in which mass loss depends only on stellar main parameters and another one for which the dependence is only due to surface phenomena? Is it not likely from one side that complex physical facts should be function of more degrees of freedom than assumed up to now by any of the present schemes? And conversely even if the present day discrepancies have put forward the suspicion that many not sufficiently considered phenomena - previously mentioned - have not enough been stressed, why should not mass loss depend both on these different possible supplementary causes, and still also on the main stellar parameters? Are surface phenomena not at least partially also dependent on these parameters ?

If, leaving now by side any abstract theoretical discussion on the

real cause of the mass loss phenomenon in itself, one just tries to express its rate in any possible parametrized form and use it as a tool to
refine the agreement between theoretical evolutionary tracks of supergiants and the observed data, is it possible to effectively find a true
interaction between stellar winds and stellar evolution, in the sense
that a given assumed rate of mass loss should lead to a better agreement
of theory with data, and conversely that such a possible agreement should
be able to put experimental constraints on the rate of mass loss ?

The answer to such a question is yes according to several research
workers in the field (Chiosi, 1980b; Maeder, 1980) the most sensible
points for the fitting of this mutual calibration being :
a) the location of the He burning phase of supergiants in the HR diagram,
which, by changing $\dot{M}$, can be moved from left to right and vice-versa and
made to coincide best with the most crowded area of data.
b) the Wolf-Rayet stars, whose position can be fitted rather well with
the H-envelope peeled off configurations obtained in the advanced evolution of large mass stars of different range.
c) the relative number of Wolf-Rayet stars and red supergiants which appears to strongly depend on the mass range and on the composition of the
stellar population whose evolution is considered.

I think that these results are important enough to support the feeling that although present day discrepancies observed are a test that the
mass loss phenomenon is much more complicated than initially surmised,
and dependent on many more physical factors than previously assumed, still
a general tendency of its basic dependence on general stellar characteristics cannot be denied, which,even without being able in the present
moment to precise its exact form, can be used at least as a first approximation for evaluating mass loss effects on the history of stars.

REFERENCES

Andriesse, C.D. : 1979, Astrophys. Space Sci. 61, 205.
Barlow, M.J., Cohen, M. : 1977, Astrophys. J. 213, 737.
Bruhweiler, F.C., Morgan, T.H., Hucht, K.A. van der : 1978, Astrophys.
        J. Letters 225, L71.
Cannon, C.J., Thomas, R.N. : 1977, Astrophys. J. 211, 910
Cassinelli, J.P., Olson, G.L., Stalio, R. : 1978, Astrophys. J. 220, 573
Castor, J.I., Abbott, D.C., Klein, R.J. : 1975, Astrophys. J. 195, 157.
Chiosi, C. : 1980a, Astron. Astrophys.
Chiosi, C. : 1980b, in Effects of Mass Loss on Stellar Evolution, IAU
        Coll. No. 59 (ed. by C. Chiosi and R. Stalio), D. Reidel, Dordrecht-
        Holland.
Chiosi, C., Greggio, L. : 1980, preprint (submitted to Astron. Astrophys.).
Chiosi, C., Nasi, E. : 1974, Astron. Astrophys. 34, 355.
Chiosi, C., Nasi, E., Sreenivasan, S.R. : 1978, Astron. Astrophys. 63,
        103.
Conti, P.S. : 1980, in Effects of Mass Loss on Stellar Evolution, IAU
        Coll. No. 59 (ed. by C. Chiosi and R. Stalio) D. Reidel,Dordrecht-

Holland.

Conti, P.S., Frost, S.A. : 1977, Astrophys. J. 212, 728.

Conti, P.S., Garmany, C.D. : 1980, Astrophys. J. 238, 190.

Czerny, M. : 1979, Acta Astron. 29, 1.

Dearborn, D.S.P., Blake, J.B., Hainebach, K.L., Schramm, D.N. : 1978, Astrophys. J. 223, 552

Deutsch, A.J. : 1956, Astrophys. J. 123, 210.

Deutsch, A.J. : 1960, in Stellar Atmospheres, Univ. of Chicago Press, Chicago, p. 543.

Fusi-Pecci, F., Renzini, A. : 1975, in Problèmes d'Hydrodynamique Stellaire, Université de Liège, p. 383.

Hearn, A.G. : 1975, Astron. Astrophys. 40, 355.

Humphreys, R.M. : 1970, Astrophys. Letters 6, 11.

Lamers, H.J.G.L.M. : 1980, in Effects of Mass Loss on Stellar Evolution, IAU Coll. No. 59, (ed. by C. Chiosi and R. Stalio), D. Reidel, Dordrecht-Holland.

Lamers, H.J.G.L.M., Paerels, F.B.S., Loore, C. de : 1980, Astron. Astrophys. 87, 68.

Lamers, H.J.G.L.M., Rogerson, J.B. : 1978, Astron. Astrophys. 66, 417.

Lighthill, M.J. : 1952, Proc. Roy. Soc. London A 211, 564.

Loore, C. de, De Grève, J.P., Lamers, H.J.G.L.M. : 1977, Astron. Astrophys. 61, 251.

Lucy, L.B., Salomon, P.M. : 1970, Astrophys. J. 159, 879.

Maeder, A. : 1980, in Effects of Mass Loss on Stellar Evolution, IAU Coll. No. 59 (ed. by C. Chiosi and R. Stalio), D. Reidel, Dordrecht-Holland.

Morton, D.C. : 1969, Astrophys. J. 147, 1017.

Morton, D.C., Jenkins, F.B., Brooks, N. : 1969, Astrophys. J. 155, 875.

Morton, D.C., Wright, A.E. : 1978, Monthly Notices Roy. Astron. Soc. 182, 478.

Proudman, L.: 1952, Proc. Roy. Soc. London A 214, 119.

Reimers, D. : 1975, in Problèmes d'Hydrodynamique Stellaire, Université de Liège, p. 369.

Weymann, R. : 1962, Astrophys. J. 136, 844.

Weymann, R. : 1963, Ann. Rev. Astron. Astrophys. 1, 97.

CONTRIBUTION TO FOURIER ANALYSIS OF THE LIGHT CURVES OF ECLIPSING VARIABLES

Zdeněk Kopal
Department of Astronomy, University of Manchester
England

ABSTRACT

The aim of the present paper will be to evaluate analytically the weights needed to express the moments A_{2m} of the light curves – which constitute the basic quantities from which all elements of the respective eclipsing system can subsequently be evaluated as weighted means of the coefficients a_n of an empirical Fourier cosine series representing the observed light curve. It will, in particular, be shown that the weight coefficients ψ_i defining even moments (A_0, A_2, A_4, A_6, ...) are expressible in a closed form in terms of simple trigonometric functions.

1. INTRODUCTION

A systematic application of Fourier methods to an analysis of the light changes of eclipsing binary systems, developed by the present writer and his school since 1974 and summarized in a recent book (Kopal, 1979a), led to the introduction of quantities generally referred to as "moments of the light curves" A_{2m}, defined by the equation

$$A_{2\mu} = \int_0^{\theta_1} (1 - \ell) \, d(\sin^{2\mu}\theta), \tag{1.1}$$

where $\ell(\theta)$ stands for the (fractional) light of the system, varying with the phase θ in the range $0 \leqslant \theta \leqslant \theta_1$; and μ is an arbitrary constant (not necessarily an integer).

The moments $A_{2\mu}$ so defined are unique and finite for any value of $\mu > -0.5$; and their analytic relationship with the geometrical elements of the eclipse which causes the light $\ell(\theta)$ to vary have been established elsewhere (cf., in particular, Chapter V of Kopal (1979a) and references quoted therein). Moreover, for distinct values of μ the $A_{2\mu}$'s are linearly independent; and a knowledge of these moments for n discrete (not necessarily equidistant) values of μ should, therefore, enable us to

233

E. G. Mariolopoulos et al. (eds.), Compendium in Astronomy, 233–245.

specify n different elements of the respective system (such as the inclination i of its orbital plane to the celestial sphere; the fractional radii $r_{1,2}$ of its components; the luminosity L_1 of the star undergoing eclipse; its degree of limb-darkening, etc.).

It will not be the aim of the present paper to detail the steps by which this can be done; for the underlying theory was already expounded elsewhere (Kopal, 1979a); and for more practical aspects of this work cf. Kopal (1982). Instead, we wish in what follows to focus attention on the first – and, in many ways the most crucial – step of such a process on which everything else depends: namely, a determination of the "empirical" values of the moments $A_{2\mu}$ of the light curves from the available observations.

In early stages of the development of our subject (e.g., Kopal, 1975a) a suggestion was put forward to resort at this stage to planimetry: i.e., to plot (on a suitable scale) $\ell(\theta)$ versus $\sin^{2\mu}\theta$ and to approximate the integral on the right-hand side of equation (1.1) by "counting squares" of the area subtended by the light curve. Such a procedure has indeed been practised extensively in the past (cf., e.g. Jurkevich et al., 1976), with fairly satisfactory results; and always works if the computer will !

However – no matter how diligent or skilful the latter may be – the outcome is bound to suffer from the following limitations inherent in the process: namely,

(i) in order to perform such quadratures, the actual observations within minima must be represented by a smooth curve drawn by free hand to follow the course of observed points; and

(ii) the areas in question may be systematically vitiated by any error in the estimated unit of light. In consequence, mere graphical work to determine the $A_{2\mu}$'s can furnish plausible values of the respective moments of the light, but cannot guarantee that these are the most probable ones; nor can it provide quantitative indications of the uncertainty within which such moments are defined by available observed data.

In the face of such a situation, the writer proposed more recently (cf. Kopal, 1979b; cf. also Section VII-3 of Kopal, 1979a) to represent the observed light changes $\ell(\theta)$ – if symmetrical with respect to $\theta=0$ – by an empirical Fourier series of the form

$$1 - \ell = \frac{1}{2}a_0 + \sum_{n=1}^{\infty} a_n \cos(n\pi\theta/\theta_1) \tag{1.2}$$

where the loss of light $1-\ell$ on the left-hand side at a time corresponding to the phase-angle θ is given to us by the observer, and the a_n's are the coefficients of terms varying with the phase as $\cos(n\pi\theta/\theta_1)$.

As many equations of condition of the form (1.2) can obviously be set up as there are observed points disclosing the light changes; and being linear in the a_n's, can be solved by the method of least squares to furnish the most probable values of the Fourier coefficients a_n, together with their uncertainty. In theory, the number of terms on the right-hand side of (1.2) is infinite to make it a true equation; but, in practice, the series can be truncated after (say N) terms if the coefficients a_n for $n>N$ turn out to be insignificant. The neglected tail of the series represents then the "noise" inherent in our problem; with the level at which it sets in depending on the quality of the underlying data (i.e., the accuracy of the observations used to specify the left-hand sides of the "equations of condition" (1.2) on which our solution for the a_n's was based).

The techniques by which such solutions can be obtained constitute a well-known part of the "calculus of observations"; and need not, therefore, be reproduced in this place. Students of eclipsing variables may find a full account of the requisite arithmetical operations e.g., in the Appendix to Chapter VI of Kopal (1959). Moreover, the writer showed recently (Kopal, 1979a,b) that the Fourier coefficients a_n on the right-hand side of equation (2) are related with the moments A_{2m} as defined by equation (1.1) by a linear form

$$A_{2m} = \sum_{n=0}^{\infty} \psi_n^{(m)}(\theta_1)\, a_n, \qquad\qquad (1.3)$$

where the $\psi_n^{(m)}(\theta_1)$'s are weight coefficients depending on the upper limit θ_1 of integration on the right-hand side of equation (1.1).

In his 1979 work the present writer established the explicit expressions for $\psi_n^{(\mu)}$ in the form of doubly-infinite expansions in trigonometric functions of θ_1, which converge none too rapidly – in fact, so slowly that a sum of more than a hundred terms proved necessary in some cases to stabilize their values correctly to 1% (for fuller details cf. Jurkevich et al., 1981). This remains, however, true only for unrestricted values of the index μ. Should, however, the latter be zero or a positive integer – the cases of primary practical importance – the writer found recently that the recalcitrant series for $\psi_n^{(m)}$ can be summed up analytically to yield expressions of transparent simplicity, obviating any need of machine computation; and the aim of the present paper will be to acquaint its readers with the results of this work.

2. RELATION BETWEEN THE MOMENTS $A_{2\mu}$ AND COEFFICIENTS a_n

Before we come to do so, let us approach our task by setting up the relation between $A_{2\mu}$ and a_n in algebraic form. At first sight, the task would appear to be easy : for by inserting in equation (1.1) for $1-\ell$ from equation (1.2) we find indeed the outcome to be of the form (1.3), where

$$\psi_n^{(m)}(\theta_1) = \frac{\varepsilon_n}{2} \int_0^{\theta_1} \cos h\theta \, d(\sin^{2\mu}\theta),$$ (2.1)

where

$$\varepsilon_0 = 1 \quad \text{but} \quad \varepsilon_n = 2 \quad \text{for } n > 0,$$ (2.2)

stands for the customary Neumann factors, and where we have abbreviated

$$h = \frac{n\pi}{\theta_1} \, .$$ (2.3)

For any value of this parameter (whether integral or fractional),

$$\cos h\theta = \frac{\sin \pi h}{\pi h} \sum_{m=0}^{\infty} \frac{(-1)^m \, \varepsilon_m \, h^2}{h^2 - m^2} \cos m\theta.$$ (2.4)

Moreover, for any value of μ (whether integral or not),

$$\sin^{2\mu}\theta = \frac{\Gamma(2\mu+1)}{2^{2\mu-1}} \sum_{j=0}^{\infty} \frac{(-1)^j \sin(2j+1)\theta}{\Gamma(\mu+j+\frac{3}{2})\Gamma(\mu-j+\frac{1}{2})} \, ;$$ (2.5

If, moreover, equations (2.3) and (2.4) are inserted in (2.1), the weight coefficients $\psi_n^{(\mu)}(\theta_1)$ can be expressed as

$$\psi_n^{(\mu)}(\theta_1) = \frac{\Gamma(2\mu+1)}{2^{2\mu-1}} \frac{\sin \pi h}{\pi h} \sum_{j=0}^{\infty} \sum_{m=0}^{\infty}$$ (2.6)

$$\frac{(-1)^{j+m} (2j+1)}{\Gamma(\mu+j+\frac{3}{2})\Gamma(\mu-j+\frac{1}{2})} \frac{\varepsilon_m h^2}{h^2 - m^2} \int_0^{\theta_1} \cos m\theta \, \cos(2j+1)\theta \, d\theta,$$

where

$$\int_0^{\theta_1} \cos m\theta \, \cos(2j+1)\theta \, d\theta =$$

$$= \frac{\sin(2j+m+1)\theta_1}{2(2j+m+1)} + \frac{\sin(2j-m+1)\theta_1}{2(2j-m+1)} \quad \text{if } m \neq 2j+1,$$

$$= \frac{\theta_1}{2} + \frac{\sin 2m \, \theta_1}{4m} \quad \text{if } m = 2j+1.$$ (2.7)

The foregoing expression for $\psi_n^{(\mu)}(\theta_1)$ as represented by equations (2.6)-(2.7), while explicit, is only of theoretical interest; being too complicated for practical work. In what follows we propose, therefore, to re-derive $\psi_n^{(\mu)}$ by a different method, which - while seemingly more complex - leads to Fourier expansions that can be summed up in a closed form.

In order to do so, let us depart from the well-known fact that the coefficients a_n of the Fourier cosine series on the right-hand side of equation (1.2) are identical with discrete values of the Fourier cosine transform

$$F(\nu) = 2 \int_0^{\theta_1} (1-\ell) \cos(2\pi\nu\theta)\, d\theta \qquad (2.8)$$

of the light changes $1-\ell$, such that

$$a_n = \frac{1}{\theta_1} F(\frac{n}{2\pi}) \equiv \frac{1}{\theta_1} F(\frac{h}{2\pi}) . \qquad (2.9)$$

On insertion for $1-\ell$ from (1.2) in (2.8) it follows that

$$F(\nu) = \sum_{n=0}^{\infty} \epsilon_n a_n \int_0^{\theta_1} \cos h\theta \cos k\theta\, d\theta, \qquad (2.10)$$

where we have abbreviated

$$k = 2\pi\nu; \qquad (2.11)$$

and where the Neumann factor ϵ_n continues to be given by equation (2.2).

On the other hand, the moments $A_{2\mu}$ of the light curves as defined by equation (1.1) can, by use of (2.5), be rewritten as

$$A_{2\mu} \equiv \int_0^{\theta_1} (1-\ell)\, d(\sin^{2\mu}\theta) \qquad (2.12)$$

$$= \frac{\Gamma(2\mu+1)}{2^{2\mu-1}} \sum_{j=0}^{\infty} \frac{(-1)^j}{\Gamma(\mu+j+\frac{3}{2})\Gamma(\mu-j+\frac{1}{2})} \int_0^{\theta_1} (1-\ell)\, d\{\sin(2j+1)\theta\};$$

and since

$$d\{\sin(2j+1)\theta\} = (2j+1)\cos(2j+1)\theta\, d\theta \qquad (2.13)$$

from (2.8) it follows that

$$\int_0^{\theta_1} (1-\ell)\cos(2j+1)\theta\, d\theta = \frac{1}{2} F(\frac{2j+1}{2\pi}). \qquad (2.14)$$

The frequency of the Fourier transform on the right-hand side of the preceding equation corresponds, by equation (2.11), to the value of

$$k = 2j + 1. \tag{2.15}$$

If so, however, it follows from equation (2.7) for $m \equiv h \equiv \pi n/\theta_1$ that, for $h \neq 2j+1$,

$$\int_0^{\theta_1} \cos h\theta \, \cos(2j+1)\theta \, d\theta = \frac{(-1)^n (2j+1)}{(2j+1)^2 - h^2} \sin(2j+1)\theta_1 \tag{2.16}$$

for $n = 0, 1, 2, \ldots$. Moreover, an insertion of (2.16) in (2.10) yields

$$F\left(\frac{2j+1}{2\pi}\right) = (2j+1) \, \sin(2j+1)\theta_1 \sum_{n=0}^{\infty} \frac{(-1)^n \varepsilon_n a_n}{(2j+1)^2 - h^2} \, ; \tag{2.17}$$

which for $h = 2j+1$ and $h\theta_1 = (2j+1)\theta_1 = n\pi$ reduces indeed to (2.9).

Lastly, of we combine equations (2.12) - (2.14) with (2.17), we find that

$$A_{2\mu} = \frac{\Gamma(2\mu+1)}{4^{\mu}} \sum_{j=0}^{\infty} \frac{(-1)^j (2j+1)^2 \sin(2j+1)\theta_1}{\Gamma(\mu+j+\frac{3}{2})\Gamma(\mu-j+\frac{1}{2})} \, \times$$

$$\times \sum_{n=0}^{\infty} \frac{(-1)^n \varepsilon_n a_n}{(2j+1)^2 - h^2} \equiv \sum_{n=0}^{\infty} \psi_n^{(\mu)}(\theta_1) \, a_n, \tag{2.18}$$

where the weight coefficients

$$\psi_n^{(\mu)}(\theta_1) = \frac{\Gamma(2\mu+1)}{4^{\mu}} \sum_{j=0}^{\infty} \frac{(-1)^{j+n} \varepsilon_n \sin(2j+1)\theta_1}{\Gamma(\mu+j+\frac{3}{2})\Gamma(\mu-j+\frac{1}{2})} \, \times$$

$$\times \frac{[(2j+1)\theta_1]^2}{[(2j+1)\theta_1]^2 - [n\pi]^2} \tag{2.19}$$

are now expressed in the form of a single Fourier sine series in odd integral multiples of θ_1.

The foregoing equations (2.18) or (2.19) are not new; they were obtained by the present writer (cf. Kopal, 1979a,b) in much the same way at an earlier date*. What has, however, not been discovered before is the fact that, for integral values of μ, the remaining series on the right-hand side of equation (2.19) can be summed up analytically; and, therefore, the entire expression for the weight coefficients $\psi^{(\mu)}(\theta_1)$ expressed in a closed form. This constitutes the task which will be carried out in the next section.

---

* Note only that in equations (2.17) - (2.19) as given in the above-quoted references the Neumann factor ε_n was inadvertently omitted.

3. EVALUATION OF THE WEIGHT FUNCTIONS $\psi_n^{(\mu)}(\theta_1)$

In evaluating the sums on the right-hand side of equation (2.19) we shall limit ourselves to the case for which the order μ happens to be a non-negative integer. The moments of the light curves of even integral orders, represented by the sequence A_0, A_2, A_4, A_6, ... , are of greatest practical importance for the solutions for the elements of eclipsing binary systems from their light curves; for they alone can be expressed in terms of these elements in a closed form (for details cf. Chapter V of Kopal, 1979a). In what follows we shall, therefore, limit ourselves to such a case; and to underline the fact that the values of μ will hereafter be restricted to the sequence 0, 1, 2, 3, ... of non-negative integers, we shall replace μ by m.

Within this restriction, equation (2.19) can be rewritten as

$$\psi_n^{(m)}(\theta_1) = \frac{(-1)^{m+2}}{\pi} \frac{(2m)!}{4^m} \sum_{j=0}^{\infty} \frac{\varepsilon_n \Gamma(j-m+\frac{1}{2})}{\Gamma(j+m+\frac{3}{2})} \times$$

$$\times \frac{(2j+1)^2 \sin(2j+1)\theta_1}{(2j+1)^2-h^2}, \tag{3.1}$$

where h continues to be given by equation (2.3). For m=0, the foregoing equation reduces, moreover, to

$$\psi_n^{(0)}(\theta_1) = (-1)^n \frac{2\varepsilon_n}{\pi} \sum_{j=0}^{\infty} \frac{(2j+1)\sin(2j+1)\theta_1}{(2j+1)^2-h^2}. \tag{3.2}$$

Since, however (cf., e.g., Oberhettinger, 1973; p. 13)

$$\sum_{j=0}^{\infty} \frac{(2j+1)\sin(2j+1)\theta_1}{(2j+1)^2-h^2} = (-1)^n \frac{\pi}{4} \tag{3.3}$$

for any value of m, and any angle θ_1 such that $0 < \theta_1 < \pi$.

Under these conditions, it follows on insertion from (3.3) in (3.2) that, for n=0,

$$\psi_0^{(0)}(\theta_1) = \frac{1}{2} ; \tag{3.4}$$

while for n > 0,

$$\psi_n^{(0)}(\theta_1) = 1. \tag{3.5}$$

Accordingly, equation (2.18) for m=0 assumes the form

$$A_0 = \frac{1}{2}a_0 + a_1 + a_2 + a_3 + ..., \tag{3.6}$$

in agreement with equation (1.2), which reduces at once to (3.6) for $\theta=0$ and $\ell(0)\equiv\lambda$; as it is known (cf., e.g., Kopal, 1976a) that $A_0=1-\lambda$.

It may also be of interest to note, in this connection, that for $m=n=0$, equation (3.2) reduces to

$$\psi_0^{(0)}(\theta_1) = \frac{2}{\pi}\{\sin\theta_1 + \frac{\sin 3\theta_1}{3} + \frac{\sin 5\theta_1}{5} + \ldots\} \qquad (3.7)$$

representing (cf. Lanczos, 1956; p. 221) the Fourier expansion of a "square wave", defined by the conditions

$$f(\theta) = \frac{1}{2} \quad \text{for } 0 < \theta < \pi ,$$

$$f(0) = f(\pi) = 0 , \qquad\qquad\qquad\qquad (3.8)$$

$$f(-\theta) = -f(\theta) ;$$

the first one of these conditions coincides with (3.4).

Note also that, for $n=0$, equation (3.1) reduces to

$$\psi_0^{(m)}(\theta_1) = \frac{1}{2}\sin^{2m}\theta_1 \qquad (3.9)$$

for any value of m (including zero); as could be easily verified by a glance at equation (2.1).

Next, let us turn our attention to the case when $m=1$. If so, equation (3.1) will assume the form

$$\psi_n^{(1)}(\theta_1) = (-1)^{n+1} 2! \frac{2\epsilon_n}{\pi} \sum_{j=0}^{\infty} \frac{2j+1}{(2j+1)^2-2^2} \cdot \frac{\sin(2j+1)\theta_1}{(2j+1)^2-h^2}. \qquad (3.10)$$

Since, however,

$$\frac{1}{(2j+1)^2-2^2} \equiv \frac{2j+1}{4}\{\frac{2j+1}{[(2j+1)^2-2^2]} - \frac{1}{2j+1}\}, \qquad (3.11)$$

Equation (3.10) can obviously be rewritten as

$$\psi_n^{(1)}(\theta_1) = \frac{1}{2}\psi_n^{(0)} - \frac{(-1)^n\epsilon_n}{\pi} \sum_{j=0}^{\infty} \frac{(2j+1)^3}{(2j+1)^2-2^2} \cdot \frac{\sin(2j+1)\theta_1}{(2j+1)^2-h^2}. \qquad (3.12)$$

If $m=2$, equation (3.1) yields

$$\psi_n^{(2)}(\theta_1) = (-1)^{n+2} 4! \frac{2}{\pi} \sum_{j=0}^{\infty} \frac{2j+1}{[(2j+1)^2-2^2][(2j+1)^2-3^2]} \frac{\sin(2j+1)\theta_1}{(2j+1)^2-h^2}.$$
$$\qquad (3.13)$$

which, since

$$\frac{1}{[(2j+1)^2-2^2][(2j+1)^2-4^2]} \equiv \frac{2j+1}{192}\left\{\frac{3}{2j+1} - \frac{4(2j+1)}{(2j+1)^2-2^2} + \frac{2j+1}{(2j+1)^2-4^2}\right\},$$

(3.14)

can be rewritten as

$$\psi_n^{(2)}(\theta_1)=\psi_n^{(1)}(\theta_1)-\frac{1}{8}\psi_n^{(0)}(\theta_1)+\frac{(-1)^n\varepsilon_n}{4\pi}\sum_{j=0}^{\infty}\frac{(2j+1)^3}{(2j+1)^2-4^2}\frac{\sin(2j+1)\theta_1}{(2j+1)^2-h^2}.$$

(3.15)

For $m = 3$, we have

$$\psi_n^{(3)}(\theta_1) = (-1)^{n+3}(6!)\frac{2}{\pi} \times$$

(3.16)

$$\times \sum_{j=0}^{\infty}\frac{\varepsilon_n(2j+1)\sin(2j+1)\theta_1}{[(2j+1)^2-2^2][(2j+1)^2-4^2][(2j+1)^2-6^2][(2j+1)^2-h^2]}.$$

Since, however,

$$\frac{1}{[(2j+1)^2-2^2][(2j+1)^2-4^2][(2j+1)^2-6^2]} \equiv \frac{2j+1}{23040}\left\{-\frac{10}{2j+1} + \frac{15(2j+1)}{(2j+1)^2-2^2} - \right.$$

$$\left. - \frac{6(2j+1)}{(2j+1)^2-4^2} + \frac{j+1}{(2j+1)^2-6^2}\right\},$$

(3.17)

it follows that

$$\psi_n^{(3)}(\theta_1) = \frac{3}{2}\psi_n^{(2)}(\theta_1) - \frac{9}{16}\psi_n^{(1)}(\theta_1) + \frac{\varepsilon_n}{64} - \frac{(-1)^n\varepsilon_n}{16\pi} \times$$

$$\times \sum_{j=0}^{\infty}\frac{(2j+1)^3\sin(2j+1)\theta_1}{[(2j+1)^2-6^2][(2j+1)^2-h^2]} ;$$

(3.18)

and similarly for higher values of m.

The remaining sums on the right-hand sides on equations (3.12), (3.15) and (3.18) are of the form

$$\sum_{j=0}^{\infty}\frac{(2j+1)^3\sin(2j+1)\theta_1}{[(2j+1)^2-(2m)^2][(2j+1)^2-h^2]} \equiv \frac{(2m\theta_1)^2}{(2m\theta_1)^2-(n\pi)^2}\sum_{j=0}^{\infty}\frac{(2j+1)\sin(2j+1)\theta_1}{(2j+1)^2-(2m)^2}$$

$$- \frac{(n\pi)^2}{(2m\theta_1)^2-(n\pi)^2}\sum_{j=0}^{\infty}\frac{(2j+1)\sin(2j+1)\theta_1}{(2j+1)^2-h^2} .$$

(3.19)

However (cf. Oberhettinger, op.cit., p. 13; equation 1.64),

$$\sum_{j=0}^{\infty} \frac{(2j+1)\sin(2j+1)\theta_1}{(2j+1)^2-(2m)^2} = \frac{\pi}{4}\cos 2m\theta_1 \tag{3.20}$$

and

$$\sum_{j=0}^{\infty} \frac{(2j+1)\sin(2j+1)\theta_1}{(2j+1)^2-h^2} = (-1)^n \frac{\pi}{4} \tag{3.21}$$

in accordance with equation (3.3.)

Therefore,

$$\sum_{j=0}^{\infty} \frac{(2j+1)^3}{(2j+1)^2-(2m)^2} \frac{\sin(2j+1)\theta_1}{(2j+1)^2-h^2} = \frac{\pi}{4}\left\{\frac{(-1)^n}{1^2-(2m/h)^2} + \frac{\cos 2m\theta_1}{1^2-(h/2m)^2}\right\}, \tag{3.22}$$

which on insertion in equations (3.12), (3.15) and (3.18) discloses that

$$\psi_n^{(1)}(\theta_1) = \frac{\varepsilon_n}{4} \frac{1-(-1)^n\cos 2\theta_1}{1^2-(h/2)^2} , \tag{3.23}$$

$$\psi_n^{(2)}(\theta_1) = \psi_n^{(1)} - \frac{\varepsilon_n}{16} + \frac{\varepsilon_n}{16}\left\{\frac{1}{1^2-(4/h)^2} + \frac{(-1)^n\cos 4\theta_1}{1^2-(h/4)^2}\right\} =$$

$$= \frac{\varepsilon_n}{4}\left\{\frac{1-(-1)^n\cos 2\theta_1}{1^2-(h/2)^2} - \frac{1}{4}\frac{1-(-1)^n\cos 4\theta_1}{1^2-(h/4)^2}\right\} ; \tag{3.24}$$

and

$$\psi_n^{(3)}(\theta_1) = \frac{3}{2}\psi_n^{(2)} - \frac{9}{16}\psi_n^{(1)} + \frac{\varepsilon_n}{64} - \frac{\varepsilon_n}{64}\left\{\frac{1}{1^2-(6/h)^2} + \frac{(-1)^n\cos 6\theta_1}{1^2-(h/6)^2}\right\} = \tag{3.25}$$

$$= \frac{15\varepsilon_n}{64}\left\{\frac{1-(-1)^n\cos 2\theta_1}{1^2-(h/2)^2} - \frac{2}{5}\frac{1-(-1)^n\cos 4\theta_1}{1^2-(h/4)^2} + \frac{1}{15}\frac{1-(-1)^n\cos 6\theta_1}{1^2-(h/6)^2}\right\} .$$

Since, moreover,

$$1 + \cos 2m\theta_1 = 2\cos^2 m\theta_1 \tag{3.26}$$

and

$$1 - \cos 2m\theta_1 = 2\sin^2 m\theta_1 . \tag{3.27}$$

Equations (3.23) - (3.25) can be rewritten as

$$\psi_{2\nu}^{(1)}(\theta_1) = \frac{\varepsilon_{2\nu}}{2} \frac{\sin^2\theta_1}{1^2-(\nu\pi/\theta_1)^2} \, , \tag{3.28}$$

$$\psi_{2\nu+1}^{(1)}(\theta_1) = \frac{\cos^2\theta_1}{1^2-[(2\nu+1)\pi/2\theta_1]^2} \, ; \tag{3.29}$$

$$\psi_{2\nu}^{(2)}(\theta_1) = \frac{\varepsilon_{2\nu}}{2} \left\{ \frac{\sin^2\theta_1}{1^2-[\nu\pi/\theta_1]^2} - \frac{\sin^2 2\theta_1}{2^2-[\nu\pi/\theta_1]^2} \right\} \, ; \tag{3.30}$$

$$\psi_{2\nu+1}^{(2)}(\theta_1) = \frac{\cos^2\theta_1}{1^2-[(2\nu+1)\pi/2\theta_1]^2} - \frac{\cos^2 2\theta_1}{2^2-[(2\nu+1)\pi/2\theta_1]^2} \, ; \tag{3.31}$$

and

$$\psi_{2\nu}^{(3)}(\theta_1) = \frac{3\varepsilon_{2\nu}}{32} \left\{ \frac{5\sin^2\theta_1}{1^2-[\nu\pi/\theta_1]^2} - \frac{8\sin^2 2\theta_1}{2^2-[\nu\pi/\theta_1]^2} + \frac{3\sin^2 3\theta_1}{3^2-|\nu\pi/\theta_1|^2} \right\} , \tag{3.32}$$

$$\psi_{2\nu+1}^{(3)}(\theta_1) = \frac{15}{16} \frac{\cos^2\theta_1}{1^2-[(2\nu+1)\pi/2\theta_1]^2} - \frac{3}{2} \frac{\cos^2 2\theta_1}{2^2-[(2\nu+1)\pi/2\theta_1]^2} +$$

$$+ \frac{9}{16} \frac{\cos^2 3\theta_1}{3^2-[(2\nu+1)\pi/2\theta_1]^2} \, ; \tag{3.33}$$

where $\nu = 0, 1, 2, \ldots$. For $\nu = 0$, the foregoing expressions (3.28), (3.30) and (3.32) reduce to (3.9).

4. RELATIONS BETWEEN THE MOMENTS A_{2m} AND FOURIER COEFFICIENTS a_n

With the aid of the results established in the preceding section we are now in a position to express the moments A_{2m} of the light curves, as defined by equation (1.1), in terms of the coefficients a_n of the Fourier expansion (1.2). Indeed, if we collect the expressions for the weight functions $\psi_n^{(m)}(\theta_1)$'s given in Section 3, we find that, in accordance with equation (1.3), for $m = 0, 1, 2, 3, \ldots$ we have

$$A_0 \equiv \sum_{n=0}^{\infty} \psi_n^{(0)}(\theta_1)a_n = \sum_{\nu=0}^{\infty} \frac{\varepsilon_\nu}{2} a_{2\nu} \, , \tag{4.1}$$

$$A_2 \equiv \sum_{n=0}^{\infty} \psi_n^{(1)}(\theta_1)a_n = \sum_{\nu=0}^{\infty} \left\{ \frac{\varepsilon_\nu}{2} \frac{a_{2\nu}\sin^2\theta_1}{1-[\nu\pi/\theta_1]^2} + \frac{a_{2\nu+1}\cos^2\theta_1}{1-[(2\nu+1)\pi/2\theta_1]^2} \right\} \, , \tag{4.2}$$

$$A_4 \equiv \sum_{n=0}^{\infty} \psi_n^{(2)}(\theta_1)a_n = \sum_{\nu=0}^{\infty} \frac{\varepsilon_\nu}{2} \left\{ \frac{\sin^2\theta_1}{1^2-[\nu\pi/\theta_1]^2} - \frac{\sin^2 2\theta_1}{2^2-[\nu\pi/\theta_1]^2} \right\} a_{2\nu} +$$

$$+ \sum_{\nu=0}^{\infty} \left\{ \frac{\cos^2\theta_1}{1-[(2\nu+1)\pi/2\theta_1]^2} - \frac{\cos^2 2\theta_1}{2^2-[(2\nu+1)\pi/2\theta_1]^2} \right\} a_{2\nu+1}, \qquad (4.3)$$

and

$$A_6 \equiv \sum_{n=0}^{\infty} \psi_n^{(3)}(\theta_1)a_n = \frac{3}{16} \sum_{n=0}^{\infty} \frac{\varepsilon_\nu}{2} \left\{ \frac{5\sin^2\theta_1}{1^2-[\nu\pi/\theta_1]^2} - \frac{8\sin^2 2\theta_1}{2^2-[\nu\pi/\theta_1]^2} + \right.$$

$$\left. + \frac{3\sin^2 3\theta_1}{3^2-[\nu\pi/\theta_1]^2} \right\} a_{2\nu} + \frac{3}{16} \sum_{\nu=0}^{\infty} \left\{ \frac{5\cos^2\theta_1}{1^2-[(2\nu+1)\pi/2\theta_1]^2} - \right.$$

$$\left. - \frac{8\cos^2 2\theta_1}{2^2-[(2\nu+1)\pi/2\theta_1]^2} + \frac{3\cos^2 3\theta_1}{3^2-[(2\nu+1)\pi/2\theta_1]^2} \right\} a_{2\nu+1}. \qquad (4.4)$$

The angle θ_1 in all the foregoing equations has been restricted (by the domain of convergence of the Fourier expansions used to obtain our results) to be less than 180°. In actual applications of these results to practical cases it will never be necessary to adopt for θ_1 a value greater than 90°. In fact, if the eclipsing system subject to analysis exhibits light variations, due to eclipses, only within the phase angles $\pm\theta_1$ around the times of conjunction, and its light $\ell(\theta)$ remains constant for $\theta > \theta_1$, no useful purpose would be served if a Fourier series of the form (1.2) were used to represent the observations beyond the angle of first contact of the eclipse, at which the variation of light will commence.

From the mathematical point of view, there is no reason not to adopt any arbitrary value of θ_1 in the arguments of periodic terms on the right-hand side of equation (1.2); for a Fourier cosine series of the form (1.2) can be used to represent, not only the variations of light within minima, but also the constancy of light between eclipses. However, should we adopt in (1.2) an arbitrary value for θ_1 well in excess of the angle of first contact of the eclipse, the asymptotic properties of the expansion on the right-hand side of equation (1.2) will worsen; and many more terms would have to be retained to obtain an equally good fit to the observed variations of light exhibited within minima, as well as its constancy between eclipses.

This all is true for systems consisting of spherical components, where the light changes are caused solely by the eclipses. For close binary systems whose components are distorted by mutual tidal action, variations of light (due to the photometric proximity effects) will persist throughout the whole cycle, and superpose upon those which may arise from eclipses. The light of such a system will never cease to vary in a continuous manner; and (cf. Kopal, 1975b, 1976b; Chapter VI of 1979a) the basic data for the solution for the elements will be moments of the light curve of the form (1.1) in which the upper limit θ_1 of integration has been extended to 90°.

If so, equations (4.1) - (4.4) will continue to express the corresponding moments A_{2m} of the light curves in terms of the Fourier coefficients a_n of the expansion on the right-hand side of equation (1.2) where we have likewise set $\theta_1 = 90^\circ$ (rendering the argument of its periodic terms to be equal to $2n\theta$). If we insert in equations (4.1) - (4.4) $\theta_1 = 90^\circ$ on their right-hand sides, these equations will assume the explicit forms given by

$$A_0 = \frac{a_0}{2} + a_1 + a_2 + a_3 + a_4 + a_5 + a_6 + \ldots , \qquad (4.5)$$

$$A_2 = \frac{a_0}{2} - \frac{a_2}{3} - \frac{a_4}{15} - \frac{a_6}{35} + \ldots , \qquad (4.6)$$

$$A_4 = \frac{a_0}{2} - \frac{a_1}{3} - \frac{a_2}{3} + \frac{a_3}{5} - \frac{a_4}{15} + \frac{a_5}{21} - \frac{a_6}{35} + \ldots , \qquad (4.7)$$

$$A_6 = \frac{a_0}{2} - \frac{a_1}{2} - \frac{a_2}{5} + \frac{3a_3}{10} - \frac{a_4}{7} + \frac{a_5}{14} - \frac{a_6}{21} + \ldots , \qquad (4.8)$$

etc.

The convergence of expansions on the right-hand sides of the preceding equations is slower than it would be for $\theta_1 \ll 90^\circ$; and the number of decimal places to which equations (4.5) - (4.8) will permit us to approximate the values of the moments A_{2m} will depend very largely on the rate of convergence of the sequence of the a_n's.

Applications of the present results to practical cases will be given in subsequent communications.

REFERENCES

Jurkevich, I., Willman, W.W., and Petty, A.F. : 1976, Astrophys. Space Sci., 44, 63.
Jurkevich, I., Petty, A.F., and Goodman, I.R. : 1981, in Photometric and Spectroscopic Binary Systems, NATO ASI Series, (ed. by E.B. Carling and Z. Kopal), D. Reidel , Dordrécht-Holland.
Kopal, Z. : 1959, in Close Binary Systems, Chapman Hall and John Wiley, London and New York.
Kopal, Z. : 1975a, Astrophys. Space Sci., 34, 431.
Kopal, Z. : 1975b, Astrophys. Space Sci., 38, 191.
Kopal, Z. : 1976a, Astrophys. Space Sci., 40, 461.
Kopal, Z. : 1976b, Astrophys. Space Sci., 45, 269.
Kopal, Z. : 1979a, in Language of the Stars, D. Reidel, Dordrecht-Holland.
Kopal, Z. : 1979b, Astrophys. Space Sci., 66, 91.
Kopal, Z. : 1982, Astrophys. Space Sci., 81, 411.
Lanczos, C. : 1956, in Applied Analysis, Prentice-Hall, New York.
Oberhettinger, F. : 1973, in Fourier Expansions, Academic Press, New York and London.

PROSPECTIVE VIEWS IN SPACE ASTROMETRY

P. Lacroute
Dijon, France

ABSTRACT

The Hipparcos project of Space Astrometry was adopted by the European Space Agency (E.S.A.) in March 1980. The corresponding satellite will be launched in 1986 and, after 2.5 years of observation and about 2 years of reductions we will be able to obtain the coordinates, the proper motions and the parallaxes of 100.000 stars with uncertainty very much smaller than now. The mean values will be :

	m < 7	m=10	m=12,5
coordinates	0."001	0."0015	0."003
proper motions	0."0014/y	0."002/y	0."0042/y
parallaxes	0."0014	0."002	0."0042

These stars will determine a very consistent sphere without systematic local errors. However, the absolute rotation by reference to the faint distant objects will be determined with other instruments.

In this way we obtain on the one hand a reference system much better than the one now available and very easy to use, and on the other hand parallaxes more numerous and more accurate which will be very useful in Astrophysics.

Hipparcos is evidently a completely new type of instrument. Our purpose is to study the foreseeable improvements and to discern the technical part of this instrumental type in the future of Astrometry.

HIPPARCOS

The description of the satellite is given in the E.S.A. report.* However we give here a short description of it which will be useful for a better understanding of the rest of the paper. The satellite measures

---

* E.S.A. S.C.I.(79) 10, December 1979.

E. G. Mariolopoulos et al. (eds.), Compendium in Astronomy, 247–251.

the angles between the stars. A complex mirror superimposes in the focal plane of a single telescope two fields of view, 68°.5 apart on the sky. In the focal plane there is a grid perpendicular to the two directions of view. By satellite rotation around an axis parallel to the grid the images of the stars cross perpendicularly the grid. The light of the stars is modulated by the grid. The comparison of the phases gives the addition to the basic angle, so as to obtain the angles between the stars. In order to eliminate the light of the stars which are not under obser- vation, we use a "dissector" limiting a sensitive diameter of 30" in the field of 54 X 54 arc minute. We move electrically the sensitive zone so as to follow the star paths. We mix quickly the measurements on the two stars observed, in order to reduce errors coming from attitude motion irregularities. The process is a continuous scanning of great circles on the sphere. In this way we obtain the five coordinates of 100.000 stars randomly distributed on the sphere. These stars form a reference system very consistent, without local systematic errors, and with abso- lute parallaxes, because we measure big angles. The results obtained will be accurate and free of magnitude effects, because we use photoelec- tric measurements on images without atmospheric distortions.

IMPROVEMENT ATTEMPTS

    1. In spite of possible smaller statistical errors on photons, the accuracy obtained for the brighter stars is not much better than the one obtained for the mean magnitude stars. This is the case because each star is put in place on the sphere only by individual angular measurements to other stars.

    On the E.S.A. report (p. 54), it is point out that after a first reduction using only the angular measurements and giving good coordinates for all the stars, it will be possible to use these coordinates and the transit times of each star in order to establish an accurate history of the attitude motion. If a smoothing of this attitude motion is available we can compute very precise attitudes for each instant. Then by using the computed attitudes and the transit times of the stars it will be pos- sible to get better information on the coordinates of the stars. In this way we connect each star with the whole of the stars observed not too far in times. We obtain accurate smoothing of the attitude motion in the case of free motion, or if we use only intermittent guiding. In the E.S.A. report some arguments are presented against an intermittent guid- ing by gas jets. However, another intermittent guiding seems possible, using wheels only, like in the present continuous guiding, if we change the actions on the wheels only suddenly at intervals of 250 sec for instance. If after detailed study this method proves to be really accept- able, we will have a real advantage because the method will be not more expensive and the weight of the results will be improved by a factor 8 for m < 6; 5 at m=7,5; 3 at m=9,5; and 2 at m=12. It is hoped that this improvement will be incorporated in the Hipparcos project.

    2. In the preceding improvement the uncertainty of the information

at each star transit is the sum of the uncertainty in the transit itself
and the uncertainty in the computed attitude. In the case of the brightest stars the last is the biggest. So we think that one could decrease
this uncertainty by lengthening the intervals of time between the guiding
actions. But then we will have to use more complex formulas for the attitude motion and moreover, especially at the end of the cycles, the gap
between the wanted program and the actual attitude motion will be too big.
At the end only some very bright stars will be improved by this method.
Therefore, it seems doubtful that the lengthening of the intervals of time
between the guiding sections will be interesting.

3. The Hipparcos random errors of proper motions are not very much
better than the best ones now available. This happens, in spite of the
better measurements, because the time base is only 2.5 years instead of,
some times, 50 years. However, these proper motions are consistent, without local systematic errors on all the sphere; that is very important.
It will be interesting and easy to reduce the uncertainty by a factor 10,
if we launch a second Hipparcos about 12 years after. This is only a
matter of money.

4. The accuracy for the faint stars is not as good as for the bright
stars. That is unavoidable consequence of the measurements in flight during continuous scanning in order to obtain inexpensively many stars.

However, by increasing the size of the optic we can move the foreseeable accuracies in the direction of faint stars. For instance an increase by a factor K in the pupil size, the number of photoelectrons increases by a factor K^2, the useful step of the grid is divided by a factor
K, and the angular uncertainties are divided by a factor K^2. But the
weights, and consequently the price increase also by a factor probably a
little smaller than K^3. So by a factor K=1,58 we move the foreseeable
accuracies of one magnitude. In the future, probably, optic a little bigger than now will be used.

5. Another way to improve the accuracy is to increase the amount of
information.

For instance, it is possible to use several dissectors working together in different parts of the focal plane. The increases of weight
and price of the satellite will be probably small. But we increase the
complexity of dissector command in the on board computer and the amount
of information to be sent to the ground. Perhaps using 2 to 4 dissectors
it will be possible to reduce the uncertainties by factors 0.7 to 0.5
or to increase the number of the stars studied.

6. Also in order to increase the amount of information it is pointed
out in the E.S.A. report (p. 38) the possibility of using image analysis
systems in the focal plane. In this way probably more numerous stars
will be studied. But the accuracy is perhaps limited by the spatial accuracy of the system and the number of transmissions to be sent to the
ground. This way will be perhaps interesting in combination with the

future technical progress.

7. In the preceding sections we have investigated the reduction of uncertainties coming from the statistic of photons; but we have also to think about the risks of systematic errors.

Taking into account the Hipparcos principles, after corrections of optical defects (field correction and color corrections) it seems that only some systematic variations of the optical device, particularly of the basic angle, are able to produce systematic errors on the coordinates. In space we have only to fear thermic variations. For instance the E.S.A. report (p. 72) draws attention to some thermic variations of the basic angle as a function of the orientation of the Sun which are able to introduce a systematic error in all the parallaxes. If we attempt to reduce the random uncertainties, we have also to improve in the best possible way the thermic stabilization. However, if we have a small systematic error, the same, in all the parallaxes, it will be possible to correct it, by a study of all the results.

Otherwise, even if the thermic stabilization is not good enough, we can study in the residuals of the first reductions the correlation between the basic angle variations and the recent history of some former thermic tests. The more we reduce the random statistical errors, the more this study is necessary. The study will be more accurate, if we use more the thermic inertia and isolation in order to obtain stability. In order to obtain only slow and regular thermic variations, which can be studied by smoothing, we have to avoid direct and frequent thermic actions. The principle will be almost the same with the principles suggested in section 1 for the attitude guiding.

8. In conclusion it seems possible to improve really in the future the foreseeable uncertainties of Hipparcos project. Some improvements are probably not too expensive (1, 5), others are probably more expensive (3, 4) and others require technical improvements (6). A reduction of the uncertainties by a factor 10 seems not utopian for the future. Such an improvement will be particulary interesting in order to improve the direct calibration of the giant stars and consequently the cosmic distance scale.

ASTROMETRIC FIELD OF TYPE HIPPARCOS INSTRUMENTS

1. Reference System

The Hipparcos - type instruments will be peerless system, rich in stars, and easy to use for the determination of an accurate reference.

However, we need other instruments in order to determine the absolute rotation by reference to the faint distant objects. This reference system will be fundamental in Solar System Mechanics.

The star proper motions without local systematic errors will be very useful in order to study the Galactic Mechanics. However, the number of data is a little too small for this statistical use and we have not faint stars. Also very often we will use indirectly the Hipparcos proper motions for elimination of systematic errors in the actual proper motions.

The Hipparcos instruments are not well adapted to observe very close stars or to study the forms of the objects (clusters, double stars, nebulae, planets).

The Hipparcos instruments are not adapted to a particular study of some stars, for instance search of planets on nearby stars, because we scan uniformly all the sphere.

2. Parallaxes

As a rule the Hipparcos type parallaxes will be peerless because they are more accurate than the others and absolute. However, other instruments are necessary for the faint stars.

All the other parallaxes by measurements of small angles are only relative parallaxes. The more we improve the accuracy the more the uncertainty in the correction to take into account the probable parallaxes of the reference stars becomes grave. However, the space-telescope will be able to give also some absolute parallaxes by direct reference to the faint distant objects.

LONG-TERM CHANGES OF THE FLARE STARS EV LAC AND BY DRA

L.N. Mavridis, G. Asteriadis, F.M. Mahmoud*
Department of Geodetic Astronomy, University of Thessaloniki
Thessaloniki, Greece

ABSTRACT

The long term changes of the quiet-state luminosity and the flare activity of the flare stars EV Lac and BY Dra during the decade 1971-80 have been studied using the photoelectric observations in the International UBV System carried out with the 30-inch Cassegrain reflector of the Department of Geodetic Astronomy, University of Thessaloniki install-ed at the Stephanion Observatory, as well as all other relevant data available. The quiet-state luminosity of the star EV Lac shows long-term fluctuations with an amplitude of $0^m.3$ and a period of about 5 years. The flare activity of this star shows similar variations, which in most of the cases run parallel to the fluctuations of the quiet-state lumi-nosity. The quiet-state luminosity of the star BY Dra shows also long-term fluctuations. The current cycle has a duration of at least 14 years and an amplitude of more than $0^m.3$. Long-term changes are also indicated for the flare activity of this star, although the observational data a-vailable are rather limited.

INTRODUCTION

The study of the flare stars of the solar neighborhood could be of considerable interest for a better understanding of the phenomena con-nected both with stellar variability and stellar evolution as well as with the production of the solar flares.

As a contribution to the discussion of the problems connected with flare stars a long-term program of photoelectric observations of flare stars of the solar neighborhood was initiated by the Department of Geodetic Astronomy, University of Thessaloniki in 1971. Among the main objectives of this program were the following:

---

* On leave of absence from the Helwan Institute of Astronomy and Geo-physics, Cairo, Egypt.

E. G. Mariolopoulos et al. (eds.), Compendium in Astronomy, 253–276.

a) Monitoring of the flare activity of the stars under consideration and determination of the characteristics of the flares observed.
b) Use of the above material for the study of an eventual variation with time of the flare activity of these stars.
c) Determination of the quiet-state luminosity of the stars observed in one or more colours.
d) Use of the above data for the study of an eventual variation with time of the quiet-state luminosity of the stars studied.

In the present paper a brief account of the results obtained during the ten years elapsed since the beginning of the program is given. On the basis of these data a more thorough study of the long-term variations of both the flare activity and the quiet-state luminosity of the stars EV Lac and BY Dra is then undertaken.

MEASUREMENTS

The observations described here have been obtained with the 30-inch Cassegrain reflector with asymmetric mount (focal ratio f/3 for the primary hyperbolic mirror and f/13.5 for the Cassegrain focus) constructed by Astro Mechanics, belonging to the Department of Geodetic Astronomy, University of Thessaloniki and installed at the Stephanion Observatory ($\lambda = -22^\circ$ 49'44", $\varphi = +37^\circ$ 45'15"). The photoelectric observations were carried out with a Johnson dual channel photoelectric photometer with offset guider unit, constructed also by Astro Mechanics. The photometer is mounted in the Cassegrain focus of the telescope and includes one RCA 1P21 and one RCA 7102 photomultipliers both of which are refrigerated during the measurements with dry ice. The measurements described here were made with the RCA 1P21 photomultiplier which was fitted with the following filters: a) one Corning filter No. 9863, 0.25 inch for the violet, b) one Corning filter No. 5030, 0.25 inch, cemented to a Schott filter No. GG13, 0.125 inch for the blue and c) one Corning filter No. 3384, 0.25 inch for the yellow. Most of the observations were carried out with the 35" diaphragm, while the rest were made with wider diaphragms. The photocurrent was measured by means of a DC amplifier and a strip chart recorder Hewlett-Packard type 7100B with a response time 0.6^s for the full range. The chart velocity was selected equal to 2.5 $cm.s^{-1}$.

Since the sky conditions at Stephanion are not very favorable to good photometry, we had to use a strictly differential method. Close to each flare star F, two comparison stars A,B of proper magnitudes and colours were selected. For the transformation of the instrumental magnitudes and colours to the International UBV System all comparison stars used were observed periodically during good nights together with stars representing the International UBV System, which were selected among the stars having high weight in Table 9 of Johnson et al.(1966).

As stated arleady the objectives of the program were manifold, including monitoring of flare activity and determination of the quiet-

state luminosity of the stars observed. As the last two objectives impose diverting requirements concerning the observational procedure, special care had to be taken for an optimum organization of the measurements. The following symmetrical observational schemes were finally used during the ten-year period under consideration:
Observational scheme No. 1: Ab,Fb(30^m), Fb(30^m), Fb(30^m), Bb, Fb(30^m), Fb(30^m), Fb(30^m), Ab and so on.
Observational scheme No. 2: Ab, Fb(30^m), Fb(30^m), Fb(30^m), Ab, Fb(30^m), Fb(30^m), Fb(30^m), Ab and so on.
Observational scheme No. 3: Ab, Bb, Ab, Fb(30^m), Fb(30^m), Fb(30^m), Ab, Bb, Ab, Fb(30^m), Fb(30^m), Fb(30^m), Ab, Bd, Ab and so on.
Observational scheme No. 4: Avbu, Fvbu, Bvbu, Fubv, Aubv and so on.

It is obvious that each observational scheme serves better different objectives. For example, observational schemes Nos. 1 and 2 can be used for flare patrol, but not for the determination of the quiet-state luminosity of the flare star. Observational scheme No. 4 on the contrary gives a very short patrol time, allows, however, the determination of the quiet-state luminosity of the flare star in the three colours of the International UBV System. Finally, observational scheme No. 3 presents a compromise between the two previous solutions, i.e. allows sufficient time for flare patrol, and at the same time gives the possibility for the determination of the quiet-state luminosity of the flare star observed in the B-colour of the International UBV System, both at the beginning and at the end of each $3x30^m$ - patrol period.

Observational scheme No. 1 was used during the years 1971-72. Observational scheme No. 2 was used during the year 1973. Observational scheme No. 3 was used during the years 1974-80. Finally, observational scheme No. 4 was used parallel to the schemes Nos. 1-3 during almost all the years 1972-80. Therefore, for all the years 1971-80 we do have patrol observations. The quiet-state luminosity of the observed flare stars in the B colour of the International UBV System can be determined for the beginning and the end of the $3x30^m$ - patrol period of each observed flare star only for the years 1974-80 during which the observational scheme No. 3 was used. Finally, the quiet-state luminosity of the observed flare stars in the three colours of the International UBV System can be determined only for the intervals of time of the years 1972-80 during which measurements according to the observational scheme No. 4 were performed.

A total of 9 flare stars of the solar neighborhood were observed during the years 1971-80. The names and some other relevant data concerning these stars are given in Table 1 (Gurzadyan, 1980). From this table we see that the V magnitudes of the stars observed range between 8.6^m and 12.95^m while, their heliocentric distances range between 2.7 and 16.3 pc.

Table 1

Flare Stars Observed at the Stephanion Observatory during the Years 1971-80

Star	α(1950)	δ(1950)	V	Spectral Type	M_v	Distance (pc)	Remarks
UV Cet	$01^h 36^m 25^s$	$-18°12'42''$	$12^m.95$	dM5.5e	$15^m.8$	2.7	Binary (the system of A and B is a close binary system with P=26.52 years).
YZ CMi	07 42 04	+03 40 48	11.2	dM4.5e	12.29	6.0	Unseen companion, astrometric binary
AD Leo	10 16 54	+20 07 18	9.43	dM4.5e	10.98	4.9	Suspected spectroscopic binary (radial velocity varies from 1 to 31 km.s^{-1}).
DT Vir	12 58 19	+12 38 42	9.79	dM2 e	9.4	12.1	
BD+16°2708	14 52 08	+16 18 18	10.20	dM0 e	10.1	10.4	
BD+55°1823	16 15 59	+55 23 48	9.96	dM1 e	8.9	16.3	Spectroscopic binary
BY Dra	18 32 45	+51 41 00	8.6	dK6 e	7.6	14.1	Visual binary, separation A,B is 2".4.
DO Cep	22 26 13	+57 26 48	11.3	dM4.5e	13.3	4.0	Optical companion with 5" separation.
EV Lac	22 44 40	+44 04 36	10.2	dM4.5e	11.65	5.1	Suspected astrometric binary.

Table 2

Total Monitoring Time (T) and Total Number of Flares Observed (N) for the
Flare Stars Studied at the Stephanion Observatory during the Years 1971–1980

Star Year	EV Lac T(h)	N	UV Cet T(h)	N	AD Leo T(h)	N	BY Dra T(h)	N	YZ CMi T(h)	N	DT Vir T(h)	N	BD +16°2708 T(h)	N	BD +55°1823 T(h)	N	DO Cep T(h)	N	Total T(h)	N
1971	5.90	2	29.60	6															35.50	8
1972	44.20	6	10.61	7							12.73	1	9.75	1					77.29	15
1973	80.70	15	47.18	20	7.95	0	42.03	0			7.05	0	23.51	0					208.42	35
1974	74.55	23	41.00	23	40.85	8	18.35	2	2.19	2	3.09	0	18.20	1	39.42	1			237.65	60
1975	82.73	13	13.12	3			22.27	1	16.37	4					46.88	0	58.25	0	239.62	21
1976	198.37	18	6.20	2	4.50	1			4.53	1			23.73	0					237.33	22
1977	164.51	10	35.43	8			5.75	0											205.69	18
1978	70.90	5	36.70	14			39.40	0											147.00	19
1979	121.72	30					25.48	0											147.20	30
1980	66.56	2					30.93	0											85.71	2
Total	910.14	124	291.84	83	53.30	9	184.47	3	23.09	7	22.87	1	75.19	2	86.30	1	58.25	0	1621.41	230

FLARES OBSERVED

Table 2 gives for each of the 9 flare stars observed the monitoring time and the number of the flares observed during each of the years 1971-80, as well as during the entire period 1971-80. From this table we see that the total monitoring time for all the 9 stars observed during the period 1971-80 was equal to $1621^{h}.41$, while the total number of the flares observed during the same period was equal to 230 flares. Out of this $1621^{h}.41$ of monitoring time, $910^{h}.14$ were devoted to the star EV Lac (124 flares), $291^{h}.84$ to the star UV Cet (83 flares), $184^{h}.47$ to the star BY Dra (3 flares) and shorter intervals of time to the remaining flare stars under consideration.

For each flare star observed during a given year following data have been determined:
a) The monitoring intervals in UT as well as the total monitoring time during each observing night. Any interruption of more than one minute has been noted.
b) The standard deviation of random noise fluctuation

$$\sigma(mag) = 2.5 \log (I_o + \sigma)/I_o , \tag{1}$$

for different times(UT) of the monitoring intervals. I_o is the intensity deflection less sky background of the quiet star and σ the standard deviation of the values of I_o during the time interval considered. The value of $\sigma(mag)$ characterizes the quality of the photometry. As a rule no fluctuation of the star's luminosity was considered as a flare unless its amplitude was at least equal to $3\sigma(mag)$.

Furthermore, for each flare observed the light curve in the b colour of our instrumental system has been plotted. On the basis of this light curve following characteristics (Andrews et al., 1969) of the flare considered were determined:
a) the date and universal time of the flare maximum,
b) the duration before and after the maximum, as well as the total duration of the flare,
c) the value of the ratio $(I_F-I_o)/I_o$ corresponding to flare maximum, where I_F is the total intensity deflection less sky background of the star plus flare,
d) the integrated intensity of the flare over its total duration, including pre-flares, if present:

$$P = \int (I_F-I_o)/I_o \, dt , \tag{2}$$

e) the increase of the apparent magnitude of the star at flare maximum

$$\Delta m(b) = 2.5 \log (I_F / I_o) , \tag{3}$$

where b is the blue magnitude of the star in our instrumental system,
f) the standard deviation of random noise fluctuation

$$\sigma(\text{mag}) = 2.5 \log (I_O + \sigma)/I_O$$

during the quiet-state phase immediately preceding the beginning of the flare,
g) the air mass at flare maximum.

For the major part of the patrol observations carried out during the years 1971-80 the above data have been already published (Asteriadis and Mavridis, 1972a,b; Contadakis and Mavridis, 1972; Asteriadis et al., 1973; Asteriadis and Mavridis, 1973; Lovell et al., 1974; Contadakis and Mavridis, 1974; Contadakis and Mavridis, 1975; Contadakis et al., 1976; Contadakis et al., 1977; Kareklidis et al., 1977a,b,c,d; Kareklidis et al., 1978; Contadakis et al., 1978a,b; Arabelos et al., 1978a,b,c,d; Contadakis et al., 1978c,d,e; Contadakis et al., 1979a,b,c; Contadakis et al., 1980; Avgoloupis et al., 1980a,b; Mahmoud et al., 1980; Asteriadis et al., 1980a,b,c; Mavridis and Varvoglis, 1980; Asteriadis et al., 1981a,b), while for the remaining observations the corresponding publications are in preparation (Contadakis et al., 1981; Asteriadis et al., 1981a,b,c,d,e; Mavridis and Varvoglis, 1981a,b; Avgoloupis et al., 1981).

QUIET-STATE LUMINOSITY

The determination of the quiet-state luminosity of the flare stars could be of interest among others from the following points of view:
a) for the study of short-period variations of the quiet-state luminosity, (BY Dra Syndrome),
b) for the study of eventual long-term changes of the quiet-state luminosity of the stars observed.

To this purpose the quiet-state luminosity of the flare stars observed has been determined on the basis of the photoelectric observations carried out according to the observational schemes Nos. 3 and 4. All values thus obtained were transformed to the International UBV System.

Tables 3a and 3b give for each of the years 1972-80 the following data concerning the quiet-state luminosity of the stars EV Lac and BY Dra respectively:
a) the interval of time covered by the quiet-state luminosity determinations carried out according to the observational schemes Nos. 3 or 4 and the corresponding mean epoch (both in JD).
b) the mean annual value of the B magnitude of the star, obtained on the basis of the observational scheme No. 3 and the corresponding total number of observations.
c) the mean annual value of the V magnitude and the (B-V), (U-B) colours of the star, obtained on the basis of the observational scheme No. 4 and the corresponding total number of observations, and
d) the mean annual value of the B magnitude of the star, obtained on the basis of both observational schemes Nos. 3 and 4 taken together and the corresponding total number of observations.

Table 3a

Photoelectric Determinations of the Quiet-State Luminosity of the Star EV Lac

Year	Time Interval JD:244	Mean Epoch JD:244	Observational Scheme No.3 $\bar{B}$	N	Observational Scheme No. 4 $\bar{V}$	N	$\overline{B-V}$	N	$\overline{U-B}$	N	Observational Schemes Nos.3+4 $\bar{B}$	N
1972	1545.427–1572.416	1561.698			$10.^{m}04$ ±0.02	8	$1.^{m}45$ ±0.02	8	$0.^{m}71$ ±0.05	6	$11.^{m}49$ ±0.02	8
1973	1929.431–1952.392	1940.314			10.00 ±0.01	19	1.39 ±0.01	19	0.63 ±0.03	14	11.39 ±0.01	19
1974	2239.472–2329.477	2282.533	$11.^{m}45$ ±0.01	89	10.07 ±0.01	31	1.36 ±0.01	31	0.60 ±0.02	29	11.45 ±0.01	120
1975	2613.359–2707.289	2661.544	11.61 ±0.01	68	10.22 ±0.01	8	1.40 ±0.01	8	0.76 ±0.02	8	11.62 ±0.01	76
1976	2981.492–3064.453	3028.084	11.42 ±0.01	221	9.97 ±0.01	40	1.44 ±0.00	40	0.60 ±0.02	31	11.42 ±0.01	261
1977	3332.455–3423.347	3375.568	11.40 ±0.01	135							11.40 ±0.01	135
1978	3743.362–3785.374	3761.964	11.36 ±0.01	61							11.36 ±0.01	61
1979	4066.433–4143.371	4100.406	11.33 ±0.01	82							11.33 ±0.01	82
1980	4445.485–4517.390	4485.235	11.58 ±0.03	73							11.58 ±0.03	73
Total			11.45 ±0.04	729	10.03 ±0.01	106	1.41 ±0.00	106	0.63 ±0.01	88	11.45 ±0.03	835

Table 3b

Photoelectric Determinations of the Quiet-State Luminosity of the Star BY-Dra

Year	Time Interval JD:244	Mean Epoch JD:244	Observational Scheme No.3 $\bar{B}$	N	Observational Scheme No. 4 $\bar{V}$	N	$\overline{B-V}$	N	$\overline{U-B}$	N	Observational Schemes Nos.3+4 $\bar{B}$	N
1973	1827.600–1864.500	1852.334			$8^{\mathrm{m}}.082$ ±0.003	35	$1^{\mathrm{m}}.227$ ±0.002	35	$0^{\mathrm{m}}.929$ ±0.006	35	$9^{\mathrm{m}}.309$ ±0.003	35
1974	2235.439–2267.477	2249.680	$9^{\mathrm{m}}.302$ ±0.009	21	8.080 ±0.005	14	1.231 ±0.003	14	0.919 ±0.008	14	9.305 ±0.006	35
1975	2584.474–2600.622	2594.367	9.338 ±0.006	25							9.338 ±0.006	25
1977	3332.402–3335.500	3333.782	9.415 ±0.021	6							9.415 ±0.021	6
1978	3698.440–3713.634	3705.255	9.419 ±0.008	27							9.419 ±0.008	27
1979	4066.413–4091.478	4079.566	9.429 ±0.010	35							9.429 ±0.010	35
1980	4426.417–4483.453	4457.131	9.496 ±0.007	38							9.496 ±0.007	38
Total			9.411 ±0.007	152	8.081 ±0.003	49	1.228 ±0.001	49	0.926 ±0.005	49	9.386 ±0.006	201

From these tables we see that a total of 835 and 201 values of the
B magnitudes of the stars EV Lac and BY Dra respectively have been determined on the basis of both observational schemes Nos. 3 and 4 taken together.

From Tables 2, 3a and 3b we see that fairly rich observational material concerning both the flare activity and quiet-state luminosity of
the flare stars observed has been accumulated during the decade 1971-80
at the Stephanion Observatory. The main advantage of this material lies
in the fact that all observations have been carried out by the same team
of observers, with the same instruments and the same observational procedure. Therefore, the corresponding data should be very homogeneous.
For this reason this material could be of considerable interest for the
study of the flare activity and the quiet-state luminosity of the stars
observed.

A first attempt to use this material for the study of long-term variations of the flare activity and the quiet-state luminosity of one of
the stars observed has been made by Mahmoud (1978). Mahmoud studied the
flare activity and the quiet-state luminosity of the star EV Lac corresponding to the years 1972-76 and concluded that the quiet-state luminosity of this star changes from year to year with a minimum during the
year 1975. During the same year a minimum of the flare activity of the
star was noticed. The amplitude of the variation of the quiet-state luminosity of the star in the B-colour during the period 1972-76 was found
equal to $0^{m}.2$.

In the present paper a discussion is made of the long-term variations of both the flare activity and the quiet-state luminosity of the
stars EV Lac and BY Dra on the basis of the material accumulated at the
Stephanion Observatory during the decade 1971-80, as well as of some
supplementary data published by other observers. The discussion of the
short-term changes of these stars, as well as the study of the material
available for the remaining seven stars observed at the Stephanion Observatory will be made in forthcoming papers.

LONG-TERM CHANGES OF THE STAR EV LAC

Long-Term Variation of the Quiet-State Luminosity

The flare star EV Lac (BD+43° 4305, Gliese No. 873) has an optical
companion five seconds of arc to the west of the flare star (Hel.Ph.Cat.
+43° 22^{h} 45^{m}, No. 115). According to Andrews and Chugainov (1969) the
magnitudes and colours of the two stars as well as of the system of both
stars are as follows:

$$
\begin{aligned}
\text{EV Lac} &: V = 10^{m}.25, & B-V &= +1^{m}.58, & U-B &= +1^{m}.06, \\
\text{Optical Companion} &: V = 12^{m}.00, & B-V &= +0^{m}.74, & U-B &= +0^{m}.21, \\
\text{EV Lac + Optical Companion} &: V = 10^{m}.05, & B-V &= +1^{m}.37, & U-B &= +0^{m}.75.
\end{aligned}
$$

The above values are the mean data of two nights (July 7-8 and 8-9, 1969). According to Andrews and Chugainov (1969) the accuracy of these observations was about $\pm 0^m.015$ for V magnitudes and (B-V) colours and ± 0.03 for (U-B) colours, and no systematic differences were found in observations between the two nights.

According to van de Kamp (1969) the flare star itself has an unseen companion, but not interpretation of the perturbations for this system was possible at that time.

As stated already our observations have been made using a diaphragm at least equal to 35". Therefore, our photoelectric data refer to the system of the two stars: EV Lac + Optical Companion. Cristaldi and Rodono (1970) have reduced their observations to the flare star itself using the following formulae:

$$I_{EV\ Lac} = 0.70\ I_{system} \quad (B\ light),$$

$$I_{EV\ Lac} = 0.83\ I_{system} \quad (V\ light).$$

No similar reduction has been made, however, in our data.

In Fig. 1 the mean annual values of the quiet-state luminosity $\overline{B}$ and the corresponding values of the colours $\overline{B-V}$ and $\overline{U-B}$ of the star EV Lac for the years 1972-80 given in Table 3a have been plotted as a function of the mean epoch of observations. From this figure we see that the mean annual values of the quiet-state luminosity $\overline{B}$ of the star EV Lac show during the years 1972-80 a clear fluctuation with a first minimum in the year 1975, noted already by Mahmoud (1978), and a second minimum in the year 1980. Furthermore, we have two maxima the first during the year 1973 and the second during the year 1979. The amplitude of this fluctuation is of the order of $0^m.3$, i.e. ten times larger than the maximum r.m.s. error of the mean annual values $\overline{B}$ given in Table 3a, which is of the order of $0^m.03$. Unfortunately, the time interval covered by our data does not allow a confirmation of the hypothesis that this fluctuation follows a five-year cycle. From Fig. 1 we see that the mean annual values of the colours $\overline{B-V}$, $\overline{U-B}$, show also a variation with time. However, as no values of the colours $\overline{B-V}$, $\overline{U-B}$ are available for the years 1977-80, we cannot check whether or not this variation is indeed correlated with the corresponding variation of the $\overline{B}$ magnitudes, as indicated by Fig. 1 (especially for the $\overline{U-B}$ colour).

Long-Term Variation of the Flare- Activity

In order to study an eventual long-term variation of the flare activity of the star EV Lac following parameters characterizing the mean annual flare activity of the star have been computed for each of the years 1971-80:
a) The average number of flares observed per hour of monitoring time, i.e.

$$P_1 = N : T, \tag{4}$$

where T is the total monitoring time (in hours) of the star during the year considered and N the corresponding total number of the flares observed (Table 2).

b) The average integrated intensity of the flares observed per hour of monitoring time, i.e.

$$P_2 = \Sigma P : T , \tag{5}$$

where P is the integrated intensity of each of the flares observed during the year considered, as defined by relation (2).

c) The average flare energy released during the flares observed per hour of monitoring time, i.e.

$$P_3 = \Sigma E : T , \tag{6}$$

where E is the flare energy i.e. the additional energy released by the star in the B-colour during each of the flares observed in the year under consideration. The values of E have been computed with the help of the formula (Cristaldi and Rodono, 1973a).

$$E = 4\pi d^2 \cdot 10^{-0.4 \, B} \cdot Q \cdot 60 \, P \quad erg , \tag{7}$$

where E is the flare energy corresponding to the B-colour of the International UBV System, d the heliocentric distance of the star in cm, B the quiet-state luminosity of the star in the B colour, Q the energy flux in the B-colour produced by a star of apparent magnitude $B = 0^m$ outside the Earth's atmosphere, and P the integrated intensity of the flare considered corresponding to the B colour, as defined by relation (2). Following Gershberg and Chugainov (1969), Q was assumed equal to $6.3.10^{-6}$ erg cm^{-2}s^{-1}. From relation (7) we see that the values of E corresponding to various flares of the same star are proportional to the values of the integrated intensities P of these flares, as the dependence on the quiet-state luminosity of the star is very weak. Therefore, the values of the parameters P_2 and P_3 should show parallel variation.

Table 4 gives the values of the parameters P_1, P_2, P_3 for the star EV Lac corresponding to each of the years 1971-80. As the total monitoring time for the year 1971 is very short ($5^h.90$) and the total number of the flares observed very limited (2 flares), the values of the parameters P_1, P_2 given in Table 4 for this year are very uncertain.

The values of the parameters P_1, P_2 given in Table 4 for the years 1972-80 have been also plotted in Fig. 1. From this figure we see that the mean annual flare activity of the star EV Lac as represented by the parameters P_1, P_2 shows a clear variation with time during the period under consideration. The average integrated intensity of the flares observed per hour of monitoring time (P_2) becomes maximum during the years 1973, 1977 and 1979 and minimum during the years 1975, 1978 and 1980. The average number of flares observed per hour of monitoring time

Table 4

Mean Annual Flare Activity of the Star EV Lac for each of the Years
1971-80 Determined on the Basis of the Patrol Observations in the B-Colour Carried out at the Stephanion Observatory.

Year	P_1 flares/hour	P_2 min./hour	P_3 erg /hour
1971	0.34	0.051	
1972	0.14	0.076	$2.29 \cdot 10^{30}$
1973	0.19	0.266	$8.75 \cdot 10^{30}$
1974	0.31	0.250	$7.85 \cdot 10^{30}$
1975	0.16	0.084	$2.08 \cdot 10^{30}$
1976	0.09	0.124	$3.78 \cdot 10^{30}$
1977	0.06	0.645	$21.10 \cdot 10^{30}$
1978	0.07	0.038	$1.30 \cdot 10^{30}$
1979	0.25	0.255	$8.78 \cdot 10^{30}$
1980	0.03	0.015	$0.54 \cdot 10^{30}$

(P_1) on the contrary becomes maximum during the years 1974 and 1979 and minimum during the years 1977 and 1980.

For a more thorough study of the long-term variation of the flare activity of the star EV Lac all patrol observations available for this star since 1969 have been used for a determination of the values of the parameters P_1,P_2. The results are given in Table 5. In this table the values of the parameters P_1,P_2 obtained on the basis of the individual observational series are given. In the same table the weighted averages of the values of P_1,P_2 corresponding to each of the years 1969-1973 are also given. The weight assigned to the values of P_1,P_2 corresponding to each observational series was taken equal to the corresponding total monitoring time T. From Table 5 we see that the values of P_1,P_2 corresponding to the years 1972, 1973 obtained on the one hand on the basis of the observations carried out at the Stephanion Observatory and on the other hand on the basis of all patrol observations available are in fairly good agreement.

The weighted averages of the parameters P_1,P_2 corresponding to the years 1969-73 given in Table 5 have been also plotted in Fig. 1. From this figure we see that the fluctuation of the values of the parameters P_1,P_2 shown by the data obtained at the Stephanion Observatory for the years 1972-80 is confirmed also by the supplementary data given in Table 5 for the years 1969-73.

Table   5

Mean Annual Flare Activity of the Star EV Lac for the Years 1969-1973
Determined on the Basis of all the Patrol Observations in the B-Colour
Available.

No.	Reference	T	N	ΣP	P_1	P_2
		1969				
1	Maslennikov and Shakhovskaya, 1969	48.25	15	42.93	0.31	0.89
2	Cristaldi and Rodono, 1969	49.78	6	2.35	0.12	0.05
3	Chugainov and Shakhovskaya, 1969	46.90	8	4.89	0.17	0.10
4	Osawa et al., 1969	33.00	12	3.86	0.36	0.12
	Weighted Average				0.23	0.30
		1970				
1	Chugainov et al., 1970	55.77	19	29.52	0.34	0.53
2	Oskanian, 1970	71.60	5	2.76	0.07	0.04
3	Grigorian and Eritsian, 1970	65.18	2	—	0.03	—
4	Cristaldi and Rodono, 1971a	21.90	4	0.66	0.18	0.03
	Weighted Average				0.14	0.22
		1971				
1	Cristaldi and Rodono, 1971b	20.05	4	3.67	0.20	0.18
2	Osawa et al., 1972	4.45	3	0.50	0.67	0.11
3	Chugainov et al., 1972	39.50	12	26.35	0.30	0.67
4	Deming and Webber, 1972	27.20	2	1.74	0.07	0.06
5	Grigorian and Eritsian, 1972	39.15	4	—	0.10	—
6	Contadakis and Mavridis, 1972	5.90	2	0.30	0.34	0.05
7	Eritsian, 1972	98.00	9	—	0.09	—
	Weighted Average				0.15	0.34
		1972				
1	Andersen and Pettersen, 1972	21.73	2	1.17	0.09	0.05
2	Cristaldi and Rodono, 1973b	17.50	3	5.92	0.17	0.34
3	Kapoor and Sinvhal, 1972	14.55	2	2.60	0.14	0.18
4	Cristaldi and Rodono, 1973a	12.40	4	3.41	0.32	0.27
5	Webber et al., 1973	5.67	1	0.07	0.18	0.01
6	Asteriadis et al., 1973	44.23	6	3.38	0.14	0.08
	Weighted Average				0.16	0.14
		1973				
1	Andersen and Pettersen, 1974	21.83	2	7.95	0.09	0.36
2	Cristaldi and Rodono, 1973c	28.00	1	4.68	0.04	0.17
3	Kapoor et al., 1973	2.05	0	0.00	0.00	0.00
4	Arabelos et al., 1978a-d	80.70	15	21.48	0.19	0.27
	Weighted Average				0.14	0.26

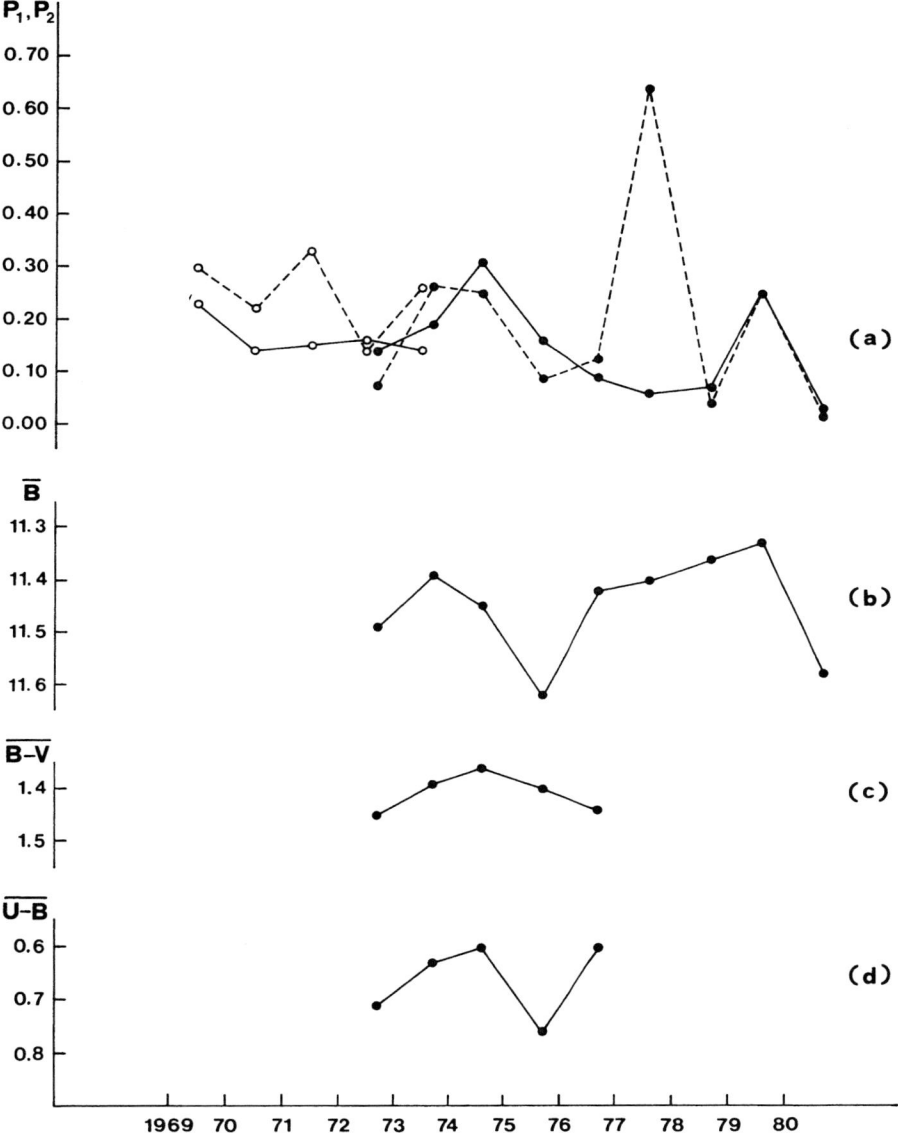

Fig. 1. Long-term variation of the flare activity and the quiet-state
luminosity of the star EV Lac. (a): Values of the parameters P_1 (.___)
and P_2 (.---) corresponding to the data obtained at the Stephanion Observatory together with the values of the parameters P_1 (○___) and P_2 (○---)
corresponding to the data given in Table 5. (b); (c) and (d): Mean annual
values $\bar{B}$, $\overline{B-V}$ and $\overline{U-B}$ of the quiet-state luminosity B and the colours
B-V and U-B respectively obtained at the Stephanion Observatory.

LONG-TERM CHANGES OF THE STAR BY DRA

Long-Term Variation of the Quiet-State Luminosity

The flare star BY Dra (BD+51°2402, Gliese No. 719) is a double-line spectroscopic binary. According to Bopp and Evans (1973) the orbital elements of the system are as follows:

$$P = 5.97599 \pm 0.00023 \text{ days },$$
$$\gamma = -25.675 \pm 0.963 \text{ km.s}^{-1},$$
$$e = 0.487 \pm 0.041 ,$$
$$i \simeq 33^{\circ} .$$

Furthermore, Bopp and Evans advanced the hypothesis that the flare star is the primary component of the system and contributes the 60% of the total light of the system. Mullan (1974) on the contrary found that the primary component should contribute as much as 70% of the total light of the system.

The quiet-state luminosity of BY Dra shows the well known small-amplitude and small-period fluctuations with both the period and the amplitude varying with time (BY Dra Syndrome). Being the prototype for this very interesting phenomenon, BY Dra has been the subject of numerous photoelectric studies over the last decades. As mentioned already in the present paper we will restrict ourselves in a discussion of the long-term variation of the quiet-state luminosity of the star. Therefore, the study of the short-term variation of its luminosity, on the basis of the data obtained at the Stephanion Observatory will be made in a forth-coming paper.

The long-term variation of the quiet-state luminosity of BY Dra has been studied by many authors in recent years (Chugainov, 1971, 1973, 1976; Vogt, 1975; Oskanyan et. al., 1977), using the photoelectric observations of this star. Phillips and Hartmann (1978) on the other hand have used the Harvard plate collection for the determination of the photographic magnitudes of the star during the years 1900-1952. Fig. 2 gives the long-term variation of the quiet-state luminosity of BY Dra as obtained by Phillips and Hartmann using their own photographic data as well as the photoelectric data discussed by Oskanyan et al. (1977). On the basis of this figure Phillips and Hartmann concluded that the quiet-state lu-minosity of BY Dra showed during the twentieth century two minima 50-60 years apart.

In order to study more thoroughly the long-term variation of the quiet-state luminosity of the star, the mean annual values $\bar{B}$, $\overline{B-V}$ of the photoelectric B-magnitudes and B-V colours of BY Dra obtained at the Stephanion Observatory (Table 3b) have been plotted in Fig. 3 to-gether with the corresponding photoelectric data published by other authors (Oskanyan et al., 1977; Chugainov, 1976). From this figure fol-lowing conclusion can be drawn:
a) There is a very good agreement between the magnitudes and colours

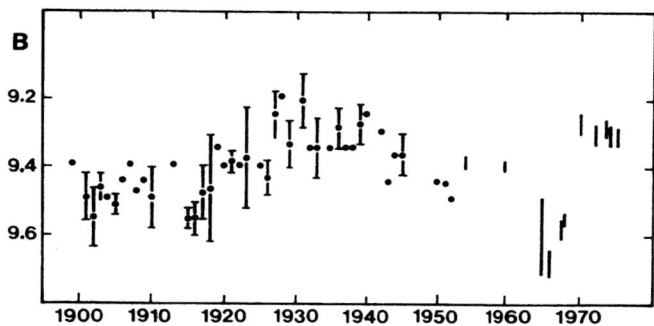

Fig. 2. Long-term variation of the quiet-state luminosity of the star
BY Dra. Dots represent the mean annual values of the photographic B magnitudes (Phillips and Hartmann, 1978); the standard deviations of these
magnitudes are indicated by the error bars; dots without error bars indicate that only one or two plates were available for that year. Lines
represent photoelectric B-magnitudes (Oskanyan et al., 1977); the height
of these lines indicates the magnitude variation of the star during the
corresponding year.

obtained at the Stephanion Observatory and the corresponding data given
by the other authors for the years in common.
b) The mean annual values $\bar{B}$ of the quiet-state luminosity of the star
BY Dra show a continuous decrease during the period 1974-80. This result
combined with the photoelectric data corresponding to the previous years,
given by the other authors, indicates that the mean annual values of
the quiet-state luminosity of the star BY Dra show definitely a long-
term variation. The period of this variation cannot get be determined
with certainty; the present cycle however, must have a period at least
equal to P = 14 years and an amplitude of more than $0^m.3$. The value P ⩾ 14
years given above for the period, does not agree with the value P = 8-9
years found by Chugainov (1973); on the contrary, this value could be
in agreement with the 7-14 year time scale found by Wilson (1978) for
the CaII-emission variability in late-type dwarfs. From Fig. 3 we see that
the mean annual values of the B-V colour of the star BY Dra show also a
long-term variation with an amplitude of the order of $0^m.1$. The period of
this variation, however, cannot be determined with certainty, due to the
fact that no observational data are available after the year 1974.

Long-Term Variation of the Flare Activity

     From Table 2 we see that only 3 flares of the star BY Dra could be
detected at the Stephanion Observatory during the years 1973-80, in
spite of the fact that the corresponding total monitoring time was equal
to $184^h.47$.

     In order to study the long-term variation of the flare activity of
the star during recent years, all published patrol observations of BY Dra

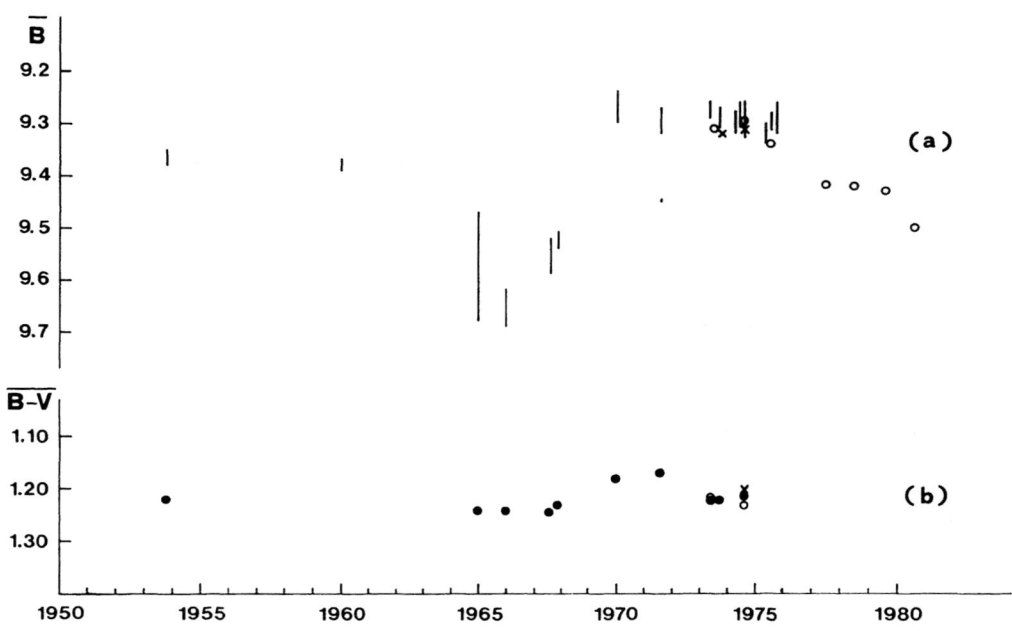

Fig. 3. Long-term variation of the quiet-state luminosity of the star
BY Dra. (a): Mean annual values of the B-magnitude, (b): Mean annual values of the B-V colour. Lines and filled circles represent the data discussed by Oskanyan et al. (1977); crosses correspond to additional data
given by Chugainov (1976); open circles give the values obtained at the
Stephanion Observatory. The height of the lines indicates the magnitude
variation of the star during the corresponding year.

since 1971 have been collected (Melkonyan et al., 1980; Slovak, 1977).
On the basis of this material and the data obtained at the Stephanion
Observatory, the values of the parameters P_1, P_2 corresponding to the
years 1972-75 (for which flares have been recorded) have been computed
and are given in Table 6. Slovak (1977) gives the characteristics of an
additional flare of the star BY Dra observed by him on June 30, 1976,
but no value of the corresponding total monitoring time is given; for
this reason, this flare could not be included in Table 6. Table 6 gives
also for each of the years 1972-75 the weighted averages of the values
of the parameters P_1, P_2. As in the case of the star EV Lac, in this
case too, the weights have been chosen equal to the corresponding total
monitoring time.

The weighted averages given in Table 6 have been also plotted in
Fig. 4. From this figure we see that the values of the parameters P_1, P_2
for the star BY Dra show an almost parallel long-term variation during
the period 1972-75. This variation is in fairly good agreement with the

Table 6

Mean Annual Flare Activity of the Star BY Dra for the Years 1971-1975
Determined on the Basis of all the Patrol Observations in the B-Colour
Available

No	Reference	T	N	ΣP	P_1	P_2
		1972				
1	Melkonyan et al., 1980	$163^h.58$	5	2.56	0.031	0.016
	Weighted Average				0.031	0.016
		1973				
1	Melkonyan et al., 1980	222.13	6	22.26	0.027	0.100
	Weighted Average				0.027	0.100
		1974				
1	Melkonyan et al., 1980	162.73	6	57.52	0.037	0.353
2	Contadakis et al., 1976	18.58	2	0.732	0.108	0.039
	Weighted Average				0.044	0.321
		1975				
1	Melkonyan et al., 1980	180.40	3	11.73	0.065	0.017
2	Mavridis and Varvoglis, 1980	22.27	1	0.21	0.045	0.009
	Weighted Average				0.020	0.059

variation of the quiet-state luminosity of the star during the same period shown in Fig. 3. However, as the total number of the flares observed during the period 1972-80 is very limited, no definitive conclusions concerning the long-term variation of the flare activity of the
star BY Dra can be reached at the present time.

CONCLUSIONS

     From the above discussion following conclusions can be drawn:
a) The quiet-state luminosity of the star EV Lac shows long-term fluctuations with a period equal to about 5 years and an amplitude equal to
about $0^m.3$. Two cycles of this fluctuation have been already covered with
the help of the photoelectric observations carried out at the Stephanion
Observatory.
b) The flare activity of the star EV Lac shows also long-term fluctuations, which in most of the cases are parallel to the ones of the quiet-
state luminosity.
c) The quiet-state luminosity of the star BY Dra shows long-term fluctuations. The current cycle which has been almost covered with photoelectric observations has a duration of at least 14 years and an amplitude
of more than $0^m.3$. This value is much shorter than the one found by

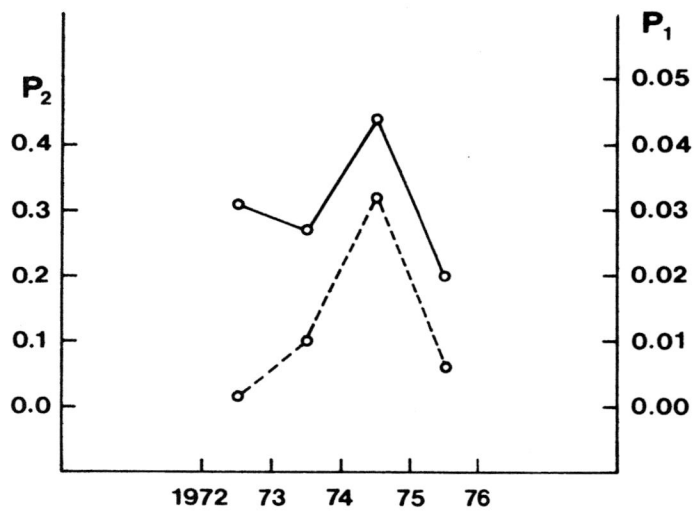

Fig. 4. Long-term variation of the flare activity of the star BY Dra; the continuous line (——) represents the values of the parameter P_1, while the dashed line (.---) represents the values of the parameter P_2.

Phillips and Hartmann for the duration of the previous cycle, on the basis of photographic (and therefore less reliable) data.
d) Long-term fluctuations are also indicated for the flare activity of the star BY Dra. However, as the total number of the flares of BY Dra observed during the last decade is very limited, no more details concerning this fluctuation can be given at the present-time. The long-term fluctuations of the quiet-state luminosity of the stars EV Lac and BY Dra mentioned above should be regarded as definitely confirmed. On the contrary, the long-term fluctuations of the flare activity of both stars, should be considered with due cautions, as the total monitoring time available for both stars and especially for the star BY Dra is very limited. This uncertainty poses serious limitations to any attempt for theoretical interpretation of the long-term fluctuations observed. On the other hand any similar attempt should take into account also the phenomena connected with the short-term changes of the same stars. For this reason the short-term changes of the stars EV Lac and BY Dra and the theoretical interpretation of the observed phenomena will be treated in forthcoming papers.

Acknowledgements

    The first of the authors would like to express his gratitude: a) to the Agency for Scientific Research and Development, Greek Ministry of Culture and Science and b) to the National Hellenic Research Foundation for their financial support during the various phases of the Flare Stars Program of the Department of Geodetic Astronomy, University of Thessaloniki. Also the third of the authors would like to thank: a) the University of Thessaloniki and b) the National Hellenic Research Foundation for their grants, which made possible his work at the Department of Geodetic Astronomy, University of Thessaloniki and his participation in the present program. Finally the authors would like to thank Dr. V. Tsikoudi for making available before publication the data concerning the flare activity and quiet-state luminosity of the star BY Dra during the year 1978, obtained on the basis of her observations at the Stephanion Observatory.

REFERENCES

Andersen, R.N. and Pettersen, B.R.: 1972, Comm. 27 IAU Inform. Bull. Variable Stars No. 723.
Andersen, B.N. and Pettersen, B.R.: 1974, Comm. 27 IAU Inform. Bull. Variable Stars No. 874.
Andrews, A.D. and Chugainov, P.F.: 1969, Comm. 27 IAU Inform. Bull. Variable Stars No. 370.
Andrews, A.D., Chugainov, P.F., Gershberg, R.E., and Oskanian, V.S.: 1969, Comm. 27 IAU Inform. Bull. Variable Stars No. 326.
Arabelos, D., Kareklidis, G., and Mavridis, L.N.: 1978a, Comm. 27 IAU Inform. Bull. Variable Stars No. 1504.
Arabelos, D., Kareklidis, G., and Mavridis, L.N.: 1978b, Proc. Acad. Athens, 52, 533.
Arabelos, D., Kareklidis, G., Mavridis, L.N., and Stavridis, D.C.: 1978c, Comm. 27 IAU Inform. Bull. Variable Stars No. 1505.
Arabelos, D., Kareklidis, G., Mavridis, L.N., and Stavridis, D.C.: 1978d, Proc. Acad. Athens, 52, 548.
Asteriadis, G. and Mavridis, L.N.: 1972a, Comm. 27 IAU Inform. Bull. Variable Stars No. 654.
Asteriadis, G. and Mavridis, L.N.: 1972b, Comm. 27 IAU Inform. Bull. Variable Stars No. 712.
Asteriadis, G. and Mavridis, L.N.: 1973, Comm. 27 IAU Inform. Bull. Variable Stars No. 816.
Asteriadis, G., Mavridis, L.N., and Stavridis, D.: 1973, Comm. 27 IAU Inform. Bull. Variable Stars No. 809.
Asteriadis, G., Kareklidis, G., and Mavridis, L.N.: 1980a, Comm. 27 IAU Inform. Bull. Variable Stars No. 1803.
Asteriadis, G., Kareklidis, G., Mavridis, L.N., and Varvoglis, P.: 1980b, Comm. 27 IAU Inform. Bull. Variable Stars No. 1804.
Asteriadis, G., Contadakis, M.E., Mahmoud, F., and Mavridis, L.N.: 1980c, Comm. 27 IAU Inform. Bull. Variable Stars No. 1806.
Asteriadis, G., Mahmoud, F., and Mavridis, L.N.: 1981a, Comm. 27 IAU Inform. Bull. Variable Stars No. 1906.
Asteriadis, G., Mahmoud, F., and Mavridis, L.N.: 1981b, Comm. 27 IAU Inform. Bull. Variable Stars No. 1907.
Asteriadis, G., Contadakis, M.E., Mahmoud, F., Mavridis, L.N., Stavridis, D., and Varvoglis, P.: 1981a, in preparation.
Asteriadis, G., Kareklidis, G., Mahmoud, F., Mavridis, L.N., Stavridis, D., Tsioumis, A.C., and Varvoglis, P.: 1981b, in preparation.

Asteriadis, G., Kareklidis, G., Mavridis, L.N., and Tsikoudi, V.:1981c,
    in preparation.
Asteriadis, G., Avgoloupis, S., Kareklidis, G., Mavridis, L.N., Stavridis,
    D., and Varvoglis, P.: 1981d, in preparation.
Asteriadis, G., Avgoloupis, S., Mavridis, L.N., and Varvoglis, P.: 1981e,
    in preparation.
Avgoloupis, S., Mavridis, L.N., and Varvoglis, P.: 1980a, Comm. 27 IAU
    Inform. Bull. Variable Stars No. 1792.
Avgoloupis, S., Phylactopoulos, P., Kareklidis, G., Mavridis, L.N., and
    Varvoglis, P.: 1980b, Comm. 27 IAU Inform. Bull. Variable Stars
    No. 1793.
Avgoloupis, S., Mavridis, L.N., and Varvoglis, P.: 1981, in preparation.
Bopp, B.W., and Evans, D.S.: 1973, Monthly Notices Roy. Astron. Soc.,
    $\underline{164}$, 343.
Chugainov, P.F. and Shakhovskaya, N.I.: 1969, Comm. 27 IAU Inform. Bull.
    Variable Stars No. 410.
Chugainov, P.F., Shakhovskaya, N.I., and Maslennikov, K.L.: 1970, Comm.
    27 IAU Inform. Bull. Variable Stars No. 485.
Chugainov, P.F.: 1971, Comm. 27 IAU Inform. Bull. Variable Stars No.520.
Chugainov, P.F., Kulapova, A.N., and Shakhovskaya, N.I.: 1972, Comm. 27
    IAU Inform. Bull. Variable Stars No. 616.
Chugainov, P.F.: 1973, Izv. Krymsk. Astrofiz., Obs., $\underline{48}$, 3.
Chugainov, P.F.: 1976, Izv. Krymsk. Astrofiz. , $\underline{54}$, 85.
Contadakis, M.E. and Mavridis, L.N.: 1972, Comm. 27 IAU Inform. Bull.
    Variable Stars No. 669.
Contadakis, M.E. and Mavridis, L.N.: 1974, Proc. Acad. Athens, $\underline{48}$, 343.
Contadakis, M.E. and Mavridis, L.N.: 1975, Comm. 27 IAU Inform. Bull.
    Variable Stars No. 1020.
Contadakis, M.E., Kareklidis, G., Mahmoud, F., Mavridis, L.N., Stavridis,
    D., and Zervaki-Zoerou, E.: 1976, Comm. 27 IAU Inform. Bull. Varia-
    ble Stars No. 1181.
Contadakis, M.E., Kareklidis, G., Mahmoud, F., Mavridis, L.N., and
    Stavridis, D.: 1977, Proc. Acad. Athens, $\underline{51}$, 721.
Contadakis, M.E., Kareklidis, G., Mavridis, L.N., Stavridis, D., and
    Zervaki-Zoerou, H.: 1978a, Comm. 27 IAU Inform. Bull. Variable Stars
    No. 1486.
Contadakis, M.E., Kareklidis, G., Mavridis, L.N., Stavridis, D., and
    Zervaki-Zoerou, H.: 1978b, Proc. Akad. Athens, $\underline{52}$, 453.
Contadakis, M.E., Kareklidis, G., Mavridis, L.N., and Stavridis, D.C:
    1978c, Proc. Acad. Athens, $\underline{53}$, 169.
Contadakis, M.E., Kareklidis, G., Mavridis, L.N., and Tsioumis, A.C.:
    1978d, Proc. Acad. Athens, $\underline{53}$, 98.
Contadakis, M.E., Kareklidis, G., Mavridis, L.N., Stavridis, D., and
    Tsioumis, A.C.: 1978e, Proc. Acad. Athens, $\underline{53}$, 148.
Contadakis, M.E., Kareklidis, G., Mavridis, L.N., and Stavridis, D.C.:
    1979a, Comm. 27 IAU Inform. Bull. Variable Stars No. 1620.
Contadakis, M.E., Kareklidis, G., Mavridis, L.N., and Tsioumis, A.C.:
    1979b, Comm. 27 IAU Inform. Bull. Variable Stars No. 1653.
Contadakis, M.E., Kareklidis, G., Mavridis, L.N., Stavridis, D., and
    Tsioumis, A.C.: 1979c, Comm. 27 IAU Inform. Bull. Variable Stars
    No. 1654.

Contadakis, M.E., Mahmoud, F., Mavridis, L.N., and Stavridis, D.: 1980, Comm. 27 IAU Inform. Bull. Variable Stars No. 1784.
Contadakis, M.E., Mahmoud, F., Mavridis, L.N., and Stavridis, D.: 1981, in preparation.
Cristaldi, S. and Rodono, M.: 1969, Comm. 27 IAU Inform. Bull. Variable Stars No. 403.
Cristaldi, S. and Rodono, M.: 1970, Astron. Astrophys. Suppl., 2, 223.
Cristaldi, S. and Rodono, M.: 1971a, Comm. 27 IAU Inform. Bull. Variable Stars No. 525.
Cristaldi, S. and Rodono, M.: 1971b, Comm. 27 IAU Inform. Bull. Variable Stars No. 600.
Cristaldi, S. and Rodono, M.: 1973a, Astron. Astrophys. Suppl., 10, 47.
Cristaldi, S. and Rodono, M.: 1973b, Comm. 27 IAU Inform. Bull. Variable Stars No. 759.
Cristaldi, S. and Rodono, M.: 1973c, Comm. 27 IAU Inform. Bull. Variable Stars No. 836.
Deming, D. and Webber, J.C.: 1972, Comm. 27 IAU Inform. Bull. Variable Stars No. 672.
Eritsian, M.A.: 1972, Soobshch. Byurakansk. Obs. Akad. Nauk Arm. S.S.R., No. 46, 23.
Gershberg, R.E. and Chugainov, P.F.: 1969, Izv. Krymsk. Astrofiz. Obs., 40, 7.
Grigorian, K.A. and Eritsian, M.A.: 1970, Comm. 27 IAU Inform. Bull. Variable Stars No. 497.
Grigorian, K.A. and Eritsian, M.A.: 1972, Soobshch. Byurakansk. Obs. Akad. Nauk Arm. S.S.R., No. 44, 104.
Gurzadyan, G.A.: 1980, Flare Stars, Pergamon Press, Oxford, New York, Toronto, Sydney, Paris, Frankfurt.
Johnson, H.L., Mitchell, R.I., Iriarte, B., and Wiśniewski, W.Z.: 1966, Commun. Lunar Planet. Lab. 4, No. 63.
Kamp, P. van den : 1969, Publ. Astron. Soc. Pacific, 81, 5.
Kapoor, R.C. and Sinvhal, S.D.: 1972, Comm. 27 IAU Inform. Bull. Variable Stars No. 750.
Kapoor, R.C., Sanwal, B.B., and Sinvhal, S.D.: 1973, Comm. 27 IAU Inform. Bull. Variable Stars No. 810.
Kareklidis, G., Mahmoud, F., Mavridis, L.N., Stavridis, D., and Zervaki-Zoerou, H.: 1977a, Comm. 27 IAU Inform. Bull. Variable Stars No. 1354.
Kareklidis, G., Mahmoud, F., Mavridis, L.N., Stavridis, D., and Zervaki-Zoerou, H.: 1977b, Proc. Acad. Athens 52, 133.
Kareklidis, G., Mahmoud, F., Mavridis, L.N., Stavridis, D., and Zervaki-Zoerou, H.: 1977c, Comm. 27 IAU Inform. Bull. Variable Stars No. 1355.
Kareklidis, G., Mavridis, L.N., and Stavridis, D.C.: 1977d, Comm. 27 IAU Inform. Bull. Variable Stars No. 1356.
Kareklidis, G., Mavridis, L.N., and Stavridis, D.C.: 1978, Proc. Acad. Athens 52, 466.
Lovell, B., Mavridis, L.N., and Contadakis, M.E.: 1974, Nature, 250, 124.
Mahmoud, F.M.: 1978, Dissertation, University of Thessaloniki, Greece.
Mahmoud, F., Mavridis, L.N., Stavridis, D., and Varvoglis, P.: 1980, Comm. 27 IAU Inform. Bull. Variable Stars No. 1799.
Maslennikov, K.L. and Shakhovskaya, N.I.: 1969, Comm. 27 IAU Inform. Bull. Variable Stars No. 401.

Mavridis, L.N. and Varvoglis, P.: 1980, Comm. 27 IAU Inform. Bull. Variable Stars No. 1891.
Mavridis, L.N. and Varvoglis, P.: 1981a, in preparation.
Mavridis, L.N. and Varvoglis, P.: 1981b, in preparation.
Melkonyan, A.S., Olah, K., Oskanyan, A.V. Jr., and Oskanyan, V.S.: 1980, Astrofizika, 16, No. 1, 107.
Mullan, D.J.: 1974, Astrophys. J., 192, 149.
Osawa, K., Noguchi, T., Okada, T., Ichimura, K., Watanabe, E., and Okida, K.: 1969, Comm. 27 IAU Inform. Bull. Variable Stars No. 399.
Osawa, K., Iscimura, K., and Shimizu, Y.: 1972, Comm. 27 IAU Inform. Bull. Variable Stars No. 608.
Oskanian, V.S.: 1970, Comm. 27 IAU Inform. Bull. Variable Stars No. 488.
Oskanyan, V.S., Evans, D.S., Lacy, C., and McMillan,R.S.: 1977, Astrophys. J. 214, 430.
Phillips, M.J. and Hartmann, L.: 1978, Astrophys. J., 224, 182.
Slovak, M.H.: 1977, Comm. 27 IAU Inform. Bull. Variable Stars No. 1271.
Vogt, S.S.: 1975, Astrophys. J., 199, 418.
Webber, J.C., Yoss, K.M., Deming, D., and Yang, K.S.: 1973, Publ. Astron. Soc. Pacific, 85, 739.
Wilson, O.C.: 1978, Astrophys. J., 226, 379.

ON THE PROBLEM OF CEPHEID AMPLITUDE DEPENDENCE ON THE STAR POSITION IN THE INSTABILITY STRIP

N.S. Nikolov and G.R. Ivanov
Department of Astronomy, University "Kliment Ohridsky",
Sofia, Bulgaria

ABSTRACT

The important problem on the Cepheid amplitude behaviour within the instability strip is reviewed. Particularly, an attempt is made to answer whether the noted differences between the photometrically and spectroscopically derived Cepheid intrinsic colours B-V are responsible for the discrepancy between the results of the investigations on the problem, and the answer is negative. The data considered separately by the authors and derived after both methods lead to the conclusion, that the amplitudes of the galactic Cepheids at a fixed period increase toward the cool edge of the instability strip.

1. INTRODUCTION

Since Sandage (1958) emphasized the role of temperature, i.e. the intrinsic colour, as a third parameter in the P-L relation, and suggested the possibility to replace the colour in the P-L-C relation with the light amplitude A, the investigation of the Cepheid amplitude dependence on the star position in the instability strip became one of the main lines in Cepheid study. The importance of these investigations increased with the attempts of practical realizations to construct a P-L-A relation, instead of the P-L-C one (Sandage and Tammann, 1971; Butler, 1976). And it is all the more a pity, that there is no agreement in the results of the various authors on this problem, because the Cepheid amplitude behaviour within the instability strip is connected not only with the Cepheid basis of the distance scale, but also with stellar evolution.

In what follows we are making an attempt to give an idea of the discrepancies existing in the different results, as well as to add new arguments in favour of the result preferred by us.

2. A BRIEF REVIEW OF THE DISCREPANCIES ON THE PROBLEM

Using data of the first photoelectric observations of a large number

277

E. G. Mariolopoulos et al. (eds.), Compendium in Astronomy, 277–284.
Copyright © 1982 by D. Reidel Publishing Company.

of Cepheids (Eggen, 1951;Eggen et al., 1957) Sandage found out in the
cited basic paper, that the small light amplitude galactic Cepheids are
concentrated at the high temperature edge of the Cepheid instability
strip. Arp's (1960) results on the Small Magellanic Cloud Cepheids were
consistent with this result. Later on Kraft (1961) comparing the colour
residual $\delta(B-V)$ from the average $<B-V>_o$ - log P relation with the defined
by him amplitude parameters

$$f_B = 10^{0.4(A_B - A_B^{max})},$$

A_B being the B-light amplitude of a Cepheid with a period P, and A_B^{max} -
the maximum amplitude observed for the same period, showed that the maximum B-light stars, i.e. those with $f_B \approx 1$, have a minimum $\delta(B-V)$. An analogous result was derived by Bahner et al.(1962), and Kraft concluded:
"When a supergiant enters the instability strip in the HR diagram, it
pulsates with a small amplitude. As the evolution proceeds, the amplitude
increases to a maximum value close to the strip centre, after which the
amplitude decreases again". (Kraft, 1961).

A different result was derived by Yakimova (1970a). She has used
the same method of comparing the δ-residuals $\delta(x) = x(P) - \overline{x(P)}$, $x(P)$
being a certain characteristic of a cepheid with the period P, and $\overline{x(P)}$
the mean x-characteristic of the same P. The difference in the method
by Yakimova was that she has stated, that the light amplitude residuals
δA_V depend not only on the colour residuals from $<B-V>_o$ - log P relation,
but in the same time also on the relation for the maximum colours -
$(B-V)_o^{max}$ - log P, and she derived correlations in the form

$$\delta A = K_1 \, \delta <B-V>_o - K_2 \, \delta(B-V)_o^{max}.$$

Using the intrinsic colours of the galactic Cepheids by Tsarevsky and
Yakimova (1970) for two groups of stars, $0.50 < \log P < 0.91$ and
$1.0 < \log P < 1.5$, Yakimova (1970a) derived positive coefficients K.
This means:at a fixed maximum temperature ($(B-V)_o^{max}$ = const), the
redder mean colour a Cepheid with a given period has, the larger its
light amplitude is; moreover, the amplitude increases with the temperature increase at maximum light. Using the Cepheid radii by Kurochkin
(1966), Yakimova has obtained the same result:the pulsation amplitudes
of Cepheids increase toward the low-temperature instability strip edge.
The same result was obtained when comparing the residuals of the absolute magnitude δM_V and δA_V (Ivanov, 1974). Qualitatively the same results
were obtained for the SMC Cepheids (Yakimova, 1970b), and Cepheids in
other galaxies (Yakimova et al., 1975).

Following Yakimova's method, the results by Yakimova (1970a) were
confirmed using Nikolov and Ivanov's (1973a,b,1975) U-B intrinsic colours
of galactic Cepheids (Nikolov and Ivanov, 1976).

A well-known paper concerning Cepheid amplitude dependence on the
position in the instability strip is by Sandage and Tammann (1971). Comparing the residuals δM_B with f_B and $\delta(B-V)$ with f_B the two authors came

to the conclusion that galactic Cepheids, as well as the Cepheids in the Small and Large Magellanic Clouds and M31 in the period ranges $0.40 < \log P < 0.86$ and $\log P > 1.30$ have the largest amplitude on the blue edge of the instability strip; the middle period Cepheids ($0.86 < \log P < 1.30$) show the opposite tendency. The data used by Sandage and Tammann are: 1) M_B from the composite of some stellar systems P-L relation (Sandage and Tammann, 1968, 1969); 2) f_B calculated from the upper envelop A_B^{max} of the diagram $A_B - \log P$ for the galactic and SMC Cepheids by Schalten-brandt and Tammann (1970), corrected for periods smaller than 4.5 days; 3) $\delta(B-V)$ residuals obtained from their relation $<B>_o - <V>_o = 0.323 \log P + 0.290$ for the galactic and Magellanic Cloud Cepheids using E_{B-V} of the galactic Cepheids, derived as simple means from E_{B-V} by Kraft (1961) and those from a formula by Tammann (1969). The result obtained by Sandage and Tammann was confirmed by Kelsall (1972) for the galactic Cepheids, and by Butler (1976) for the Small Magellanic Cloud Cepheids on the basis of photographic photometry.

In a review by Yakimova et al. (1975) as a main cause for the difference between Yakimova's and our results and the one by Sandage and Tammann (1971) was pointed out in the fact, that Sandage and Tammann have not studied the Cepheids in different systems separately, as it was done by Yakimova and us, but as one single complex; in this way the individual δ-residual of a single Cepheid is a superposition of two residuals, i.e. the δ-residual of this star from the mean $\overline{x}(P)$ for its stellar system, and the residual of this mean $\overline{x}(P)$ from the general one for all considered stellar systems: our Galaxy, the Magellanic Clouds, and M31. In particular, this explains Sandage and Tammann's result of the largest amplitudes at the blue edge of the instability strip for comparatively small periods, up to $\simeq 7.5$ days, since there is an excess of Cepheids with small periods and large amplitudes in SMC in comparison with those in the Galaxy, and besides one has in mind that the SMC Cepheids are bluer than the galactic ones (Gascoigne, 1969).

A confirmation of Yakimova's result follows from Fernie's (1970) equations for the classical Cepheid intrinsic colour in maximum and minimum light, as Madore (1976a) drew the attention on it. Moreover, this author, comparing the residuals from the reddening-free function W (van den Bergh, 1968) for the zero-amplitude line $W_{A=0} = -4.1 \log P$ and B-amplitudes of the galactic and Magellanic Cloud Cepheids, showed a correlation between these two values, in the same sense that at a fixed period the pulsation amplitudes increase monotonically toward the cool edge of the instability strip (Madore, 1976b). A return to the opposite behaviour of the Cepheid amplitudes when $\delta(B-V)$ taken from the period-colour relation by Dean et al. (1978) are compared with f_B by Nikolov and Momchev (1972), was indicated by Efremov and Nikolov (1979).

An attempt to reveal the reasons for the discrepancies in the investigations by Yakimova (1970a) and by Sandage and Tammann (1971) was made by Nikolov and Ivanov (1976) and Ivanov and Nikolov (1979), but no unique cause either in the methods, or in the data used in these investigations have been elucidated. But it is worth paying attention to Ivanov and

Nikolov's (1979) result, that if we treat the same data used by Sandage and Tammann (1971) in a different way, namely if we take into consideration $(B-V)_o^{max}$, it proves that Cepheid amplitudes at a fixed period increase toward the low-temperature strip edge.

3. ARE THE DIFFERENCES IN THE PHOTOMETRICALLY AND SPECTROSCOPICALLY DERIVED CEPHEID INTRINSIC COLOURS RESPONSIBLE FOR THE DIFFERENT RESULTS OF THE CEPHEID AMPLITUDE BEHAVIOUR WITHIN THE INSTABILITY STRIP?

It is well-known that there is a disagreement between the intrinsic colours of the Cepheids found out on the basis of photometric (the basic work by Kron, 1958) and spectroscopic (the basic work by Kraft, 1961) methods and that these colours are of great importance for the problem in consideration.

Mianes (1963) was the first to mention a systematic difference between the long-period Cepheid colour excesses on the basis of the six-colour photometry and the spectroscopical ones by Kraft (1961). Efremov (1966) confirmed the existence of such a difference and showed that it reaches about $0^m.20$ at P > 30 days. He pointed out the possible influence of the luminosity on Kraft's intrinsic colours for Cepheids with long periods. Later on, Tsarevsky and Yakimova (1970) concluded that there is no such systematic difference. Once more, the problem was discussed by Canavaggia et al. (1975), who have shown that the long-period excesses derived by Tsarevsky and Yakimova (1970) after Kraft's (1961) method are systematically greater than the six-colour ones with about $0^m.2$ and pointed out again that the influence of the star luminosity on Kraft's (1960) Γ-photometry may be responsible for the systematic increase in their excesses. Dean et al. (1978) reconfirmed this systematic difference and Efremov and Nikolov (1979) considered once more the luminosity effect as it possible cause.

In order to check up whether the differences in the Cepheid intrinsic colours derived after the two methods may influence the result of the dependence of the Cepheid amplitude on the star position in the instability strip, in this section we consider separately the original data obtained by various authors. Our hypothesis is that, if the difference in the data of the two methods is decisive, the cause of the inconsistent results lies in the procedures of the two methods. In this aspect we would like to mention the paper by Efremov and Nikolov (1979), where the photometrically derived Cepheid colours by Dean et al. (1978) show (with the exception of the small amplitude s-Cepheids) a tendency of an increase of Cepheid amplitudes toward the blue edge of the instability strip, in contrast to the results obtained by Yakimova (1970), Yakimova et al. (1975), and Nikolov and Ivanov (1976), on the basis of the spectroscopically derived Cepheid intrinsic colours. If one obtains independently of the method for deriving the Cepheid colours (at least in the majority of cases) the same result for the behaviour of the Cepheid amplitudes across the instability strip, the method and its inherent errors are not decisive for the final results.

We have calculated the correlation coefficients r between δA_v and δ(B-V) for the maximum, mean and minimum light, derived from the original data by different authors. The residuals δ(x) were obtained from the hand drawn x-log P relations. We accept the 95 per cent confidence as significant. If r is positive, the amplitude increases toward the red edge of the instability strip, in contrast to the case of a negative r. The results are presented in Table I. The last three lines are for the Large and Small Magellanic Clouds after the observations by Gascoigne (1969) and Madore (1975). The results for the SMC are given for the short- and large-period Cepheids (the cut of the period is accepted at log P =1),

Table I

		max		mean phase		min		
Author	Method	r	confid	r	confid	r	confid	n
Mianes (1963)	6 col ph	-0.46	98	0.31	90	0.53	99	26
Williams (1966)	uvby ph	-0.47	99	0.12	-	0.97	98	48
Tsarevsky & Yakimova (1968)	G ph	-0.43	99	0.20	98	0.43	99	155
Makarenko (1970)	UBV	-0.16	95	0.01	-	0.53	99	185
Fernie (1970)	Ir ph	-0.64	99	0.50	95	0.77	99	17
Sandage & Tammann (1971)	G ph	-0.56	99	0.00	-	0.55	99	46
Schmidt (1972)	H. line	-0.57	99	0.04	-	0.34	80	19
Kelsall (1972)	uvby ph	-0.49	95	0.06	-	0.31	80	22
Parsons & Bell (1975)	6 col ph	-0.58	99	0.05	-	0.44	99	43
Dean et al. (1978)	6 col ph	-0.58	99	0.03	-	0.52	99	99
Nikolov & Ivanov (1975)	mean E_{B-V}	-0.64	99	0.45	99	0.62	99	42
Gascoigne (1969) Madore (1975) LMC		-0.65	99	-0.08	-	-0.16	-	26
Gascoigne (1969) SMC short-period		-0.62	98	-0.39	80	-0.07	-	16
Madore (1975) large-period		-0.30	-	0.52	95	0.68	95	16

because of the problem, discussed not once, of the probable different nature of the small-period and large-amplitude galactic cepheids, in comparison with those from the SMC Cepheids.

A survey of Table I shows negative correlation, with a confidence equal to and more than 95 per cent, in almost all cases, for the relation of δA_v with $\delta(B-V)$ at maximum light. We have a similar situation for the relation of δA_v with $\delta(B-V)$ at minimum light, but the correlation coefficients are positive. And only for the relation of δA_v with $\delta(B-V)$ at mean light, which the authors usually regard as a criterion for the Cepheid amplitude behaviour in the instability strip, one sees a not so homogeneous picture. In this case, there is a positive r for all authors investigating the galactic Cepheids. The positive r indicates that at a constant period the cepheid amplitude increases monotonically from the blue to the red edge of the instability strip. But only in two cases, namely for the spectroscopic data by Tsarevsky and Yakimova (1970), and for the infrared photometric data by Fernie (1970) we have a significant correlation coefficient r.

Such results obviously signify that the method for deriving $(B-V)_o$-colours does not affect the result of the amplitude behaviour on the instability strip. The same conclusion follows from the last line for the galactic Cepheids in Table I. The correlation coefficient r in this line, was calculated for a sample of 42 Cepheids with the best colour excesses E_{B-V} from the catalogue by Nikolov and Ivanov (1975), with mean quadratic errors of the $E_{B-V} \leqslant 0.02$. This catalogue contains the Cepheids colour excesses E_{B-V}, each of which is derived as a mean from all determinations by different authors, who have used the photometric as well as the spectroscopic method. The positive coefficient r and its sufficient significance in the case of precisely determined E_{B-V} and respectively $(B-V)_o$ independently of the method, probably shows that the errors of $(B-V)_o$ affect mainly the result of the Cepheid amplitudes in the instability strip, and consequently the method used for the deriving of $(B-V)_o$ may have a secondary contribution. On the other hand the highest confidence of the coefficient r in the case when the Cepheids with the most precisely determined $(B-V)_o$ are selected and the positive sign of this coefficient of correlation between δA_v and $\delta<B-V>$ along with the positive sign of r in all other cases for the galactic Cepheids, gives an assurance on the conclusion, that in the Galaxy the Cepheid amplitudes show a general tendency toward an increase from the blue to the red edge of the instability strip.

4. CONCLUSION

Independently of the conclusion drawn in the previous section on the increase of galactic Cepheid amplitudes toward the red edge of the instability strip, the correct general conclusion is, that the problem of the Cepheid amplitude dependence on the star position in the instability strip, is still under consideration. Up to now the attempts made to reveal some reasons about the discrepancy in the results obtained by various authors did not give clear outcomes. In order to proceed the efforts in this direction we shall ask several questions, which, according to our opinion are important. One of them is: what is the reason for introducing the $(B-V)_o$ in the maximum light or along those in the mean light, when resolving the Cepheid amplitude dependence on the star position within the instability strip? Another very important problem concerns the possible difference in the physical nature of the various stellar system Cepheids: is there such a difference, what is its kind, to what limits does it reach, and what are the restrictions which it imposes on our investigations?

ACKNOWLEDGEMENTS

Our thanks are due to Prof. L.N. Mavridis for his kind proposal, which gives us the possibility to publish this paper in such a distinguished volume. We are obliged to many colleagues and especially to Dr. Yu. N. Efremov for discussions on the problem in consideration. We express our special thanks to Mrs. S. Pomenova, who edited the paper.

REFERENCES

Arp, H.C.: 1960, Astron. J. 65, 404.
Bahner, H., Hiltner, W.A. and Kraft, R.P.: 1962, Astrophys. J. Suppl. 6, 319.
Bergh, S. van den : 1968, J. Roy. Astron. Soc. Can. 62, 145.
Butler, C.J.: 1976, Astron. Astrophys. Suppl. 24, 299.
Canavaggia, R., Mianes, P., and Rousseau, J.: 1975, Astron. Astrophys. 43, 275.
Dean, J.F., Warren, P.R., and Cousins, A.W.J.: 1978, Monthly Notices Roy. Astron. Soc., 183, 569.
Efremov, Yu.N.: 1966, Perem. Zvezdy 16, 18.
Efremov, Yu.N. and Nikolov, N.S.: 1979, Astrophys. Space Sci. 63, 193.
Eggen, O.J.: 1951, Astrophys. J. 153, 195.
Eggen, O.J., Gascoigne, S.C.B. and Burr, E.G.: 1957, Monthly Notices Roy. Astron. Soc. 117, 406.
Fernie, J.D.: 1970, Astrophys. J. 161, 679.
Gascoigne, S.C.B.: 1969, Monthly Notices Roy. Astron. Soc. 146, 1.
Ivanov, G.R.: 1974, Dissertation, University of Sofia.
Ivanov, G.R. and Nikolov, N.S.: 1979, Astrophys. Space Sci, 60, 329.
Kelsall, T.: 1972, Goddard Space Flight Center, Preprint.
Kraft, R.P.: 1960, Astrophys. J. 131, 330.
Kraft, R.P.: 1961, Astrophys. J. 134, 616.
Kron, G.E.: 1958, Publ. Astron. Soc. Pacific, 70, 561.
Kurochkin, N.E.: 1966, Perem. Zvezdy 16, 10.
Madore, B.M.: 1976a, Observatory No. 1015.
Madore, B.M.: 1976b, Roy. Greenwich Obs. Bull. No. 182.
Mianes, P.: 1963, Ann. Astrophys. 26, 1.
Nikolov, N.S. and Ivanov, G.R.: 1973a, Astron. Zh. 51, 132.
Nikolov, N.S. and Ivanov, G.R.: 1973b, Perem. Zvezdy 19, 207.

Nikolov, N.S. and Ivanov, G.R.: 1975, Astrophys. Invest. Bulgar. Acad.
    Sci. 1, 25.
Nikolov, N.S. and Ivanov, G.R.: 1976, Astrophys. Invest. Bulgar. Acad.
    Sci. 3, 1.
Nikolov, N.S. and Momchev, G.K.: 1972, Bull. Sect. Astron., Acad. Bulgare
    Sci. 5, 73.
Sandage, A.R.: 1958, Astrophys. J. 127, 513.
Sandage, A.R. and Tammann, G.A.: 1968, Astrophys. J. 151, 531.
Sandage, A.R. and Tammann, G.A.: 1969, Astrophys. J. 157, 683.
Sandage, A.R. and Tammann, G.A.: 1971, Astrophys. J. 167, 293.
Schaltenbrandt, R. and Tammamm, G.A.: 1970, Astron. Astrophys. 7, 289.
Tammann, G.A.: 1969, in The Spiral Structure of Our Galaxy, IAU Symp.
    No. 38 (ed by W. Becker and G. Contopoulos) p. 237.
Tsarevsky, G.S. and Yakimova, N.N.: 1970, Perem. Zvezdy, 17, 120.
Yakimova, N.N.: 1970a, Perem. Zvezdy 17, 253.
Yakimova, N.N.: 1970b, Astron. Circ. No. 595, 5.
Yakimova, N.N., Nikolov, N.S. and Ivanov, G.R.: 1975, in Variable Stars
    and Stellar Evolution, IAU Symp. No. 67 (ed. by V.E. Sherwood and
    L. Plaut) p. 202.

EARLY-TYPE EMISSION-LINE STARS

S.N.Svolopoulos
National University of Athens
Athens, Greece

ABSTRACT

Early-type emission-line stars keep the continuous interest of the scientists. As it is believed the emission lines are formed in the extensive envelopes surrounding these stars. The various objects of this sort could be generally grouped into the following classes:
a) the Of stars; b) the Wolf-Rayet stars; c) the Be-stars; d) the Shell stars; e) the B-type supergiants (α Cygni and P Cygni); f) the Stars with composite spectra and g) the Symbiotic stars.

In this paper after a general discussion of some of the contemporary problems of the previous objects, we are briefly referring to our endeavours for a related research in the Laboratory of Astrophysics at the National University of Athens.

THE Of STARS

The spectrum of these stars is of type O, displaying, in addition, the emission lines HeII $\lambda4686\overset{\circ}{A}$, NIII $\lambda4634\overset{\circ}{A}$, $\lambda4640\overset{\circ}{A}$ and $\lambda4611\overset{\circ}{A}$. Some other lines, like H_α, appear in weak emission.

In the case of the Of stars, a star of high surface temperature is surrounded by a gaseous envelope of small density and low temperature.

THE WOLF-RAYET STARS

These systems are similar to the objects of the previous group. Again, a very hot star is surrounded by an envelope.

Broad emission bands are the main features of the Wolf-Rayet spectra. These spectra are classified either as WC-spectra, or as WN-spectra (Underhill,1960). In the WC-spectra strong lines of CII, CIII, CIV and of OII, OIII, OIV, OV, OVI appear, in addition, to the emission lines of

285

E. G. Mariolopoulos et al. (eds.), Compendium in Astronomy, 285–293.
Copyright © 1982 by D. Reidel Publishing Company.

H, HeI and HeII. On the contrary in the WN-spectra, the emission lines of NII, NIII, NIV and NV are prominent.

THE Be STARS

The Be stars are stars with B-type spectra containing emission lines. Struve (1931) was the first to prove that emission lines are often observed in the spectra of rapidly rotating B-type stars. According to Slettebak (1949, 1955) in the average the Be stars are rotating faster than the normal B stars.

Consequently, the origin of the emission lines is related to the rotational speed of the star. The equilibrium in the equatorial zone of the rapidly rotating B stars is disturbed, and therefore material escapes from the star to the surrounding space, where gaseous rings are formed.

Generally, H_α is the strongest emission line in Be spectra. The emission is weaker at the higher members of the Balmer series. Many years ago Curtiss (1923) gave a formula which shows the relation between the width of the emission lines (Δ_λ) and the wave length (λ). If H_β is the observed width of H_β in Å, the formula gives:

$$\Delta_\lambda = 0.000628(\lambda-3270) (H_\beta-2.61) + 2.61$$

In the majority of the Be stars, emission is observed only at the early members of the Balmer series. However, there are cases, where emission has been observed to extend to very high members of the Balmer series. In a very few cases emission appears even at higher members than the Balmer absorption breaks. For example, there are spectra containing absorption lines of the Balmer series to about H_{15}, while the corresponding emission lines extend up to H_{22}.

Be spectra, with very bright emission at the lines of the Balmer series, also, show emission in certain lines of FeII. In some of these stars emission is observed at the forbidden lines of FeII, too.

The spectra of the rings are of type A, while the spectra of the central stars are similar to B3. This means that the ionization within the rings is much smaller than the ionization in the layers of the star.

Many Be spectra show emission at M_gII $\lambda4481$Å and at the strong lines of SiII. Also, emission is suspected at some lines of CrII and NiII.

Most of the emission lines are double and only in a very few cases their two components are of the same brightness.

Burbidge (1951, 1953, 1956) after simple assumptions determined the size of the emitting regions. From the observed line intensities, it was found that in average the radius of the emitting regions is ten times greater than the radius of the central star. Therefore, a ring, around an average Be star, consisting, mainly, of hydrogen at a density of 10^{-11} gr/cm^3 and in a temperature of about 12000°K, must have a mass e-qual to 10^{-8} solar masses.

One important property of the Be stars is the variability of the e-mission lines. The ratio V/R, of the intensity V of the violet component of the emission to the intensity R of the red component, varies in an impressive way. McLaughlin (1938) had proved that the variation of the ratio V/R consists of many years cycles. For the explanation of this variation, he suggested that the gaseous rings not only rotate around the central star, but also the rings pulsate continuously.

The importance of an expanding gaseous envelope was pointed out by Gerasimovic (1934). According to him the centrifugal acceleration due to rotation will continue to expand the envelope, while a new part of the envelope is formed below. When a gaseous envelope has become very dense, then the Lyman-α radiation radiated from the envelope towards the star, prevents the ejection of more atoms from the stellar surface. This pro-cess continues by the moment, when the envelope following its expansion becomes very thin. Then, the intensity of Lyman-α radiation from the en-velope towards the star has been considerably decreased and gaseous masses start again to flow from the star to the surrounding space.

Though the processes mentioned by Gerasimovic, to some extend, take place within the rings, it is true that the radial velocity curves do not support Gerasimovic hypothesis. Actually, with the observed radial veloc-ity data we can determine the motions of the gases, and therefore to check the theory.

A simple model for the variation of the V/R ratio was given by Huang (1975). According to this model the rotating gaseous ring is elliptical and the great axis of its elliptical shape rotates slowly. Accordingly the position of the ring relative to the observer changes continuously. Consequently, the variations of V/R follow.

Rottenberg (1952) has proved that, if the motions of the gases and the radiation scattering are both given, then it is possible to determine the profiles of the various emission lines in the Be spectra.

A much more detailed examination of the same problem was done by Marlborough (1969, 1970). Marlborough assumed a steady state and that the ejection of the gases was a result of the high rotational velocity of the star. In addition he assumed certain density distributions and then he constructed a series of models. For each one of these models he determined the excitation and the ionization of hydrogen atoms at various parts of the gaseous rings. After that he computed the line profiles for each model. The comparison of the computed profiles with the observed ones gave information about the emitting regions of Be stars.

After a study of a great number of models, Poeckert and Marlborough (1978), demonstrated that the inclination of the rings influences the

form of the emission line profiles and the intensity of the continuous
spectrum too. Also, they proved that the profiles and the continuum spectrum are influenced by changes in the envelope's density and temperature.
However, a small set of observational data is not enough in order to determine the exact model for a certain Be star.

According to Stalio (1980) the variation of the luminosity of a Be
star together with the H_α-profile and the profile of ultraviolet resonance lines give information about the motions, the temperature and the
density in the exterior layers.

SHELL STARS

The shell stars have a structure similar to that of the Be stars.
However, the envelope of the shell stars is much more massive than the
envelope of the Be stars. Therefore, the absorption in the shell stars
is much more greater than in the Be stars and it forms deep absorption
lines. The shell spectrum reminds of the spectrum of an A-type supergiant
with strong lines of FeII, CrII, TiII etc, while the lines of the neutral
metals are weak. Also, the forbidden lines appear strong in comparison
to the appearance of the permitted lines. This happens because there is
not thermodynamic equilibrium in the shells and the radiation field is
diluted (Struve and Wurm, 1938). For this same reason the permitted lines
MgII $\lambda4481\text{Å}$, SiII $\lambda4128\text{Å}$, and $\lambda4130\text{Å}$ are abnormally weak in comparison
with the strength of the lines from the ionized metals.

Furthermore the most important difference between the spectra of the
supergiants and the spectra of the shell stars is that the continuous
opacity is greater in the supergiants than in the shell stars. Actually,
in the supergiants we can not see the photosphere below their extended
atmospheres.

The shell's density is very high near the stellar surface, while
further away the density drops and the material there is very thin. Within the shell the gases move with various velocities, as it can be seen
from the profiles of certain lines, especially of FeII and of H.

When the shell is active, the range of the observed radial velocities is very great (Struve, 1942; Merrill and Sanford 1944; Underhill,
1966). Then some absorption lines become asymmetric and the observed by
the Balmer lines radial velocities become more negative with higher series member. This phenomenon is usually called the Balmer progression and
it can be explained by the fact that material is ejected from the star
with a high speed which is decelerated, as the material moves away from
the star.

There are some cases exhibiting a reverse line asymmetry and then
the Balmer progression is reversed. Many times various disturbances take
place within the shells because of abnormal changes in the radiation of
the central star.

B-TYPE SUPERGIANTS (α CYGNI AND P CYGNI)

The spectra of the supergiants of luminosity class Ia of spectral types between O9 and B9 display a complicated profile at H_α. It consists of a sharp emission peak and a violet-displaced absorption component. Some other lines, such as H_β, HeI $\lambda5876\text{Å}$ and HeI $\lambda6678\text{Å}$, often exhibit a similar profile.

Usually, this type of profile is called P Cygni-profile and it can be explained by the hypothesis of a rapidly expanding envelope. The violet-displaced absorption line is formed in the section of the envelope that is found on the line of sight and which moves towards the observer. Some typical P Cygni-profiles are illustrated in Figure 1.

STARS WITH COMPOSITE SPECTRA

This group contains systems with spectra resulting from the super-position of two stellar spectra. This is the case of the close binaries. In their spectra lines of H, HeI, HeII, even in some cases of CaII, are observed in emission. These emission lines are formed in gaseous concentrations, that are arranged in rings, or shells, or streams in the vicinity of the binary. Many times the gaseous concentrations cover totally the one of the two stars.

The variety of the gaseous formations in a close binary is quite great. Very often the emission lines reveal gaseous concentrations at Lagrange L_1, in the region between the two stars. For example, emission lines are observed in phases following the main eclipse of an ecliptic variable.

In Figure 2 the profiles of HeI $\lambda4472\text{Å}$ in the spectra of β Lyrae are given for certain phases. β Lyr is a well known system. Gaseous streams are situated in the area of Lagrange L_1 of the system, while gases escape from the region of Lagrange L_2 and form an expanding envelope surrounding the whole system. The expanding envelope around β Lyr is fairly evident from the observation of strong absorpion lines of HeI which are affected by dilution. The triplets are stronger than the singlets. HeI $\lambda3888\text{Å}$ is the strongest line of all HeI-lines.

However, no emission is formed in the expanding envelope of β Lyr. Dadaev (1979) found that all the emission lines of β Lyr can be interpreted without any assumption of a common gaseous envelope surrounding the whole system. The emission has its origin in the streams moving in the neighbourhood of the two stellar components.

The orbital elements of β Lyrae have been found changed in the course of three years (Struve, Svolopoulos and Zebergs, 1960). It is possible that the varying effects of the gaseous streams may produce real dynamic changes in the orbital elements of one of the two components.

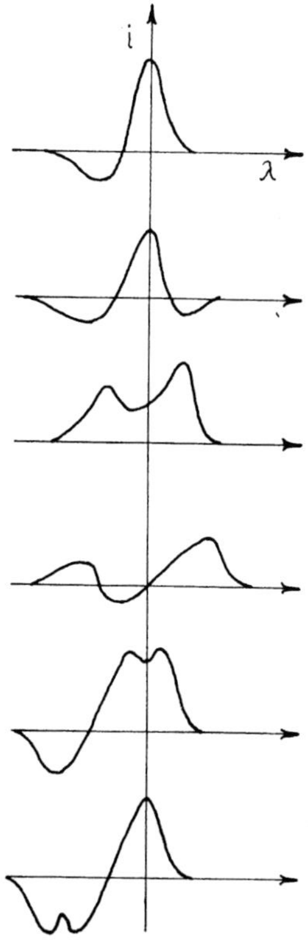

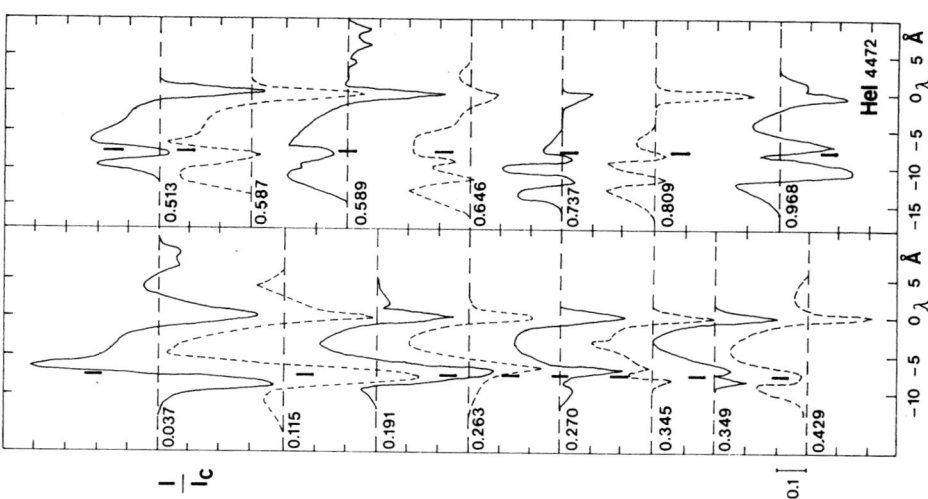

Fig. 2. Profiles of HeI λ4472Å in the spectra of β Lyrae for various phases.

Fig. 1. Typical P Cygni-profiles

SYMBIOTIC STARS

This group shows many similarities with the previous one, since the
spectra of the symbiotic stars display the superposition of two sorts of
features normally requiring two widely different excitation conditions.
It could be the case of the superposition of the spectra of two stars of
largely different surface temperature. However, the binary hypothesis
does not seem universally acceptable. Consequently, the spectra of the
symbiotic stars are often called combination spectra.

Z Andromedae is considered to be the prototype of the group. It is
a variable with a semi-regular light-curve of a period of 714 days and
with an amplitude less than about 3 mag. When the star is bright, the
features of a B-type shell star is observed in its spectrum. On the contrary, when the star becomes faint, its spectra are dominated by TiO-
bands.

It seems that the symbiotic stars are objects of semi-periodic erup-
tions, a phenomenon resembling to the recurrent novae. This view is supported by the fact, that in the combination spectra we observe the spectroscopic changes of a slow nova, superimposed on the spectral features
of a late type star.

As an example of the present day research on this subject, we are
mentioning a recent paper by Paczyński and Rudak (1980). Their basic assumptions are that the symbiotic stars are binaries and also that the
one component of the pair is a degenerate dwarf. The degenerate dwarf ac-
cretes matter from a wind of the secondary component. Finally, two models are suggested. In type I model the luminosity is produced in a stably
burning hydrogen shell. Variations of the accretion rate can cause variations in radius and effective temperature. In type II model the accretion is smaller than a critical value. In that case the hydrogen burning
proceeds through flashes. Therefore, the type II model could be related
to slow recurrent novae.

AX MONOCEROTIS

In the Laboratory of Astrophysics at the University of Athens a research survey dealing, with studies of certain B type stars with bright
emission lines has been undertaken since some years ago (Svolopoulos 1966,
1967, 1970, 1973, 1975, 1976). In October 1980 E.Danezis, Scientific Assistant at this Laboratory, took several new spectra by using the 152 cm
telescope at Haute Provence Observatory. Among those spectra there were
some of AX Monocerotis having a dispersion of 20 Å/mm.

AX Mon is a very interesting object with a variable composite spectrum. Merrill (1923) had classified some of its spectra as spectra of
type M. However, TiO bands have never been observed since then. After the
modern extensive studies by Cowley (1964) and Peton (1974), it is believed that AX Mon consists of a pair of stars. The one component is a

K-type giant and the other is a rapidly rotating B-type star, while in the surrounding space there is an extensive gaseous envelope. Though Cowley classified the hotter star as a star of B 3nn-type, the absence of the line HeII λ4200Å in the spectra studied by Peton gives more weight in classifying it as a B 0.5-star.

The most probable elements of the AX Mon are the following:

Spectrum: B 0.5 + K 2II

$P(\text{Binary})$: $232^d,5$ $P(\text{shell}) = 12$ years

$e = 0,02$, $a.\sin i = 1.64 \times 10^8$ km, $K = 51.5$ km/sec

$\gamma = +6.5$ km/sec, $f(m) = 3.26$ $m_B : m_K = 2.5$

The general changes in the spectrum of AX Mon during the 232.5 days-period are the following:

In the interval between the phases 0.9 and 0.1 a strong emission is observed at the lines of FeII and a narrow emission in the centre of the stellar absorption lines of H, He and MgII. Between the phases 0.1 and 0.4 the metallic lines are of a medium strength and they exhibit a double central absorption. Between the phases 0.4 and 0.6 the metallic lines are very strong. The emission of FeII lines is considerably strong. Lines of TiII, CrII and NiII are observed, too. Finally, between the phases 0.6 and 0.9 the spectra are similar to the spectra appearing in the interval between the phases 0.1 and 0.4.

The importance of the studies of the early-type emission-line stars can be easily seen from all that were mentioned previously. It is hoped that by these studies the processes that take place in the space surrounding the hot stars will be clarified.

REFERENCES

Beals, C.S.: 1950, Publ. Dominion Astrophys. Obs., 9, 1.
Bidelman, W.P.: 1976, in Be and Shell Stars, IAU Symp. No. 70 (ed. by A. Slettebak) Dordrecht-Holland, p. 457.
Burbidge, E.M. and Burbidge, G.R.: 1951, Astrophys. J., 113, 84.
Burbidge, E.M. and Burbidge, G.R.: 1953, Astrophys. J., 117, 407.
Burbidge, E.M. and Burbidge, G.R.: 1953, Astrophys. J., 118, 252.
Burbidge, E.M. and Burbidge, G.R.: 1956, Vistas Astron., 2, 1446.
Cowley, A.P.: 1964, Astrophys. J. 159, 817.
Curtiss, R.H.: 1923, Publ. Univ. Obs. Michigan, 3, 1.
Dadaev, A.N.: 1979, Izv. Glav. Astron. Obs. Pulkovo, No. 196, 28.
Gerasimovic, P.B.: 1934, Monthly Notices Roy. Astron. Soc., 94, 737.
Huang, S.S.: 1975, Sky Telesc., 49, 359.
Marlborough, J.M.: 1969, Astrophys. J., 156, 135.
Marlborough, J.M.: 1970, Astrophys. J., 159, 575.
McLaughlin, D.R.: 1938, Popular Astronomy, 46, 368.
Merrill, P.W.: 1923, Publ. Astron. Soc. Pacific 35, 303.
Merrill, P.W. and Sanford, R.F.: 1944, Astrophys. J., 100, 1.

Paczyński, B. and Rudak, B.: 1980, Astron. Astrophys., 82, 349.
Peton, A.: 1974, Astrophys. Space Sci., 30, 481.
Poeckert, R. and Marlborough, J.M.: 1978, Astrophys. J. Suppl. 38, 229.
Polidan, R.S.: 1976, in Be and Shell Stars, IAU Symp. No. 70 (ed. by A.
    Slettebak) Dordrecht-Holland, p. 401.
Rosseland, S.: 1936, Theoretical Astrophysics, Oxford, p. 284.
Rottenberg, J.A.: 1952, Monthly Notices Roy. Astron. Soc., 112, 125.
Sahade, J.: 1960, Star and Stellar Systems, 6, 466.
Slettebak, A.: 1949, Astrophys. J., 110, 498.
Slettebak, A.: 1955, Astrophys. J., 121, 653.
Stalio, R.: 1980, in Variability in Stars and Galaxies, Proc. 5th European
    Reg. Meet., IAU, Liege.
Struve, O.: 1931, Astrophys. J., 73, 94.
Struve, O. and Wurm, K.: 1938, Astrophys. J., 88, 84.
Struve, O.: 1942, Astrophys. J., 95, 134.
Struve, O., Svolopoulos, S.N., Zebergs V.: 1960, Astrophys. J., 131, 111.
Svolopoulos, S.N.: 1966, Ann. Astrophys. 29, 29.
Svolopoulos, S.N.: 1967, Astron. Nachr., 290, 155.
Svolopoulos, S.N.: 1970, Contr. Lab. Astronomy, Univ. Ioannina, 4.
Svolopoulos, S.N.: 1973, B.A.C., 24, 167.
Svolopoulos, S.N.: 1975, Astron. Astrophys. 41, 199.
Svolopoulos, S.N.: 1976, Astron. Nachr., 297, 87.
Underhill, A.: 1960, Stars and Stellar Systems, 6, 411.
Underhill, A.: 1966, in The Early Type Stars, D. Reidel, Dordrecht-Holland,
    p. 231.

PART VII

GALAXIES

THE STRUCTURE OF THE SMALL MAGELLANIC CLOUD

M. T. Brück
Department of Astronomy, University of Edinburgh
Edinburgh, U.K.

ABSTRACT

A survey of the distribution of various types of object in the Small Magellanic Cloud is used to build up a picture of the stellar content of its dominant features and to reconstruct in broad outline the evolutionary history of this galaxy as a whole.

1. Introduction

The two Magellanic Clouds, the nearest external galaxies, are of exceptional interest for a variety of reasons, of which we mention three in particular. First, the Clouds are the examples par excellence of their individual galaxy types : the LMC (type SBm) falls between the barred spirals and the true irregulars, typified by the SMC (type IBm), which has less recognisable spiral structure but the same basic components of bar and short γ-shaped bright arms. Secondly, the Clouds serve as an example of an interacting system, in the case of which meaningful models of their motion relative to the Galaxy may be constructed. Thirdly, being nearby, it is possible to study individual stars within them, estimate their ages, the circumstances of their formation and in this way perhaps illuminate the general question of star and galaxy formation everywhere.

For certain aspects of these problems the Small Cloud has some special advantages. Being the least massive of the triple system of the Galaxy and the Magellanic pair, it is expected to be the predominant victim of local gravitational and tidal effects. On the north side it adjoins the Magellanic Stream, which is a hint that the latter may represent débris torn from the SMC as the Magellanic system passed through supposed perigalacticon between 10^8 and 10^9 years ago.

Like all late-type galaxies, the Magellanic Clouds contain a substantial gas (HI) component. Normally, HI is found in conjunction with young stellar populations, in spiral arms and galactic disks. In the case of the Magellanic Clouds, their HI envelopes are so extensive as

297

E. G. Mariolopoulos et al. (eds.), Compendium in Astronomy, 297–313.
Copyright © 1982 by D. Reidel Publishing Company.

to merge into each other, leaving the entire system encompassed by one
large wrapping of HI extending over 30° of sky, equivalent to 30 kpc in
metric measure. Such a huge structure has the dimensions of an entire
large galaxy. An interesting question for discussion is whether it is
indeed one large unit, with stars as well as gas filling the inter-Cloud
regions, or whether the Clouds, as far as stars are concerned, are independent individual objects.

2. Basic Data

Though there have been various determinations of the distances to the
Magellanic Clouds, they are not so discrepant as to affect a general discussion of the structure of the SMC, and we adopt the usually accepted
figures of 50 and 65 kpc for the distances respectively of the Large and
Small Clouds. Their angular separation of 20° thus corresponds to 20 kpc.
Their masses, obtained from their rotation curves are 1×10^{10} and
1.5×10^9 solar masses, corresponding very roughly to 1/10 and 1/100 of
the mass of the Galaxy.

3. The Structure of the SMC

The photograph in Plate 1 of the SMC and the diagram on Figure 1
illustrate the main features, which are the asymmetrically placed bar
with the short luminous arms emerging from one end. The SMC also possesses
a faint section of an outer spiral (the outer spiral is a conspicuous
characteristic of the earlier type SBm galaxies like the LMC) and also
a unique feature of its own, namely, a wing protruding to the east in
the general direction of the Large Cloud. Further out in the same direction is another appendage, the outer wing which is less luminous than
the inner one, and this is followed, at an even greater distance, by a
small concentration of stars and gas (shown on Figure 4) which may not
be an integral part of the SMC. On short exposure photographs, the bar,
inner arms and inner wing dominate the picture, especially those taken
in blue light or through an Hα filter, because of the luminous blue associations and HII regions located there. On longer exposure photographs,
the outline of the SMC becomes more extended, taking on an elliptical
shape which is normally identified with a flattened tilted disk. Deeper
photography reveals the SMC as a rather smooth spheroid, reminiscent in
shape of a dwarf elliptical galaxy. Finally, on long exposure photographs,
it has been possible to extend the limits of the SMC to an approximately
circular contour of 6° radius, identified tentatively as a halo.

It is reasonable to envisage each successive shape as displaying
progressively older populations. For the purposes of analysing the structure of the SMC it is convenient to treat these subdivisions separately,
in order of increasing age : the bar, arms and wing; the disk; and the
halo. The outer wing, outer arm and inter-Cloud bridge will also be discussed separately.

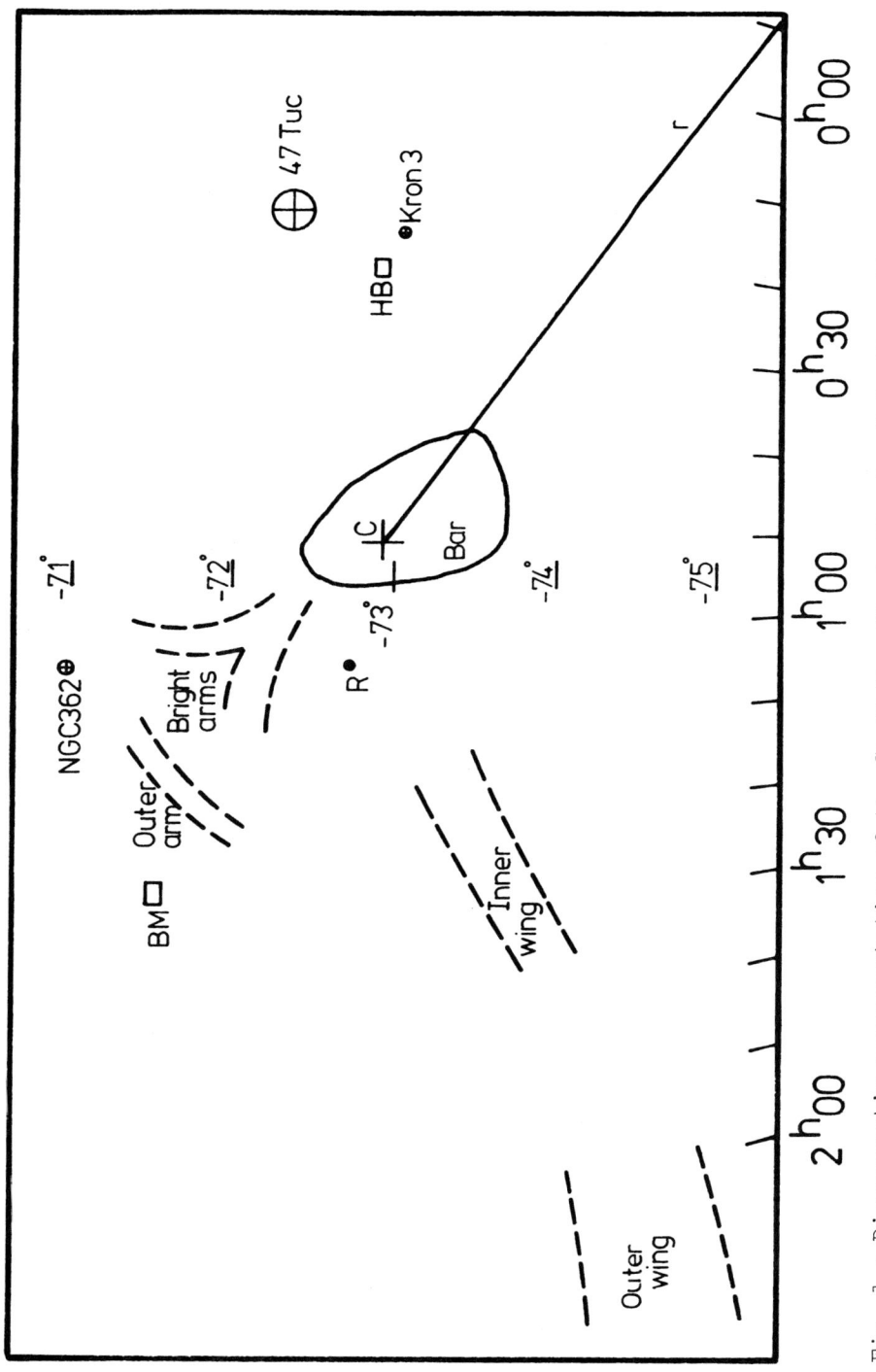

Fig. 1 : Diagrammatic representation of the Small Magellanic Cloud. C is the optical centre; R the radio rotation centre. The line from C is the radial direction to which the profiles in Figures 4 and 5 refer. The globular clusters 47 Tuc and NGC 362 (galactic) and Kron 3(SMC) are shown for references. The fields BM (Brück and Marsoglou, 1978) and HB (Hawkins and Brück, 1981) have been photometrically observed and are referred to in the text.

Plate 1: The Small Magellanic Cloud. Photograph taken with the 1.2 metre U.K. Schmidt Telescope. The photograph shows the central parts of the Cloud and the inner wing , with the galactic globular cluster 47 Tucanae on the right.

4. The Bar, Arms and Inner Wing

Though these conspicuous features were noted by earlier observers, serious discussion of their interpretation must begin with two classic papers, the star counts of de Vaucouleurs (1955) and the high resolution mapping in 21-cm radio of Hindman (1967). De Vaucouleurs performed counts to limiting magnitudes (m_{pg}) 14 and 16 ($M = -5^m$ and -3^m) and produced isopleths which showed (to -3^m) three main centers of concentration in the arms, the bar and the wing or "prominence", as he labelled it. The more luminous stars (to -5^m) were predominantly found in the arms, some in the wing and none in the bar.

A very similar pattern to the -5^m isopleths emerges in the distribution of HII regions (Davies et al., 1976) which represents the location of the very youngest stars, aged a few times 10^6 years. In HII, the arms are the most dominant, followed by the wing and the bar.

Hindman's HI contours showed a striking resemblance to de Vaucouleurs' -3^m counts and left in no doubt the strong correlation between gas and young ($\leq 10^7$ years) stars. This correlation has recently been placed on a quantitative basis by Azzopardi and Vigneau (1977) who find a tight relation of the form $\rho \propto B^n$ between the surface density ρ of these stars and the HI brightness B, with n taking the value 1.4. This formula was first applied by Sanduleak (1969) with respect to OB stars for which he found $n = 1.84$.

Azzopardi and Vigneau have also identified supergiants in the SMC whose distribution also favours the same key regions of the bar, arms and wing, and whose surface density is also correlated fairly well with HI. The correlation index n is not everywhere the same. Even in the bar there is an apparent difference between their relative densities in the north and south ends. It is interesting to look back at Hindman's work and find that he had already noted, with regard to de Vaucouleurs' -3^m counts, a higher proportion of stars in the northern part of the bar relative to the HI content. Thus, though the numbers of all young stars are correlated with the HI gas density, there are regional variations in the exact form which that correlation assumes, and a clear distinction in this respect between the various sub-structures.

Photometry of stars in the bar by Basinski, Bok and Bok (1967) shows - from their position on the colour-magnitude diagram - that these are young stars of age 5.10^6 years, while the age of the wing is given as 10^7 years by Westerlund (1961).

All evidence points to current or very recent ($10^6 - 10^7$ years) star formation in the bar, arms and wing. This does not mean that there have not been periods of star formation at other times. It is possible to get an idea of the duration of the star-formation epoch by looking at the distribution of other, less young, objects. The distribution of cepheid variable stars by Payne-Gaposckin and Gaposckin (1966) and subdivided by period (i.e. age) shows a tendency for younger ($\sim 5.10^7$ years) objects to outline the arms, bar and wing, while the older ones ($\sim 10^8$ years) form a gradually more diffuse distribution until, for objects older than 3.10^8 years these special regions are no longer recognisable.

In a similar manner the distribution of clusters, also classified by age, can be used for the same purpose. Obviously the most effective way of attaching an age to a cluster is to observe its colour-magnitude diagram. This information is available for some clusters but the vast majority of them (~ 300) have not been analysed in this way. Instead, it has been possible to provide an effective classification by overall appearance and colour based on intercomparison of UBV photographs using clusters with well-observed colour magnitude diagrams as calibrators (Brück, 1975). Compact populous blue clusters have been identified showing a distribution which favours the bar and arms, while luminous loose blue associations are found throughout the three regions. The age of the populous variety is estimated at 10^7 years (Janes and Carney, 1980, value for NGC 330); the others are younger. An older group, the small blue clusters, judged to be similar in size and luminosity to the Pleiades and hence having an age of $\sim 5.10^7$ years, are distributed smoothly in a pattern which does not in any way highlight the arm, bar or wing. Thus, though these three regions contain clusters older than 5.10^7 years, it is only as part of a wider structure described in connection with the disk in the next section. It may be concluded that these small blue clusters had already existed before the star formation events occurred which were responsible for the present appearance of the three luminous regions.

A very definite higher limit may be placed on the age of the bar, arms and wing by noting that clusters of intermediate age and older ($> 3.10^9$ years) are completely absent in these regions; indeed these objects are congregated in an entirely different part of the SMC, well to the west of the bar. It is noteworthy that planetary nebulae are found in the bar (Westerlund, 1968) but not at all in the arms and wing. Planetary nebulae represent later stages of stellar evolution with ages of $\sim 10^9$ years. We can therefore state with confidence that star formation did not start specifically in the bar earlier than 10^9 years ago; in the arms and wing it began considerably more recently.

Combining all this information allows us to set upper and lower limits to the era during which star formation occurred either continuously or in bursts. It is $\sim 5.10^6$ to $\sim 10^9$ years ago – with some more recent activity – in the bar; from 5.10^7 ago onwards and still continuing at a high level in the arms, and only currently in the wing. The longer period involved in the case of the bar may explain its higher star density, and the lower gas density in the wing presumably is the reason why this region has not produced the compact young groups characteristic of the bar and arms.

5. The Outer Wing, Outer Arm and Bridge

These features, less luminous and further out on the eastern side of the SMC, have been observed photometrically by Westerlund (1961),

Brück and Marsoglu (1978) and Kunkel (1980) respectively. Their CM diagrams are similar with main sequence termination corresponding to an age
of $2-6.10^7$ years in the case of the outer arm and to 10^7 years in the outer
wing and bridge. It is tempting to ascribe a common birth time to all
three, but it must be remembered that the bridge may not belong to the
SMC and that its apparent alignment with the outer wing may be a fortui-
tous projection effect.

6. The Disk

Models of barred galaxies, built up from a study of many known examples, have been constructed mainly with a view to understanding their
complex dynamical structure. The topic is reviewed by Freeman (1975)
and also by de Vaucouleurs and Freeman (1972). The characteristic feature of the Magellanic barred variety is the asymmetry of the bar with
respect to the disk of the galaxy. The bar, shaped like a prolate spheroid, lies in the plane of a flat exponential disk, in such a manner that
its centre is displaced by about 1 kpc from the centre of the disk and
also from the centre of rotation which does not necessarily coincide
with the centre of the disk. Following de Vaucouleurs' (1960) interpretation of his own star count observations combined with the photoelectric
surface brightness isophotes of Elsässer (1958), de Vaucouleurs and
Freeman have applied this model to the case of the SMC and have derived
the following data : b/a (ratio of minor to major axes of the apparent
disk) 0.5 leading to $i = 60^{\circ}$, p (position angle of the line of nodes)
45°; ϑ (distance of the end of the bar from the node) 20°. This small
value of ϑ means that the bar is being viewed almost end on, so that
one end - assumed to be the broad end - is closer to us than the narrow
end from which the arms protrude.

The disk, in this analysis, is recognised from the presence of disk-
like objects, which means objects of normal Population I such as define
the disk of our own Galaxy. Among such indicators are the small blue
clusters of the Pleiades type, which are very plentiful in the SMC. It
has already been mentioned in the previous section that these particular
clusters are not associated with the bar, arms and wing, but that they
occupy a more extended area. Like the -3^{m} stars, they form an elliptical structure, as do the moderately old Cepheids. Using those clusters,
similar values of b/a (0.5) and p (55°) are found (Brück, 1975) to those
of de Vaucouleurs. Hindman's HI brightness contours are also roughly
elliptical in shape and yield b/a = 0.33, p = 55°. The position angle 55°
is also the direction of maximum HI velocity gradient, denoting a rotating
body and confirming the orientation of its direction of rotation. The
centre of symmetry of the rotation curve, which is the centre of rotation
of the SMC, is displaced by $0^{\circ}.7$ from the centre of the SMC disk, taken
to be the centroid of distribution mapped out by the star counts and located within the bar.

The disk, then, defined as the region occupied by typical Population
I material, is established as having an apparent elliptical shape with
major axis position angle of 45° to 55°, b/a = 1/2 to 1/3, corresponding

to $i = 60°$ to $70°$.

The value of i obtained from b/a rests on the very simple assumption that the apparent shape is due solely to foreshortening of a perfectly flat, perfectly circular disk. How valid is this assumption ? There are a number of points which lead one to query whether the reality is quite so uncomplicated : (1) A finite thickness of the disk would increase the value of b/a, making the value derived for i a minimum (for irregular galaxies in general the true axial ratio is ~0.3 (Hodge and Hitchcock, 1966)). (2) Hindman's analysis of the HI velocity pattern shows "very little evidence of a decrease in velocity with distance from the major axis" contrary to what would be expected of a flattened system seen at small inclination. Hindman, on the basis of these two considerations, suggests a large value of i, possibly $90°$, i.e. an edge-on view. (3) Extended HI maps show contours of decreasing ellipticity. The value of i based on contour gets smaller with increasing size, the ultimate contours being virtually round, indicating a face-on view, i = 0. The decreasing ellipticity is also evident in the outer photoelectric isophotes and even in de Vaucouleurs' -3^m isopleths. (4) The shape of the outer arm (Figure 1) - a characteristic feature of SBm galaxies of which only a small faint portion is discernible in the SMC - is in the form of a circular arc. Since the outer arm, according to the model, lies in the plane of the disk, its circular shape would suggest that it is unforeshortened, i.e. viewed face on (Brück and Marsoglu, 1978). (5) The distribution of objects of increasing age, discussed in the previous section, shows a transition towards the elliptical outline of the disk, in which the bar loses its identity as it merges into the wider distribution. Since the bar is a three dimensional structure, it is likely that its contribution introduces a bias into the picture of the disk. (6) Star counts to considerably fainter limits (21^m, involving a ~10^9 years old population as described in the next section) have distinctly circular outer contours, suggestive in shape of a halo in the sense of a three dimensional spherical corona. Their outer outline imitates the outer contours of HI - but the HI component cannot simultaneously represent a three dimensional halo and a supposed flat disk.

These apparently contradictory pieces of evidence cannot be reconciled in terms of a flat disk; it is necessary to postulate a roughly spherical or ellipsoidal volume, rotating as before, but without a flat disk, the Population I disk-type objects inhabitating instead the denser equatorial regions while the older objects fill the entire volume. It appears that in the SMC the term "disk" refers to a type of population rather than a flat geometrical entity. The orientation of the outer arm (point (4) above) remains unexplained; perhaps it is a projection of a spherical shell. Additional evidence that the SMC disk has a complex spatial structure is found in the argument put forward by Florsch, first in 1972 and recently in considerable detail (Florsch et al., 1981). The basis of Florsch's work is the distinct difference in the apparent distance modulus between Population I objects at the north-west and south-east ends of the major axis diameter. Even when allowance is made for estimated variable extinction in various regions of the Cloud, there still remains a discrepancy of the order of $0^m.3$ interpreted as indicating that the extremities of the "disk" are at different distances in the line of sight. Florsch also suggests that the SMC is very deep, possibly as deep as 23 kpc at the south-west extremity of the bar.

7. The Halo

Just as the word "disk" is used to describe a type of object as well as a particular shape, so too the term "halo" may denote Population II objects, as found in the halo of our Galaxy, or may define a geometrical corona-like three dimensional structure. In the former sense, i.e. of a population description, the SMC was accorded a 7° diameter in Westerlund's (1974) well known table of the substructures of the SMC. Its presence is based on the observation of characteristic Population II objects, namely, red globular clusters and RR Lyrae stars. In an important investigation of RR Lyrae stars in a field about 2° to the north of the SMC bar, Graham (1975) identified a high density of such stars, higher by a factor of five, he suggests, than the corresponding density in the halo of our Galaxy.

The SMC also contains several red globular-shaped clusters, whose centroid is substantially displaced from the centroid of the Population I clusters. These clusters in turn may be subdivided into at least two categories on the basis of colour, either through their colour-magnitude diagrams, by integrated photometry (Kron and Guetter, 1976), or even visually, as explained earlier. One group is typified by the cluster Kron 3 with age 3.10^9 years, the other by Lindsay 1 with age at least 4.10^9 years, both ages established by Gascoigne (1980) from the morphology of their colour-magnitude diagrams. The CM diagram of the younger (3.10^9 years) cluster has a main sequence turn-off at about 21^m which is within the scope of measurement of currently available instruments. Unfortunately, at present, turn-off points fainter than this (i.e. belonging to groups older than $\sim 4.10^9$ years) cannot be established. However, it is surmised that the observation of RR Lyrae stars is in itself a guarantee that a population of age $\sim 10^{10}$ years is certainly present, perhaps in considerable numbers.

A valuable survey of CM diagrams in the SMC by Kontizas (1980) reveals that what may be termed the "intermediate age" population (i.e. $\sim 10^9$ years) is common, not only in the clusters themselves but in the surrounding fields. The same age group has been found by others (Brück and Marsoglu, 1978; Hardy et al., 1980; Stewart, 1980) and has been recently confirmed by high quality electronographic photometry of a field well clear of the bar and the cluster domain (Hawkins and Brück, 1981). The CM diagram, shown in Figure 2, has been interpreted on the basis of modern evolutionary models (Ciardullo and Demarque, 1977) as belonging to a 3 ± 1.10^9 year old population with a possible mixture of stars of ages up to 10^{10} years. In the CM diagram the concentration of points around 21^m represents the main sequence turn-off of the 3.10^9 year old (intermediate) stars; the abrupt upper limit of this grouping denotes a very definite lower age limit to the sample. The concentration of brighter

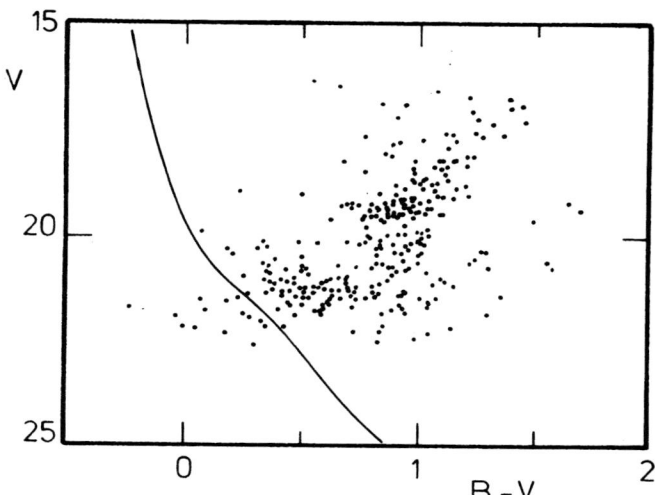

Fig. 2 : Colour magnitude diagram of a typical halo field in the SMC (Hawkins and Brück, 1981)

red stars at 19^m in the CM diagram is made up of the evolved "clump" giants of the same population plus the horizontal branch component of a possible older group of stars.

The conglomerate population represented in Figure 2 has been found to occupy an approximately circular area surrounding the SMC and for this reason we refer to it henceforth as the "halo", meaning the word in the geometrical sense. Certainly, as described above, it may well contain a proportion of halo-type, i.e. Population II, stars, but from what is learned from photometric analysis, a significant constituent is of intermediate age, much younger than the conventional members of the halo of our own Galaxy.

Counts to $21^m.5$ of UK Schmidt Telescope III AJ photographs performed by COSMOS reveal that this halo extends to a distance of about 6° from the SMC rotation centre (Brück, 1978, 1980). To reach the halo limits requires the measurement of at least six survey ($6^o \times 6^o$) fields of the UK Schmidt Telescope, and because of the uncertainty of photometric calibration at the faint end of the magnitude range, there is liable to be a difference of 20 percent or more in star density between adjacent fields. As a result, the overall contour, made up by combining contours from separate fields, may suffer a certain amount of distortion. Two contours, one at a level of 6 counts per mm^2 to $21^m.5$ (~17,000 stars per square degree) and the other the ultimate detectable contour, are shown, smoothed, in Figure 3. In spite of photometric difficulties, it is evident that the shape tends to roundness, and follows approximately the shape of the blue photoelectric isophotes, which are also shown on the diagram, and also those of the HI contours of the Magellanic system (Mathewson et al.,

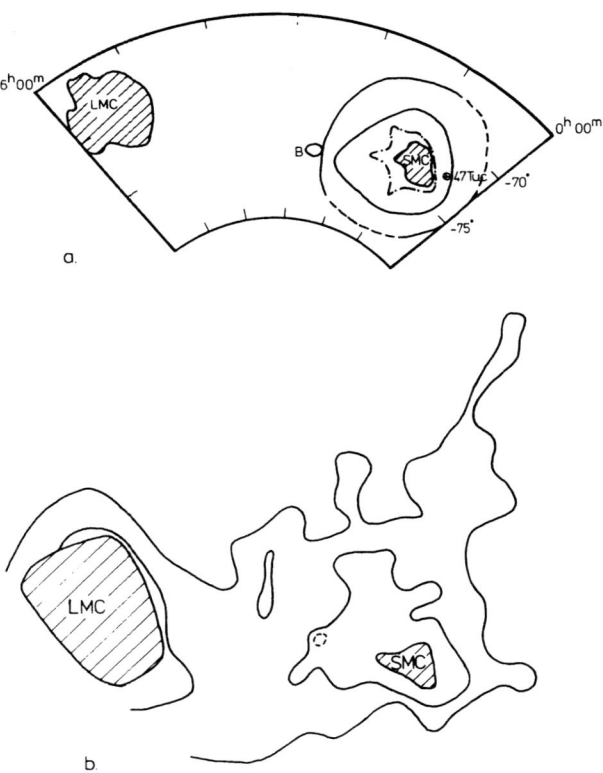

Fig. 3:(a) Map showing the Large and Small Magellanic Clouds. The hatched
areas are the central parts of the Clouds as delineated on the Smithsonian Astrophysical Observatory star charts. Contours surrounding the
SMC are :
Solid lines (————) : isopleths to $21^m.5$ from star counts using U.K.
Schmidt telescope photographs. Inner contour represents approximately
6 stars per mm^2 (1 mm = 1.1 arcminute); outer contour is the extreme measurable contour with approximately 2 stars per mm^2. The outer contour
may be distorted (see text); dashed line (-----) represents interpolation
in fields not recorded. Dotted (-.-.-.) contour : the outermost contour
photoelectrically observed (Elsässer, 1958)
B is the intercloud bridge (Kunkel, 1980; Brück, 1978)
(b) HI radio contours by Mathewson et al. (1974) of the Magellanic system.
Two contours only have been included, to 10 and 2 brightness units
(1 unit = 2 x 10^{19} atoms cm^{-2}). The outer contour shows an extension in
the direction of the Magellanic Stream and also encloses both Clouds.
The dotted circle is a region of higher radio brightness coincident with
the bridge B. (The HI map is on a galactic coordinate projection - coordinates not reproduced here - and is not exactly on the same scale
as (a)).

1974) confirming the tendency in contours of all kinds towards increasing roundness as one moves away from the central irregular features of the arms, bar and wing. In fact the halo contours obliterate even the outer wing (Brück, 1978) so that the description of the wings as "prominences" or "tidal arms" is not appropriate : these outer regions are not extraneous objects but integral parts of the SMC itself.

One degree beyond the limiting halo contour falls a group of stars belonging to the inter-cloud bridge. This group shows up on COSMOS maps and is referred to as "patch D" in the first paper in the series (Brück, 1978). It has also been investigated by Kunkel (1980) who incorporates it into his model of a past Magellanic-Galaxy encounter. It is quite possible, however, that this bridge may belong to the SMC halo : a major difficulty in determining the ultimate limit of the halo hinges on a correct value for the non-Magellanic images on the photographs. There may be a low density of Magellanic stars present, spreading right across the inter-Cloud gap and coextensive with the HI envelope, which defies detection because the present uncertainty in the foreground galactic star counts exceeds the level of the possible Magellanic star density.

8. The Oldest Population

While it has been established that an older (4.10^9 years upwards, probably Population II) component is present in the halo, there is only one type of object belonging to this component for which a complete sample is available. This is the red globular clusters, most of which fall in the extreme western part of the halo. Their distribution is quite unrelated to the features of the main body of the SMC, with a centroid displaced ~1 kpc from the centre of the disk and ~2 kpc from the centre of rotation. These clusters appear to be even more displaced than those of intermediate age.

The asymmetrical distribution of the halo as a whole is evident from the fact that COSMOS counts are about 5 times higher in the west than in the east (Brück, 1980). An additional asymmetry in the older component would be expected to influence the relative densities of stars around ~19^m and ~21^m in different parts of the halo, since the former group, as Figure 2 illustrates, includes the oldest stars whereas the latter represents the intermediate population only. However, the photometric consistency between photographs of different regions is not sufficient to allow such a comparison to be made at present. Within the limits of error, the luminosity functions of the two extremities of the halo are the same (Brück, 1980). Another test of the behaviour of the older stars is described in relation to the radial profile of the halo in the next section which suggests that there may be a small difference between their distribution and that of the intermediate age members.

9. Luminosity Functions

The relation between star density and HI gas density plays an important part in the discussions so far. Describing it in the form

$\rho \propto B^n$ (the Schmidt formula) or a linear relation between log ρ and log B (where ρ is star density and B is HI brightness) with slope n, gives a value of n - the correlation index - which is found to vary with the type of object observed, and in some cases also with location. Details of the various values obtained in the SMC have been assembled by Azzopardi and Vigneau (1977) and by Brück (1980). There has been some controversy as to the physical meaning of the index n, originally introduced as a coefficient of star formation rate, which need not be discussed in the present context. Suffice it to remark that variations in n are equivalent in practice to variations in luminosity function, whatever the physical meaning of the index may be. For the SMC the index varies from 3 for the most luminous young objects to 1 for 19^m stars (M≤0). It is also higher in the bright arms than in the bar, confirming de Vaucouleurs' (1955) assertion, based solely on data available at the time (visually performed counts to $M = -5^m$ and -3^m) that the luminosity function is not the same in these two regions.

De Vaucouleurs' data may be extended by adding counts to fainter limits within the same population. Figure 2 shows that if counts are terminated in B at 19^m the prolific intermediate and older halo population is effectively excluded. Such counts and their variation with distance along the apparent major axis of the SMC in the south-west (the bar side) direction are shown in Figure 4, together with de Vaucouleurs' counts to 16^m along the same axis. On account of the very high star density the inner zones are unmeasurable on the UK Schmidt telescope photographs and therefore there is only a small overlap between the two curves. However, it is clear that the two curves represent the same "disk" stellar population; by displacing them vertically they have been made to coincide to form one continuous curve. The vertical shift necessary to achieve coincidence is 1.71 in log N, corresponding to a gradient in the cumulative luminosity function d logNm/dm of 0.6 per magnitude. This is the same figure which de Vaucouleurs obtained for the more luminous stars indicating that it is constant at this value over a range of magnitudes between 0 and -5^m. The gradient is close to the corresponding value for the solar neighbourhood (0.55) and much higher than for Population I samples (0.2) in our Galaxy.

The luminosity function gradient is equivalent, of course, to a mass function gradient but not necessarily to an initial mass function as the existing stars may be the accumulation of more than one era of star formation or of even a continuous process of star formation. The high gradient means that there are relatively large numbers of lower luminosity (or mass) stars among the SMC "disk" population compared with, for example, zero-age galactic clusters.

As regards the halo, having postulated that this is composed of an intermediate age probably together with an older population, we may attempt to determine whether both have the same profile or whether they behave differently as regards their spatial distribution. Knowing that the faintest stars observed in the present case (≥21^m) represent mainly intermediate age stars, whereas the stars between 19^m and 21^m represent

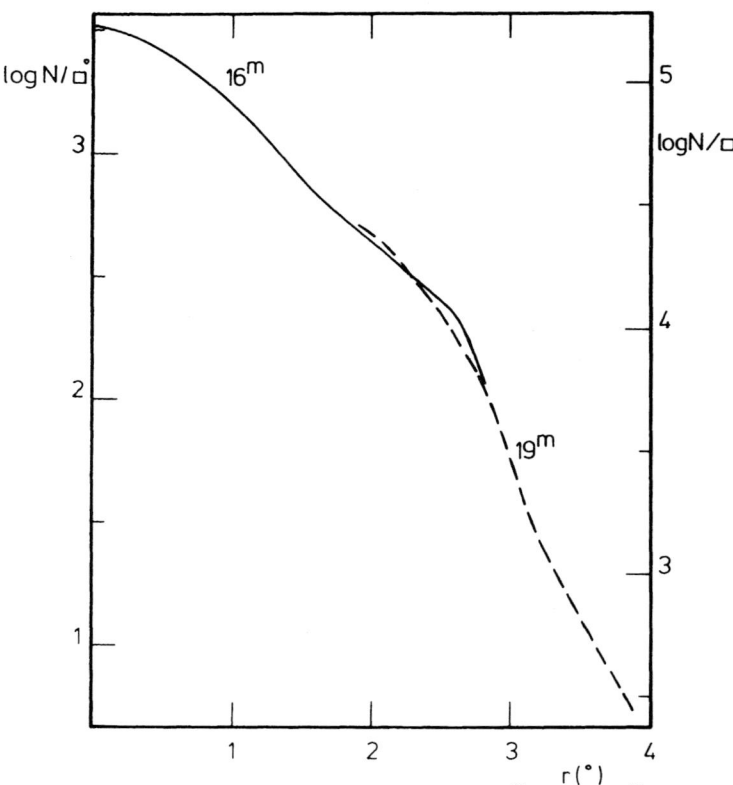

Fig. 4 : Variation in total star counts to 16^m and 19^m with radial distance from the SMC optical centre in the South-west direction (shown on Figure 1). Left ordinate refers to 16^m counts and right ordinate to 19^m counts.

a mixture of evolved intermediate and older stars, a segregation into these magnitude groupings might be expected to reflect differences in distribution between the two populations. In Figure 5, profile (a) represents the intermediate age only while (b) includes both age groups. From the close similarity in the outer regions between the two curves it would appear that there the intermediate age population of profile (a) also dominates profile (b), in other words that the intermediate age population forms the major contributor to the mixture. There appears to be some divergence between the curves in the outer regions, (b) falling off slightly sooner, suggesting that the older population has relatively higher density near the edge of the halo than elsewhere. Quantification of the relative densities is beyond the scope of present data.

10. Motions in the SMC

Reference has already been made to Hindman's study of HI which also includes a detailed analysis of the velocity fields. In it the author emphasizes the baffling complexity of the motions within the SMC which

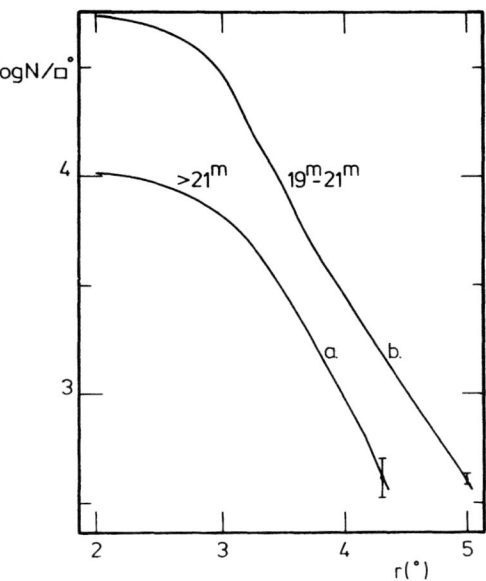

Fig. 5 : Variation in star counts along the same radial direction as Figure 4 (a) Stars fainter than 21^m including main sequence stars of intermediate age population. (b) Stars between 19^m and 21^m including horizontal branch and clump stars of intermediate and older populations.

defy explanation in terms of a simple rotating disk. This analysis has never been superseded or extended. A similar study of optical radial velocities by Ardeberg and Maurice (1979) reveals once again the diffi- culty of interpreting the motions and adds a further complication in the form of a distance differential between the various parts of the Cloud, as postulated also by Florsch (Section 6). Particularly interesting in connection with the present discussion is their suggestion that the outer wing of the SMC is part of a warp which crosses the main body and is also in recession. This contradicts the popular and somewhat intuitive view that, because the wings seen in projection on the sky point in the direc- tion of the LMC, they must therefore be lined up in that direction in space, and thus lie on the nearer side.

The complex nature of SMC motions has also been demonstrated by Florsch et al. (1981) who divides the main body of the SMC into two groups with quite distinct radial velocity patterns. Mention has already been made (Section 6) of Florsch's suggestion that the SMC has considerable depth as well as having a distance gradient along its major axis.

All these observations combine to reinforce the view that the struc- ture of the SMC cannot be made to conform to a simple model of a rotating flatdisked body.

11. The History of the SMC

Reviewing the information on the distribution of objects of various
ages, the distribution of gas (HI) and the correlation index relating
star to gas density, one may attempt to draw some inferences regarding
the past history of the SMC in more detail than had been possible in earlier assessments (Walker et al., 1969; van den Bergh, 1975).

Its first generation of stars appear to be less numerous than the
second intermediate ($\sim 3.10^9$ years) age generation which is widespread.
These fill a halo, coextensive with the HI gas but not directly correlated
with it in density; in fact the intermediate and older stars are asymmetrically situated with respect to the gas, which appears to have subsequently built up in the equatorial regions of the Cloud, forming the
apparent disk where the next generation (5.10^7-10^8 years) of stars were
born. The pervasive nature of the intermediate age stars indicates that
there was a lull in star formation between these two events, though it
is difficult to assert the exact upper limit of the age of the disk stars;
small clusters, older than 10^8 years, are quite undetectable.

Meantime the bar, with high concentration of gas, built up and began
star production which appears to have gone on continuously for the past
$\sim 10^9$ years. The star density there is high, but the rate of star formation at present (the correlation index $n \approx 1$) is lower than in the arms
and wing. These two regions are the youngest in the SMC where stars began
forming only $\sim 10^7$ years ago, first in the arms and later in the wing,
and still continues at a higher rate than in the bar (correlation index
2-3 for the arms). The wing is distinctive in that it has not managed
to produce the populous clusters found elsewhere.

The outer arm and wing represent regions where a burst of star formation occurred about 1-6.10^7 years ago, but has since ceased.

12. Conclusion

In the last few years, thanks to large instruments and automatic
star counting facilities, it has been possible to extend observations
of faint stars to wider spatial limits in the SMC, so that it is in fact
the first and only external galaxy in the case of which the stellar and
gaseous content are found to coincide in extent. Is it in this respect
a typical normal small galaxy, or has the proximity of its large neigh-
bour affected its behaviour now or in the past ? Is it an independent
unit, or a "siamese twin" of the Large Cloud, attached to it by a bridge
of stars and gas ? Is it perhaps even linked by stars and gas to our own
Galaxy ? What is the role of the Magellanic Stream ?

The Small Magellanic Cloud has not yet yielded up all its secrets;
there are still fascinating questions awaiting the enquiring astronomer.

Photography by Photolabs, Royal Observatory, Edinburgh. Original
negative by U.K. Schmidt Telescope Unit.

REFERENCES

Ardeberg, A. and Maurice, E.: 1979, Astron. Astrophys. 77, 277.
Azzopardi, A. and Vigneau, J.: 1977, Astron. Astrophys. 56, 151.
Basinski, J.M., Bok, B.J., and Bok, P.F.: 1967, Monthly Notices Roy.
    Astron. Soc. 137, 55.
Bergh, S. van den: 1975, Ann. Rev. Astron. Asrophys. 13, 217.
Brück, M.T.: 1975, Monthly Notices Roy. Astron. Soc. 173, 327.
Brück, M.T.: 1976, Occ. Rep. Roy. Obs. Edinburgh No. 1.
Brück, M.T.: 1978, Astron. Astrophys. 68, 181.
Brück, M.T.: 1980, Astron. Astrophys. 87, 92.
Brück, M.T. and Marsoglu, A.: 1978, Astron. Astrophys. 68, 193.
Ciardullo, R.B. and Demarque, P.: 1977, Trans. Yale Univ. Obs. 33.
Davies, R.D., Elliott, K.H., and Meaburn, J.: 1976, Mem. Roy. Astron.
    Soc. 81, 89.
Elsässer, H.: 1958, Z. Astrophys. 45, 24.
Florsch, A.: 1972, Publ. Obs. Strasbourg, 2, fasc. 1.
Florsch, A., Marcout, J., and Fleck, E.: 1981, Astron. Astrophys. 96, 158.
Freeman, K.C.: 1975, Stars and Stellar Systems 9, 409.
Gascoigne, S.C.B.: 1980, in Star Clusters, IAU Symp. No. 85 (ed. by J.
    E. Hesser), D. Reidel, Dordrecht-Holland, p. 305.
Graham, J.A.: 1975, Publ. Astron. Soc. Pacific, 87, 641.
Hardy, E., Melnick, J., and Reheault, C.: 1980, in Star Clusters, IAU
    Symp. No. 85, (ed. by J.E. Hesser) D. Reidel, Dordrecht-Holland, p. 343.
Hawkins, M.R.S. and Brück, M.T.: 1981, Monthly Notices Roy. Astron. Soc.
    (in press).
Hindman, J.V.: 1967, Australian J. Phys. 20, 147.
Hodge, P.W. and Hitchcock, J.L.: 1966, Publ. Astron. Soc. Pacific, 78, 79.
Janes, K.A. and Carney, B.W.: 1980, in Star Clusters, IAU Symp. No. 85
    (ed. by J.E. Hesser), D. Reidel, Dordrecht-Holland, p. 349.
Kontizas, M.: 1980, Astron. Astrophys. Suppl. Ser. 40, 151.
Kron, G.E. and Guetter, H.H.: 1976, Astron. J., 81, 817.
Kunkel, W.: 1980, in Star Clusters, IAU Symp. No. 85 (ed. by J.E. Hesser),
    D. Reidel, Dordrecht-Holland, p. 353.
Mathewson, D.S., Cleary, M.N., and Murray, J.D.: 1974, in Galactic Radio
    Astronomy, IAU Symp. No. 60 (ed. by F.J. Kerr and S.C. Simonson,
    III), D. Reidel, Dordrecht-Holland, p. 617.
Payne-Gaposchkin, C. and Gaposchkin, S.: 1966, Smithsonian Contrib.
    Astrophys., No. 9.
Sanduleak, N.: 1969, Astron. J. 74, 47.
Stewart, N.: 1980, Ph. D. Thesis, Univ. of Edinburgh.
Vaucouleurs, G. de: 1955, Astron. J. 60, 219.
Vaucouleurs, G. de: 1960, Astrophys. J. 131, 574.
Vaucouleurs, G. de and Freeman, K.C.: 1972, Vistas Astron. 14, 163.
Walker, M.F., Blanco, V.M., and Kunkel, W.E.:1969, Astron. J. 74, 44.
Westerlund, B.E.: 1961, Uppsala Astron. Obs. Ann. 5, No. 2.
Westerlund, B.E.: 1968, in Planetary Nebulae, IAU Symp. No. 34 (ed. by
    D.E. Osterbrock and C.R. O'Dell), D. Reidel, Dordrecht-Holland, p.23.
Westerlund, B.E.: 1974, in Galaxies and Relativistic Astrophysics, Proc.
    First European Astron. Meeting (ed. by B. Barbanis and J.D. Hadji-
    demetriou), Springer-Verlag, Berlin-Heidelberg-New York, p. 39.

LARGE SCALE DISTRIBUTION OF ELEMENTS AND GRAVITY

Wilhelmina Iwanowska
Institute of Astronomy
Nicolaus Copernicus University
Torun, Poland

ABSTRACT

Present perspectives for the idea of the gravitational separation of elements as a primordial factor generating gradients of the chemical composition of matter in gravitating systems are discussed.

More than ten years ago radial and vertical gradients of metallicity have been found in our and other galaxies (Morgan and Mayall 1957; Arp, 1965; McClure 1969). Later on, it appeared more and more evident that these gradients increase with the atomic number of elements. These two facts led the present author to think that gravitational separation of elements could be at work in gaseous media of gravitating systems. I expressed this supposition at the IAU meeting in Brighton in 1970 and in 1972 presented a joint paper by E. Basinska and myself (Basinska and Iwanowska,1974, Paper I) at the First European Astronomical Meeting in Athens. Dr. Basinska pursued this topic further (Basinska-Grzesik,1974, 1978, Papers II and III). In the present note I intend to reconsider the arguments for this idea, taking into account those papers as well as new data published since then.

In Paper I an attempt has been made to estimate the rate of gravitational diffusion of HeI or FeI atoms (component 2) relatively to HI atoms (component 1) in a column of gas perpendicular to the plane of the Galaxy at the solar distance from its center. A static model of the Galaxy was assumed with the present distribution of mass as given in the model by Perek (1962). The evolutionary process consisted in a continuous conversion of gas into stars changing the ratio of stellar to gaseous components from zero up to its present value. The relative velocity of diffusion ($v_1 - v_2$) was calculated with the classical formula for neutral two-component gas given by Chapman and Cowling (1952, 1970)

$$v_1 - v_2 = -D_{12}\left[\frac{1}{c_1 c_2}\frac{\partial c_1}{\partial z} + \frac{m_2 - m_1}{c_1 m_1 + c_2 m_2}\frac{1}{p}\frac{\partial p}{\partial z} - \frac{m_1 m_2 (F_1 - F_2)}{kT(c_1 m_1 + c_2 m_2)} + \frac{\alpha_{12}}{T}\frac{\partial T}{\partial z}\right] \quad (1)$$

with

E. G. Mariolopoulos et al. (eds.), Compendium in Astronomy, 315–322.

$$D_{12} = \frac{3}{8n\sigma_{12}^2} \left[\frac{kT(m_1+m_2)}{2\pi\,m_1 m_2}\right]^{1/2}, \quad \sigma_{12} = \frac{1}{2}(\sigma_1 + \sigma_2).$$

c_1, c_2 and m_1, m_2 are concentrations and mass values of hydrogen and helium or iron atoms respectively;
σ_1, σ_2 - the diameters of the respective atoms;
F_1, F_2 - any external forces per unit mass acting on the particles;
T is the temperature of gas, assumed to be 100 K everywhere;
p - gas pressure.
Assuming hydrostatic equilibrium and taking into account the pressure term only, as we have done, one can write the formula as follows:

$$\upsilon_1 - \upsilon_2 = d_{12}(m_2-m_1)\frac{g}{n(kT)^{1/2}}, \tag{2}$$

with

$$d_{12} = \frac{3}{8\,\sigma_{12}^2}\left(\frac{m_1+m_2}{2\pi m_1 m_2}\right)^{1/2},$$

wherefrom one can see that the diffusion velocity is proportional to the atomic mass difference (m_2-m_1), as well as to the gravity acceleration (g) and inversely proportional to the number density of gas (n) and to the square root of temperature (T).

The calculated velocities increase with the height above the galactic plane ($|z|$) and with time (t), owing to the diminishing gas density (n). They reach at the present moment up to 1 km s^{-1} or 10 km s^{-1} for He and Fe atoms respectively at the height $|z| = 3$ kpc.

From numerical integration of the equation of continuity

$$\partial\rho/\partial t + \upsilon\,\mathrm{grad}\,\rho + \rho\,\mathrm{div}\,\upsilon = 0 \tag{3}$$

- calculated backwards - Dr. Basinska derived the changes of relative abundances n(He)/n(H) and n(Fe)/n(H) with time (t) and height above the galactic plane ($|z|$). The mean gradients of these abundances obtained for $0 < |z| < 300$ pc and $t = 10^{10}$ yrs (the time of the solar birth), expressed in the usual logarithmic scale, are

$$\frac{\partial[He/H]}{\partial z} = -0.07 \text{ kpc}^{-1}, \quad \frac{\partial[Fe/H]}{\partial z} = -0.64 \text{ kpc}^{-1}.$$

Mayor (1977) stated from photometric measurements of F,G,K stars performed by several authors that [Fe/H] value at $|z| = 500$ pc falls down to half of its value at $z = 0$. This means that

$$\frac{\partial[Fe/H]}{\partial z} = -0.60 \text{ kpc}^{-1}$$

over this interval of z at the time of birth of these stars. The excellent agreement with our calculated value is accidental, however, since

our model cannot be considered as physically real. Our accepted temperature value 100 K for the gas is not compatible with our accepted $n(z)$
distribution for hydrostatic equilibrium*. To bring these parameters to
local hydrostatic equilibrium the temperature should increase with $|z|$
up to a maximum value of 10^6 K at $|z|$ = 1500 pc what should diminish our
$\upsilon_1 - \upsilon_2$ values by a factor proportional to $(kT)^{1/2}$, or up to two orders
of magnitude in this extreme case. However, our accepted $n(z)$ distribution contains all matter of the dynamical model of the Galaxy including
its "unseen" component. We still don't know exactly what kind of matter
it is, yet, it seems reasonable to exclude this component from the gas
phase over the considered lifetime of the Galaxy of $14 \cdot 10^9$ yrs. We have
then to divide our accepted n value for $z = 0$ in the solar neighbourhood
by a factor of about 10 and maybe by factors increasing with height since
the "dark halo" is a very extended distribution. Therefore, these two
corrections nearly balance each other, leaving our calculated velocities
correct up to an order of magnitude. There remains, however, another
unknown factor - turbulence and other motions mixing galactic gas and
hampering more or less the process of gravitational diffusion. As we have
remarked, however, turbulence cannot destroy the systematic process of
gravitational sedimentation of heavy elements over distance ranges surpas-
sing by orders of magnitude the "mixing lengths" of turbulence. It slows
down the process of diffusion in a similar way as does increasing temperature. It is difficult, however to estimate at present to what degree
turbulent motions have slown down the process of diffusion in different
phases of interstellar gas.

In Papers II and III Dr. Basinska recalculated the changes of
$n(Fe)/n(H)$ with time also along the z-coordinate in our Galaxy, using a
two-phase model for interstellar gas, consisting of cold dense clouds
and warm intercloud medium. For this last component she obtained values
of the relative diffusion velocity very close to those obtained in Paper
I. Dr. Basinska calculated also changes with time of the radial gradients
of $n(Fe)/n(H)$ for stationary polytropic models of spherical systems meant
as globular protoclusters and elliptical protogalaxies evolving in the
gas-to-star conversion process. She obtained small changes in central
regions but significant one in the haloes of these models, increasing
with the total mass. Altogether she tested numerically what could be
expected from formula (2) that gravitational separation of elements can
be effective in two extreme cases: (a) in tenuous media with very low
density and (b) in dense media with very strong gravity acceleration.
Low temperature is a propitious environment in all cases. These statements invite one to look for a possible effectiveness of the process of
gravitational separation of elements at an earlier phase of Universe
whatever the process of galaxy formation was: a collapse of a tenuous
cloud (case a) or an explosive fragmentation of very dense material and
its expansion after the Big-Bang (case b).

Collapsing models of galaxies with incorporated star formation and

---

* Dr. B. Paczynski has called my attention to this inconsistency.

nucleosynthesis in stars have been extensively investigated by many
authors, to mention Larson, Tinsley, Truran, Cameron, Talbot, Arnett,
Pagel, Pachett a.o. (to which references are given in Paper III). With
some flexible assumptions metal enrichment progressing with time and
metallicity gradients could be adjusted by these authors to those observed in galaxies. Recently such adjustment of theory to the $[Fe/H]$
values obtained from uvbyβ photometry for F-type stars of different ages
was performed by Twarog (1980) with satisfactory fitting when Larson's
model of collapsing Galaxy with generous infall of gas was taken.

However, not any theory known to the present author explains the
mentioned fact of the dependence of abundances on the atomic mass of
elements. This correlation found by the present author from spectra of
long-period variables (Iwanowska, 1968), confirmed by Kuroczkin (1974)
for super-metal-rich stars, can be seen among main sequence stars studied
by Khokhlova (1976, 1977).

More recently many studies of the chemical composition of giants
in globular clusters have been made with high dispersion spectra (Griffin,
1979; Cohen, 1979, 1980; Pilachowski, a.o. 1980a,b, a.o.). All or nearly
all of their results show the effect of atomic mass (or atomic number).
Let us see e.g. the abundances obtained by Griffin for the giant L 973
in the globular cluster M 13 and those for α Boo taken as a comparison
star. The run of elemental deficiency with atomic mass (Fig. 1) is clearly seen in both stars, being steeper for the more metal poor one. This
phenomenon is more or less evident also in other abundance determinations,
which adversely do not reflect the relations expected from the theory
of nucleosynthesis in stars (Cohen, 1979, 1980). The correlation of elemental abundances with atomic mass appears as a finger print of gravity
which has produced abundance gradients in the prestellar gas. The field
of our Galaxy seems to have mastered abundance gradients over distances
up to 100 kpc, within the reach of our globular clusters. Each cluster
has worked out its own abundance gradient (connected with its own individual field) as is seen in colour gradients of globular clusters, most
clearly in 47 Tucanae from spectroscopic investigations.

Both these gradients, galactic and globular, seem to appear in a
curious way in Fig. 2 which presents the $[m/H]$ data collected by Harris
and Canterna (1979) for globular clusters (with some minor changes: 47
Tuc and M71 are shifted down to the values of $[m/H]$ = -1.27 according
to a revision by Cohen, 1980). As was noted by Alcaino (1980) there is
a gap in $[m/H]$ values between -0.9 and -1.2 for clusters close to the
galactic center. The upper group (a) of 12 globulars exceeds in abundance
the lower one (b), contained in the same distance limits 0-8 kpc, by
$\Delta[m/H] = 1.06 \pm 0.08$. Avoiding drastic assumption that all abundance determinations for 12 globulars of group (a) are erroneous by one order
of magnitude, I prefer to see in this group those globular clusters which
have been stripped out of their outer shells by tidal action of the galactic field and exhibit their internal, more metal rich stellar population.

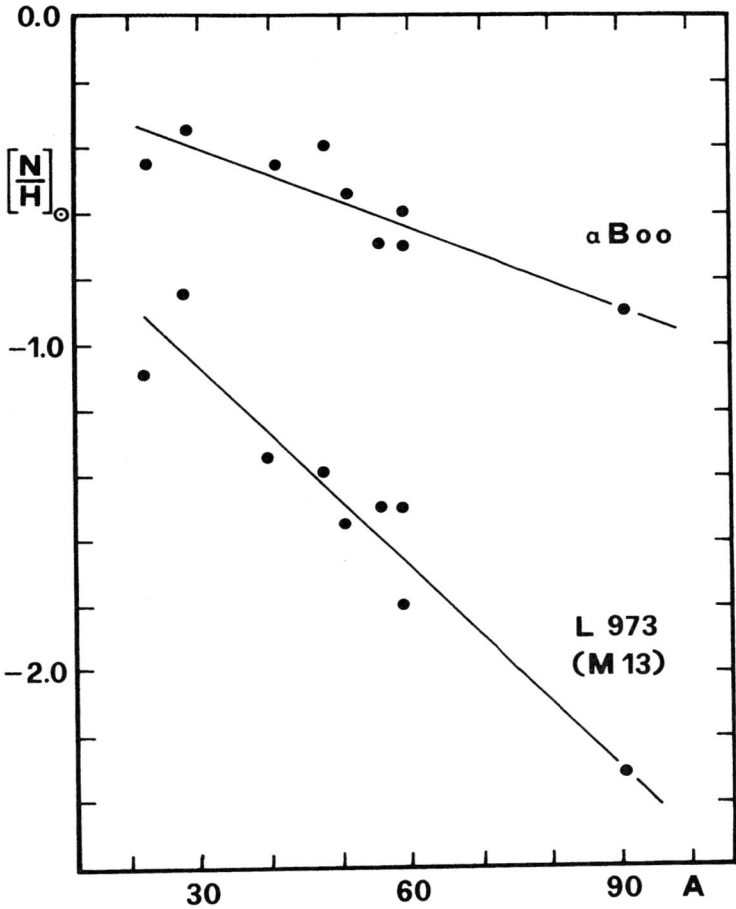

Fig. 1. Abundances of elements vs. atomic mass in α Bootis and L 973, a late type giant in the globular cluster M 13 according to Griffin (1979).

As can be seen in Table I the two groups differ also in diameters and luminosities, group (a) containing smaller and fainter clusters with lower $|z|$-coordinate and slower radial velocity $|V_r|$. All these differences have proper signs for the supposition of stripping in group (a), the level of statistical significance of these differences can be estimated from their mean errors given in Table I. Alcaino (1979) notes four more "metal rich" clusters close to the galactic center, namely: NGC 5927, 6366, 6528 and 6539. Their [m/H] values given by Kukarkin (1974) are positive but lower than 0.1. They are not included in our Fig. 2 nor in the data of Table I. Their inclusion would increase the values and significance levels of the mean parameter differences given in Table I.

Tidal disruption of globular clusters was investigated by Tremaine,

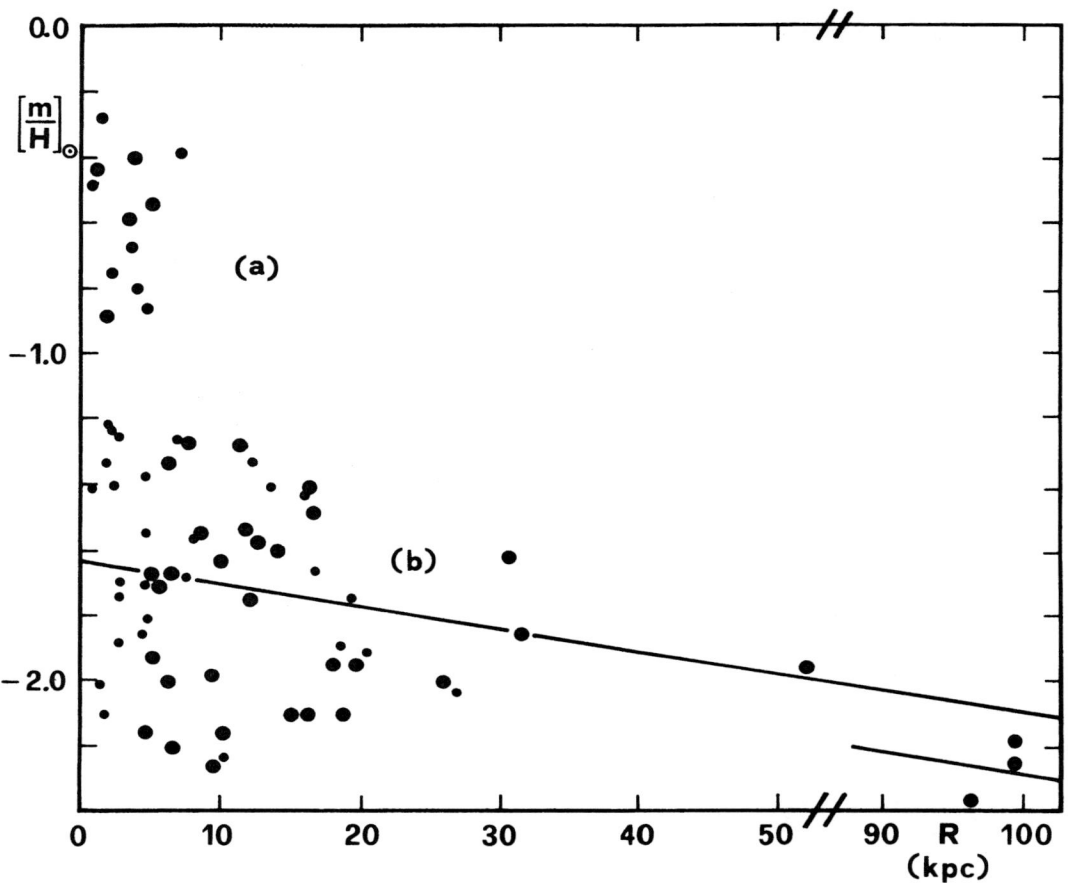

Fig. 2. Metallicity vs. galactocentric distance for globular clusters
according to Harris and Canterna (1979). Big dots - primary data, small
dots - secondary data.

Ostriker and Spitzer (1975) and several other authors. Alladin and
Parthasarathy (1978) calculated the time of disruption for a model of
globular cluster moving in circular orbits at different distances from
the galactic center. They obtained times of disruption as short as 5.4×10^5 yrs for R = 0.32 kpc and 5.7×10^9 yrs for R = 8 kpc. According to these
results it is very likely that some globulars contained in these distance
limits are more or less stripped.

      For all the lower points in Fig. 2 a straight line is drawn which
represents the galactic halo abundance gradient $\partial [m/H] / \partial R = -0.007$ kpc^{-1}
which is lower by one order of magnitude than the gradient found in the
galactic disk, namely - 0.05 for old stars and to -0.10 kpc^{-1} - -0.20 kpc^{-1}
for the younger ones (Mayor, 1977; Luck and Bond, 1980).

Table I

Mean parameter values for groups (a) and (b) of globular clusters.

Parameter	(a)	(b)	(a) - (b)
$[m/H]_{\odot}$	-0.59	-1.65	+1.06 ± 0.082
Diameter (pc)	16.75	24.35	-7.60 ± 2.40
Luminosity ($10^5 L_{\odot}$)	1.18	1.65	-0.47 ± 0.43
$|z|$-coordinate (kpc)	1.86	2.14	-0.28 ± 0.44
$|V_r|$ (km s^{-1})	76.2	88.1	-11.9 ±20.3

Thus, our actual state of knowledge about the large-scale distribution of elements can be briefly characterized by two statements:
1. Abundance gradients are present in all gravitating systems: star clusters, galaxies and clusters of galaxies.
2. Abundance gradients increase with atomic mass of elements.

In view of these facts it seems worth while to investigate the role of gravity in establishing such distribution of elements, in particular, to see, if gravitational separation of elements could affect this distribution at an early phase of evolution of gravitating systems. This evolution should be modelled according to two schemes: (a) gravitational collapse of systems, (b) explosive fragmentation and expansion of fragments (with possible later collapse), forming systems of all hierarchies. For the case (a) ample choice of collapsing models have been elaborated. They could be used for a study of element distribution and its evolution. Case (b) is on the level of hypotheses rather than theory. They go back to the ideas of Ambartsumian about superdense matter and its explosive expansion. The idea seems attractive in view of some recent developments. One of them is the "swiss-cheese like" distribution of galaxies and their clusters discovered recently, another are the explosive structures in double radio galaxies and quasars. One can hope that further progress in the theory of superdense matter will enable at some time to construct physical models of expanding protogalaxies.

Speaking about the distribution of elements at an early phase of universe one admits that the elements were already there, in some amount, that they have been formed in pregalactic material, presumably in the big-bang or some "minor bangs". Remembering how helpful the big-bang was for creating the excess of helium, we can only hope that seemingly insuperable difficulties for creation of other heavy elements through explosive synthesis will be alleviated in a better understanding of the big-bang processes and we shall feel more comfortable with subsequent slow production of these elements in stars.

REFERENCES

Alcaino, G.: 1979, Astron. Astrophys. 29, 281.
Alcaino, G.: 1980, Astrophys. Space Sci. 70, 363.
Alladin, S.M. and Parthasarathy, M.: 1978, Monthly Notices Roy. Astron.
     Soc. 184, 871.
Arp, H.: 1965, Astrophys. J. 141, 43.
Basinska, E. and Iwanowska, W.: 1974, in Stars and the Milky Way System
     (ed. by L.N. Mavridis), Springer Verlag, p. 165 (Paper I).
Basinska-Grzesik, E.: 1974, Estratto Mem. Soc. Astron. Ital. 45, 109
     (Paper II).
Basinska-Grzesik, E.: 1978, Astrophys. Space Sci. 58, 389 (Paper III).
Chapman, S. and Cowling, T.G.: 1952, The Mathematical Theory of Non-
     Uniform Gases, Cambridge Univ. Press; 1970, 3rd ed.
Cohen, J.: 1979, Astrophys. J. 231, 751.
Cohen, J.: 1980, Astrophys. J. 241, 981.
Griffin, R.: 1979, Monthly Notices Roy. Astron. Soc. 187, 269.
Harris, W.E. and Canterna, R.: 1979, Astrophys. J. 231, L 19.
Iwanowska, W.: 1968, Astrophys. Space Sci. 2, 128.
Khokhlova, V.L.: 1976, Papers from Astron. Zh. 3, No. 1.
Khokhlova, V.L.: 1977, in The Chemical and Dynamical Evolution of Our
     Galaxy, IAU Coll. No. 45 (ed. by E. Basinska-Grzesik and M. Mayor),
     p. 177.
Kuroczkin, D.: 1974, Bull. Astron. Obs. Torun 54.
Luck, B.E. and Bond, H.E.: 1980, Astrophys. J. 241, 218.
Mayor, M.: 1977, in The Chemical and Dynamical Evolution of Our Galaxy,
     IAU Coll. No. 45 (ed. by E. Basinska-Grzesik and M. Mayor), p. 213.
McClure, R.D.: 1969, Astrophys. J. 74, 50.
Morgan, W.W. and Mayall, N.U.: 1957, Publ. Astron. Soc. Pacific 69, 291.
Perek, L.: 1962, Adv. Astron. Astrophys. 1, 165.
Pilachowski, C.A., Canterna, R., and Wallerstein, G.: 1980, Astrophys.
     J. 235, L 21.
Pilachowski, C.A., Wallerstein, G., and Leep, E.M.: 1980, Astrophys. J.
     236, 508.
Tremaine, S.D., Ostriker, J.P., and Spitzer, L.: 1975, Astrophys. J.
     196, 407.
Twarog, B.A.: 1980, Astrophys. J. 242, 242.

ACTIVE GALACTIC NUCLEI AND PARTICLE ACCELERATION IN ACCRETION DISKS AROUND MASSIVE BLACK HOLES

Minas Kafatos*, Maurice M. Shapiro[+] and R. Silberberg[+]
#Department of Physics, George Mason University, Fairfax, U.S.A
*Laboratory for Astronomy and Solar Physics, NASA/Goddard Space Flight Center, U.S.A.
†Laboratory for Cosmic Ray Physics, Naval Research Laboratory, Washington, DC, U.S.A.

ABSTRACT

The nuclei of active galaxies are compact power sources, mainly non-thermal. Much of the energy is emitted in the form of X-rays and γ-rays, and also of synchrotron radiation at radio and infrared wavelengths. From these galactic nuclei, radio lobes, jets and high-speed clouds are expelled. The central power source has been interpreted using models that invoke an ultra-massive black hole or, alternatively, a spinar. We examine here the acceleration processes near a black hole. An accretion disk around supermassive black hole in the center of an active galactic nucleus is shown to be a likely site of particle acceleration. Electrons can be accelerated to relativistic energies by various electromagnetic processes as well as by purely gravitational processes (e.g. Penrose pair production). Protons are accelerated to very high energies by the following mechanisms: 1) stochastic Fermi acceleration, 2) betatron acceleration, and possibly 3) shock wave acceleration. We find that, depending on the confinement time for the protons, different galactic nuclei would contribute to different parts of the rigidity spectrum. Examining the magnetic field strengths in different disk models we find that the betatron process can boost particle momenta by some 4 orders of magnitude. Finally, if shock waves can be sustained in an accretion disk (for example if the disk has a corona around it, or if supersonic turbulence can be sustained), such shock waves would also accelerate particles to cosmic-ray energies. Neutrinos would be a useful experimental probe of these models, whereas gamma-ray observations would be more useful for electron acceleration processes.

INTRODUCTION

Various models have been proposed to explain the activity of powerful galaxies and quasars. Recent high-energy observations in the X-ray and γ-ray regions of the electromagnetic spectrum indicate that many active nuclei (AGN) are X-ray sources and some of them are powerful γ-ray sources. Active galactic nuclei are characterized by (a) very high lumi-

323

E. G. Mariolopoulos et al. (eds.), Compendium in Astronomy, 323–345.

nosities (quasars, in particular, which require up to tens of solar masses per year to be converted into energy), and (b) short time scales of
variability, ranging from a few hours to days for the shortest ones. These
features argue for a compact, efficient power source. High-resolution
radio observations with VLA and VLBI also imply small sizes, of arc-milli-
seconds or less (Wittels et al., 1975; Kellermann et al., 1977).

Most recent observations, therefore, seem to favor supermassive objects as the power sources. These compact, supermassive objects could be
(i) magnetized, rotating plasma masses, e.g. of the "spinar" or "magnetoid"
type (Cavaliere and Morrison, 1970; Woltjer, 1971; Sturrock and Barnes*,
1972; Ginzburg and Ozernoy, 1977), or else (ii) massive black holes accreting matter; in this case the accretion disk itself might contain cha-
otic magnetic fields, or it could set up large-scale fields as it rotates
(Lovelace, 1976; Eardley and Lightman, 1975; Pringle, Rees and Pacholczyk,
1973; Lynden-Bell, 1969; Fabian et al., 1976; Kafatos and Leiter, 1979;
Kafatos, 1980).

The nature of the central source (Silberberg and Shapiro, 1979),
might be determined from gamma-ray observations and high energy neutrino
measurements (Berezinsky and Ginzburg, 1981). How long the ultimate collapse of a spinar will be delayed is unknown at present. Moreover, a su-
per-massive black hole seems to provide a more natural model for acceleration, since it is as compact as possible, and it generally permits higher
efficiencies for converting accreting mass into radiation. Moreover, some
spinar models yield too much energy in the form of γ-rays, in excess of
10^{48} erg s^{-1} (Sturrock, 1971). Although it might be possible to develop
spinar models that are consistent with existing observations, we concentrate here, for the reasons cited, on particle acceleration mechanisms
in the vicinity of black holes.

By black hole we mean objects which possess an event horizon but do
not necessarily contain a mathematical singularity at the center. We do
not single out a particular disk model but examine a variety of processes
that would accelerate electrons and protons to cosmic-ray energies. Furthermore, we concentrate on processes in the vicinity of the black hole.
Relativistic beam models (Blandford and Rees, 1978; Blandford and Königl,
1979) could be very important in explaining the compact, extragalactic
variable sources. Beam models will be examined here if they are specifically tied to supermassive black holes (Lovelace, 1976; Lovelace, MacAuslan
and Burns, 1979). We find that, owing to severe electron losses in the
accretion disk, most models predict escape of the electrons perpendicular
to the disk, i.e. in beams. We mention, too, the possibility (Cavaliere,
private communication) that a black hole-spinar model may also be compatible with the observations. In this model the "spinar" would be a magnetized region surrounding a black hole core.

---

* Actually, the Sturrock-Barnes model is also consistent with the combination of a black hole and corona.

TYPES OF ACTIVE GALAXIES AND CONDITIONS IN AND OUTSIDE THE CENTRAL NUCLEI

Active galaxies have large concentrations of mass within $\sim$100 pc from the central "power engine", and are characterized by the emission of radio lobes, jets and high-speed clouds. Among the types of active galaxies are radio galaxies, Seyferts of class 1 and 2, and quasars. Seyfert galaxies (both class 1 and 2) have bright compact nuclei; much of their luminosity is radiated in the infrared. Both have broad Balmer lines (broader for class 1) that imply motions of $\sim$10^3 km/sec (500-1000 km/sec for class 2, and a few thousand km/sec for class 1). In addition, however, those of class 1 are powerful X-ray emitters. They exhibit variability in radio emission and have narrower forbidden lines than class 2, while the allowed and forbidden lines of class 2 have the same width (Khachikian and Weedman, 1974; Stein and Weedman, 1976). (The Balmer lines are formed in dense filaments close to the nucleus with $\sim 10^8$ atoms/cm^3, while the forbidden lines arise in more rarefied regions, $\sim$10^3 atoms/cm^3).

Jones, O'Dell and Stein (1974a,b) have developed methods to determine the size of the regions of the compact synchrotron and Compton photon emission and the magnetic fields in those regions. The method is based on using the spectra of the synchrotron and inverse Compton radiation, the frequency at which synchrotron self-absorption becomes important (Ginzburg and Syrovatskii, 1969), and polarization. Jones and Stein (1975) considered the origin of the infrared emission from the nucleus of the class 2 Seyfert NGC 1068. The nuclei of class 2 Seyferts are so heavily obscured by dust and gas that the nature of the underlying source is not established: it could be non-thermal like other active galactic nuclei, or it could be thermal emission from early-type supergiants that satisfy the small mass-to luminosity ratio M/L < 0.05, or even < 0.003 (Telesco, Harper and Loewenstein, 1976). If the source is non-thermal, one can estimate by combining the data in Table I of Jones and Stein (1975) and the more recent measurements of Telesco, Harper and Loewenstein (1976), that the non-thermal source is exceedingly small, $\sim$10^{14} cm, and the magnetic field is $\sim$400 to 1000 gauss. These values, if the source is non-thermal, would fit best the black hole model.

In the production of neutrinos, a crucial quantity is the path length of material (e.g., in g/cm^2) traversed by protons in the active galaxies. In the class 2 Seyfert galaxy, NGC 1068, the mass of the gas in the nuclear dust cloud is $M_{gas} \simeq 10^8 M_{\odot}$ in θ=2.5", i.e. in 250 pc (Jones and Stein, 1975). This corresponds to a mean density $\sim$10^2 atoms/cm^3. In the case of the class 1 Seyfert galaxy 3C 120, one can infer from the O III line that n_e is 1500/cm^3 in a volume $V \simeq 5$x10^{58} cm^3, i.e. $n \simeq 1500$ atoms/cm^3 in $\sim$12 pc, and that it contains clouds or filaments with $n \simeq 10^8$ atoms/cm^3 and $M \simeq 200 M_{\odot}$ (Shields, Oke and Sargent, 1972). From the absorption of low-energy X-rays the column densities of matter along the direction of sight to the central nuclei of the class 1 Seyfert NGC 4151, radio galaxy Cen A, and quasar 3C 273 have been measured-they are about 0.1 or 0.2 g/cm^2 (Ives, Sanford and Penston, 1976; Sanford and Ives, 1976). Recent estimates for 3C 273 are lower; maybe its column density is variable.

This mass is concentrated at the center; at the sites of the H_β, H_γ and H_δ absorption lines of NGC 4151, the column density is ~10^8 times lower.

The energy spectra of photons from active galactic nuclei of various types of galaxies are shown in Fig. 1. These are given in terms of power per log E interval. All of these differ greatly from the spectra of normal galaxies like ours. In our galaxy, the energy of thermal emission at optical wave lengths exceeds that at X-rays by ~10^4 while in the active galaxies (with the exception of class 2 Seyferts) the energy in X-rays exceeds that in the thermal optical photons. Our galactic nucleus, however, has at least four discrete X-ray sources with powers near 10^{37} ergs/sec at energies 2-10 KeV (Cruddace et al., 1978). The spectrum of the quasar 3C 273 (up to 500 MeV) is taken from Swanenburg et al. (1978) and references therein, while the upper limit at 3×10^{11} eV is from Weekes (1978). The latter limit would be ~3 x higher when corrected for γ-ray absorption by the 3°K microwave radiation.

Some quasars have now been observed at X-ray wave lengths: MK 2251-178 (Ricker et al., 1978; Canizares, McClintock and Ricker, 1978) and 0241 + 622 (Apparao et al., 1978). Many more quasars have been observed recently with the Einstein X-ray Observatory (Giacconi, 1979). The spectrum of the class 1 Seyfert NGC 4151 is based on Baity et al. (1975) and references therein, Auriemma et al. (1978), and Schönfelder (1978), Zanrosso et al. (1979), and references therein. The infrared measurements of NGC 4151 and of the other sources mentioned in this paragraph are largely based on Rieke and Low (1972,1975).The flux between 0.1 and 10 MeV appears too high for the synchrotron self-Compton process, and Leiter and Kafatos (1978) and Kafatos (1980) proposed that it originates in the Penrose Compton process in the ergosphere of the black hole. Schönfelder (1978) proposed that such flux from class 1 Seyferts could explain the diffuse flux of gamma rays from 1-100 MeV;this is an alternative explanation to that of Stecker, Morgan and Bredekamp (1972) who proposed redshifted gamma rays from cosmological p̄-p annihilation.

The Ariel 5 catalog (Cooke et al., 1978) and the fourth UHURU catalog (Forman et al.,1978) listed 18 X-ray sources that are class 1 Seyfert galaxies. Elvis et al. (1978) and Tananbaum et al. (1978) analyzed the data of these sources. Ward et al. (1978) identified 3 more Seyferts as X-ray sources. Ward et al. (1978) and Bradt et al. (1978) discussed 6 more Seyfert-like sources which also emit X-rays; these, however, have narrower emission lines (~ 1000 km/sec). Seyferts of class 1 are stronger X-ray emitters than those of class 2. The spectrum of the radiogalaxy Cen A (or NGC 5128) is based on Mushotzky et al. (1978) and references therein; the high-energy gamma-ray data, (E > 300 GeV) are based on Grindlay et al. (1975). The X-ray flux can also be interpreted as due to inverse Compton interactions by hot (10^4 K) blackbody radiation produced at the galactic nucleus (Beall et al. 1978). Cen A has been found to vary rapidly, on a time scale of a day, at mm wave lengths (Kellermann, 1974), and over 2 to 5 hours at X-ray wave lengths (Delvaille, Epstein and Schnopper, 1978), implying a highly compact nucleus for this radio galaxy. The radio, infrared, and X-ray spectra from the central, compact source in the radio

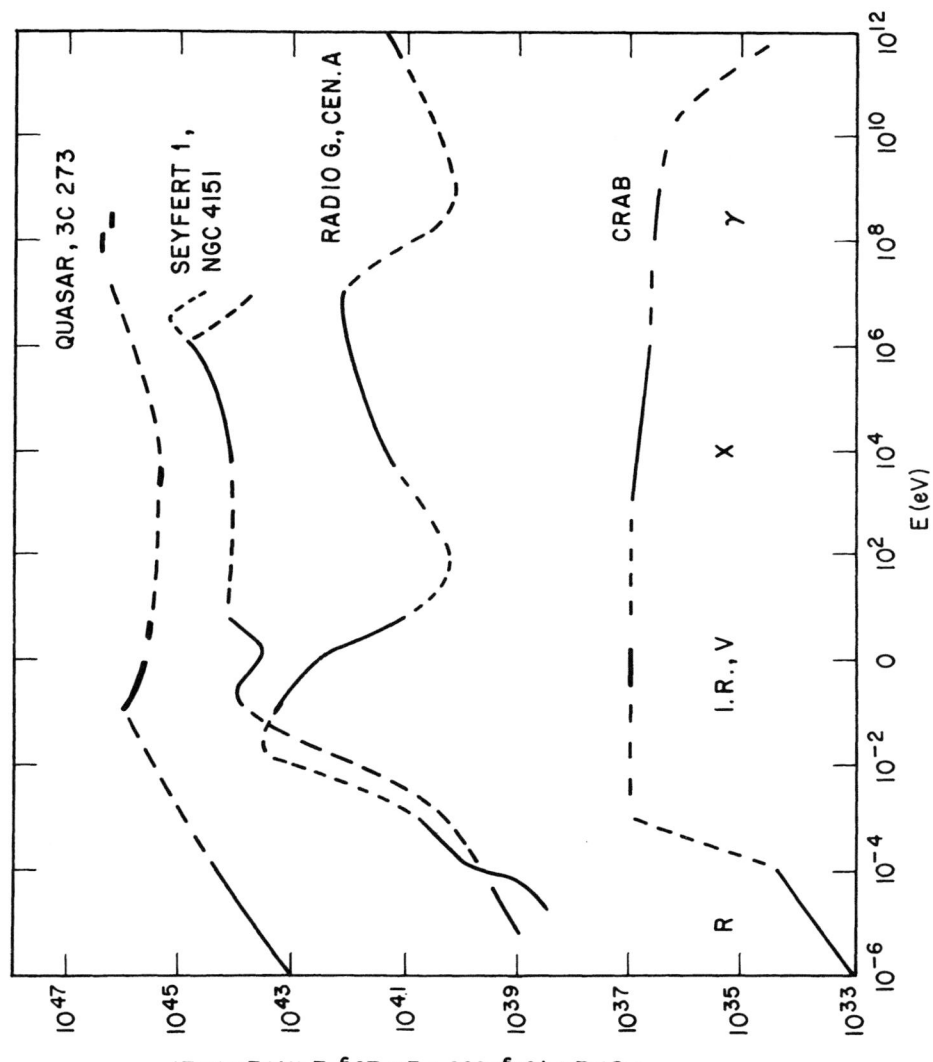

Fig. 1. The luminosities (per log-energy interval) of some active galactic nuclei, at various photon energies. For comparison, the luminosity of the Crab supernova remnant is also displayed.

Table 1a

Luminosities of Some Active Galaxies and Quasars
(in erg/sec)

Object	Radio	Infrared	Optical	X-rays	γ-rays	Type of Object, Distance	Reference
NGC4151			3×10^{42}	3.8×10^{42} (2-10 keV)	10^{45} (<3 MeV)	Seyfert 20 Mpc	Schönfelder, 1978, Giacconi, 1978, Gursky and Schwartz, 1977.
3C120	2.5×10^{42}	6.5×10^{44} (1-10 μ)	4×10^{43}	4×10^{44} (2-10 keV)		Radio, optical variable Seyfert 190 Mpc	Giacconi, 1978, Gursky and Schwartz, 1977, Sandage, 1972, O'Dell et al., 1978.
M87	9×10^{41}	$<10^{43}$ (1-300 μ)	$<10^{44}$	5×10^{42} (2-10 keV)		Giant elliptical galaxy, Virgo A 22 Mpc	Gursky and Schwartz, 1977, Sandage, 1972, Neugebauer, 1978.
NGC5128	10^{41}	7×10^{42} (3-300 μ)		$(0.4-2.5) \times 10^{42}$ (2-10 keV)	10^{44} (<10 MeV) 300 GeV detected	Giant elliptical galaxy with dust, Cen A, 5 Mpc	Gursky and Schwartz, 1977, Sandage, 1972, Neugebauer, 1978, Hall et al., 1976.

Table 1a (continued)

Object	Radio	Infrared	Optical	X-rays	γ-rays	Type of Object, Distance	Reference
CYGNUS A	10^{44}	3×10^{44} (8-13 μ)	8×10^{44} (thermal?)	2.5×10^{45} (6-7.5 keV)		Radio galaxy 323 Mpc	Kafatos, 1978.
3C273	2.2×10^{44}	1×10^{47} (1-300 μ)	7×10^{46}	7×10^{45} (2-10 keV)	3×10^{46} (50-500 MeV)	Quasar 950 Mpc	Swanenburg et al., 1978, Gursky and Schwartz, 1977, Sandage, 1972, Neugebauer, 1978.

Total Luminosities of Bright Extragalactic Objects

Q 0420-388	Most luminous Quasar	2×10^{48} ergs/sec
3C232	Luminous Quasar	8×10^{47}
3C273	Luminous Quasar	2×10^{47}
Mkn 231	Most luminous Galaxy	1.5×10^{46}

$(6 \times 10^{46}$ ergs/s $= 1 \, M_{\odot} \, c^2$/year$)$

galaxy Cygnus A (nearly as powerful as 3C 271 from radio to X-ray frequencies) has been explored by Kafatos (1978), who considered alternative
models like single and double compact radio clouds. He found that only
the double cloud model can be fitted to the X-ray observations. Indications of variability of the X-ray flux suggest that these, too, originate
in a compact source.

For the purposes of comparison, the spectrum of a galactic non-thermal source, the Crab Nebula and pulsar, is also shown. It is surprisingly
similar to that of the quasar 3C 273, though its luminosity is $\sim 10^9$ times
lower. The data for the spectrum of the Crab are from Apparao (1973) and
references therein. Above 500 MeV, the γ-ray flux shown is that from the
pulsar (McBreen et al., 1973), and near 10^6 MeV, that of Grindlay,
Helmken and Weekes (1976) and references therein. Similar results at
E > 500 GeV were recently obtained by Gupta et al. (1976). The energy output of some young supernovae in relativistic particles could be much
higher. Marscher and Brown (1978) estimate a radio luminosity of $\sim 2 \times 10^{39}$
ergs/sec for the recent supernova SN 1970g, about 1 year after the
supernova burst.

Another class of compact nuclei and variable radio sources are the
BL Lacertae objects located possibly in giant elliptical galaxies. The
BL Lac objects are characterized by absence of strong emission lines.
Recently two or three of these were observed by their X-ray emission,
which appears strongly variable (Mushotzky et al., 1978); see also
Schwartz et al. (1978). Recent "instantaneous" spectral energy curve measurements for MK 501 (Kondo et al., 1981) indicate that the total luminosity of this BL Lac object has previously been underestimated by a factor
of about 3 or 4.

Table 1a shows the luminosities of some active galaxies and quasars.
Table 1b shows the variability time scales for some of these objects.

GAS, EJECTA, AND ACCELERATION OF PARTICLES IN THE OUTER REGION

Strong X-ray emission has been observed from clusters of galaxies;
recent studies are presented by Mushotzky et al. (1978); Helmken et al.
(1978) and Jones and Forman (1978). X-ray spectral lines of highly ionized
iron (Fe XVII to Fe XX IV) are observed from such sources (Fabricant et
al., 1978 and references therein). These observations seem to be consistent with thermal emission corresponding to kT = 2 to 10 KeV. The appreciable abundance of iron, Fe/H $\simeq 1.4 \times 10^{-5}$ — about half of the Fe/H ratio
in the solar system — (Cameron, 1973) implies that the galaxies of the
cluster (in which nucleosynthesis takes place) are the source of this gas.
Superclusters also appear to be X-ray sources (Murray et al., 1978).

Cloudlike structures escape at high velocities (100 to 1000 km/sec)
from Seyfert galaxies; the radio lobes of radio galaxies also have high
velocities. While the processes giving rise to escape of gas from active
galaxies are not yet understood, there are various proposals, e.g. radi-

Table 1b

Variability of Some Active Galaxies and Quasars

Object	Time scale	Comments	Reference
NGC4151	1.5d as short as 10 min?	Seyfert; variability in X-Rays (2-60 KeV)	Mushotzky, Holt and Serlemitsos, 1978.
3C120	0.5 yr	Seyfert; radio, optical variability	Burbidge, Jones and O'Dell, 1974.
NGC5128 (Cen A)	<0.5 yr as short as a few days	Giant elliptical radio galaxy; radio, X-ray variability	Beall et al., 1978.
Cyg A	1 yr?	Radio galaxy; X-ray variability?	Kafatos, 1978.
AO 0235+164	A few weeks (in Oct. 1975)	BL Lac; it flared in optical and IR, radio was slower; it achieved 0.2 magn./day in the optical >10^{48} erg/s.	Burbidge, 1978.
OJ287	0.5 yr or less in radio. periodic? 30d, 26d, 8d, 39 min possible periodicity	BL Lac; variable in radio, optical and IR; varies by a few tenths of magn./day	Stein, O'Dell and Strittmatter 1976.
BL Lac	Several weeks ν>10 GHz a few months ν~1 GHz a few yr ν<1 GHz	Rapid variations in radio, optical and IR	Stein, O'Dell and Strittmatter 1976.
3C279	~1 month in 1926-1937, 4 yr in radio	QSO. Optical variability in 1936-1937	Burbidge, Jones and O'Dell, 1974.
CTA 102	2.3 yr Components in radio	QSO. This object dissipates energy at a rate of 100-1000 M$\odot$ c²/yr!	Burbidge, Jones and O'Dell, 1974.
3C273 C D	1.5 yr Components in radio 4.6 yr Components in radio 5 yr ~days in optical and X-rays	QSO. Primarily radio, optical, X-ray variability. Constant in IR.	Burbidge, Jones and O'Dell, 1974.

ation pressure, and plasma interactions. Under certain circumstances, such as gas densities > 10^4 atoms/cm^3, cosmic rays can act as the driving force (Eilek, 1977). At lower densities, the forbidden lines would be too strong.

An early review of the emission of gas in Seyfert galaxies and of energy input requirements was given by Burbidge, Burbidge and Sandage (1963). Walker (1968) presented a study of the motions of the class 2 Seyfert galaxy NGC 1068. Near the nucleus of this galaxy there are 3 or 4 clouds about 500 pc in diameter (these data are revised to correspond to a Hubble constant of 55 km/sec/Mpc; for 80 km/sec Mpc the diameter would be $\sim$300 pc). The masses of the clouds are $\sim10^7$ M$_\odot$. The clouds have large internal turbulent velocities, and velocities with respect to the center up to $\sim10^3$ km/sec. The age of these clouds is 10^5 to 10^6 years, and there is evidence for an earlier generation of violent events $\sim3 \times 10^6$ years ago. The kinetic energy required to accelerate the clouds is $\sim10^{55}$ ergs. Ulrich (1973) discusses the motions of the four large clouds of the class 1 Seyfert Galaxy NGC 4151. Anderson (1974) suggests that the motions of the clouds in NGC 4151 may be only rotational, rather than radial. A theoretical study of the flow of hot winds (T $\simeq$ 5 x 10^6 $^\circ$K) from the galaxies of Seyfert nuclei (with velocities up to $\sim$800 km/sec), cloud formation, and outward expansion of spiral arms is given by Wolfe (1974).

Radio galaxies frequently have two lobes symmetric about the central galaxy. These have given rise to several fascinating interpretations. An early review, with illustrative photographs, is given by Maltby, Matthews and Moffet (1963). A recent review by De Young (1976) discusses both the observational features and the explanatory models. The linear extent of radio sources varies from a few tens of kiloparsecs to 6 Mpc. The radio luminosities vary from 10^{40} to 10^{46} ergs/sec. The integrated radio spectra generally fit a power law S$_\nu$ $\propto$ ν^α with $\alpha = -0.6$ to -1.2, and corresponding electron energy spectra N(E) dE = KE$^{-(1-2\alpha)}$ dE. The matter density and electron density in the radio lobes is low: n$_e$=3x10^{-6} to 3x10^{-3} cm^{-3}, but due to the large volume, the mass is still 10^4 to 10^8 M$_\odot$. Extended radio sources are generally associated with elliptical galaxies. Data on polarization imply that the magnetic vector is generally aligned with the major axis of the source. X-rays ($\sim$4x10^{41} ergs/sec at 2 - 10 KeV) have been observed from the radio lobes of Cen A (Cooke, Lawrence and Perola, 1978). The minimum total energy present in relativistic particles and magnetic field ranges from 10^{56} to 10^{61} ergs, and the magnetic field, assuming equipartition, ranges from 10^{-6} to 10^{-3} gauss. There is a problem with the lifetime of electrons in the source; B $\propto$ R^{-2}, hence B is $\sim10^{-5}$ G; the field strength as the source emerges from the galactic nucleus is 1 to 10 G. The synchrotron lifetime of electrons radiating at 10 GHz in this field is at most 1 year. This has led to the suggestion that radio sources must replenish the supply of relativistic electrons. Various models have been proposed:

(a) The nuclear region of the galaxy is the source of relativistic fluid that breaks forth along the rotation axis and becomes collimated into beams. The shock front of the beam and the circumgalactic medium is the source of the highly relativistic electrons (and of protons as well) (Scheuer, 1974; Blandford and Rees, 1974). This model was modified by Wiita (1978a,b), who considered a hotter and denser gas in the galactic nucleus, and applied the model to Cygnus A.

(b) Massive condensed objects having masses of 10^6 to 10^9 $M_\odot$ are ejected from the galactic nucleus by a gravitational slingshot mechanism, that results from a three- or four-body interaction. These provide an in-situ energy supply, (Burbidge, 1967; Saslaw, Valtonen and Aarseth, 1974; Harrison, 1977). These massive objects could be black holes with accretion disks (Rees and Saslaw, 1975), or massive "spinars" with radii 10^{16} to 10^{18} cm (Flasar and Morrison, 1976). The latter apply this model to Cygnus A. Such a model has been applied by Valtonen (1976) to the radio tail galaxies.

(c) A cloud of thermal plasma, relativistic particles and magnetic field is ejected from the galactic nucleus and confined by ram pressure. (De Young and Axford, 1967). Sanders (1976) carried out two-dimensional hydrodynamic calculations and found that adequate collimation is possible when there is a condensed object of 5×10^8 $M_\odot$ at the center of the galactic disk. Ejection of plasmoids along the axis of rotation can occur after shearing of rotating magnetic fields (Sturrock and Barnes, 1972).

For the large collinear double sources there is an increase in brightness near the head of the source and the maximum of the brightness distribution lies along the axis. De Young (1977) concludes that the source region within the radio lobe (whether ejected spinar, ejected black hole, plasmoid or beam-medium interface) must be a source of hot thermal plasma ($> 10^6$ $^\circ$K) in addition to relativistic particles and magnetic fields.

The flux, structural and temporal variations, and superluminar velocities of compact extragalactic radio sources have been interpreted by means of a model in which a relativistic blast-wave (Marscher, 1978a,b) moves toward the observer. In this model, a highly relativistic blast wave, with an energy of $\sim 10^{57}$ ergs strikes a "screen" or shell-like structure, e.g., one consisting of filaments. The shocked plasma of the screen in which particles (electrons) are accelerated becomes a time-dependent source of a synchrotron radiation, with rapid movement of the shock front along the screen. Various models of superluminar velocities, including one resembling the above, have been proposed by Blandford, McKee and Rees (1977).

The acceleration of particles by betatron and Fermi processes at the turbulent interface of fast moving plasmoids (emitted from radio galaxies) and the intergalactic medium has been studied by Pacholczyk and Scott (1976a,b). The first of these papers deals with the radio-tail galaxies. Ram pressure confinement and deceleration of plasmoids leads to Rayleigh-Taylor and Kelvin-Helmholtz instabilities, with resulting turbulence. The magnetic field in the plasmoid is amplified by turbulence, resulting in betatron acceleration and synchrotron emission. Also Fermi acceleration occurs; the turbulent vortices act as magnetic scattering centers. This model has been further developed by Christiansen, Pacholczyk and Scott (1977) and Christiansen, Scott and Vestrand (1978).

The small-scale structure and time variations of extragalactic radio sources, especially of quasars, has been studied with very-long-baseline interferometry (VLBI), (Wittels et al., 1975; Kellermann et al., 1977 and references therein). Structures as small as a tenth of an arcsecond could be explored, and time variations on scales of days and months were observed. Several of these sources were found to exhibit apparent super-light expansion: the Seyfert 3C 120 and quasars 3C 273, 3C 279 and 3C 345. Two components of the latter were found to increase their separation from ~ 0.9 or 1.0 milliarcseconds in 1971 to 1.7 milliarcseconds in 1976 (Wittels et al., 1976; Cohen et al., 1976). The apparent velocity of separation was ~ 2.5 c between 1971 and 1974, and ~ 8 c from 1974 to 1976.

Various models for explaining such apparent superlight velocities were proposed by Blandford, McKee and Rees (1977) and Marscher (1978). These involved, for example, a broad and relativistic blast-wave striking a filament at nearly right angles, and generating a turbulent shock across a broad front that is formed at different parts of the filament in a time $< L/c$. Here L is the distance on the "screen" from the point where the signal was first formed to the point of the emission maximum at time t. The flux thus varies on a time scale short relative to the light travel time across the source. The observations of these sources are discussed by Cotton (1976) and references therein. The theoretical implications (ultra relativistic blast waves) and the energy input requirements are explored by Jones and Tobin (1977).

The rapid variability time scale presents a problem: The corresponding estimated source size ct_v would imply a brightness temperature that exceeds the "inverse Compton" limit of $\sim 10^{12}$ K, (Kellermann and Pauliny-Toth, 1969). (The energy density of synchrotron radiation would be so high as to cause too rapid energy losses of the relativistic electrons by the inverse Compton process). Blandford and Königl (1979) present a solution of this problem in terms of a relativistic jet moving from the nucleus of an active galaxy with a Lorentz factor $\gamma < 10$ towards the observer. High energy input requirements are also alleviated by the beaming of radiation toward the observer, i.e. the observer is more likely to see sources beaming toward him. This is supported by the observation of Readhead et al. (1978), who find greater c-shaped distortion in jets of distant quasars; beaming toward us renders these more easily observable, and small distortions appear magnified when looked at head-on. Blandford and Königl (1979) suggest that the optically-violent variable quasars (many with apparent super-luminar velocities) and Lacertids correspond to objects beaming toward our galaxy. Readhead, Cohen and Blandford (1978) have described a radio galaxy NGC 6251 that has well aligned jets, one in the nucleus at 1.7 pc, another extending to 100 kpc, and radio lobes separated by 3 Mpc.

The problem of axisymmetric emission of radio lobes in the case of radio galaxies, generally elliptical ones, versus emission of clouds (not along a line of symmetry) in Seyfert galaxies, mostly spiral ones, has not been resolved. A possible explanation is the following: in an elliptical galaxy, the magnetic field is aligned with the axis of rotation and axis of ejection or particle beam; in spiral galaxies, on the other hand, the total magnetic field could have a component along the spiral arms. There is then an angle between the axis of the net magnetic field and that of rotation. The magnetic field axis (and presumably the particle beam) would then rotate about the spin axis, with the power of the beam spread out over an extended cone, resulting in plasma interactions and the ejection of clouds. In the case of class 2 Seyferts, the beam could strike the accretion disk, dissipate therein, but cause sufficient remote energy output so as to yield broad emission lines in an extended region about the galactic nucleus. A corresponding model could be constructed with a massive spinar: axisymmetric in elliptical galaxies, non-axisymmetric in spiral ones.

ACTIVE GALACTIC NUCLEI: MODELS AND ACCELERATION MECHANISMS

Various models have been proposed to explain the huge energy output (10^{42} to 10^{47} ergs/sec or 10^8 to 10^{13} times the solar luminosity) from active galactic nuclei. These are: (1) ultra-massive black holes of $\sim 10^7$ or 10^8 $M_\odot$ with accretion disks (Lynden-Bell, 1969; Lynden-Bell and Rees, 1971; Hills, 1975); (2) ultra-massive stars or magnetoplasmic bodies or "spinars" or "oblique rotators" (Hoyle and Fowler, 1963a,b; Fowler, 1964a,b; Morrison, 1969; Cavaliere and Morrison, 1970; Ozernoy and Usov, 1971; Ginzburg and Ozernoy, 1977), and (3) compact young star clusters with frequent supernovae (Spitzer and Saslaw, 1966; Spitzer and Stone, 1967; Colgate, Calvin and Petschek, 1975 and Arons, Kulsrud and Ostriker, 1975).

Recently there have been several studies of the rate of energy production by mass accretion at the black holes. Hills (1975) developed a quantitative approach to the problem, but overestimated the mass build-up rate of black holes by tidal disruption of stars in the Roche lobe. Peebles (1972) and Bahcall and Wolf (1976) carried out a pioneering work on the distribution and rate of consumption of stars around a black hole. Subsequent studies (Frank and Rees, 1976; Lightman and Shapiro, 1977 and Dokuchaev and Ozernoy, 1977) treated the important "loss cone" effects diffusion of stars into the loss cone by gravitational interactions, and critical radius r_c, outside of which the diffusion of orbits does not lead to the loss cone by stars. Taking into account these results, the evolution and power output of a black hole and its environment was calculated by Young, Shields and Wheeler (1977), Young (1977), Shields and Wheeler (1978) and Ozernoy and Reinhardt(1978).

Kafatos and Leiter (1979) have calculated the density of gas just outside the ergosphere of a black hole and inside the ergosphere; for a mass of $\sim 10^8$ $M_\odot$; they estimate about 10^{11} to 10^{13} atoms/cm^3.

The fluxes and energy spectra of X-rays (up to ~ 1 MeV) from black

holes have been calculated by Eardley et al. (1978). They considered
three types of accretion disk models: the standard disk model, the two-
temperature model with ion temperature exceeding that of electrons by
one or two orders of magnitude, and the optically thin model. References
that describe these models in greater detail are given by Eardley et al.
(1978). The gamma-ray fluxes and spectra have been estimated by Maraschi
and Treves (1977), Leiter and Kafatos (1978), Kafatos and Leiter (1979)
and Kafatos (1980). The last three papers are based on the Penrose Compton
and pair production processes in the ergosphere of the black hole. The
Penrose Compton process yields gamma rays of up to $\sim$5 MeV, and the Penrose
pair production to $\sim10^3$ MeV.

For the multiple pulsar model of active galactic nuclei, (with an
energy output rate of $\sim10^{45}$ ergs/sec), one requires approximately ten
supernovae per year in a star cluster of radius $\sim$1 pc. Each supernova
produces a kinetic energy output 10^{51} to 10^{52} ergs. Numerous mechanisms
for acceleration near a pulsar or in a supernova remnant have been proposed: (a) in the vicinity of the light cylinder, when the angular momentum vector and that of the magnetic field are parallel (Goldreich and
Julian, 1969); by electromagnetic waves from an oblique magnetic dipole
rotator (Pacini, 1967; Gunn and Ostriker, 1969); by electric field at
the pulsar caps (Pacini and Rees, 1970; Sturrock, 1971); (b) farther out
in the supernova remnant, various processes of acceleration have been
studied. Kulsrud, Ostriker and Gunn (1972) have considered the electro-
magnetic-wave mechanism in the remnant during the decade following the
outburst. Scott and Chevalier (1975) and Chevalier, Robertson and Scott
(1976) have discussed the effects of moving magnetic knots and the statistical Fermi mechanism in supernova remnants $\sim$100 years old. Axford, Leer
and Skadron (1977), and Blandford and Ostriker (1978) have investigated
the role of wave turbulence and shocks at the boundary of supernova remnants. Finally, Chevalier (1977) has suggested that betatron acceleration
occurs in compressed shocked regions. In an assembly of merging supernova
remnants, there is further stochastic acceleration (Kulsrud and Arons,
1975).

A supermassive rotator could have acceleration mechanisms like a
pulsar. However, Sturrock (1971) calculated that too much of the energy
output would then be in the form of high-energy gamma rays between 100
and 1000 GeV. Sturrock and Barnes (1972) proposed a rotating-magnetoid
model with transitions between open and closed magnetic field configurations. The open configurations would contain current sheets. Reconnection
of magnetic field lines across the current sheets gives rise to acceleration of particles and ejection of plasmoids. Ozernoy and Uzov (1971) estimate the dimensions of the supermassive rotator to be of the order of
10^{16} cm.

The neighborhood of a black hole is likely to be an efficient accel-
erator: as gas and turbulent clouds spiral in, magnetic irregularities
approach each other, and the magnetic field builds up, perhaps to $\sim10^3$
gauss in the accretion disk. The Fermi acceleration process, the betatron
process, and acceleration by wave turbulence and shocks (see references

just above) should be rather efficient here. For example, compression of old supernova remnants as they approach the black hole neighborhood could increase magnetic fields from $\sim 10^{-6}$ to $\sim 10^{-3}$ gauss with corresponding acceleration of cosmic rays in these compressed supernova remnants. The mechanism of Sturrock and Barnes (1972) may also operate here. Furthermore, the magnetic accretion disk can act as an electric dynamo and accelerate oppositely directed beams of particles up to energies of $\sim 10^{19}$ or 10^{20} eV (Lovelace, 1976).

While experimental data have not yet permitted a clear conclusion as to the best model, some observations are more consistent with the ultra-massive black hole model. Tananbaum et al.(1978) found the X-ray flux from the class 1 Seyfert galaxy NGC 4151 to be rapidly variable, with rise times of ~ 700 sec, implying a size of the emission region $< 2 \times 10^{13}$ cm*. Such increases in flux occur too frequently to fit the multiple supernova model and the power constraints of the galaxy. The gravitational radius of a black hole with mass M_h is $r_g = GM_h/c^2 =$ $=1.5 \times 10^{12} M_h/10^7 M_\odot$ cm, while the emission peaks at a radius of about $1.5 \times 10^{13} (M_h/10^7 M_\odot)$ cm (Lightman, Giacconi and Tananbaum, 1978). These authors conclude that the observations of Tananbaum et al. (1978) fit a black hole model with $M_h \lesssim 4 \times 10^6 M_\odot$. The radio galaxy Cen A has a central radio component whose spectrum cuts off at ~ 30 GHz (Kellermann, 1974). If the cutoff results from synchrotron self-absorption, the size of the emitting region is $\sim 10^{16}$ cm (Fabian et al., 1976). Based on these considerations, and on relatively rapid X-ray variability (with a time scale of 6 days, observed by Winkler and White, 1975), Fabian et al. (1976) conclude in favor of the black hole model for Cen A.

More recently, Mushotzky et al. (1978) have found even more rapid variability in X-rays, on a time scale of a day, and Delvaille, Epstein and Schnopper (1978), even of 2 to 5 hours. A study of the photometric profile of the radio galaxy Virgo A (M 87), with a mass of $10^9 M_\odot$ in 100 pc, has been found to fit the black hole model (Young et al., 1978; and Sargent et al., 1978). Finally, recent observations with the "Einstein" satellite of X-ray variability in 3C 273, of the order of a day, would also favor a black hole model.

It is, of course, possible to invoke a model in which the variable X-ray component of emission occupies a small part of the total size of the region, as would be the case with a model featuring a "hot spot" on an accretion disk, or with a flare model.

MATTER AND MAGNETIC FIELDS IN THE BLACK HOLE ACCRETION DISK

The density of matter in the accretion disk is about a trillion times higher than the average density of gas in our galactic disk. The

---

*However, the observation is at $< 3\sigma$ level. Fully established time variations are as short as a couple of hours.

density of atoms (averaged over the whole solid angle about the black hole) is (Kafatos, 1980)

$$n \simeq 2 \times 10^{12} \ (\dot{M}_1/M_8^2) \ r_*^{-3/2} \ \alpha^{-1} cm^{-3} \ ,$$

where $\dot{M}_1$ is the mass per year accreted by the black hole in units of solar mass $M_\odot$ per year, M_8 is the dimensionless mass of the black hole in units of $10^8 \ M_\odot$, $r_* = r/r_g$ where r is the radial distance, and r_g is the gravitational radius $= 1.5 \times 10^{13} \ M_8$ cm and α is the viscosity parameter ~ 0.1. We note that the density is large close to the black hole, at $r \simeq r_g$, and falls off as $r^{-3/2}$. Near the equatorial plane of the disk, the density may be higher by an order of magnitude, while it is much smaller (maybe by 3 orders of magnitude) at large zenith angles. The column length of matter is

$$\bar{x} = m_H \int_{1.2r_g}^{100r_g} n \ dr \ g/cm^2 \simeq 3.2 \times 10^{12} \ m_H (\dot{M}_1/M_8^2) \alpha^{-1} \ r_g \ g/cm^2 \ ,$$

where m_H is the mass of a hydrogen atom; thus $\bar{x} \simeq 80 \ (\dot{M}_1/M_8) \alpha^{-1} \ g/cm^2$. For plausible values $\dot{M}_1/M_8 \simeq 0.1$, and $\alpha = 0.1$, $\bar{x}$ is greater than the mean free path of protons ($\lambda_p \simeq 50 \ g/cm^2$), when viewed in the plane of the accretion disk.

The magnetic field near a black hole has in general poloidal (B_r, B_z) and toroidal B_ϕ components; here ϕ is the azimuthal angle. These components are given by Takahara (1979)

$$B_\phi \simeq 2.5 \times 10^4 \ r_*^{-3/4} \ M_8^{-\frac{1}{2}} \ gauss,$$

$$B_r \simeq 0.1 \ B_\phi \qquad\qquad gauss,$$

$$B_z \simeq 1.7 \times 10^4 \ r_*^{-7/4} \ \dot{M}_1 \ M_8^{-3/2} \ gauss.$$

ACCELERATION AND ENERGY-LOSS PROCESSES NEAR A BLACK HOLE

Since high-energy neutrinos are produced mostly by interactions of protons, we shall discuss electrons here rather briefly, noting, however, that they are especially important for gamma-ray production processes. The acceleration of electrons by a dynamo and magnetic flare mechanism has been studied by Sturrock and Barnes (1972), Harrison (1976), Takahara (1979) and Lovelace et al. (1979). The production of electron pairs by breakdown of vacuum and acceleration in magnetic fields has been explored by Blandford and Znajek (1977). Production of electron pairs by the Penrose pair production process has been investigated by Kafatos and Leiter (1979) and Kafatos (1980).

Electrons are particularly prone to suffer energy losses by generating photons or increasing the energy of photons. These losses occur through bremsstrahlung, inverse-Compton, and synchrotron processes. The energy loss time scale (Kafatos, 1980) is shortest for the synchrotron process:

$$t_s \simeq \frac{4.1 \times 10^5}{B_\perp^2 \; E_e(GeV)} \; sec \; .$$

Thus an electron of 1 GeV would lose half of its energy in ~ 0.4 second in a magnetic field of 10^3 gauss.

(a) The Fermi Acceleration Process

Protons can be accelerated by various processes in or near a black hole ergosphere. In turbulent magnetic field inhomogeneities, stochastic Fermi acceleration can occur. The mean rate of energy increase is:

$$\frac{dE}{dt} = a_F \; \frac{v}{c} \; E,$$

where v is the velocity of the particle and

$$a_F = \frac{u^2}{v\lambda} \; .$$

Here λ is the mean free path of particles while colliding with magnetic inhomogeneities of cell size ℓ, and u is the velocity of these inhomogeneities. λ is estimated from $\lambda > \ell \simeq 0.1 \; h \; V_A/c_s$, where h is the half thickness of the accretion disk, V_A is the Alfven speed, and c_s is the speed of sound; $V_A < c_s$ and $h \simeq 10^{14} M_1$ cm. We also have $u \simeq u_r$ where the radial drift velocity $u_r \simeq \alpha u_\phi$ (Kafatos and Leiter, 1979) and α is the viscosity parameter $\simeq 0.1$ (Shapiro et al., 1976). Here $u_\phi = (GM/r)^{\frac{1}{2}} = cr_*^{-\frac{1}{2}}$. Hence $u \simeq c \; \alpha \; r_*^{-\frac{1}{2}}$, and

$$a_F = 10^{-13} \; c\alpha^2 \; r_*^{-1} \; \dot{M}_1^{-1} \; sec^{-1} \; ,$$

when $v \simeq c$.

The rigidity spectrum is

$$\frac{dJ}{dR} \propto \frac{1}{R^{1+(a_F\tau)^{-1}}} \; ,$$

where τ is the confinement time in the acceleration region. A confinement time of $\sim 10^5$ sec in the ergosphere (about 1 day) is enough for efficient acceleration yielding an exponent like that of the galactic cosmic-ray spectrum corrected for leakage. However, the rate of energy loss by nuclear collision could be competitive with the rate of energy gain due to acceleration, except in regions sufficiently remote from the accretion disk, or if the drift velocity is $\sim u_\phi$ rather than u_r.

The magnetic field and dimensions of the accretion disk limit the energies obtainable by the stochastic Fermi process to $E < 10^{19}$ eV; the gyro-radius cannot exceed the dimensions of the source.

If small coronal loops are formed in the accretion disk (Galeev et al., 1979) one can consider either first-order Fermi acceleration as a reflecting wave moves up the loop, or second-order due, e.g., to random reconnection of loops. In the case of the first-order process,

$$a_F = \frac{v_\perp^2}{c^2} \frac{1}{\ell} \frac{d\ell}{dt} \, ,$$

where ℓ is the distance between reflections, $\ell \simeq z_o/\alpha^{1/3}$, where z_o is the convective cell size; also $\ell \simeq R_E \, r/3\alpha^{1/3}$, where R_E = ratio of luminosity L to Eddington luminosity L_E (Kafatos and Leiter, 1979; Kafatos, 1980). Here $R_E \simeq 10^{-2}$. Hence $\ell \simeq 3 \times 10^{-3} \, r\alpha^{-1/3}$. For $d\ell/dt \simeq u_r = \alpha u_\phi$, we obtain $d\ell/dt \stackrel{=}{\equiv} \alpha \, c \, r_*^{-\frac{1}{2}}$. If $v^2/c^2 \simeq 1/2$, then $a_F \simeq 0.3\alpha^{4/3} \, M_8^{-1} \, r_*^{-3/2} \, \phi \sec^{-1}$. For $d\ell/dt = u_\phi$, $a_F \simeq 0.3\alpha^{1/3} \, M_8^{-1} \, r_*^{-3/2} \sec^{-1}$. Thus, acceleration occurs on a time scale of minutes. In the case of the second-order Fermi acceleration,

$$a_F(r) \simeq 0.6 \, \alpha^{7/3} \, M_8^{-1} \, r_*^{-2} \sec^{-1},$$

$$a_F(\phi) \simeq 0.6 \, \alpha^{4/3} \, M_8^{-1} \, r_*^{-2} \sec^{-1}.$$

Acceleration up to $\sim 10^{18}$ eV is possible in this process.

(b) The Betatron Process

The momentum of a particle in a rising magnetic field increases according to the equation

$$P^2(t) = \frac{1}{3} \, P^2(0) \{ 1 + 2 \frac{B(t)}{B(0)} \} \quad .$$

If the original magnetic field $B(0) \simeq 10^{-5}$ gauss, and

$$B(t) \simeq 10^4 \, M_8^{-\frac{1}{2}} \, (r_*/10)^{-3/4} \text{ gauss,}$$

as in the accretion disk, then $P(t)/P(0) \simeq 3 \times 10^4$. Thus, if the accretion disk picks up protons from a hot component of the interstellar medium ($T \simeq 10^6$ K), then the betatron process will accelerate them to nearly 10 GeV in the disk. This process could serve as an injector for the Fermi mechanism, as could the Penrose process.

(c) The Shock Wave Acceleration and the Dynamo Processes

The shock wave acceleration process (e.g. Axford et al., 1977; Blandford and Ostriker, 1978; and Eichler, 1979) should produce a large number of high energy particles, especially if the ion temperature is high ($T_i \sim 10^{12}$K).

Lovelace (1976) has proposed an electromagnetic dynamo model that could accelerate beams of particles up to 10^{19} eV.

We conclude that a black-hole ergosphere (where magnetic irregularities approach each other, or "wind up") provides a powerful machine for accelerating particles in various ways.

REFERENCES

Anderson, K.S. : 1974, Astrophys. J. 187, 445.
Apparao, K.M.V. : 1973, Astrophys. Space Sci. 25, 3.
Apparao, K.M.V., Bignami, G.F., Maraschi, L., Helmken, H., Margon, B., Hjellming, R., Bradt, H.V., and Dower, R.G. : 1978, Nature 273, 450.
Arons, J., Kulsrud, R.M., and Ostriker, J.P. : 1975, Astrophys. J. 198, 687.
Auriemma, G., Angeloni, L., Belli, B.M., Bernardi, A., Cardini, D., Costa, E., Emanuele, A., Giovannelli, F., and Ubertini, P. : 1978, Astrophys. J. 221, L7.
Axford, W.I., Leer, E., and Skadron, G. : 1977, 15th Int. Cosmic-Ray Conf., Plovdiv, Bulgaria, 2, 273.
Bahcall, J.N. and Wolf, R.A. : 1976, Astrophys. J. 209, 214.
Baity, W.A., Jones, T.W., Wheaton, W.A., and Peterson, L.E. : 1975, Astrophys. J. 199, L5.
Beall, J.H., Rose, W.R., Graf, W., Price, K.M., Dent, W.A., Hobbs, R.W., Conklin, E.K., Ulich, B.L., Dennis, B.R., Crannell, C.J., Dolan, J. F., Frost, K.J., and Orwig, L.E. : 1978, Astrophys. J. 219, 836.
Berezinsky, V.S. and Ginzburg, V.L. : 1981, Monthly Notices Roy. Astron. Soc. 194, 3.
Blandford, R.D. and Rees, M.J. : 1974, Monthly Notices Roy. Astron. Soc. 169, 395.
Blandford, R.D., McKee, C.F. and Rees, M.J. : 1977, Nature 267, 211.
Blandford, R.D. and Znajek, R.L. : 1977, Monthly Notices Roy. Astron. Soc. 179, 433.
Blandford, R.D. and Ostriker, J.P. : 1978, Astrophys. J. 221, L29.
Blandford, R.D. and Rees, M.J. : 1978, Pittsburg Conf. p. 328.
Blandford, R.D. and Königl, A. : 1979, Astrophys. J. 232, 34.
Bradt, H.V., Burke, B.F., Canizares, C.R., Greenfield, P.E., Kelley, R. L., McClintock, J.E., Paradijs, J. van, and Koski, A.T. : 1978, Astrophys. J. 226, L111.
Burbidge, G.R., Burbidge, E.M., and Sandage, A.R. : 1963, Rev. Mod. Phys. 35, 947.
Burbidge, G.R. : 1967, Nature 216, 1287
Burbidge, G.R., Jones, T.W., and O'Dell, S.L. : 1974, Astrophys. J. 193, 43.
Burbidge, G.R. : 1978, Phys. Scripta 17.
Cameron, A.G.W. : 1973, Space Sci. Rev. 15, 121.
Canizares, C.R., McClintock, J.E., and Ricker, G.R. : 1978, Astrophys. J. 226, L1.
Cavaliere, A. and Morrison, P. : 1970, Astrophys. J. 162, L133.
Chevalier, R.A., Robertson, J.W., and Scott, J.S. : 1976, Astrophys. J. 207, 450.
Chevalier, R.A. : 1977, Astrophys. J. 213, 52.
Christiansen, W.A., Pacholczyk, A.G., and Scott, J.S. : 1977, Nature 266, 593.

Christiansen, W.A., Scott, J.S.,and Vestrand, W.T.: 1978, Astrophys. J.
    223, 13.
Cohen, M.H., Moffet, A.T., Romney, J.D., Schilizzi, R.T., Seielstad, G.
    A., Kellermann, K.I., Purcell, G.H., Shaffer, D.B., Pauliny-Toth,
    I.I.K., Preuss, E., Witzel, A.,and Rinehart, R. : 1976, Astrophys.
    J. 206, L1.
Colgate, S.A., Calvin, J.D.,and Petschek, A.G. : 1975, Astrophys. J. 197,
    L105.
Cooke, B.A., Ricketts, M.M., Maccacaro, T., Pye, J.P., Elvis, M., Watson,
    M.G., Griffiths, R.E., Pounds, K.A., McHardy, I., Maccagni, D., Se-
    ward, F.D., Page, C.G., and Turner, M.J.L. : 1978, Monthly Notices
    Roy. Astron. Soc. 182, 489.
Cooke, B.A., Lawrence, A.,and Perola, G.C. : 1978, Monthly Notices Roy.
    Astron. Soc. 182, 661.
Cotton, W.D. : 1976, Astrophys. J. 204, L63.
Cruddace, R.G., Fritz, G., Shulman, S., Friedman, H., McKee, J., and
    Johnson, M. : 1978, Astrophys. J. 222, L95.
Delvaille, J.P., Epstein,A.,and Schnopper,H.W. : 1978, Astrophys.J. 219, L81.
Dokuchaev, V.I.,and Ozernoy, L.M. : 1977, J. Exp. Teor. Fiz. 73, 1587.
Eardley, D.M. and Lightman, A.P. : 1975, Astrophys. J. 200, 187.
Eardley, D.M., Lightman, A.P., Payne, D.G., and Shapiro,S.L. : 1978,
    Astrophys. J. 224, 53.
Eichler, D. : 1979, Astrophys. J. 232, 106.
Eilek, J.A. : 1977, Astrophys. J. 212, 278.
Elvis,M., Maccacaro,T., Wilson, A.S., Ward,M.J., Penston,M.V., Fosbury,R.
    A.E.,and Perola, G.C.:1978, Monthly Notices Roy. Astron.Soc. 183, 129.
Fabian, A.C., Maccagni D., Rees, M.J.,and Stoeger, W.R. : 1976, Nature
    260, 683.
Fabricant, D., Topka, K., Harnden, F.R., Jr., and Gorenstein, P. : 1978,
    Astrophys. J. 226, L107.
Flasar, F.M.,and Morrison, P. : 1976, Astrophys. J. 204, 352.
Forman, W., Jones, C., Cominsky, L., Julien, P., Murray, S., Peters, G.,
    Tananbaum, H., and Giacconi, R. : 1978, Astrophys. J. Suppl. 38, 357.
Fowler, W.A. : 1964, Rev. Mod. Phys. 36, 545.
Fowler, W.A. : 1964, in Quasars and High-Energy Astronomy, 2nd Texas Symp.
    on Relativistic Astrophysics, Gordon and Breach, New York, p. 151.
Frank, J. and Rees, M.J. : 1976, Monthly Notices Roy. Astron. Soc. 176,
    633.
Galeev, A.A., Rosner, R.,and Vaiana, G.S. : 1979, Astrophys. J. 229,318.
Giacconi, R. : 1978, Phys. Scripta 17, 159.
Giacconi, R. : 1979, 16th Int. Cosmic Ray Conf., Kyoto, 14, 19.
Ginzburg, V.L. and Syrovatskii, S.I. : 1969, Ann. Rev. Astron. Astrophys.
    7, 375.
Ginzburg, V.L., and Ozernoy, L.M. : 1977, Astrophys. Space Sci. 50, 23.
Goldreich, P.,and Julian, W.H. : 1969, Astrophys. J. 157, 869.
Grindlay, J.E., Helmken, H.F., Hanbury Brown, R., Davis, J., and Allen,
    L.R. : 1975, Astrophys. J. 197, L9.
Grindlay, J.E., Helmken, H.F., and Weekes, T.C. : 1976, Astrophys. J.
    209, 592.
Gunn, J.E. and Ostriker, J.P. : 1969, Phys. Rev. Letters 22, 728.
Gupta, S.K., Ramana Murthy, P.V., Sreekantan, S.V., and Tonwar, S.C. :

1976, Astrophys. J. 221, 268.
Gursky, H., and Schwartz, D.A. : 1977, Ann. Rev. Astron. Astrophys. 15, 541.
Hall, R.D., Meegan, L.A., Wolraven, G.D., Djuth, F.T., and Harrison, E. R. : 1976, Nature 264, 525.
Harrison, E.R. : 1976, Nature, 264, 525.
Harrison, E.R. : 1977, Astrophys. J. 213, 827.
Helmken, H., Delvaille, J.P., Epstein, A., Geller, M.J., Schnopper, H.W., and Jernigan, J.G. : 1978, Astrophys. J. 221, L43.
Hills, J.G. : 1975, Nature 254, 295.
Hoyle, F., and Fowler, W.A. : 1963, Monthly Notices Roy. Astron. Soc. 125, 169.
Hoyle, F. and Fowler, W.A. : 1963, Nature 197, 533.
Ives,J.C., Sanford,P.W.,and Penston,M.V. : 1976, Astrophys. J. 207, L159.
Jones, C. and Forman, W. : 1978, Astrophys. J. 224, 1.
Jones, T.W., O'Dell, S.L., and Stein, W.A. : 1974, Astrophys. J. 188, 353.
Jones, T.W., O'Dell, S.L., and Stein, W.A. : 1974, Astrophys. J. 192, 261.
Jones, T.W. and Stein, W.A. : 1975, Astrophys. J. 197, 297.
Jones, T.W. and Tobin, W. : 1977, Astrophys. J. 215, 474.
Kafatos, M. : 1978, Astrophys. J. 225, 756.
Kafatos, M. : 1980, Astrophys. J. 236, 99.
Kafatos, M. and Leiter, D. : 1979, Astrophys. J. 229, 46.
Kellermann, K.I., and Pauliny-Toth, I.I.K. : 1969, Astrophys. J. 155, L71.
Kellermann, K.I. : 1974, Astrophys. J. 194, L135.
Kellermann, K.I., Shaffer, D.B., Purcell, G.H., Pauliny-Toth, I.I.K., Preuss, E., Witzel, A., Graham, D., Schilizzi, R.T., Cohen, M.H., Moffet, A.T., Romney, J.D., and Niell, A.E. : 1977, Astrophys. J. 211, 658.
Khachikian, E.Y. and Weedman, D.W. : 1974, Astrophys. J. 192, 581.
Kondo, Y., et al. : 1981, Astrophys. J. in press.
Kulsrud, R.M., Ostriker, J.P., and Gunn, J.E. : 1972, Phys. Rev. Letters 28, 636.
Kulsrud, R.M. and Arons, J. : 1975, Astrophys. J. 198, 709.
Leiter, D. and Kafatos, M. : 1978, Astrophys. J. 226, 32.
Lightman, A.P. and Shapiro, S.L. : 1977, Astrophys. J. 211, 244.
Lightman, A.P., Giacconi, R., and Tananbaum, H. : 1978, Astrophys. J. 224, 375.
Lovelace, R.V.E. : 1976, Nature 262, 649.
Lovelace, R.V.E., McAuslan, J., and Burns, M. : 1979, in Particle Acceleration Mechanisms in Astrophysics, (ed. by J. Arons, C. McKee, and C. Max), p. 399.
Lynden-Bell, D. : 1969, Nature 223, 690.
Lynden-Bell, D., and Rees, M.J. : 1971, Monthly Notices Roy. Astron. Soc. 152, 461.
Maltby, P., Matthews, T.A. and Moffet, A.T. : 1963, Astrophys.J. 137, 153.
Maraschi, L. and Treves, A. : 1977, Astrophys. J. 218, L113.
Marscher, A.P. and Brown, R.L. : 1978, Astrophys. J. 220, 474.
Marscher, A.P. : 1978, Astrophys. J. 219, 392.
Marscher, A.P. : 1978, Astrophys. J. 224, 816.
McBreen, B., Ball, S.E., Jr., Campbell, M., Greisen, K., and Koch, D. : 1973, Astrophys. J. 184, 571.

Morrison, P.: 1969, Astrophys. J. 157, L73.

Murray, S.S., Forman, W., Jones, C., and Giacconi, R.: 1978, Astrophys. J. 219, L89.

Mushotzky, R.F., Serlemitsos, P.J., Becker, R.H., Boldt, E.A., and Holt, S.S.: 1978, Astrophys. J. 220, 790.

Mushotzky, R.F., Boldt, E.A., Holt, S.S., Pravdo, S.H., Serlemitsos, P.J., Swank, J.H., and Rothschild, R.H.: 1978, Astrophys. J. 226, L65.

Mushotzky, R.F., Holt, S.S., and Serlemitsos, P.J.: 1978, Astrophys. J. Letters 225, L115.

Mushotzky, R.F., Serlemitsos, P.J., Smith, B.W., Boldt, E.A., and Holt, S.S.: 1978, Astrophys. J. 225, 21.

Neugebauer, G.: 1978, Phys. Scripta 17, 149.

O'Dell, S.L., Puschell, J.J., Stein, W.A., Owen, F., Porcas, R.W., Mufson, S., Moffett, T.J., and Ulrich, M.-H.: 1978, Astrophys. J. 224, 22.

Ozernoy, L.M. and Usov, V.V.: 1971, Astrophys. Space Sci. 13, 3.

Ozernoy, L.M. and Reinhardt, M.: 1978, Astrophys. Space Sci. 59, 171.

Pacholczyk, A.G. and Scott, J.S.: 1976, Astrophys. J. 203, 313.

Pacholczyk, A.G. and Scott, J.S.: 1976, Astrophys. J. 210, 311.

Pacini, F.: 1967, Nature 216, 567.

Pacini, F. and Rees, M.J.: 1970, Nature, 226, 622.

Peebles, P.J.E.: 1972, Astrophys. J. 178, 371.

Pringle, J.E., Rees, M.J., and Pacholczyk, A.G.: 1973, Astron. Astrophys. 29, 179.

Readhead, A.C.S., Cohen, M.H., Pearson, T.J., and Wilkinson, P.N.: 1978, Nature 276, 768.

Readhead, A.C.S., Cohen, M.H., and Blandford, R.D.: 1978, Nature 272, 131.

Rees, M.J. and Saslaw, W.C.: 1975, Monthly Notices Roy. Astron. Soc. 171, 53.

Ricker, G.R., Clark, G.W., Doxsey, R.E., Dower, R.G., Jernigan, J.G., Delvaille, J.P., MacAlpine, G.M., and Hjellming, R.M.: 1978, Nature 271, 35.

Rieke, G.H. and Low, F.J.: 1972, Astrophys. J. 176, L95.

Rieke, G.H. and Low, F.J.: 1975, Astrophys. J. 200, L67.

Sandage, A.: 1972, Astrophys. J. 178, 25.

Sanders, R.H.: 1976, Astrophys. J. 205, 335.

Sanford, P.W. and Ives, J.C.: 1976, Proc. Roy. Soc. London A 350, 491.

Sargent, W.L.W., Young, P.J., Boksenberg, A., Shortridge, K., Lynds, C.R. and Hartwick, F.D.A.: 1978, Astrophys. J. 221, 731.

Saslaw, W.C., Valtonen, M.J., and Aarseth, S.J.: 1974, Astrophys. J. 190, 253.

Scheuer, P.A.G.: 1974, Monthly Notices Roy. Astron. Soc. 166, 513.

Schönfelder, V.: 1978, Nature, 274, 344.

Schwartz, D.A., Bradt, H.V., Doxsey, R.E., Griffiths, R.E., Gursky, H., Johnson, M.D., and Schwarz, J.: 1978, Astrophys. J. 224, L103.

Scott, J.S. and Chevalier, R.A.: 1975, Astrophys. J. 197, L5.

Shapiro, S.L., Lightman, A.P., and Eardley, D.M.: 1976, Astrophys. J. 204, 187.

Shields, G.A., Oke, J.B., and Sargent, W.L.W.: 1972, Astrophys. J. 176, 75.

Shields, G.A. and Wheeler, J.C.: 1978, Astrophys. J. 222, 667.

Silberberg, R. and Shapiro, M.M.: 1979, Conf. Papers, 16th Int. Cosmic

Ray Conf., Kyoto, 10, 357.
Spitzer, L. and Saslaw, W.C.: 1966, Astrophys. J. 143, 400.
Spitzer, L. and Stone, M.E.: 1967, Astrophys. J. 147, 519.
Stecker, F.W., Morgan, D.L., and Bredekamp, J.: 1972, Phys. Rev. Letters
    27, 1969.
Stein, W.A., O'Dell, S.L., and Strittmatter, P.A.: 1976, Ann. Rev. Astron.
    Astrophys. 14, 173.
Stein, W.A. and Weedman, D.W.: 1976, Astrophys. J. 205, 44.
Sturrock, P.A.: 1971, Astrophys. J. 170, 85.
Sturrock, P.A.: 1971, Astrophys. J. 164, 529.
Sturrock, P.A. and Barnes, C.: 1972, Astrophys. J. 176, 31.
Swanenburg, B.N., Bennett, K., Bignami, G.F., Caraveo, P., Hermsen, W.,
    Kanback, G., Masnou, J.L., Mayer-Hasselwander, H.A., Paul, J.A.,
    Sacco, B., Scarsi, L., and Wills, R.D.: 1978, Nature 275, 298.
Tananbaum, H., Peters, G., Forman, W., Giacconi, R., Jones, C., and Avni,
    Y.: 1978, Astrophys. J. 223, 74.
Takahara, F.: 1979, Prog. Theor. Phys. 62, 629.
Telesco, C.M., Harper, D.A., and Loewenstein, R.F.: 1976, Astrophys. J.
    203, L53.
Ulrich, M.-H.: 1973, Astrophys. J. 181, 51.
Valtonen, M.J.: 1976, Astrophys. J. 209, 35.
Walker, M.F.: 1968, Astrophys. J. 151, 71.
Ward, M.J., Wilson, A.S., Penston, M.V., Elvis, M., Maccacaro, T., and
    Tritton, K.P.: 1978, Astrophys. J. 223, 788.
Weekes, T.C.: 1978, in Proc. of the 1978 DUMAND Summer Workshop, Vol. 2
    (ed. by A. Roberts), DUMAND Scripps Inst. of Oceanography, Code
    A-010, La Jolla, CA 92093, p. 313.
Wiita, P.J.: 1978, Astrophys. J. 221, 41.
Wiita, P.J.: 1978, Astrophys. J. 221, 436.
Winkler, P.F. Jr. and White, A.E.: 1975, Astrophys. J. 199, L139.
Wittels, J.J., Knight, C.A., Shapiro, I.I., Hinteregger, H.F., Rogers,
    A.E.E., Whitney, A.R., Clark, T.A., Hutton, L.K., Marandino, G.E.,
    Niell, A.E., Rönnäng, B.O., Rydbeck, O.E.H., Klemperer, W.K., and
    Warnock, W.W.: 1975, Astrophys. J. 196, 13.
Wittels, J.J., Cotton, W.D., Counselman III, C.C., Shapiro, I.I., Hinte-
    regger, H.F., Knight, C.A., Rogers, A.E.E., Whitney, A.R., Clark,
    T.A., Hutton, L.K., Rönnäng, B.O., Rydbeck, O.E.H., and Niell, A.E.:
    1976, Astrophys. J. 206, L75.
Wolfe, A.M.: 1974, Astrophys. J. 188, 243.
Woltjer, L.: 1971, in Nuclei of Galaxies (ed. by D.J.K. O'Connell),
    Doubleday, New York, p. 477.
Young, D.S. de and Axford, W.I.: 1967, Nature 216, 125.
Young, D.S. de : 1976, Ann. Rev. Astron. Astrophys. 14, 447.
Young, D.S. de : 1977, Astrophys. J. 211, 329.
Young, P.J., Shields, G.A., and Wheeler, J.C.: 1977, Astrophys. J. 212,
    367.
Young, P.J.: 1977, Astrophys. J. 215, 36.
Young, P.J., Westphal, J.A., Kristian, J., Wilson, C.P., and Landauer,
    F.P.: 1978, Astrophys. J. 221, 721.
Zanrosso, E.M., Long, J.L., Zych, A.D., Gibbons, R., White, R.S., and
    Dayton, B.: 1979, 16th Int. Cosmic Ray Conf., Kyoto, 12, 26.

PART VIII
COSMOLOGY AND RELATIVITY

NON LINEAR EINSTEIN-MAXWELL DIFFERENTIAL EQUATIONS

G. Antonacopoulos
Department of Astronomy, University of Patras, Patras Greece

C.G. Kostakis
Department of Mathematics, Hellenic Air-Force Academy,
Dekelia, Attica, Greece

ABSTRACT

Starting with the Einstein-Maxwell field equations in general relativity we construct the general differential equations govering the components of the metric tensor. These equations allow us to find h_{ij} in various orders.

1. INTRODUCTION

The general method of obtaining equations of motion in the theory of relativity introduced by Einstein et al. (1938) and Einstein and Infeld (1940, 1949) has been successful for equations of motion for charged particles (cf. Bażański, 1956, 1957; Bertotti, 1955; Chase, 1954; Infeld and Wallace, 1940; Dionysiou, 1980, 1981b, 1981c). We made a great simplification in the procedure of obtaining the equations of motion by the use of Dirac's δ-function in the energy-momentum tensor.

Instead of treating point masses and charges for mathematical convenience we consider them to be distributed continuously in space. Then we can introduce the "mass density" and the "charge density" μ and ρ such that μdV and ρdV are the masses and the charges contained in the volume dV. The densities μ, ρ are in general functions of the coordinates and the time. The integrals of μ and ρ over a certain volume are the masses and the charges contained within volume.

We note that masses and charges are actually "pointlike" so that the densities μ,ρ are zero every where except at points where the masses and charges are located, so

$$\int_V \mu \, dV = \sum_{\nu=1}^{n} m_\nu, \tag{1.1}$$

E. G. Mariolopoulos et al. (eds.), Compendium in Astronomy, 349–359.

$$\int_V \rho dV = \sum_{\nu=1}^{n} e_\nu . \tag{1.2}$$

Therefore μ and ρ can be expressed with the help of the δ-function in the following form :

$$\mu(\vec{r},t) = \sum_{\nu=1}^{n} m_\nu \delta(\vec{r}-\vec{r}_\nu(t)), \tag{1.3}$$

$$\rho(\vec{r},t) = \sum_{\nu=1}^{n} e_\nu \delta(\vec{r}-\vec{r}_\nu(t)), \tag{1.4}$$

where $\vec{r}_\nu(t)$ are the position vectors of masses and charges respectively.

2. GENERAL FORMULATION OF THE EQUATIONS

We consider the gravitational and electromagnetic fields of n charged particles, with masses m_1, m_2, ..., m_n with charges e_1, e_2, ..., e_n and with position vectors $\vec{r}_1(t)$, $\vec{r}_2(t)$, ..., $\vec{r}_n(t)$ respectively. Let the coordinate of the point of the word line of the i-th particle be $\xi_i^\alpha(t)$, where $\alpha = 0,1,2,3$ and $i = 1,2,3,...,n$. The gravitational and electromagnetic fields of such particles are defined by Einstein-Maxwell differential equations :

$$R_{ij} = \frac{8\pi G}{c^4} (T_{ij} + E_{ij} - \tfrac{1}{2} g_{ij} T), \tag{2.1}$$

$$F^{ij}_{;j} = \frac{4\pi}{c} J^i \tag{2.2}$$

$$F_{ij;k} + F_{jk;i} + F_{ki;j} = 0 , \tag{2.3}$$

where $J^0 = c\rho$ is the charge density, $(J^1, J^2, J^3) \equiv \vec{J}$ is the current density; R_{ij} is the Ricci tensor;

$$F_{ij} = A_{i;j} - A_{j;i} = A_{i,j} - A_{j,i} \tag{2.4}$$

is the electromagnetic field tensor which satisfies equation (2.3).

Also using equation (2.4), equation (2.3) can be written

$$F_{ij,k} + F_{jk,i} + F_{ki,j} = 0 , \tag{2.3a}$$

that is semi colon (covariant derivative) goes to comma (partial derivative); g_{ij} is the metric tensor of the combined field; G is the gravitational constant, c is the velocity of light and T_{ij}, E_{ij} are the energy
momentum tensors of the gravitational and electromagnetic fields respectively. It should be noted that Latin indices range from 0 to 3, while
Greek ones from 1 to 3.

We must now introduce another assumption

$$f,\alpha \equiv \frac{\partial f}{\partial x^{\alpha}} , \quad f,o \equiv \frac{\partial f}{\partial x^{o}} = \frac{1}{c} \frac{\partial f}{\partial t} , \tag{2.5}$$

where the operation of $\partial/\partial x^{o}$ on any quantity lowers its order by one.

The energy-momentum tensor of the gravitational field is :

$$T^{ij}(\vec{r},t) = \sum_{\nu=1}^{n} \frac{m_{\nu}c}{\sqrt{-g}} \frac{dx_{\nu}^{i}}{ds} \frac{dx_{\nu}^{j}}{dt} \; \delta(\vec{r}-\vec{r}_{\nu}(t)), \tag{2.6}$$

where $\vec{r}_{\nu}$ are the position vectors of the particles, and the summation
extends over all the particles of the system. The energy-momentum tensor
of the electromagnetic field E_{ij} is :

$$E_{ij} = \frac{1}{4\pi} (-F_{ik}F^{k}_{j} + \frac{1}{4} g_{ij}F_{ke}F^{ke}), \qquad (I)$$

or

$$E^{ij} = \frac{1}{4\pi} (-F^{ik}F^{j}_{k} + \frac{1}{4} g^{ij}F_{ke}F^{ke}), \qquad (II) \tag{2.7}$$

or

$$E^{i}_{j} = \frac{1}{4\pi} (-F^{ik}F_{jk} + \frac{1}{4} \delta^{i}_{j} F^{km}F_{km}), \qquad (III)$$

where it is clear that

$$E^{i}_{i} = 0 \quad \text{(traceless)} \quad \text{and} \quad E_{ij} = E_{ji} \quad \text{(symmetric)}. \tag{2.8}$$

Avoiding the form of the matter tensor the field equations become (Bertotti, 1955)

$$R_{ij} = \frac{8\pi G}{c^{4}} \; E_{ij}, \tag{2.9}$$

since with the aid of equation (2.8) we have the scalar curvature R
of the space zero, i.e., R=0. The explicit formula for E_{ij} is, from
equation (2.7)

$$E_{\alpha\beta} = \frac{1}{4\pi} (F_{\alpha\gamma} F_{\beta\gamma} - F_{o\alpha} F_{o\beta} + \tfrac{1}{2} \delta_{\alpha\beta} F_{o\gamma} F_{o\gamma} - \tfrac{1}{4} \delta_{\alpha\beta} F_{\gamma\epsilon} F_{\gamma\epsilon}) , \quad (I)$$

$$E_{o\alpha} = \frac{1}{4\pi} F_{\beta o} F_{\beta\alpha} , \qquad\qquad\qquad (II) \quad (2.10)$$

$$E_{oo} = \frac{1}{4\pi} (\tfrac{1}{2} F_{\alpha o} F_{\alpha o} + \tfrac{1}{4} F_{\alpha\beta} F_{\beta\alpha}) , \qquad\qquad (III)$$

where $E_{\alpha\beta}$ is the Maxwell stress tensor, $cE_{o\alpha}$ is the Poynting vector and E_{oo} is the energy density. With regard to the matter tensor, equation (2.6), we can write

$$T^{ij} = \sum_{\nu=1}^{n} m_{\nu}(t) \frac{dx^i}{dt} \frac{dx^j}{dt} \delta(\vec{r} - \vec{r}_{\nu}), \qquad\qquad (2.11)$$

where the relativistic mass $m_{\nu}(t)$ is defined by

$$m_{\nu}(t) = \frac{m_{\nu}(0)}{\sqrt{-g}} \frac{dx^o}{ds} , \qquad\qquad\qquad (2.12)$$

$m_{\nu}(0)$ = the mass and the time $x^o = ct$.

From the field equations (2.1), we have

$$(T^{ij} + E^{ij})_{;j} = 0. \qquad\qquad\qquad (2.13)$$

Thus the equation (2.13) is essentially contained in the field equations (2.1). On the other hand, the equation (2.13), expressing the law of conservation of energy and momentum, contains the equation of motion of the physical system to which the energy-momentum tensor under consideration refers.

From equations (2.13) we can obtain the equations of motion for ν-th particle by integrating them over V_{ν}, where V_{ν} is the three dimensional region surrounding the ν-th particle. Thus we get :

$$\int_{V_{\nu}} (T^{ij} + E^{ij})_{;j} d\vec{r} = 0 \qquad \nu = 1,2,3,\ldots,n. \qquad (2.14)$$

According to (2.7 III) and (2.2.) we get :

$$E^{ij}_{;j} = \frac{1}{c} F^i_{\ j} J^j, \qquad\qquad\qquad (2.15)$$

where the current four-vector is defined by

$$J^i = \sum_{\nu=1}^{n} \frac{e_\nu c}{\sqrt{-g}} \frac{dx^i}{dx^o} \delta(\vec{r}-\vec{r}_\nu) \tag{2.16}$$

(cf. Landau and Lifshitz, 1975).
Thus instead of (2.14), we have

$$\int_{V_\nu} (T^{ij}_{;j} + \frac{1}{c} F^i_j J^j)d\vec{r} = 0, \tag{2.17}$$

$\nu = 1,2,3,\ldots,n$ (Dionysiou, 1979, 1980, 1981a).

We can write out these equations only if we solve before the field equations (2.1), (2.2), and (2.3), where the Einstein's equation (2.1) is hyperbolic partial differential equation.

The Einstein et al. (1938) and Einstein and Infeld (1940, 1949) theory provides a model for the solution of the field equations by successive approximation based on expansions in inverse powers of c.

According to this theory we shall write the coefficients of the metric tensor as

$$g_{ij} = g_{ij}^{(o)} + h_{ij}, \qquad g^{ij} = g^{(o)ij} + h^{ij}, \tag{2.18}$$

where $g_{ij}^{(o)}$, $g^{(o)ij}$ denote the flat metric (1, -1, -1, -1) and the deviations from it by h_{ij}, h^{ij}. We develop the quantities h_{ij}, h^{ij} in series of inverse powers of c, that is to say

$$g_{oo} = 1 + \frac{1}{c^2} {}_2g_{oo} + \frac{1}{c^4} {}_4g_{oo} + \frac{1}{c^5} {}_5g_{oo} + 0(c^{-6}) \qquad (I)$$

$$g_{o\alpha} = \frac{1}{c^3} {}_3g_{o\alpha} + 0(c^{-5}), \qquad (II) \tag{2.19}$$

$$g_{\alpha\beta} = -\delta_{\alpha\beta} + \frac{1}{c^2} {}_2g_{\alpha\beta} + \frac{1}{c^4} {}_4g_{\alpha\beta} + \frac{1}{c^5} {}_5g_{\alpha\beta} + 0(c^{-6}) \qquad (III)$$

and

$$g^{oo} = 1 + \frac{1}{c^2} {}_2g^{oo} + \frac{1}{c^4} {}_4g^{oo} + \frac{1}{c^5} {}_5g^{oo} + 0(c^{-6}), \qquad (I)$$

$$g^{o\alpha} = \frac{1}{c^3} {}_3g^{o\alpha} + 0(c^{-5}) \qquad (II) \tag{2.20}$$

$$g^{\alpha\beta} = -\delta_{\alpha\beta} + \frac{1}{c^2} {}_2g^{\alpha\beta} + \frac{1}{c^4} {}_4g^{\alpha\beta} + \frac{1}{c^5} {}_5g^{\alpha\beta} + 0(c^{-6}), \qquad (III)$$

where subscripts denote the orders of the corresponding quantities,

$$_2 g_{oo} = - 2U \tag{2.21}$$

is demanded by the principle of equivalence (Chandrasekhar and Esposito, 1970), and by virtue of the laws of the conservation of mass and linear momentum we have (Ohta et al., 1973, 1974; Dionysiou, 1976, 1977).

$$_3 g_{oo} = _2 g_{o\alpha} = _1 g_{\alpha\beta} = _3 g_{\alpha\beta} = _4 g_{o\alpha} = 0. \tag{2.22}$$

The Ricci tensor is defined by

$$R_{ij} = g^{lm} R_{limj} , \tag{2.23}$$

where R_{limj} is the curvature or Riemann–Christoffel tensor

$$R_{limj} = \tfrac{1}{2} (g_{lj,im} + g_{im,lj} - g_{lm,ij} - g_{ir,lm}) +$$

$$+ g^{np}(\Gamma_{n,im} \Gamma_{p,jl} - \Gamma_{n,ij} \Gamma_{p,lm}) \tag{2.24}$$

and

$$\Gamma_{p,ij} = \tfrac{1}{2} (g_{pi,j} + g_{pj,i} - g_{ij,p}) \tag{2.25}$$

are the well known Christoffel symbols of the first kind.

Then from equation (2.1) we get (Dionysiou, 1980)

$$\square\, h_{ij} = -g^{(o)lm}(h_{lj,im} + h_{im,lj} - h_{lm,ij}) -$$

$$- h^{lm}(h_{lj,im} + h_{im,lj} - h_{lm,ij} - h_{ij,lm}) -$$

$$- 2g^{lm} g^{np}(\Gamma_{n,im}\Gamma_{p,jl} - \Gamma_{n,ij}\Gamma_{p,lm}) + \frac{16\pi G}{c^4} (T_{ij} + E_{ij} - \tfrac{1}{2} g_{ij}T). \tag{2.26}$$

This is the differential equation for defining h_{ij} on the Einstein–Maxwell differential equations.

But

$$g = |g_{ij}| = g_{oo}g_{11}g_{22}g_{33} + O(c^{-6}), \tag{2.27}$$

$$\frac{dx^o}{ds} = (g_{oo} + \frac{2}{c} g_{o\lambda}v_\lambda + \frac{1}{c^2} g_{\lambda\mu}v_\lambda v_\mu)^{-1/2}, \tag{2.28}$$

$$T_{lm} = g_{il}g_{jm}T^{ij} \tag{2.29}$$

and

$$T = T_i^i = g_{ij} T^{ij} = \sum_{\nu=1}^{n} \frac{m_\nu c^2}{\sqrt{-g}} \frac{ds}{dx^o} \delta(\vec{r} - \vec{r}_\nu).$$ (2.30)

Next, we consider the retarded potentials at the vector positions $\vec{r}_i$ as

$$\varphi(\vec{r}_i, t) = \int_V \frac{\rho(\vec{r}, t - R/c)}{R} dV + \text{const.}$$ (I)

(2.31)

$$\vec{A}(\vec{r}_i, t) = \frac{1}{c} \int_V \frac{\vec{J}(\vec{r}, t - R/c)}{R} dV + \text{const.},$$ (II)

where $R = |\vec{r}_i - \vec{r}|$ is the distance from the volume element $dV = d\vec{r}$ to the field point $\vec{r}_i$ at which redetermine the potentials and

$$\vec{J}(\vec{r}, t) = \rho(\vec{r}, t)\vec{\upsilon}(\vec{r}).$$ (2.32)

If the motion of the charges is sufficiently slow and smooth we can expand (2.31) in inverse powers of c (Dionysiou and Vaiopoulos, 1979; Bertotti, 1955; Eddington, 1953) as :

$$\varphi(\vec{r}, t) = A_o = \sum_{j=1}^{n} \frac{e_j}{|\vec{r}_i - \vec{r}_j|} + \frac{1}{2c^2} \sum_{j=1}^{n} e_j \frac{\partial^2}{\partial t^2} |\vec{r}_i - \vec{r}_j| -$$

$$- \frac{1}{6c^3} \sum_{j=1}^{n} e_j \frac{\partial^3}{\partial t^3} |\vec{r}_i - \vec{r}_j|^2 + 0(c^{-4})$$ (2.33)

for the scalar potential and

$$\vec{A}^i(\vec{r}_i, t) = -\vec{A}_i(\vec{r}_i, t) = -\frac{1}{c} \sum_{j=1}^{n} \frac{e_j \vec{v}_j}{|\vec{r}_i - \vec{r}_j|} +$$

$$+ \frac{1}{c^2} \sum_{j=1}^{n} e_j \vec{v}_j - \frac{1}{2c^3} \sum_{j=1}^{n} e_j \frac{\partial^2}{\partial t^2} (\vec{v}_j |\vec{r}_i - \vec{r}_j|) + 0(c^{-4})$$ (2.34)

for the vector potential, where the four-vector A_i is defined by

$$A_i = (\varphi, -\vec{A})$$

with the conditions that A^i, $i = 0$ and $A_i = g_{ij}^{(o)} A^j$; The differentiation $\partial/\partial t$ is done with respect to $\vec{r}_j$; the velocities $\vec{v}_j$ of the charges e_j, the distance $|\vec{r}_i - \vec{r}_j|$ from the charges e_j to the field point $\vec{r}_i$ are unretarded - that is, they refer to the same instant as $\varphi(\vec{r}_i, t)$ and $\vec{A}(\vec{r}_i, t)$.

Thus, we finally get (Dionysiou, 1980, 1981b)

$$F_{o\alpha} = \frac{\partial}{\partial x_i^\alpha} \left(\sum_{j=1}^{n} \frac{e_j}{|\vec{r}_i - \vec{r}_j|} \right) + \frac{1}{c^2} \left[\frac{1}{2} \sum_{j=1}^{n} e_j \frac{\partial}{\partial x_i^\alpha} \frac{\partial^2}{\partial t^2} |\vec{r}_i - \vec{r}_j| + \right.$$

$$\left. + \frac{\partial}{\partial t} \left(\sum_{j=1}^{n} \frac{e_j v_j^\alpha}{|\vec{r}_i - \vec{r}_j|} \right) \right] + \frac{1}{c^3} \left[-\frac{1}{6} \sum_{j=1}^{n} e_j \frac{\partial}{\partial x_i^\alpha} \frac{\partial^3}{\partial t^3} |\vec{r}_i - \vec{r}_j|^2 - \frac{\partial^2}{\partial t^2} \sum_{j=1}^{n} e_j v_j^\alpha \right] +$$

$$+ 0(c^{-4}), \tag{2.35}$$

$$F_{\alpha\beta} = \frac{1}{c} \left[\frac{\partial}{\partial x_i^\alpha} \left(\sum_{j=1}^{n} \frac{e_j v_j^\beta}{|\vec{r}_i - \vec{r}_j|} \right) - \frac{\partial}{\partial x_i^\beta} \left(\sum_{j=1}^{n} \frac{e_j v_j^\alpha}{|\vec{r}_i - \vec{r}_j|} \right) \right] + \frac{1}{c^3} \left[\frac{1}{2} \sum_{j=1}^{n} e_j \right.$$

$$\left. \left(\frac{\partial}{\partial x_i^\alpha} \frac{\partial^2}{\partial t^2} v_j^\beta |\vec{r}_i - \vec{r}_j| - \frac{\partial}{\partial x_i^\beta} \frac{\partial^2}{\partial t^2} v_j^\alpha |\vec{r}_i - \vec{r}_j| \right) \right] + 0(c^{-4}), \tag{2.36}$$

where $\partial/\partial t$ is done with respect to r_j and

$$E_{\alpha\beta} = \frac{1}{4\pi} \left[-\frac{\partial}{\partial x_i^\alpha} \left(\sum_{j=1}^{n} \frac{e_j}{|\vec{r}_i - \vec{r}_j|} \right) \frac{\partial}{\partial x_i^\beta} \left(\sum_{j=1}^{n} \frac{e_j}{|\vec{r}_i - \vec{r}_j|} \right) + \right.$$

$$\left. \frac{1}{2} \delta_{\alpha\beta} \frac{\partial}{\partial x_i^\gamma} \left(\sum_{j=1}^{n} \frac{e_j}{|\vec{r}_i - \vec{r}_j|} \right) \frac{\partial}{\partial x_i^\gamma} \left(\sum_{j=1}^{n} \frac{e_j}{|\vec{r}_i - \vec{r}_j|} \right) \right] + 0(c^{-4}), \tag{2.37}$$

$$E_{o\alpha} = \frac{1}{4\pi c} \left\{ \frac{\partial}{\partial x_i^\beta} \left(\sum_{j=1}^{n} \frac{e_j}{|\vec{r}_i - \vec{r}_j|} \right) \left[\frac{\partial}{\partial x_i^\alpha} \left(\sum_{j=1}^{n} \frac{e_j v_j^\beta}{|\vec{r}_i - \vec{r}_j|} \right) - \right. \right.$$

$$\left. \left. - \frac{\partial}{\partial x_i^\beta} \left(\sum_{j=1}^{n} \frac{e_j v_j^\alpha}{|\vec{r}_i - \vec{r}_j|} \right) \right] \right\} + 0(c^{-3}), \tag{2.38}$$

and

$$E_{oo} = \frac{1}{8\pi} \left[\nabla \left(\sum_{j=1}^{n} \frac{e_j}{|\vec{r}_i - \vec{r}_j|} \right) \right]^2 + 0(c^{-2}), \tag{2.39}$$

where $\nabla \equiv \frac{\partial}{\partial x^\alpha}$.

The available information about the term g_{ij} of various orders is summarized in the following table :

Table 1

Einstein-Maxwell Equations *

$$g_{ij} = g_{ij}^{(G)} \oplus g_{ij}^{(E)}$$

Equation of motion	Orders of the metric coefficients needed		
	$g_{\alpha\beta} = g_{\alpha\beta}^{(G)} \oplus g_{\alpha\beta}^{(E)}$	$g_{0\alpha} = g_{0\alpha}^{(G)} \oplus g_{0\alpha}^{(E)}$	$g_{00} = g_{00}^{(G)} \oplus g_{00}^{(E)}$
Newtonian	$_0 g_{\alpha\beta}^{(G)}=\delta_{\alpha\beta}$, $_0 g_{\alpha\beta}^{(E)}=0$	$_1 g_{0\alpha}^{(G)} = {}_1 g_{0\alpha}^{(E)}=0$	$_2 g_{00}^{(G)}=-2U$, $_2 g_{00}^{(E)}=0$
½-P.N.A.	$_1 g_{\alpha\beta}^{(G)}={}_1 g_{\alpha\beta}^{(E)}=0$	$_2 g_{0\alpha}^{(G)}={}_2 g_{0\alpha}^{(E)}=0$	$_3 g_{00}^{(G)}={}_3 g_{00}^{(E)}=0$
1-P.N.A.	$_2 g_{\alpha\beta}^{(G)}=-2U\delta_{\alpha\beta}$ $_2 g_{\alpha\beta}^{(E)}=0$	$_3 g_{0\alpha}^{(G)}=P_\alpha$, $_3 g_{0\alpha}^{(E)}=0$	$_4 g_{00}^{(G)}=2(U^2-2\varphi)$, $_4 g_{00}^{(E)}$
1½-P.N.A.	$_3 g_{\alpha\beta}^{(G)}={}_3 g_{\alpha\beta}^{(E)}=0$	$_4 g_{0\alpha}^{(G)}={}_4 g_{0\alpha}^{(E)}=0$	$_5 g_{00}^{(G)}(t)$, $_5 g_{00}^{(E)}=0$
2-P.N.A.	$_4 g_{\alpha\beta}^{(G)}$, $_4 g_{\alpha\beta}^{(E)}$	$_5 g_{0\alpha}^{(G)}$, $_5 g_{0\alpha}^{(E)}$	$_6 g_{00}^{(G)}$, $_6 g_{00}^{(E)}$
2½-P.N.A.	$_5 g_{\alpha\beta}^{(G)}(t)$, $_5 g_{\alpha\beta}^{(E)}$	$_6 g_{0\alpha}^{(G)}$, $_6 g_{0\alpha}^{(E)}=0$	$_7 g_{00}^{(G)}$, $_7 g_{00}^{(E)}$

where

1) $U = G \sum_{i=1}^{n} \frac{m_i}{|\vec{r} - \vec{r}_i|}$,

2) $\varphi = \frac{3}{4} G \sum_{i=1}^{n} \frac{m_i v_i^2}{|\vec{r}-\vec{r}_i|} - \frac{1}{2} G^2 \sum_{i=1}^{n}\sum_{j=1}^{n} \frac{m_i m_j}{|\vec{r}-\vec{r}_i|\,|\vec{r}_i-\vec{r}_j|}$, $i \neq j$,

(Chandrasekhar and Esposito, 1970; Tsoupakis, 1979; Antonacopoulos and Tsoupakis, 1979; Antonacopoulos, 1979).

3) $P_\alpha = {}_3 g_{0\alpha}^{(G)} = \frac{G}{2} \sum_{i=1}^{n} \frac{m_i}{|\vec{r}-\vec{r}_i|} [7\upsilon_{\alpha i}+(\vec{n}.\vec{\upsilon}_i)n_{\alpha i}]$,

where

$\vec{n} = \frac{\vec{r}-\vec{r}_i}{|\vec{r}-\vec{r}_i|}$,

* Where $g_{ij}^{(G)} \equiv$ gravitational terms, $g_{ij}^{(E)} \equiv$ electromagnetic terms and the symbol $\oplus \equiv$ the mixing of the metric in the Einstein-Maxwell differential equations.

4) $\displaystyle {}_5 g^{(G)}_{00}(t) = \frac{4}{3} G \frac{d^3 I_{\alpha\alpha}}{dt^3}$, $\displaystyle {}_5 g^{(G)}_{\alpha\beta}(t) = \frac{2G}{c^3} \frac{d^3}{dt^3} D_{\alpha\beta}$,

where

$$I_{\alpha\beta} = \sum_m m\, x_\alpha\, x_\beta$$

is the dipole moment tensor and

$$D_{\alpha\beta} = 3 I_{\alpha\beta} - \delta_{\alpha\beta} I_{\gamma\gamma}$$

is the quandrupole moment tensor (Dionysiou, 1976, 1977).

5) $\displaystyle {}_4 g^{(G)}_{\alpha\beta}$, $\displaystyle {}_5 g^{(G)}_{0\alpha}$, $\displaystyle {}_6 g^{(G)}_{00}$

these terms are defined by Ohta et al. (1973, 1974).

6) $\displaystyle {}_5 g^{(G)}_{\alpha\beta}$, $\displaystyle {}_6 g^{(G)}_{0\alpha}$, $\displaystyle {}_5 g^{(G)}_{00}$

these are the lowest order terms that derive from the imposition of the Sommerfeld radiation condition (namely that at infinity there is only outgoing radiation) at infinity.

7) $\displaystyle {}_4 g^{(E)}_{\alpha\beta}$, $\displaystyle {}_5 g^{(E)}_{0\alpha}$, $\displaystyle {}_6 g^{(E)}_{00}$, $\displaystyle {}_4 g^{(E)}_{00}$

these terms are defined by Kostakis (1981).

3. CONCLUSIONS

In this paper we have used the Einstein-Infeld-Hoffmann (1938) and Einstein-Infeld (1940, 1949) theory in order to find the differential equations for defining h_{ij} on the Einstein-Maxwell field equations. Also, we give informations about the terms g_{ij} of various orders.

REFERENCES

Antonacopoulos, G. and Tsoupakis, E. : 1979,   Astrophys. and Space
    Sci., 62, 183
Antonacopoulos, G. : 1979, Astrophys. and Space Sci., 62, 217
Bažański, S. : 1956, Acta Phys. Pol. 15, 363.
Bažański, S. : 1957, Acta Phys. Pol. 16, 423.
Bertotti, B. : 1955, Nuovo Cimento, 2, 231.
Chandrasekhar, S. and Esposito, F.P. : 1970, Astrophys. J. 160, 153.
Chase, D.M. : 1954, Phys. Rev. 95, 243.
Dionysiou, D.D.: 1976, Nuovo Cimento, 33B, 519; 35B, 363.

Dionysiou, D.D.:   1977, Nuovo Cimento Letters, 19, 383.
Dionysiou, D.D.:   1979, Nuovo Cimento, 52B, 56.
Dionysiou, D.D.:   1980, Astrophys. and Space Sci., 73, 295.
Dionysiou, D.D.:   1981a, Astrophys. and Space Sci., 76, 513.
Dionysiou, D.D.:   1981b, Astrophys. and Space Sci., 77, 383.
Dionysiou, D.D.:   1981c, Int. J. of Theor. Phys., 20, 1.
Dionysiou, D.D. and Vaiopoulos, D.A.: 1979, Nuovo Cimento Letters, 26, 5.
Eddington, A.S.: 1953, Fundamental Theory, Cambridge Univ. Press,
     Cambridge.
Einstein, A., Infeld, L., and Hoffmann, B.: 1938, Ann. Math. 39, 65.
Einstein, A. and Infeld, L.: 1940, Ann. Math. 41, 797.
Einstein, A. and Infeld, L.: 1949, Can. J. Math. 1, 209.
Infeld, L. and Wallace, P.R.: 1940, Phys. Rev. 57, 797.
Kostakis, C.G.: 1981, Ph. D. Thesis. Univ. of Patras.
Landau, L.D. and Lifshitz, E.M.: 1975, The Classical Theory of Fields,
     Pergamon Press, Oxford.
Ohta, T., Okamura, H., Kimura, T., and Hiida, K.: 1973, Prog. Theor.
     Phys. 50, 492.
Ohta, T., Okamura, H., Kimura, T., and Hiida, K.: 1974, Prog. Theor.
     Phys. 51, 1220.
Tsoupakis, E.: 1979, Ph. D. Thesis, Univ. of Patras.

RECENT RESULTS FOR THE MOON'S SECULAR ACCELERATION AND THEIR IMPLICATION
FOR THE POSSIBLE VARIATION OF G IN DIRAC'S LARGE NUMBER HYPOTHESIS

L.V. Morrison
Royal Greenwich Observatory
Herstmonceux Castle, Hailsham, East Sussex, U.K.

ABSTRACT

   Recent data on the secular acceleration of the Moon imply a rate of
variation of $\underline{G}$ between zero and -6 parts in 10^{11} per year with Dirac's
additive creation hypothesis.

   In order to reconcile the cosmological predictions of the Large
Number Hypothesis with Einstein's general theory of relativity, Dirac
(1938) proposed that the universal constant of gravitation, $\underline{G}$, should
remain constant with respect to the Einstein unit of time, but vary with
respect to the atomic unit of time. Dirac's hypothesis also requires
the continuous creation of matter and he considers two cases : the addition and multiplication of matter (Dirac, 1974). In the additive case
matter is created uniformly throughout space, and in the multiplicative
case matter is created where it already exists, in proportion to the
amount existing there. Roxburgh (1977) has shown that other cosmological
models are possible which reconcile the Large Number Hypothesis with the
theory of general relativity, but I shall restrict my discussion to the
two cases proposed by Dirac.

   On the hypothesis that $\underline{G}$ decreases with atomic time, Dirac deduces
that with additive creation orbits in the Solar System (and elsewhere)
would contract with atomic time and that with multiplicative creation
they would expand. The contraction of any orbit would produce a secular
acceleration in the orbital motion on atomic time, whereas the expansion
would produce a deceleration. By measuring this acceleration on atomic
time we can obtain a measure of the rate of change of $\underline{G}$, once all the
other known physical causes producing secular accelerations have been
removed. The relationships between the rate of change of $\underline{G}$ and the
secular acceleration for the Dirac cosmologies are

$$\frac{\dot{G}}{G} = - \frac{\text{secular acceleration}}{\text{mean motion}} \quad \text{(additive case)},$$

E. G. Mariolopoulos et al. (eds.), Compendium in Astronomy, 361–366.
Copyright © 1982 by D. Reidel Publishing Company.

$$\frac{\dot{G}}{G} = + \frac{\text{secular acceleration}}{\text{mean motion}} \quad \text{(multiplicative case)}$$

In this paper I shall consider the implications of recent measurements of the Moon's secular acceleration. The secular acceleration is the rate of change of mean motion in longitude and is expressed in seconds of arc per century per century (" cy^{-2}).

There are two known causes of the observed secular acceleration of the Moon, besides the possible cosmological one. The first is due to the attractions of the planetary system. Strictly, the effect is a very long-period oscillation, but it can be treated as a secular acceleration over a period of a few thousand years. I shall assume that its value, (+14".28 cy^{-2}) in Einstein units with constant $\underline{G}$, is known with certainty from dynamical theory and the masses of the planets. I shall consider this contribution as having been eliminated in all further discussion of the secular acceleration. The second known cause of the secular acceleration is the tidal interaction between the Earth and Moon. This contribution cannot be calculated accurately from dynamical theory and has to be determined from observation. If there were a cosmological acceleration due to varying $\underline{G}$, it would be combined with the tidal acceleration in the observed secular acceleration when measured on atomic time. These two contributions are separated by measuring the tidal part using one of the following methods.

The first method is to measure the secular acceleration in the Moon's motion on the time-scale inferred from the orbital motions of other bodies in the Solar System in which $\underline{G}$ is treated as a constant. The second is to measure the periodic perturbations on the orbits of artificial satellites caused by the lunar tides, and from their amplitudes and phase-lags to deduce the tidal acceleration. The third method is to construct numerical models of the ocean tides and to deduce the tidal acceleration of the Moon from the rate of dissipation of tidal energy required in the model to represent the actual tides. In all three methods, the tidal acceleration is measured in the natural (or Einstein) unit of time implicit in the dynamical system. I shall refer to the time-scale having this unit of time as dynamical time.

Secular Acceleration on Dynamical Time

1. The acceleration of the Moon relative to the Sun is determined by analysing the recorded occurrences of ancient and medieval total solar eclipses. The motion of the Sun (i.e. the reflected orbital motion of the Earth) defines the dynamical time-scale. By this method Muller (1976) finds the tidal acceleration of the Moon to be (-30 ± 3) " cy^{-2}. This result is likely to be an improvement on the earlier ones of Muller and Stephenson (1975) and Newton (1970, 1972) because it removes a bias in these earlier results due to an error in the motion of the nodes of the Moon's orbit.

2. The acceleration of the Moon relative to the stellar background is determined from timings of occultations of stars made in the 17th century to the present. The dynamical time-scale is deduced from the timings of the transits of Mercury across the Sun. Using this method, Morrison and Ward (1975) found the tidal acceleration of the Moon to be $(-26 \pm 2)"$ cy^{-2}. This should be an improvement over the previous result of $(-22\pm1)"$ cy^{-2} obtained from this method by Spencer Jones (1939) and Clemence (1948), because Morrison and Ward used a longer time-span of observations.

3. The acceleration of the Moon is obtained relative to dynamical time from meridian-circle observations of its position over several decades. The dynamical time-scale is deduced from meridian-circle observations of the planets made over the same period. From such observations in the period 1913-1968 Oesterwinter and Cohen (1972) found the tidal acceleration of the Moon to be $(-38 \pm 8)"$ cy^{-2}. This result may be somewhat confused for our present purpose because they used atomic time in the period 1955-1968.

4. From the observed perturbations of the mean inclination and nodes of the orbits of two artificial satellites due to the M_2 tide, Goad and Douglas (1978) find the tidal acceleration of the Moon to be $(-26 \pm 3)"$ cy^{-2}. In deriving this result, they assume that there is no contribution from the solid-Earth tides. An uncertainty of about $2"$ cy^{-2} appears to arise from the unmeasured contributions from the N_2 and O_1 tides which have to be estimated from tidal models.

5. From a comparison of several numerical models giving the amplitudes and lags of ocean tides, Lambeck (1977) estimates the tidal acceleration to be $(-30.6 \pm 3.1)"$ cy^{-2}. Uncertainties arise from lack of knowledge of the ocean-continent boundary conditions, solid-Earth tides, yielding of the Earth, and dissipation mechanisms.

Secular Acceleration and Atomic Time

1. From an analysis of eight years of lunar laser ranging data, Ferrari et al. (1980) find the acceleration on atomic time to be $(-23.8\pm3.1)"$ cy^{-2}. Independent analyses by Calame and Mulholland (1978) and King (1979) give results which fall within the range of uncertainty of the result of Ferrari et al.. The separate integration of the rotational and orbital models for the Moon, rather than their simultaneous integration, may contribute a systematic error to all of these results.

2. The most recent result obtained by van Flandern (1979) from the analysis of timings of occultations of stars on the atomic scale is $(-21.5 \pm 3.2)"$ cy^{-2}. Shortcomings in the planetary part of the analytical lunar theory and systematic errors in the star catalogues have produced biassed estimates of the acceleration in previous analyses; however, Van Flandern now believes that these have been substantially eliminated.

Comparison of Secular Accelerations on Dynamical and Atomic Time

The representative results for the Moon's acceleration from each
independent method discussed above are listed in Table 1 and plotted in
Figure 1 with the range of uncertainty shaded. Overlapping solutions
produce strips of darker shading. The diagonal line, along which the
accelerations on dynamical and atomic time are equal, denotes the solution $\dot{G}/G = 0$. Solutions above this line imply $\underline{G}$ varying with additive
creation, since the acceleration on atomic time is greater than on dynamical time, thus giving a residual acceleration and contracting orbits.
Solutions below the line $\dot{G}/G = 0$ give a residual deceleration, expanding
orbits, and hence $\underline{G}$ varying with multiplicative creation.

Sandage and Tammann's (1975) value of 55 km s^{-1} Mpc^{-1} for the Hubble
constant, H_o, gives the age of the universe, t_o, from the relation
$t_o = H_o^{-1}$, as 1.8 x 10^{10} yr. In Dirac's cosmology, $- \dot{G}/G = t_o^{-1}$, and hence
$\dot{G}/G = -5.6$ x 10^{-11} yr^{-1}. With this rate of change for $\dot{G}/G$ and the value
1."73 x 10^7 yr^{-1} for the mean motion of the Moon, we find the residual
acceleration on atomic time with additive creation to be + 9."7 cy^{-2}.
The line $\dot{G}/G = -5.6$ x 10^{-11} yr^{-1} is shown in Figure 1, where the indications are that $\dot{G}/G$ lies between this value and zero.

The evidence for the variability of $\underline{G}$, however, is far from conclu-
sive, because systematic errors in any of the methods used to derive the
secular acceleration of the Moon could alter the solution considerably.
It would appear from Figure 1 that a rate for $\dot{G}/G$ as great as
-5.6 x 10^{-11} yr^{-1} with multiplicative creation is unlikely. We note,
on the other hand, that additive creation meets with severe criticism
on the grounds of stellar evolution (van den Berg, 1976; Maeder, 1977a,b).

The Large Number Hypothesis has also been found to be irreconcilable
with the 3K background radiation (Steigman, 1978), but this objection
has been circumvented (Canuto and Hsieh, 1978; Canuto et al., 1979;
Canuto and Hsieh, 1979), by introducing a gauge function $\beta(t)$ such that
$G\beta^2 = 1$ with $\underline{G} \sim t^{-1}$ and $\beta \sim t^{\frac{1}{2}}$, where t is time in atomic units. This
leads to the relationships :

$$\frac{\text{secular acceleration}}{\text{mean motion}} = \frac{\dot{\beta}}{\beta} = + \tfrac{1}{2}t_o^{-1}.$$

Using the value of 55km s^{-1} Mpc^{-1} for the Hubble constant, we find

$$\frac{\text{secular acceleration}}{\text{mean motion}} = + 2.8 \text{ x } 10^{-11} \text{ yr}^{-1} ;$$

i.e. residual secular acceleration = + 4."8 cy^{-2}.

This result is consistent with the residual secular acceleration on
atomic time indicated by the measurements shown in Figure 1.

Table 1

Secular acceleration of the Moon

On dynamical time

Method	Author	Result ($"cy^{-2}$)
Total solar eclipses	Muller (1976)	-30 ± 3
Transits of Mercury	Morrison & Ward (1975)	-26 ± 2
Meridian-Circle observations	Oesterwinter & Cohen (1972)	-38 ± 8
Perturbations of artificial satellites	Goad & Douglas (1978)	-26 ± 3
Numerical tide models	Lambeck (1977)	-30.6 ± 3.1

On atomic time

Method	Author	Result ($"cy^{-2}$)
Lunar laser ranging	Ferrari et al. (1980)	-23.8 ± 3.1
Lunar occultations of stars	van Flandern (1979)	-21.5 ± 3.2

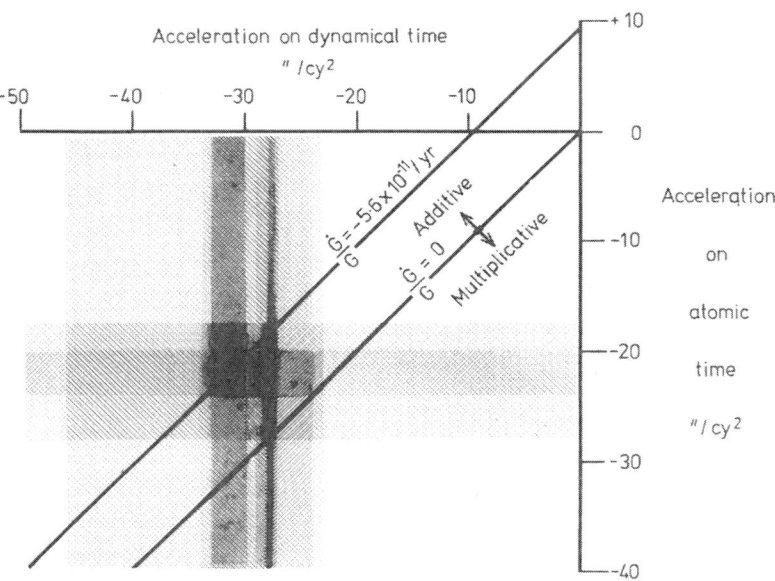

Fig. 1 : Plot of the observational results for the secular acceleration of the Moon listed in Table 1. The shaded bands cover the ranges of uncertainty of the results. Overlapping results produce darker bands. See the text for the significance of the diagonal lines.

A further increase in the accuracy of the determination of the Moon's secular acceleration on atomic time is to be expected in the next few years from the lunar laser ranging experiment. This will improve the reliability of the observational evidence from the Moon's secular acceleration for or against the variability of $\underline{G}$ in atomic units.

REFERENCES

Berg, D.A. van den : 1976, Monthly Notices Roy. Astron. Soc. <u>176</u>, 455

Calame, O. and Mulholland, J.D.: 1978, in Tidal Friction and the Earth's Rotation, (ed. by P. Brosche and J. Sündermann), Springer-Verlag, Berlin, p. 43.

Canuto, V. and Hsieh, S.-H. : 1978, Astrophys. J. <u>224</u>, 302

Canuto, V. and Hsieh, S.-H. : 1979, Astrophys. J. Suppl. <u>41</u>, 243

Canuto, V., Hsieh, S.-H.,and Owen, J.R. : 1979, Monthly Notices Roy. Astron. Soc. <u>188</u>, 829

Clemence, G.M. : 1948, Astron. J. <u>53</u>, 169

Dirac, P.A.M. : 1938, Proc. Roy. Soc. London A <u>165</u>, 199

Dirac, P.A.M. : 1974, Proc. Roy. Soc. London A <u>338</u>, 439

Ferrari, A.J., Sinclair, W.S., Sjogren, W.L., Williams, J.G., and Yoder, C.F. : 1980, J. Geophys. Res. <u>85</u>, 3939

Flandern,.T.C. van : 1979, private communication

Goad, C.C. and Douglas, B.C. : 1978, J. Geophys. Res. <u>83</u>, 2306

King, R.W. : 1979, private communication

Lambeck, K. : 1977, Phil. Trans. Roy. Soc. London A <u>287</u>, 545

Maeder, A. : 1977a, Astron. Astrophys. <u>56</u>, 359

Maeder, A. : 1977b,Astron. Astrophys. <u>57</u>, 125

Morrison, L.V. and Ward, C.G. : 1975, Monthly Notices Roy. Astron. Soc. <u>173</u>, 183

Muller, P.M. :1976, Determination of the Cosmological Rate of Change of G and the Tidal Accelerations of Earth and Moon from Ancient and Modern Astronomical Data. Spec. Publ. No. 43-46, Jet Propulsion Lab.,Pasadena, Calif.

Muller, P.M. and Stephenson, F.R.: 1975, in Growth Rhythms and the History of the Earth's Rotation (ed. by G.D. Rosenberg and S.K. Runcorn), John Wiley and Sons, London, p. 459.

Newton, R.R. : 1970, Ancient Astronomical Observations and the Accelerations of the Earth and Moon, The Johns Hopkins Press, Baltimore & London

Newton, R.R. : 1972, Medieval Chronicles and the Rotation of the Earth, The Johns Hopkins Press, Baltimore & London

Oesterwinter, C. and Cohen, C.J. : 1972, Celes. Mech. <u>5</u>, 317

Roxburgh, I.W. : 1977, Nature <u>268</u>, 504

Sandage, A. and Tammann, G.A. : 1975, Astrophys. J. <u>197</u>, 265

Spencer Jones, H. : 1939, Monthly Notices Roy. Astron. Soc. <u>99</u>, 541

Steigman, G. : 1978, Astrophys. J. <u>221</u>, 407

CLASSICAL PHASE SPACE FROM A RELATIVIST'S POINT OF VIEW

S.Persides
Department of Astronomy, University of Thessaloniki
Thessaloniki, Greece

ABSTRACT

A new formulation of phase space of classical mechanics is proposed based on a known four-dimensional formulation of the space-time. With space-time represented by a four-dimensional manifold S (with affine structure and stratified by Euclidean three dimensional spaces) the phase space $\bar{\Gamma}$ for a single particle is defined as the cotangent bundle of S. Thus $\bar{\Gamma}$ is an eight dimensional manifold and a vector bundle with symplectic structure and tensor fields transferred from S. Some remarks on the constants of motion are made. In the general case of a system with n degrees of freedom $\bar{\Gamma}$ is a manifold with 2n+2 dimensions.

I. INTRODUCTION

Although four centuries have passed from the time Newton laid the foundations of classical mechanics, not only the research interest on the subject has not diminished but in the last thirty years has grown and has been extended to many related fields.

At the beginning of the twentieth century classical mechanics had attained a rather complete and well polished structure, that of analytical dynamics (Whittaker, 1965). Then for half a century no essential progress has been made. Perhaps two were the underlying causes for this standstill. First, the existing mathematics (pure and applied) did not help any theoretical or computational progress. Second, quantum theory and general relativity absorbed most of the interest of physicists. After 1950, however, the situation has changed. Advances made in pure mathematics (mainly in topology and differential geometry) in the first half of the twentieth century started to have a profound impact on the formulation and progress of classical mechanics. At the same time applied mathematics and technology created the electronic computer, a tool that in scientific research appears to be as important as the telescope in Astronomy. Quantum theory and specially general relativity have benefited from this progress. But the most profound change has been made in the formulation and devel-

E. G. Mariolopoulos et al. (eds.), Compendium in Astronomy, 367–377.
Copyright © 1982 by D. Reidel Publishing Company.

opment of classical mechanics for the simple reason that this branch of Physics had already a coherent formulation before the introduction of modern mathematics. Thus substantial progress has been made in the theory and the applications of classical mechanics highlighted by the Kolmogorov-Arnold-Moser theorem and extensive numerical calculations of complicated mechanical systems. Finally, the whole subject of modern classical mechanics has been presented in textbook form (Abraham and Marsden, 1978; Arnold, 1978; Thirring, 1978; Wang, 1979) as a well founded and complete branch of mathematical physics.

There is, however, one element or point of view which has not been taken into account, perhaps justifiably. That is the fact that the space-time we live in is four dimensional. In relativity (special or general) this fact is not just a point of view or something the theory can do without it, but a fundamental property of the physical world and the physical theory. It is a basic assumption that all physical phenomena can be described as taking place in a four dimensional manifold with a locally Minkowskian metric structure. This manifold represents space-time and there is no essential distinction or unique separation between space and time.

On the other hand in classical mechanics we consider space as a three dimensional Euclidean manifold and consider time as a parameter in describing the change of physical systems. Thus for a relativist two obvious questions are raised: Is there an equivalent four dimensional formulation of classical mechanics? If there is one, is it useful in some way? The answer to the first question has been given essentially more than fifty years ago (Cartan, 1923, 1924; Friedrichs, 1927) and elaborated further after 1960 (Trautman, 1963, 1965, 1966; Havas, 1964). However, investigations along these lines have stopped at this point, the formulation has not been extended to the phase space and the second question has not been answered. There is a justification to that. Since the four dimensional formulation is equivalent to the three dimensional one, no essentially new information can be extracted from the four dimensional formulation. This argument, however, does not rule out the possibility that the four dimensional formulation will bring out some features of classical mechanics which are difficult to be discovered in the three dimensional formulation. The purpose of this paper is to extend the four dimensional formulation to phase space and set up the framework in which some questions concerning mechanical systems can be reformulated and perhaps answered. Emphasis is given to the structure of phase space as determined by the structure of the four dimensional space-time and to the first integrals of motion.

In Sec. 2 we review the four-dimensional formulation of classical mechanics and add a few remarks on its structure motivated from recent advances in general relativity. In Sec. 3 we construct the phase space for a single particle and examine some of its features. Finally, in Sec. 4 we indicate how the construction can be carried out for a mechanical system with constraints and close with some general remarks.

2. THE SPACE-TIME OF CLASSICAL MECHANICS

Space, time and space-time are undoubtedly fundamental concepts in any physical theory (Ehlers, 1973). The human race and every single human being grow up learning the meaning of space-time. Scientists build models of space-time using mathematical concepts. The twentieth century mathematics has offered a remarkable concept, that of a manifold, for the description of almost all models of space-time that appeared in human history up to now. Aristotle, Galileo, Leibniz and Newton did not describe their space-times as manifolds and it is not always clear what they meant by the terms "space" and "time". We can, however, describe the evolution of the space-time concept in a way that brings out the characteristic similarities and differences of the various space-time concepts. In all definitions, including those of Minkowski and Einstein, space-time is a four dimensional manifold S with the topology of R^4 (the topology of an Einsteinian space-time may be different). The differences among the various space-times lie on the structure imposed on S. Aristotelian space-time can be considered as a product $E^3 x E^1$ with two Euclidean metrics, one on E^3 and one on E^1. Between any two events there is a unique spatial separation and a unique time difference (Penrose, 1966, 1968), i.e., there is absolute space and absolute time. Although Galileo and Newton considered space-time more or less as Aristotelian, their works implied different structures. We can say that in Galilean and Newtonian space-times between two general events only a time-difference can be defined, while a spatial separation between two events can be defined only if their time difference is zero. Furthermore, the geodesics (straight lines) of the space-time are determined by the "free" particles, where free particle for Galilean and Newtonian space-times means respectively not including and including the gravitational field. Leibniz's space-time has no preferred straight lines and parallel translation of a vector in time is not defined (Ehlers, 1973). Minkowskian and Einsteinian space-times differ radically from the previous ones in that they impose a four-metric structure on S with Lorentzian signature.

Now, what is the structure of space-time of classical mechanics? The answer to this delicate question comes from Newton's first law, combined with the fact that translational motion of a box <u>cannot</u> be detected from experiments inside the box, while rotational motion <u>can</u> be detected. This means that the world lines of free particles are geodesics of the space-time and that the axes of a (Cartesian) coordinate system can be parallelly propagated in time. Thus the concept of an "inertial" coordinate system is defined as a coordinate system whose spatial axes are parallelly propagated along a (timelike) geodesic. In modern terms this means that S is an affine space.

After the introduction of general relativity by Einstein in 1915 Cartan (1923, 1924) and Friedrichs (1927) presented a four-dimensional non-relativistic formulation of the space-time structure including the gravitational field. The essential change was that "free" particles were those moving under the influence of a gravitational field only, i.e., in this model the geodesics are determined by the gravitational field. When

the gravitational field vanishes this formulation gives a four-dimensional
formulation of the structure of space-time of classical mechanics in modern mathematical language. From recent versions of this theory (Trautman
1965, 1966; Havas, 1964; Misner, Thorne and Wheeler, 1973) we obtain the
following axiomatic definition of the space-time of classical mechanics:
1) Space-time is a four dimensional (differentiable) manifold S homeomor-
   phic to R^4.
2) S is endowed with a symmetric affine connection which defines a covariant derivative operator ∇_a.
3) On S there are a scalar field t and a symmetric contravariant tensor
   field h^{ab} of signature 0+++.
4) On S the following conditions are satisfied:
    a. $\nabla_a t \neq 0$ everywhere,
    b. $\nabla_a \nabla_b t = 0$,
    c. $h^{ab}\nabla_b t = 0$,   (1)
    d. $\nabla_c h^{ab} = 0$,
    e. $R^a{}_{bcd} = 0$
($R^a{}_{bcd}$ is the Riemann tensor defined from ∇_a).

    The agreement of this formulation with the usual one where space and
time are separate is easily obtained. Conditions 1 and 2 specify the dimensionality of the space-time and the geodesic structure as determined
from the priviledged class of motions (i.e., the motion of free particles). The hypersurfaces t = const. do not have common points (since
$\nabla_a t \neq 0$) and specify a stratification of the space-time, each stratum or
slice t = const. being a <u>simultaneity section</u>. Since $\nabla_a \nabla_b t = 0$, t is an
affine parameter along timelike geodesics (i.e., geodesics which do not
lie in a simultaneity section) and can be used as a time coordinate. This
is the <u>absolute Newtonian time</u> and is determined up to a linear transfor-
mation $t \rightarrow \kappa t + \lambda$. Since h^{ab} has signature 0+++ and $h^{ab}\nabla_b t = 0$, the tensor
field h^{ab} defines an invertible tensor field on each hypersurface t=const.
Thus we have a positive definite metric on each simultaneity section.
Finally, the conditions $\nabla_c h^{ab} = 0$ and $R^a{}_{bcd} = 0$ imply that this metric
is flat, so each simultaneity section is isometric to the Euclidean three
dimensional space E^3.

    The space-time of classical mechanics is presented in Fig. 1.
The value of t increases upwards. Three simultaneity sections for $t = t_1$,
$t = t_2$ and $t = t_3$, have been drawn. A and B are timelike geodesics or
world lines of free particles. C is a world line of a particle subjected
to acceleration. The coordinate system 0 has its origin moving along
A and its spatial axes parallelly propagated. It is an inertial coordinate
system. Another inertial coordinate system (moving with respect to 0
with constant velocity) could be set up using the geodesic B. The distance of points (or events) 1 and 4 or 2 and 5 or 3 and 6 has an invariant meaning (i.e., independent of the coordinate system). However, the
distance particle B (or C) moved from $t = t_1$ to $t = t_2$ (i.e., the spatial
distance of 1 and 2) does not have an invariant meaning (i.e., it is not
independent of the coordinate system used).

    We consider now a few consequences of the structure given to the

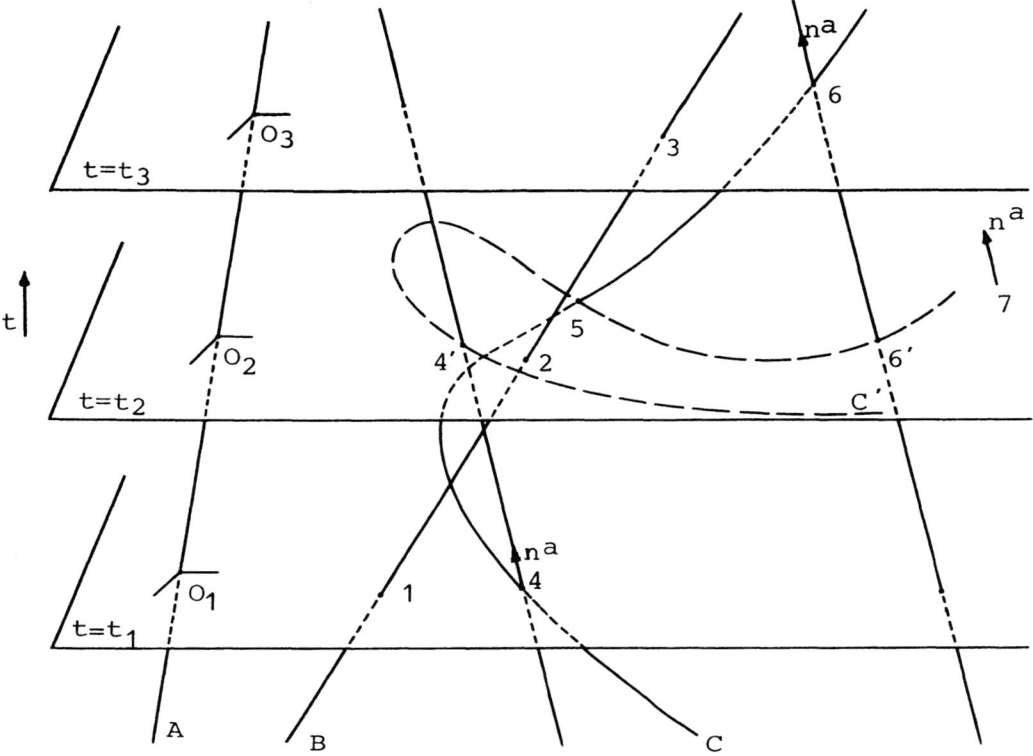

Fig. 1.

space-time S. If $t_a = \nabla_a t$ then we define the tensor field

$$\Gamma^{ab}{}_{cd} = h^{ab} t_c t_d. \tag{2}$$

We can easily prove the following properties (parentheses and brackets denote the symmetric and antisymmetric part respectively)

$$\Gamma^{ab}{}_{cd} = \Gamma^{ba}{}_{cd} = \Gamma^{ab}{}_{dc} = \Gamma^{(ab)}{}_{(cd)} \neq 0,$$

$$\Gamma^{ab}{}_{c[d}\Gamma^{ef}{}_{g]h} = \Gamma^{ab}{}_{cb} = 0. \tag{3}$$

There is a class of coordinate systems in which the non-zero components of $\Gamma^{ab}{}_{cd}$ are

$$\Gamma^{11}{}_{00} = \Gamma^{22}{}_{00} = \Gamma^{33}{}_{00} = 1. \tag{4}$$

We can raise indices using h^{ab}. Since, however, h^{ab} is not invertible, we cannot define h_{ab} uniquely and lower tensor indices. The relation

$$h^{ab}h_{bc}h^{cd} = h^{ad} \tag{5}$$

defines h_{ab} up to addition of a term of the form $t_{(a}s_{b)}$ where s_b is an arbitrary covariant vector field. In some cases the ambiguity in h_{ab} does not affect the results. For example for any contravariant vector v^a for which $t_a v^a = 0$ there is no ambiguity in $h_{ab}v^a v^b$ and $h_{a[b}t_{c]}v^a$. However, $\nabla_c h_{ab}$ is not in general the same with $\nabla_c(h_{ab}+t_{(a}s_{b)})$. Consequently we may ask that $\nabla_c h_{ab} = 0$.

The infinitesimal symmetries of S can be investigated by requiring that some of the structure of S is preserved under an infinitesimal coordinate transformation $x^\mu \to x'^\mu = x^\mu + \xi^\mu(x^\nu)$. Thus we can search for the vector fields ξ^μ which satisfy $\mathcal{L}_\xi t_a = 0$, $\mathcal{L}_\xi \Gamma^{ab}{}_{cd} = 0$, etc. If we ask that $\mathcal{L}_\xi t_a = 0$, $\mathcal{L}_\xi h^{ab} = 0$ and the Christoffel symbols remain zero, we obtain the Galilean group of transformations ($\mathcal{L}_\xi$ denotes the Lie derivative with respect to ξ^a).

It should be emphasized that the structure of S and the structure of each simultaneity section T_t are closely related. In modern mathematical language, since there is a unique mapping $f_t : T_t \to S$ for each simultaneity section T_t, the metric tensor n_{ab} on T_t is the pullback of h_{ab} (the ambiguity of h_{ab} is not transferred to n_{ab} since $f_t^* t_a = 0$). Also h^{ab} is the pushforward of n^{ab}. Since I, i.e., the null boundary of an asymptotically Minkowskian space-time in general relativity, and S have metrics with signatures 0++ and 0+++ respectively, some of the techniques used to study I (Geroch, 1977) can also be used to study S.

3. THE PHASE SPACE OF CLASSICAL MECHANICS

The first step towards the construction of the configuration space and then the phase space is the mapping of the whole four dimensional space-time S onto a single simultaneity section, i.e., a three dimensional Euclidean space. Such a mapping is not determined uniquely by the structure of S as described in the previous section and consequently some additional structure has to be imposed on S. This additional structure can be specified by choosing a contravariant vector n^a at some arbitrary point 7 of the space-time manifold (see Fig. 1.), such that $n^a t_a \neq 0$ (i.e., n^a does not lie on a simultaneity section). Then this vector can be normalized so that $n^a t_a = 1$ and parallelly transported to any point of the manifold. The geodesics passing from each point of S and having n^a as a tangent vector give a mapping of S onto each simultaneity section. Thus, e.g., point 6 is mapped to point 6', point 4 to 4' etc. The "spatial distance of 4 and 5" and the "distance travelled by particle C from t_2 to t_3" mean now the distances 4'5 and 56' respectively. The world line C is mapped onto C' and becomes a trajectory in E^3 parametrized by t.

The usual procedure is now to obtain from E^3 (on which S is mapped) the configuration space M and then the cotangent bundle T*M. In the case of time-dependent mechanical systems the extended configuration space M_e = MxR is taken instead of M and the cotangent bundle T^*M_e is constructed (Thirring, 1978; Wang, 1979). T^*M or T^*M_e is in each case the phase space Γ. Its structure is determined from the way the cotangent bundle is constructed and the structure transferred from M or M_e.

A relativist, however, who abides by the principle of equal treatment of space and time, will not follow this procedure, but will construct the configuration space directly from S and then consider the cotangent bundle of the configuration space. This procedure does not make any distinction between time-independent and time-dependent systems and has in some cases certain advantages. In what follows we will construct the phase space for a single particle and determine its structure.

For a single particle the space-time S is also the configuration space. At each point q of the configuration space we construct the tangent space T_qS, that is the vector space with elements the tangent vectors of curves of S at q. Let T_q^*S be the dual vector space of T_qS, that is the cotangent space at qϵS. The union of all T_q^*S, with each pϵT_q^*S considered as a pair (q, p)ϵT^*S, is the cotangent bundle T^*S and is called the phase space $\bar{\Gamma}$ of S. The phase space is an eight dimensional manifold (since S is four dimensional) and a vector bundle. Its structure is determined by its construction procedure and from the transfer of the structure of S. The construction procedure endows $\bar{\Gamma}$ with a <u>symplectic structure</u>, that is an antisymmetric invertible tensor field Ω_{ab} for which $D_{[a}\Omega_{bc]} = 0$. Furthermore, there is a preferred covariant vector field $A_{\underline{a}}$ (on $\bar{\Gamma}$) and a mapping (projection) $\pi:\bar{\Gamma}\rightarrow S$ that maps (q, p)$\epsilon\bar{\Gamma}$ to qϵS. If S is considered as a four dimensional hypersurface of $\bar{\Gamma}$ we have also the mapping $\sigma:S\rightarrow\bar{\Gamma}$ that maps qϵS to (q, 0)$\epsilon\bar{\Gamma}$. Using these mappings we can transfer all the structure of S to $\bar{\Gamma}$.

A mechanical system is described in the standard formulation of phase space by its Hamiltonian function H which can be considered as a scalar field on the phase space Γ. For a single particle let $p_a\epsilon T_q^*S$ be its momentum at q. Then the kinetic energy for a particle of unit mass

$$T = \frac{1}{2} h^{ab}p_a p_b \tag{6}$$

can be considered as a scalar field on Γ. The potential energy V(q, p) of the particle can be also considered as a scalar field on Γ (or on S, if it does not depend on p). Thus the Hamiltonian for a single particle is

$$H = T + V. \tag{7}$$

From this scalar field (assumed to be C^∞) on Γ we obtain the vector field

$$H^a = \Omega^{ma}D_m H, \tag{8}$$

where D_m is an arbitrary derivative operator on Γ. The integral curves of H^a are the trajectories that describe the physical system in phase space. The coordinates q^i, p_i of a point on a trajectory satisfy the well known Hamilton equations

$$\frac{dq^i}{dt} = \frac{\partial H}{\partial p_i} \quad , \quad \frac{dp_i}{dt} = \frac{\partial H}{\partial q^i} \quad , \qquad (i = 1,2,3) \ . \tag{9}$$

In the formulation, however, presented in this paper the phase space $\bar{\Gamma}$ is eight dimensional, p_0 (the zero component of p_a) does not appear in the kinetic energy T and H is not adequate for the description of the motion in $\bar{\Gamma}$. The missing term seems to be provided by the vector field n^a on S. Let us define a <u>generalized Hamiltonian</u> for a single particle

$$\bar{H} = H + p_a n^a = \frac{1}{2} h^{ab} p_a p_b + V(q, p) + p_a n^a \ . \tag{10}$$

Then we consider the vector field $\bar{H}^a = \Omega^{ma} D_m \bar{H}$ where Ω^{ma} and D_m are now in the eight dimensional phase space $\bar{\Gamma}$. The claim is that the integral curves of this vector field are the trajectories which describe the motion in $\bar{\Gamma}$. To establish this claim we have to prove equations (9).

Let us use an inertial Cartesian coordinate system on S in which $h^{ab} = \text{diag} [0, 1, 1, 1]$ and $n^a = [1, 0, 0, 0]$ (such a system can always be easily found). Then $T = \frac{1}{2} [(p_1)^2 + (p_2)^2 + (p_3)^2]$, $p_a n^a = p_0$ and let us assume for simplicity that the potential energy is $V = V(q)$, i.e., it does not depend on p_a. We have

$$\bar{H} = H + p_0 = \frac{1}{2} [(p_1)^2 + (p_2)^2 + (p_3)^2] + V(q) + p_0. \tag{11}$$

The integral curves of $\bar{H}^a$ are described by $q^i = q^i(\tau)$, $p_i = p_i(\tau)$ (τ is a parameter along each integral curve and $i = 0, 1, 2, 3$) which satisfy the equations

$$\frac{dq^i}{d\tau} = \frac{\partial \bar{H}}{\partial p_i} \quad , \quad \frac{dp_i}{d\tau} = - \frac{\partial \bar{H}}{\partial q^i} \quad , \qquad (i = 0, 1, 2, 3). \tag{12}$$

First we observe that $d\bar{H}/d\tau = 0$, hence $\bar{H}$ remains constant along a trajectory (even when V depends on q^0). Furthermore $\partial \bar{H}/\partial p_0 = 1$, $\partial \bar{H}/\partial q^i = \partial H/\partial q^i$ and equations (12) give for $i = 0$

$$\frac{dq^0}{d\tau} = 1, \quad \frac{dp_0}{d\tau} = - \frac{\partial H}{\partial q^0} \ . \tag{13}$$

Since $dq^0/d\tau = 1$ or $t \equiv q^0 = \tau + \text{const.}$, t can be used as a parameter along a trajectory in $\bar{\Gamma}$. For $i = 1, 2, 3$ we replace in equations (12) $d\tau$ by dt, $\partial \bar{H}/\partial q^i$ by $\partial H/\partial q^i$ and $\partial \bar{H}/\partial p_i$ by $\partial H/\partial p_i$. Thus we obtain equations (9). Hence the motion in the eight dimensional phase space $\bar{\Gamma}$ is governed by equations (9) with $\bar{H} = \text{const.}$ and $dp_0/dt = -\partial H/\partial t$. This last equation gives $p_0 = \text{const.}$ when H is independent of t, in which case $E = H - p_0 = \text{const.}$ is the total energy of the particle.

We examine now some implications of the present formulation for the integrals of motion. If F is a scalar field on $\bar{\Gamma}$ and $dF/d\tau = 0$, then F is a <u>constant of motion</u>. Note that $\bar{H}$ is always a constant of motion. Generally $dF/d\tau = \Omega^{ma} (D_m \bar{H})(D_a F) = [\bar{H}, F]$, where $[\bar{H}, F]$ is the Poisson bracket. Hence a constant of motion commutes with $\bar{H}$. If F is a constant of motion then the vector field $\xi^a = \Omega^{ma} D_m F$ is an <u>infinitesimal symmetry</u> of the system in phase space, since we can easily prove that $\mathcal{L}_\xi \Omega_{ab} = 0$ and $\mathcal{L}_\xi \bar{H} = 0$. Hence, in order to find constants of motion we can search for (infinitesimal) symmetries. This is a problem in which the present formulation seems to have an advantage over the standard formulation of phase space. Since ξ^a is a vector field on an eight dimensional space (i.e., the phase space $\bar{\Gamma}$), it is possible that it cannot be transferred to a lower dimensional space Γ (such as the phase space of the standard formulation). In mathematical language this means that the pushforward of ξ^a from $\bar{\Gamma}$ to Γ may not exist. In such a case the symmetry represented by ξ^a (and the corresponding constant of motion) cannot be found by searching in Γ for an infinitesimal symmetry. A more general problem would be to find a tensor field (of higher valence than ξ^a) on $\bar{\Gamma}$ that contains the information of ξ^a and can be transferred to Γ. Such a tensor field would be a generalization of the symmetry concept similar to that of a Killing tensor field on a Riemannian manifold.

To illustrate the above we consider briefly the two-dimensional harmonic oscillator with kinetic and potential energy

$$T = \frac{1}{2} \left[(p_1)^2 + (p_2)^2 \right] , \quad V = \frac{1}{2} \left[(\omega_1 q^1)^2 + (\omega_2 q^2)^2 \right] \tag{14}$$

respectively, with $q^o = t$, $q^1 = x$, $q^2 = y$. The generalized Hamiltonian is

$$\bar{H} = H + p_o = \frac{1}{2} \left[(\omega_1 x)^2 + (p_1)^2 \right] + \frac{1}{2} \left[(\omega_2 y)^2 + (p_2)^2 \right] + p_o. \tag{15}$$

The equations of motion can be solved easily and give x, y, p_1, p_2 as functions of τ or t and some arbitrary constants. Solving with respect to these constants we find

$$
\begin{aligned}
t' &= t - \tau, \\
p' &= p_o, \\
A &= x\sin\omega_1 t + \frac{p_1}{\omega_1} \cos\omega_1 t, \\
B &= \frac{p_1}{\omega_1} \sin\omega_1 t - x\cos\omega_1 t, \\
C &= y\sin\omega_2 t + \frac{p_2}{\omega_2} \cos\omega_2 t, \\
D &= \frac{p_2}{\omega_2} \sin\omega_2 t - y\cos\omega_2 t .
\end{aligned}
\tag{16}
$$

Since these constants of motion depend on t, the corresponding ξ^a cannot be transferred to a lower dimensional time-independent manifold. For some combinations, however, this transfer is possible, e.g. for ξ^a which correspond to $A^2 + B^2$ and $C^2 + D^2$ (if ω_1/ω_2 is rational there is a third combination of A, B, C, D which gives a global constant of motion). Since in the general case there always exist six constants of motion (or five independent of τ) on $\bar{\Gamma}$, the problem of finding constants of motion in the standard formulation on Γ is that of finding these combinations whose ξ^a can be transferred. Thus the present formulation has many constants of motion but the corresponding surfaces extend to infinite t. Any attempt

to define new such surfaces of finite extent in phase space gives much
fewer and more complicated surfaces.

4. GENERAL REMARKS

For a more complicated mechanical system we have to construct from
S the configuration space $\bar{M}$ and then the phase space $\bar{\Gamma}$. Below we describe
roughly two alternative procedures for constructing $\bar{\Gamma}$ in the case of
holonomic constraints or no constraints at all.

We can define generalized (independent) coordinates q^1, q^2,..., q^n
on each simultaneity section T_t. Thus we obtain an n dimensional manifold
M and a mapping $\varphi : M \rightarrow E^3 \times E^3 \times ...\times E^3$. This mapping can be used to transfer the structure of a simultaneity section to M and define a metric
tensor on M from which the kinetic energy $T = 1/2\ h_{ij}\dot{q}^i\dot{q}^j$ can be obtained.
Then the configuration space $\bar{M}$ can be defined as a fiber bundle with base
the (one dimensional) time manifold E^1 and copies of M as fibers. Another
procedure to obtain $\bar{M}$ would be to construct $S \times S \times ...\times S$, where the number
of factors equals the number N of particles. Since each S is a fiber
bundle with base a time axis, there is a unique mapping between any two
bases and the dimensionality of $S \times S \times ...\times S$ is essentially 3N+1. Taking
into account the constraints we lower still the dimensionality and define
again generalized coordinates q^0, q^1,..., q^n on $\bar{M} \subset S \times S \times ...S$. In each
procedure the phase space $\bar{\Gamma}$ will be defined again as the cotangent bundle
of $\bar{M}$. If the mechanical system has n degrees of freedom, then the phase
space $\bar{\Gamma}$ has 2n+2 dimensions.

In the formulation of phase space presented in this paper, space
and time are treated on the same basis as far as this is possible, given
the differences in the structure of S with respect to space and time.
The phase space $\bar{\Gamma}$ has two more dimensions from the phase space Γ of the
standard formulation. This increase of dimensionality is due to the time
coordinate and the corresponding component of the momentum. A study of
a mechanical system in a higher dimensional space will probably increase
the apparent complexity of the problem but simplify the geometry of symmetries and surfaces defined by the constants of motion. It would be
interesting to study in this framework the existence of a third (global
or not) integral of motion, the possibility of defining approximate constants of motion (i.e., quantities which vary very little for a long time
interval), the transition from integrable to ergodic systems etc.

Acknowledgments

I would like to thank Dr. B. Xanthopoulos for helpful discussions
on the subject.

REFERENCES

Abraham, R., Marsden, J.E.: 1978, Foundations of Mechanics, Benjamin/
    Cummings, Reading.

Arnold, V.I.: 1978, Mathematical Methods of Classical Mechanics, Springer-Verlag, Berlin.

Cartan, E.: 1923, Ann. Ec. Norm. Sup. 40, 325.

Cartan, E.: 1924, Ann. Ec. Norm. Sup. 41, 1.

Ehlers, J.: 1973, in The Physicist's Conception of Nature, (ed. by J. Mehra), D. Reidel, Dordrecht-Holland.

Friedrichs, K.: 1927, Math. Ann. 98, 566.

Geroch, R.: 1977, in Asymptotic Structure of Space-Time, (ed. by F.P. Esposito, L. Witten), Plenum Press, New York.

Havas, P.: 1964, Rev. Mod. Phys. 36, 938.

Misner, C.W., Thorne, K.S., Wheeler, J.A.: 1973, Gravitation, Freeman, San Francisco.

Penrose, R.: 1966, An Analysis of the Structure of Space-Time, Adams Prize Essay, Cambridge Univ.

Penrose, R.: 1968, in Battelle Rencontres, (ed. by C.M. DeWitt and J.A. Wheeler), Benjamin, New York.

Thirring, W.: 1978, Classical Dynamical Systems, Springer-Verlag, New York.

Trautman, A.: 1963, Compt. Rend. Acad. Sci. Paris, 257, 617.

Trautman, A.: 1965, in Lectures on General Relativity, Prentice-Hall, Englewood Cliffs.

Trautman, A.: 1966, in Perspectives in Geometry and Relativity, Indiana Univ. Press, Bloomington.

Wang, C.-C.: 1979, Mathematical Principles of Mechanics and Electro-magnetism, Plenum Press, New York.

Whittaker, E.T.: 1965, Analytical Dynamics of Particles and Rigid Bodies, Cambridge Univ. Press, Cambridge.

PART IX
EXTRATERRESTRIAL INTELLIGENCE

THE COLONIZATION OF THE GALAXY -- A KEY CONCEPT IN THE SEARCH FOR EXTRA-TERRESTRIAL INTELLIGENCE

Michael D. Papagiannis
Chairman, Department of Astronomy, Boston University
Corresponding Member, Academy of Athens

ABSTRACT

The Drake Equation, which precludes stellar colonization and assumes an independent origin for each stellar civilization, predicts, according to most of its backers, a number $N = 10^5 - 10^6$ of advanced technological civilizations in the Galaxy, i.e., roughly one per million stars. The concept, on the other hand, of the colonization of the Galaxy predicts, either $N = 10^{10} - 10^{11}$ if the colonization has already occurred, with space colonies in orbit around every well behaved star of the Galaxy, or $N = 10^{-1} - 10^0$ if the colonization has not yet taken place, because the reason it has not yet occurred must simply be that throughout the long history of the Galaxy there were too few civilizations capable of initiating the colonization process. These conditions simplify considerably the search for extraterrestrial intelligence because we now need to investigate only a small number of stars in our own vicinity, rather than millions of faraway stars as required by the Drake equation. If a systematic astronomical search in our own solar system, and in particular in the asteroid belt, as well as in 10-20 of our nearby stars would prove negative, we would have to conclude that the colonization of the Galaxy has not yet occurred and therefore we must be one of the very few if not the only advanced civilization in the entire Galaxy.

1. THE DRAKE EQUATION

The Drake equation (Drake, 1965) has provided most of the theoretical background in our search for extraterrestrial intelligence from almost the inception of the program, i.e., since Cocconi and Morrison (1959) proposed the search for radio signals at the 21 cm line of atomic hydrogen, and Drake in 1960 with project Ozma (Drake, 1961) conducted the first radio observations of ε-Eridani and τ-Ceti. The Drake equation tries to estimate the number N of technological civilizations currently present in the Galaxy, and hence the chances of success for any given search, from the expression (Papagiannis, 1978a),

$$N = R.P.L , \qquad (1)$$

E. G. Mariolopoulos et al. (eds.), Compendium in Astronomy, 381–390.

where R is the rate at which stars are being formed in the Galaxy, P is
the probability that a given star will possess the necessary conditions
(planets at the right distances, adequate life-time, etc.) for the origin
and subsequent evolution of life to an advanced civilization, and L is
the average life-span of advanced technological civilizations. Though
there is considerable diversity of opinion in the literature about the
values of these three factors and in particular about the values of P and
L (Sturrock, 1980), the proponents of the Drake equation favor in general
values in the ranges, $R \simeq 10\text{-}20$ stars/years, $P \simeq 10^{-1} \text{-}10^{-3}$, and
$L \simeq 10^5 \text{-}10^7$ years, which when introduced in (1) yield values for N
(Shklovskii and Sagan, 1966; Sagan and Drake, 1975; Goldsmith and Owen,
1980) of the order,

$$N \simeq 10^5 \text{-}10^6. \tag{2}$$

It must be emphasized that the Drake equation assumes that life originated and evolved independently in each solar system to the level of an
advanced civilization. The reason the Drake equation excludes the colonization of the Galaxy as an important factor in the estimate of N, is
basically historic. When the equation was first formulated in the early
60's, it was believed that interstellar traveling could represent a realistic possibility only if a 5-10 light-year trip to another star could
be completed within a reasonable fraction of a human life, i.e., in
10-20 years. This, however, necessitates starship velocities of the order
of $V \simeq 0.5c$, which are practically inconceivable within our present knowledge of physics. As a result the possibility of interstellar travel was
left aside. As we will see, however, in Section 3, our ideas have now
changed substantially, and multigeneration interstellar trips at $V \simeq 0.02c$
are now considered realistic possibilities for advanced civilizations.

2. IMPLICATIONS OF THE DRAKE EQUATION

Since the Galaxy has about 2×10^{11} stars, the result obtained in
(1) implies that there must be roughly one advanced civilization per
million stars. In order, therefore, to assure a high probability of discovering another advanced civilization, we must search several million
stars. It was this type of deduction that led to proposals of grandiose
schemes, such as Project Cyclops (Oliver and Billingham, 1972) which
envisions thousands of large radio antennas working in unison in an intensive cosmic radio search program, because without such projects it would
take thousands of years to complete a radio search of several million
stars.

Another important implication of the Drake equation is the large
number of advanced civilizations that must have appeared throughout the
long history of the Galaxy. Since the age of the Galaxy ($\sim 12 \times 10^9$ years)
is much longer than the assumed average life-span of advanced civilizations ($\sim 10^6$ years), it follows that the total number N_T of advanced
civilizations that blossomed in the Galaxy must be given by the expression
(Freeman and Lampton, 1975; Papagiannis, 1978a),

$$N_T = R.P.T , \tag{3}$$

where T is the period over which advanced civilizations must have been appearing in the Galaxy. Allowing 2-3 billion years for the production of heavy elements and the formation of type-I stars, and 4-5 billion years for the appearance of an advanced civilization in a new solar system, we still find that the first advanced civilizations could have appeared about 5 billion years ago and hence we have $T \simeq 5$ billion years. Using in (3) this value of T and the values for R and P we used in (1), we find,

$$N_T \simeq 10^9 , \tag{4}$$

which simply states that if there are now 10^5-10^6 advanced civilizations in the Galaxy, during its long history the Galaxy must have housed close to 10^9 advanced civilizations (Papagiannis, 1978a).

3. THE FEASIBILITY OF INTERSTELLAR TRAVEL AND COLONIZATION

The change of our views about interstellar travel was significantly advanced by the new ideas about human colonies in space developed by O'Neill (1977) and others. In just 20 years of manned space exploration, we have managed to establish semi-permanent scientific stations in space where groups of astronauts and cosmonauts have spent several months at a time. It seems totally reasonable to expect that in the next century there will be large habitats in space housing thousands of people on a permanent basis. These space colonies would consist probably of inter-locking modules, each one about 100 meters in diameter, instead of a single colossal unit 1-10 km in diameter as it has often been implied. The reason is that a modular structure allows for a staged construction and occupancy which makes it much more practical, and in addition it is far less expensive since the smaller units require thinner walls and hence a smaller total amount of building materials (Papagiannis, 1980a). Figure 1 shows a symmetric modular arrangement in the form of a double helix, proposed by Papagiannis (1980a), that can be built in stages and makes provision for continuous expansion.

These space colonies will in time become totally independent of the Earth and will set themselves in separate orbits around the Sun that will bring them closer to their sources of energy and raw materials. These space colonies would be perfectly capable of undertaking long trips to other stars at speeds of V = 0.01-0.05c, which are totally feasible with the use of nuclear fusion (Hart, 1975; Papagiannis, 1978a,b; Martin and Bond, 1980). These voyages would take several centuries and hence many generations, but this will be of no concern to their inhabitants since their everyday lives will continue to be the same whether their colony is going in circles around the Sun or is on its way to another star. Our own planet, after all, is a large spaceship that orbits the Galaxy every 250 million years and no one seems to mind this endless cosmic voyage.

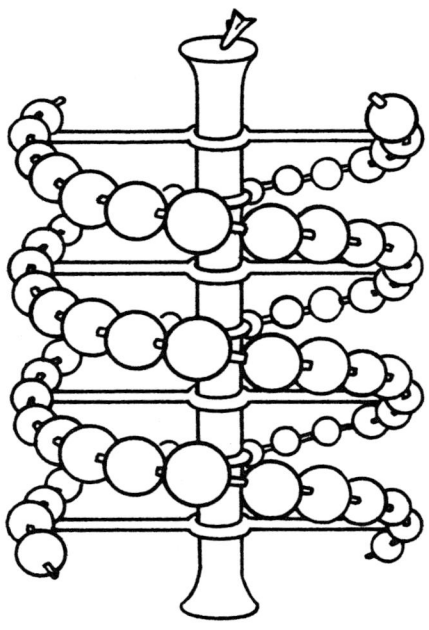

Fig. 1. A Space Colony with a Double-Helix design, which provides for staged construction and occupancy, structural symmetry, and needs far less material to build it than an equivalent large single unit.

     Space colonies will also make the colonization of the Galaxy much easier and much more complete, because when they will reach another solar system their inhabitants will continue to live in their space habitats, to which they must have become totally accustomed after a many-generation-long interstellar trip. This will alleviate the need to search among a large number of stars to find a suitable planet to settle. It also will allow them to populate with space colonies essentially every well behaved star of the Galaxy, because all they would need from a solar system would be energy and raw materials to build and sustain more space habitats, all of which should be readily available in practically every solar system. Planets, on the other hand, where they could land and breathe freely the atmosphere are likely to be extremely rare (Papagiannis, 1981a), despite of many assertions to the contrary by popular movies and television programs. The colonization, therefore, of the Galaxy based exclusively on the colonization of suitable planets would have been extremely difficult, since they would have to travel probably hundreds of light years to find a suitable planet.

4. THE INEVITABILITY OF GALACTIC COLONIZATION

Life in general seems to obey the following four basic principles, expounded by Papagiannis (1980a):
I.   Tends to expand, like a gas, to occupy all available space.
II. Adapts to the additional requirements of every available space.
III. Evolves continuously to levels of higher organization.
IV. The evolution proceeds faster as the level of organization increases.

Life, e.g., developed lungs to move out of the sea and onto the land, and wings so that it can fly in the air. Man has continued this relentless expansion into new spaces, except that by substituting technology for biological evolution, he has managed to accelerate immensely the adaptation process. Man is now capable of going from hot to cold, or more impressively from underwater to outer space, by simply changing his clothes. Life in every form has always found the means to expand into new turf. How did, e.g., ants and spiders manage to be present on every little island of the Earth? Did they evolve independently on each and every island? Obviously not. They did find nevertheless a way, possibly by clinging to the feathers of a bird or by burying their eggs inside a floating log, to reach every unoccupied territory. It is inconceivable, therefore, that man, having the technical means to travel to other stars, would ever refrain from doing what seems to be a natural tendency for all life. Even on philosophical grounds, a Galaxy populated by beings capable of appreciating the grandeur of Nature would be a far more desirable cosmic entity than a practically barren Galaxy. We, therefore, are likely to proceed in this direction.

Several prominent scientists, such as Drake (1980), Newman and Sagan (1981), and others, have questioned the feasibility of galactic colonization for a variety of reasons, including technical, economic, social, etc. An analysis of these problems by Papagiannis (1980a), indicates that it is conceivable that some of these difficulties could prevent some galactic civilizations at some stage of their development from engaging in interstellar travel. It is virtually impossible, however, to find a universal reason that would prevent a billion of galactic civilizations, each with an average life-span of a million years and a combined existence of five billion years, from travelling to their neighboring stars, as it is stringently required by the Drake equation. The burden, therefore, is now on those who still oppose the colonization concept to convince us that whatever reason they have chosen to espouse has a universal and iron-clad application to all galactic civilizations all of the time, because it would only take a small group, during one generation, of a single stellar civilization to defy this rule and the colonization would rapidly spread throughout the entire Galaxy.

5. IMPLICATIONS OF THE COLONIZATION CONCEPT

Assuming a velocity $V = 0.02-0.03c$ for interstellar travel and allowing for a build-up period of several centuries before new colonies can initiate their own missions to other stars, say 500 years for a trip of 10 light years and 500 years for a build-up period, leads to a colonization wave velocity V_c of about 1% of the speed of light (Hart, 1975; Jones, 1976; Papagiannis, 1978a,b; Kuiper and Morris, 1977; Drake, 1980)

$$V_c \simeq 0.01c \,. \tag{5}$$

This implies that the entire Galaxy, which has a diameter of about 100,000 light years, can be colonized in less than 10 million years. This, however, is a very short period relative to the nearly 12 billion year age of the Galaxy, or the 5 billion years that the Galaxy could have been housing advanced civilizations.

The conversion, therefore, of the Galaxy from a practically barren desert to a buzzing ecoumenopolis teeming with intelligent life, will register on the cosmic clock as an almost instantaneous event, something like a person going from kindergarten to a Ph.D. in a single week. The probability, therefore, that the colonization of the Galaxy is now in progress is less than 0.2%, which makes such a coincidence very unlikely. It also points out that since the colonization of the Galaxy could have occurred at any time during the past 5 billion years, if the Galaxy has already been colonized this cosmic event in all probability must have taken place hundreds of millions of years ago.

From the above it follows that the Drake equation retains its validity only in the pre-colonization period, because the colonization of the Galaxy changes drastically the entire picture and renders the Drake equation inapplicable. With the notion of space colonies, we can expect to find habitats in orbit around every well behaved star of the Galaxy and therefore after the completion of the colonization we would expect to have,

$$N = 10^{10} - 10^{11} \tag{6}$$

advanced stellar civilizations in the Galaxy, closely linked in a galactic network of interstellar communications.

If, on the other hand, the colonization has not yet taken place, it simply means that the number N_T of advanced civilizations that could have initiated the colonization process throughout the past 5 billion years of galactic history, must have been very small. This is contradictory, however, to the value $N_T = 10^9$ we obtained from the Drake equation in (4), and in order to bring N_T in line with the conclusions of the colonization theory we must reduce it by a factor of at least 10^6 to $N_T \simeq 10^3$. Since neither R nor T in (3) can be reduced by such a colossal factor, it follows that the discrepancy must be in the probability P.

A possible explanation has been proposed by Papagiannis (1978a, 1981b). It is generally assumed that once life originates on a hospitable planet, it will almost inevitably evolve to higher intelligence and then to an advanced technological civilization. The reason being that intelli-

gence and technology give a great advantage to those that possess them
and hence are bound to be favored by the forces of evolution. This is
basically a correct conclusion, but there is another important factor
that is usually ignored. This is that the evolution from primitive life
to an advanced civilization has taken in the case of the Earth close to
4 billion years and in general it is likely to represent a very long
period, even in cosmic terms.

The balance, however, that must be maintained on a planet between
runaway glaciation and a runaway greenhouse effect so as to sustain liquid water on its surface is a very precarious one (Papagiannis, 1981a).
It would not be surprising, therefore, if only one in a million of the
planets where life originates was capable of sustaining the liquid water
conditions for 4-5 billion years to allow life to complete the painfully
slow evolutionary process from primitive microorganisms to a civilization
with advanced technology. If an evolutionary term

$$f_e \simeq 10^{-6} \tag{7}$$

was to be included in the Drake equation, as proposed by Papagiannis
(1978a, 1981b), the probability P would become $P \simeq 10^{-7}$-10^{-9} instead of
$P \simeq 10^{-1}$-10^{-3}, and would lead to $N_T \simeq 10^3$ instead of $N_T \simeq 10^9$, which
would bring the results of the Drake equation in line with the conclusions
of the colonization concept. The use of this very low probability in (1)
would also produce a very low value for N,

$$N \simeq 10^{-1}\text{-}10^0 \tag{8}$$

which would indicate that in all probability we must be the only advanced
civilization present in the entire Galaxy.

It should be pointed out, however, that though, as suggested by (8),
we might not be able to find any other advanced civilization in the
Galaxy, there are still likely to be millions of planets with life at
different stages of their long biological evolution. It is conceivable,
therefore, that some day we might be able to identify some of these life-
carrying worlds from the spectroscopic signature of life (Owen, 1980).
The presence, e.g., of large quantities of free oxygen in the atmosphere
of a planet strongly suggests a biological origin. The reason is that
oxygen tends to combine readily with many other chemical elements and
can exist free in large quantities in a planetary atmosphere only if it
is continuously replenished by some mechanism. The dissociation of water
vapor by ultra-violet radiation from the central star can produce some
free oxygen on a planet, but not in the amounts currently present in the
terrestrial atmosphere. A biological origin, therefore, would be the most
likely explanation if we were to detect strong oxygen or ozone lines in
a planet of another solar system. Such measurements, however, are still
20-30 years in the future because they require the development of new
observational techniques and the availability of large telescopes in
space.

6. CONCLUSIONS

From the discussion above it follows that the colonization concept leads to either of the following two extremes:
I.   The colonization of the Galaxy has already occurred and every well behaved star of the Galaxy must be populated with colonies.
II.   The colonization has not yet taken place, in which case we are probably the only advanced civilization in the Galaxy.
The distillation of the colonization concept into these two extreme alternatives, makes also the search for exraterrestrial intelligence much easier. The reason is that if the colonization has already occurred, we would expect to find space colonies in orbit around most of the stars in our neighborhood, including our own Sun. Alternatively, if we do not find them in our own vicinity, there is little hope that we would be able to find them anywhere else in the Galaxy.

Let us now carry this analysis one step further. If there were such space colonies in our own solar system, where would they be? One would expect to find them primarily either near their sources of energy or near their sources of raw materials. If these colonies, e.g., were dependent on solar energy, we would expect to find them in orbits closer to the Sun. These people, however, have managed to undertake long interstellar trips staying away from any star for several centuries. Consequently, they must possess alternative sources of energy, such as nuclear fusion, which they will probably continue to use when they colonize a new solar system. As a result they are more likely to settle in orbits closer to their sources of raw materials.

The best source of raw materials in our own solar system seems to be the asteroid belt where practically everything, from metals to organic compounds, is readily available and can be launched into space at a minimal expenditure of energy due to the practically negligible gravitational forces of the asteroids. Frozen water and nuclear fuel for fusion can also be easily obtained from Jupiter and its outer Galilean satellites, Ganymede and Callisto. Hence the asteroid belt is the most logical place to search for space habitats. Such space colonies could have easily escaped detection from the Earth up to now, lost among the thousands of natural asteroids. A careful astronomical search, however, might be able to discover some physical characteristics of certain asteroids, such as an abnormally high infrared temperature or an unexpected radio emission, that would indicate an artificial origin. There are already several known asteroids with peculiar properties, such as No. 624 Hector, which seems to have a dumbbell shape, and No. 532 Herculina, which seems to have several small satellites in orbit around it. All these peculiar asteroids would certainly be among the first to be examined in a comprehensive search of the asteroid belt.

A systematic astronomical search for exraterrestrial intelligence can now be undertaken with primary targets our own solar system and in particular the asteroid belt (Papagiannis, 1978b, 1981b), and 10-20 nearby stars. Such a concerted effort (Papagiannis, 1980b) would allow us to ascertain within a reasonable period, probably before the turn of the century, whether our stellar neighborhood has been colonized or not. If the results are positive, it will open a completely new era for mankind in which we will probably be invited to join as junior partners an already well-established galactic society. If, on the other hand, the results would turn out to be negative, we must begin to consider seriously that we might be one of the few if not the only advanced civilization in the Galaxy. This initially might be a deep disappointment, but in the long run it will develop into a great challenge because it would be up to us to infuse life with cosmic consciousness in the entire Galaxy, a unique opportunity and a very special privilege for mankind.

REFERENCES

Cocconi, G. and Morrison, P.: 1959, Nature, 184, 844.
Drake, F.D.: 1961, Phys. Today, 14, 40. Also in The Quest for Extraterrestrial Life (ed. by D. Goldsmith) Univ. Sci. Books, Mill Valley, CA, USA, 1980.
Drake, F.D.: 1965, Current Aspects of Exobiology (ed. by G. Mamikunian and M.H. Briggs) Pergamon Press, New York.
Drake, F.D.: 1980, in Strategies for the Search for Life in the Universe (ed. by M.D. Papagiannis) D. Reidel, Dordrecht-Holland.
Freeman, J. and Lampton, M.: 1975, Icarus, 25, 368.
Goldsmith, D. and Owen, T.: 1980, The Search for Life in the Universe. Benjamin/Cummings Publ. Co., Menlo Park, CA.
Hart, M.H.: 1975, Quart. J. Roy. Astron. Soc., 16, 128.
Jones, E.M.: 1976, Icarus, 28, 421.
Kuiper, T. and Morris, M.: 1977, Science, 196, 616.
Martin, A.R. and Bond, A.: 1980, in Strategies for the Search for Life in the Universe (ed. by M.D. Papagiannis) D. Reidel, Dordrecht-Holland.
Newman, W.J. and Sagan C.: 1981, Icarus (in press).
Oliver, B.M. and Billingham, J.: 1972, Project Cyclops, NASA CR114445 NASA Ames Res. Cent., Moffett Field, CA, USA.
O'Neill, G.K.: 1977, The High Frontier-Human Colonies in Space. William Morrow and Co., New York.
Owen, T.: 1980, in Strategies for the Search for Life in the Universe (ed. by M.D. Papagiannis) D. Reidel, Dordrecht-Holland.
Papagiannis, M.D.: 1978a, in Origin of Life (ed. by H. Noda) Center Acad. Publ., Tokyo, Japan.
Papagiannis, M.D.: 1978b, Quart. J. Roy. Astron. Soc., 19, 277.
Papagiannis, M.D.: 1980a, in Strategies for the Search for Life in the Universe (ed. by M.D. Papagiannis) D. Reidel, Dordrecht-Holland.
Papagiannis, M.D.: 1980b, in Strategies for the Search for Life in the Universe (ed. by M.D. Papagiannis) D. Reidel, Dordrecht-Holland.
Papagiannis, M.D.: 1981a, in Origin of Life (ed. by Y. Wolman) D. Reidel, Dordrecht-Holland.
Papagiannis, M.D.: 1981b, in Extraterrestrials-Where Are They? (ed. by M. Hart and B. Zuckerman) Pergamon Press.
Sagan, C. and Drake, F.: 1975, Sci. Am. 233, (May) 83.

Shklovskii, J.S. and Sagan, C.: 1966, Intelligent Life in the Universe,
    Holden Day, San Francisco, USA.
Sturrock, P.A.: 1980, in Strategies for the Search for Life in the Uni-
    verse (ed. by M.D. Papagiannis) D. Reidel, Dordrecht-Holland.

PART X
ASSOCIATED FIELDS

NUMERICAL TREATMENT OF THE MAGNETOHYDRODYNAMIC FLOW BOUNDED BY AN INFINITE POROUS PLATE

C.L. Goudas, G.A. Katsiaris, M.A. Drymonitou, and A.J. Vernardis
University of Patras
Patras, Greece

ABSTRACT

The scheme and some numerical results obtained for the problem of one dimensional flow and heat transfer of an electrically conducting fluid, subject to an external transverse and constant magnetic field, when the fluid moves on the one side of an infinite non-conducting and porous plate which in general moves in the same direction with the fluid, is exposed. The results produced were found to be satisfactory by comparison with other results obtained from analytical treatment of cases corresponding to simplified body forces and conditions of motion.

INTRODUCTION

The problem to be tackled here is the unsteady motion of an electrically conducting, viscous and incompressible fluid subject to a homogeneous magnetic field. The flow is confined on the one side of an infinite and non-conducting plate which is moving in a fixed direction in its own plane with a velocity that is a general function of the time. The plate is porous and through its pores fluid of the same kind is injected or sucked. The fluid moves with a velocity parallel to that to the plate. This velocity at large distances from the plate is in general another function of the time only, while at smaller distances it is a function of the distance from the plate and the time. The externally applied magnetic field is normal to the plate and produces an induced field that affects the motion. Finally, the plate and the fluid acquire different temperatures and as a result the plate is either heated or cooled. We assume that the plate always preserves its temperature. This implies that when the fluid is warmer the plate can absorb any amount of heat. Also, when the fluid is cooler, the plate can provide all the heat needed and still retain its temperature.

The quantities to be evaluated here are the viscous, the magnetic and the thermal boundary layers of the fluid as functions of the time, of the various types of fluid parameters and of the plate, free stream,

E. G. Mariolopoulos et al. (eds.), Compendium in Astronomy, 393–414.
Copyright © 1982 by D. Reidel Publishing Company.

and suction or injection velocities.

The numerical treatment of unsteady boundary layers is of considerable interest, partly as an approach to the description of the difficult features of the unsteady boundary layers, and partly because of its importance in certain practical problems.

The subject has recently been reviewed by Riley (1975) and by Sears and Telionis (1975).

Particular cases of this problem, in some cases under more general conditions, have being considered by other investigators. Hasimoto (1964), e.g., studied the flow of an electrically conducting medium along an infinite plate moving impulsively in the presence of a transverse magnetic field, while Dukowicz et al. (1968) treated numerically the same problem. However, the porosity of the plate and the heat transfer aspect were not considered.

Standard numerical procedures for computing the flow field have been used by many authors in studying two or three dimensional viscous, thermal and hydromagnetic boundary layers and studying their behaviour.

Rubbert (1965) and Rubbert and Landahl (1967) developed the method of parametric differentiation for the solution of a class of non-linear two-point boundary value problems occurring in transonic flow. The existence and uniqueness of the solution by this method have been proved by Yakovlev (1965).

This method was also applied successfully to incompressible and compressible boundary layer problems (Yakovlev, 1965; Tan and Dibiano, 1972; Narayana and Ramamoorthy, 1972; Nath 1973).

Takhar and Soundalgekar (1976) studied the similarity solution of the energy equation for the thermal boundary layer of an incompressible, viscous fluid past a semi-infinite plate using the Merson's method and Newton's iteration.

Axisymmetric steady flow of a viscous, incompressible and electrically conducting liquid between two non-conducting coaxial circular discs is solved numerically by Khare (1977) in presence of a transverse magnetic field using a series expansion method.

A boundary layer solution for the flow of an electrically conducting fluid over a semi-infinite flat plate in the presence of a transverse magnetic field and taking into account the heat due to viscous dissipation and stress-work has been presented by Soundalgekar and Takhar (1977) but they omitted some of the terms in the equations. Ingham (1979) solved the same problem with full boundary layer equations and showed that the problem can be solved quickly and accurately with the finite differences technique.

Cebeci et al. (1979) studied numerically the response of a stagnation point boundary layer to a change in the external velocity. They also showed that their solutions are in excellent agreement with the known analytical one.

Three-dimensional boundary layer problems were solved using exact finite differences methods by Der and Raetz (1962), Hall (1967), Dwyer (1968), and Der (1971).

In this paper we shall first present the techniques employed for the numerical treatment of the problem described in the first part of this introduction and then proceed to giving some of the results obtained and their significance.

THE EQUATIONS OF MOTION

Under the conditions described above and for u', v', denoting the components of the fluid velocity along the Ox' and Oy' axes, respectively, we arrive (see Drymonitou, 1981) to the following equations governing the motion, the magnetic field and the energy of the fluid:

$$\frac{\partial u'}{\partial t'} + v'\frac{\partial u'}{\partial y'} = -\frac{1}{\rho}\frac{\partial p'}{\partial x'} + \nu\frac{\partial^2 u'}{\partial y'^2} + \frac{\mu_o}{\rho}H_{y'}\frac{\partial H_{x'}}{\partial y'}, \text{(momentum)}, \qquad (1)$$

$$\frac{\partial H_{x'}}{\partial t'} + v'\frac{\partial H_{x'}}{\partial y'} = H_{y'}\frac{\partial u'}{\partial y'} + \frac{1}{\sigma\mu_o}\frac{\partial^2 H_{x'}}{\partial y'^2}, \text{(magnetic field)}, \qquad (2)$$

$$\frac{\partial T'}{\partial t'} + v'\frac{\partial T'}{\partial y'} = \frac{\kappa}{\rho c_p}\frac{\partial^2 T'}{\partial y'^2} + \frac{\nu}{c_p}(\frac{\partial u'}{\partial y'})^2 + \frac{1}{\sigma\rho c_p}(\frac{\partial H_{x'}}{\partial y'})^2, \text{(energy)}, \qquad (3)$$

$$\frac{\partial v'}{\partial y'} = 0, \text{(continuity)}, \qquad (4)$$

where ρ denotes the density of the fluid, t' the time, x', y' the position coordinates, ν the coefficient of kinematic viscosity, μ_o the magnetic permeability, σ the electrical conductivity of the fluid, T' the temperature, $H_{x'}$, $H_{y'}$, the corresponding components of the magnetic field, κ the thermal conductivity, p the pressure and c_p the specific heat of the fluid under constant pressure. The second and third terms of the right hand side of the energy equation (3) represent the amount of heat produced by friction and the Joule heating, respectively.

We shall seek solution of this system of equations for the following boundary conditions:
i. For the velocity and temperature we shall assume that

$$u' = u_o', \quad v' = -v_o, \quad T' = T_w' \quad \text{for} \quad y' = 0, \qquad (5)$$

$$u' \rightarrow U'(t'), \qquad T' \rightarrow T'_\infty \qquad \text{for} \quad y' \rightarrow \infty, \tag{5}$$

where U' and v_o are the time-dependent free-stream velocity along the Ox' - axis and the constant velocity of suction $(v_o > 0)$ along the Oy'-axis, respectively, T'_w the temperature of the plate and T'_∞ the temperature of the fluid at large distances from the plate.

ii. For the magnetic field we shall assume that

$$H_{x'} = 0, \qquad H_{y'} = H_o \qquad \text{for} \quad y' = 0,$$
$$H_{x'} = 0, \qquad H_{y'} = H_o \qquad \text{for} \quad y' \rightarrow \infty. \tag{6}$$

DIMENSIONLESS FORM OF THE EQUATIONS

Equation (4) gives that

$$v' = -v_o, \qquad (v_o > 0).$$

Equation (1), on the other hand, applied for the free-stream region of the fluid gives that

$$-\frac{1}{\rho} \frac{\partial p}{\partial x'} = \frac{dU'}{dt'}. \tag{7}$$

On account of relation (7) and the fact that $H_y = H_o = \text{constant}$, equations (1)-(3) become

$$\frac{\partial u'}{\partial t'} - v_o \frac{\partial u'}{\partial y'} = \frac{dU'}{dt'} + \nu \frac{\partial^2 u'}{\partial y'^2} + \frac{\mu_o}{\rho} H_o \frac{\partial H_{x'}}{\partial y'}, \tag{8}$$

$$\frac{\partial H'_{x'}}{\partial t'} - v_o \frac{\partial H_{x'}}{\partial y'} = H_o \frac{\partial u'}{\partial y'} + \frac{1}{\sigma \mu_o} \frac{\partial^2 H_{x'}}{\partial y'^2}, \tag{9}$$

$$\frac{\partial T'}{\partial t'} - v_o \frac{\partial T'}{\partial y'} = \frac{\kappa}{\rho c_p} \frac{\partial^2 T'}{\partial y'^2} + \frac{\nu}{c_p} \left(\frac{\partial u'}{\partial y'}\right)^2 + \frac{1}{\sigma \rho c_p} \left(\frac{\partial H_{x'}}{\partial y'}\right)^2. \tag{10}$$

We introduce now the following dimensionless variables and parameters:

$$y = (v_o/\nu)y', \qquad \text{for the distance from the plate,}$$

$$t = (v_o^2/4\nu)t', \qquad \text{for the time,}$$

$$u = u'/U_o, \qquad \text{for the velocity,}$$

$$U = U'/U_o, \qquad \text{for the free-stream velocity,}$$

$$\alpha = (\sigma\mu_o)^{-1}, \qquad\qquad \text{for the diffusion of the magnetic field,}$$

$$P_m = \nu\sigma\mu_o = \nu/\alpha , \qquad \text{for the Prandle magnetic number,}$$

$$P = \rho\nu c_p/\kappa , \qquad\qquad \text{for the Prandle number,}$$

$$E = \frac{U_o^2}{c_p(T_w'-T_\infty')} , \qquad \text{for the Eckert number,}$$

$$\theta = \frac{T'- T_\infty'}{T_w'- T_\infty'} , \qquad\qquad \text{for the temperature,}$$

$$H = (\mu_o/\rho)^{1/2} \cdot \frac{H_{x'}}{U_o} , \qquad \text{for the induced magnetic field,}$$

$$M = (\mu_o/\rho)^{1/2} \cdot \frac{H_o}{v_o} , \qquad \text{for the magnetic parameter,}$$

$$\omega = (4\nu/v_o^2)\omega', \qquad\qquad \text{for the angular velocity,*}$$

$$\tau_w = -\rho v_o U_o (\frac{\partial u}{\partial y})_y = 0, \text{ for the skin friction on the wall,}$$

where U_o is the mean free-stream velocity and M (the parameter of the magnetic field) the ratio of the Alfvén wave velocity over the velocity of suction. The magnetic Prandle number P_m, on the other hand, is the ratio of the viscous over the magnetic diffusion and is the same with the usual Prandle number, i.e. the ratio of the viscous over the thermal diffusion. The coefficient P_m is generally small and is used as a measure of the ratio of the viscous over the magnetic boundary layer thickness.

By introducing the above dimensionless variables and parameters, equations (8) - (10) become:

$$\frac{\partial u}{\partial t} = 4 \frac{\partial^2 u}{\partial y^2} + 4 \frac{\partial u}{\partial y} + 4M \frac{\partial H}{\partial y} + \frac{dU}{dt} , \qquad\qquad (11)$$

$$\frac{\partial H}{\partial t} = \frac{4}{P_m} \frac{\partial^2 H}{\partial y^2} + 4 \frac{\partial H}{\partial y} + 4M \frac{\partial u}{\partial y} , \qquad\qquad (12)$$

---

* The angular velocity ω', and hence the dimensionless variable ω, will appear later in the definition of the free-stream velocity U(t) which will be taken as a periodic function of the time.

$$\frac{\partial \theta}{\partial t} = \frac{4}{P} \frac{\partial^2 \theta}{\partial y^2} + 4 \frac{\partial \theta}{\partial y} + 4E (\frac{\partial u}{\partial y})^2 + \frac{4E}{P_m} (\frac{\partial H}{\partial y})^2 \ . \tag{13}$$

The boundary conditions, on the other hand, for the same dimension-less variables are:

$$u = u_o : \quad \theta = 1, \quad H = 0, \qquad \text{when} \quad y = 0 \ ,$$
$$y \rightarrow \infty : \ u \rightarrow U(t), \quad \theta \rightarrow 0, \quad H \rightarrow 0, \qquad \text{when} \quad y \rightarrow \infty, \tag{14}$$

while their initial values (t = 0) are

$$u = u(y,0), \quad H = H(y,0), \quad \theta = \theta(y,0). \tag{15}$$

NUMERICAL TREATMENT USING FINITE DIFFERENCES

Let us first give to equations (11) - (13) the form

$$\frac{\partial u}{\partial t} = a_o \frac{\partial^2 u}{\partial y^2} + a_1 \frac{\partial u}{\partial y} + a_2 \frac{\partial H}{\partial y} + a_3 \ , \tag{16}$$

$$\frac{\partial H}{\partial t} = b_o \frac{\partial^2 H}{\partial y^2} + b_1 \frac{\partial H}{\partial y} + b_2 \frac{\partial u}{\partial y} \ , \tag{17}$$

$$\frac{\partial \theta}{\partial t} = c_o \frac{\partial^2 \theta}{\partial y^2} + c_1 \frac{\partial \theta}{\partial y} + c_2 (\frac{\partial u}{\partial y})^2 + c_3 (\frac{\partial H}{\partial y})^2 \ , \tag{18}$$

where a_i, b_i, c_i, i = 0,1,2,3 have obvious definitions. This constitutes a system of non-linear partial differential equations of the parabolic type.

We shall now set up a scheme for the numerical evaluation of its solutions by transforming it into a system of simultaneous equations in terms of the finite differences of the variables involved. To do so we shall employ the forward differences for all partial derivatives with respect to time and the central differences for all partial derivatives with respect to position. Thus, if we denote by Q(y,t) any of the varia-bles u,H,θ, then the partial derivatives with respect to position and time at y = iΔy and t = jΔt, with Δy and Δt taken fixed, are

$$(\frac{\partial Q}{\partial y})_{i,j} = \frac{Q_{i+1,j} - Q_{i-1,j}}{2\Delta y} \ , \tag{19}$$

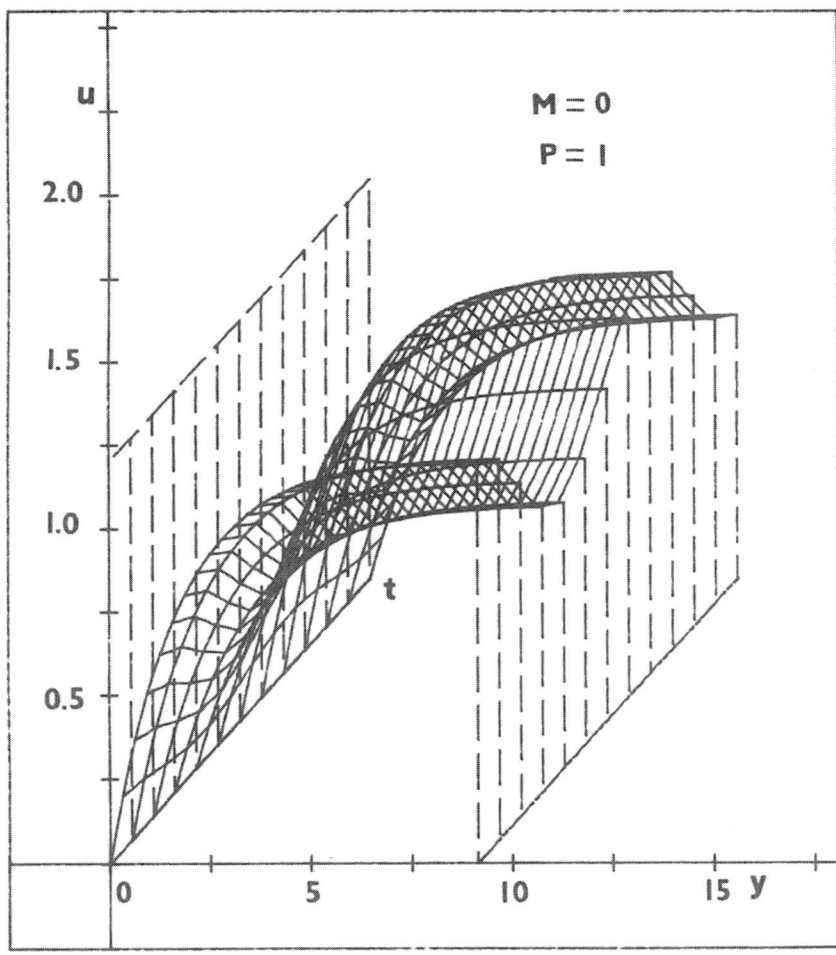

Fig. 1. Three-dimensional presentation of the velocity u of the fluid
as a function of the distance y from the plate and the time t, for M=0
and P=1. The plate is motionless and the free-stream velocity varies
periodically with the time.

$$\left(\frac{\partial^2 Q}{\partial y^2}\right)_{i,j} = \frac{Q_{i+1,j} - 2Q_{1,j} + Q_{i-1,j}}{(\Delta y)^2} , \tag{20}$$

$$\left(\frac{\partial Q}{\partial t}\right)_{i,j} = \frac{Q_{i,j+1} - Q_{i,j}}{\Delta t} , \tag{21}$$

where the notation $Q(i\Delta y, j\Delta t) = Q_{i,j}$ is used. By substituting in equations (16) - (18) the expressions (19) - (21) of the derivatives we obtain the following linear algebraic system:

$$u_{i,j+1} = u_{i-1,j}\left(\tau_u - \frac{a_1 k}{2h}\right) + u_{i,j}(1 - 2\tau_u) + u_{i+1,j}\left(\tau_u + \frac{a_1 k}{2h}\right) +$$

$$+ \frac{ka_2}{2h}\left(H_{i+1,j} - H_{i-1,j}\right) + a_3 , \tag{22}$$

$$H_{i,j+1} = H_{i-1,j}\left(\tau_H - \frac{b_1 k}{2h}\right) + H_{i+1,j}\left(\tau_H + \frac{b_1 k}{2h}\right) + H_{i,j}(1-2\tau_H) +$$

$$+ \frac{kb_2}{2h}\left(u_{i+1,j} - u_{i-1,j}\right) , \tag{23}$$

$$\theta_{i,j+1} = \theta_{i-1,j}\left(\tau_\theta - \frac{c_1 k}{2h}\right) + \theta_{i,j}(1-2\tau_\theta) + \theta_{i+1,j}\left(\tau_\theta + \frac{c_1 k}{2h}\right) +$$

$$+ \frac{c_2 k}{2h}\left(u_{i+1,j} - u_{i-1,j}\right)^2 + \frac{c_3 k}{2h}\left(H_{i+1,j} - H_{i-1,j}\right)^2 , \tag{24}$$

in which $k = \Delta t$, $h = \Delta y$ and τ_u, τ_H and τ_θ stand for the following quantities:

$$\tau_u = \frac{a_o k}{h^2} , \tag{25}$$

$$\tau_H = \frac{b_o k}{h^2} , \tag{26}$$

$$\tau_\theta = \frac{c_o k}{h^2} . \tag{27}$$

The choice of values for τ_u, τ_H, τ_θ, is critical, since the criterion of stability of the solution to be obtained depends on it (Milne, 1970, p. 119). We shall distinguish the following two cases:

    Case 1:

$$a_o = b_o = c_o \tag{28}$$

Then, obviously, $\tau_u = \tau_H = \tau_\theta = \tau$ and the stability criterion is satisfied for $\tau \leqslant 0.5$, while optimal convergence is achieved for

$$\tau = \frac{1}{6} . \tag{29}$$

We remind that the error in the derivatives is $O(k) + O(h^2)$.

Case 2:

$$a_o \neq b_o \neq c_o . \tag{30}$$

In this case the selection of the steps h and k is made so that the necessary conditions for stability, $\tau_u \leqslant 0.5$, $\tau_H \leqslant 0.5$, $\tau_\theta \leqslant 0.5$, will be satisfield while at the same time the values of τ_u, τ_H, and τ_θ corresponding to the selected values of h and k will be as close as possible to 1/6, so that optimal convergence for the three unknown functions will be achieved. In all cases the values of h and k should be selected as large as the above conditions allow.

The differences equations (22) - (24) are then solved by the explicit method (Richtmyer and Morton, 1967) and the procedure of calculation follows this scheme: The numerical values of u,H and θ corresponding to the space-time point y = ih, t = (j+1)k are calculated using their already known values corresponding to the points y = (i-1)h, ih, (i+1)h, t = jk and the cycle goes on to cover the cases i = 2,3,...,n-1, j = 1,2,...,m-1, where n and m define the space-time grid over which the calculation will be performed. We remark that the values of u,H and θ for i = 1,2,...,n-1, j = 1 are known from the initial conditions while the values of the same variables for i = 2, n, j = 1,2,..., m-1 are known from the boundary conditions.

THE COMPLETE SCHEME OF CALCULATIONS

The scheme of calculations for the complete solution of the problem in hand that will produce the numerical tables for the velocity distribution, the magnetic field and the temperature distribution is divided into the following phases:

Phase 1: Definition of the space-time grid at the nodes of which the values of u,H,θ will be computed. This phase starts with the definition of h and k.

Phase 2: Calculation of the values of the unknown functions at the boundary nodal points. This is done using the boundary and initial conditions of the solutions seeked.

Phase 3: Calculation of the values of the unknown functions u,H and

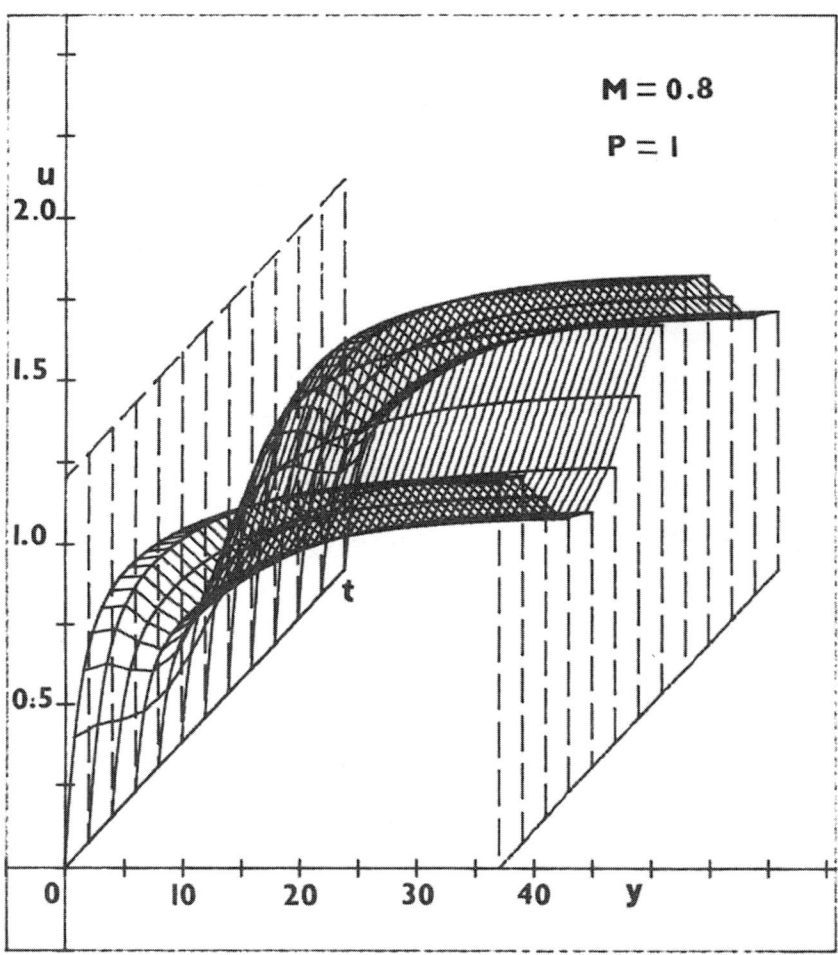

Fig. 2. Three dimensional presentation of the velocity u of the fluid
as function of y and t, for M=0.8 and P=1.

θ at the rest of the nodes of the grid.

Obviously, all three phases of calculation are done in loops. Concerning phase 1, we first calculate the step sizes h and k and then the parameters n and m that define completely the grid. Subsequently, we execute phase 2 which involves calculation of the quantities $u_{i,j}$, $H_{i,j}$, $\theta_{i,j}$ for i = 1,2,..., n, j = 1 and for j = 1,2,..., m-1, i = 1 and i = n. We then proceed to phase 3 which implies the calculation of the values of $u_{i,j}$, $H_{i,j}$, $\theta_{i,j}$ for i = 2,3,..., n, j = 2,3,..., m-1. This implies solution of the following algebraic equations:

$$j = 1: u_{i,2} = u_{i-1,1}(\tau_u - \frac{a_1 k}{2h}) + u_{i,1}(1 - 2\tau_u) + u_{i+1,1}(\tau_u + \frac{a_1 k}{2h}) +$$

$$+ \frac{ka_2}{2h}(H_{i+1,1} - H_{i-1,1}) + a_3 ,$$

$$H_{i,2} = H_{i-1,1}(\tau_H - \frac{b_1 k}{2h}) + H_{i,1}(1 - 2\tau_H) + H_{i+1,1}(\tau_H + \frac{b_1 k}{2h}) +$$

$$+ \frac{kb_2}{2h}(u_{i+1,1} - u_{i-1,1}),$$

$$\theta_{i,2} = \theta_{i-1,1}(\tau_\theta - \frac{c_1 k}{2h}) + \theta_{i,1}(1 - 2\tau_\theta) + \theta_{i+1,1}(\tau_\theta + \frac{c_1 k}{2h}) +$$

$$+ \frac{c_3 k}{2h}(H_{i+1,j} - H_{i-1,j})^2 + \frac{c_2 k}{2h}(u_{i+1,1} - u_{i-1,1})^2 ,$$

$$j = 2: u_{i,3} = u_{i-1,2}(\tau_u - \frac{a_1 k}{2h}) + u_{i,2}(1 - 2\tau_u) + u_{i+1,2}(\tau_u + \frac{a_1 k}{2h}) +$$

$$+ \frac{ka_2}{2h}(H_{i+1,2} - H_{i-1,2}) + a_3 ,$$

$$H_{i,3} = H_{i-1,2}(\tau_H - \frac{b_1 k}{2h}) + H_{i,2}(1 - 2\tau_H) + H_{i+1,2}(\tau_H + \frac{b_1 k}{2h}) +$$

$$+ \frac{kb_2}{2h}(u_{i+1,2} - u_{i-1,2}) ,$$

$$\theta_{i,3} = \theta_{i-1,2}(\tau_\theta - \frac{c_1 k}{2h}) + \theta_{i,2}(1 - 2\tau_\theta) + \theta_{i+1,2}(\tau_\theta + \frac{c_1 k}{2h}) +$$

$$+ \frac{c_2 k}{2h}(u_{i+1,2} - u_{i-1,2})^2 + \frac{c_3 k}{2h}(H_{i+1,2} - H_{i-1,2})^2 ,$$

. . . .

$$j = m-1: u_{i,m} = u_{i-1,m-1}(\tau_u - \frac{a_1 k}{2h}) + u_{i,m-1}(1 - 2\tau_u) + u_{i+1,m-1}(\tau_u + \frac{a_1 k}{2h}) +$$

$$+ \frac{ka_2}{2h}(H_{i+1,m-1} - H_{i-1,m-1}) + a_3 \; ,$$

$$H_{i,m} = H_{i-1,m-1}(\tau_H - \frac{b_1 k}{2h}) + H_{i,m-1}(1-2\tau_H) + H_{i+1,m-1}(\tau_H + \frac{b_1 k}{2h}) +$$

$$+ \frac{kb_2}{2h}(u_{i+1,m-1} - u_{i-1,m-1}) \; ,$$

$$\theta_{i,m} = \theta_{i-1,m-1}(\tau_\theta - \frac{c_1 k}{2h}) + \theta_{i,m-1}(1-2\tau_\theta) + \theta_{i+1,m-1}(\tau_\theta + \frac{c_1 k}{2h}) +$$

$$+ \frac{c_2 k}{2h}(u_{i+1,m-1} - u_{i-1,m-1})^2 + \frac{c_3 k}{2h}(H_{i+1,m-1} - H_{i-1,m-1})^2 \; ,$$

for i = 2,3,...,n.

The scheme presented here and the computer programme based on it are general and can be applied for any system of parabolic partial differential equations in two independent variables and for whatever boundary and initial conditions.

In order to check the accuracy of the results and in addition to the fulfilment of the necessary conditions for stability and convergence of the solutions computed, the programme was tested against the numerical values of known analytical solutions corresponding to one special case of the equations under study. The solutions in question were given by Kafousias (1976). The comparison showed that in all cases the numerical values were in agreement to the ones corresponding to the analytical solution up to the fifth significant figure. For the objectives of this study, i.e. the calculation of the boundary layer thickness, this accuracy is more than satisfactory since the boundary layers are well defined when more than two significant figures are correct.

Another independent check of the accuracy of the results is the size of the viscous boundary layer, which in the absence of magnetic field must be equal to the thickness of the thermal boundary layer. This was found in all cases to be as expected.

CALCULATION OF THE BOUNDARY LAYERS

A special subroutine was added to the programme in order to calculate the thickness of the viscous, magnetic and thermal boundary layers. This calculation was done on the basis of the following definition of the boundary layers, expressed only for the viscous one. The definition of the other two boundary layers is strictly analogous.

Case 1: $U_o = 0$.

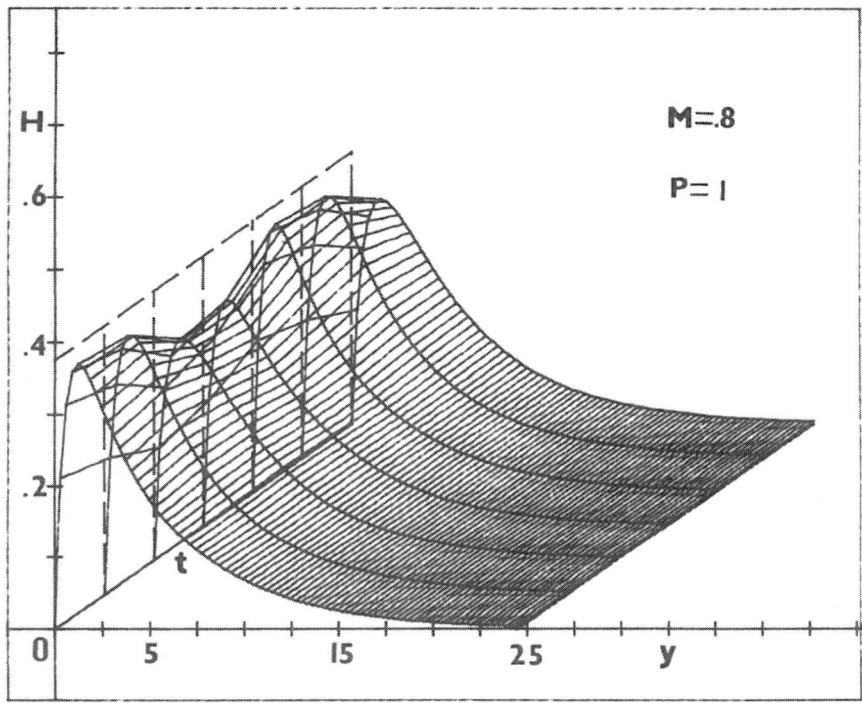

Fig. 3. Three-dimensional presentation of the induced field intensity H
as function of y and t for M = 0.8 and P = 1.

    The motion of the boundary surface is responsible for the
formation of a layer of the fluid adjacent to it, that
moves parallel to the boundary with velocities $u(y,t)$. If
$u_{max} = \max[u(y,t)]$ then the boundary layer thickness δ is
defined from the relation $u(\delta,t) = u_{max}/100$.

Case 2: $U_o \neq 0$.
    The combined motion of the free-stream and the boundary
surface produce in the vicinity of the surface a layer of
thickness δ which is determined from the condition

$$[u(\delta,t) - U_o] = U_o/100, \quad U_o > 0.$$

    The actual calculation of the boundary layer thickness proceeds
for each case as follows:

    In case 1 the integration scheme gives for every value of $t_j = jk$
the corresponding value of $u(y_i,t_j)$, where $y_i = ik$ and $i = 0,1,2,...,n$. A
simple subroutine defines the three consecutive nodes $\ell-1,\ell,\ell+1$, between
which the max $|u(y,t)|$ occurs and calculates this maximum and the corre-
sponding location by quadratic interpolation. Then the subroutine pro-

ceeds to calculate the boundary layer thickness at this instance of time
by first defining the consecutive nodes between which the quantity
$|u(y_i,t)|$ $-u_{max}/100$ changes sign. The value of δ is estimated through
linear interpolation between the above two points. Once δ is calculated
the control is transferred to the main routine for the calculation of the
values of $u(y_i,t)$ for $t = (j+1)k$. In certain cases ambiguity may rise as
to where the max $|u(y,t)|$ occurs, especially when at two or more dis-
tinct distances y from the boundary the quantity $|u(y,t)|$ receives com-
parable values. In such cases there is need to carry out quadratic inter-
polations in every group of three consecutive nodes among which the true
$\max|u(y,t)|$ may occur.

In case 2 the procedure of calculation is only slightly different
since, for non-moving boundary, the maximum velocity is U_o and the calculation of δ is done by linear interpolation between the nodes at which
the value of the quantity $99U_o/100 - u(y_i,t_i)$ changes sign. For moving
plate the boundary layer thickness is more complicated since we may encounter multiple boundary layers separated by layers moving with velocity
almost equal to that of the free-stream. In such cases we must determine
all the active layers by position and thickness. This is done by identi-
fying the pairs of nodal points where the quantity $|u(y,t) - U_o| - U_o/100$
changes sign. Obviously, between the pair of nodes where transition from
positive to negative sign occurs there exists a terminal point of an
active layer and the start of a non-active one. The term active implies
different kinetic or other behaviour from the space of the free-stream.
Transition, on the other hand, from negative to positive sign implies
that between the said two nodes there is a terminal point of a non-active
and the start of an active layer. The calculation of the terminal points
is then made by linear interpolation. Thus the active and non-active
layers are identified by position and thickness. Once this is done the
control is transferred to the main routine for the calculation of the
same quantities for the next time-step.

In both cases the boundary layer thickness is time-dependent if
the free-stream velocity and/or the wall velocity are time-dependent.
In all cases it is of interest to calculate the envelope of the boundary
layers, i.e. the part of the fluid that at some time instance can become
active in the above sense. This is done by the same routine.

The calculation of the magnetic and thermal boundary layer thickness
is made by subroutines structured to function on the same line as the
one described already for the calculation of the viscous boundary layer.

CALCULATION OF THE SKIN FRICTION

The skin friction τ_w is given from the expression

$$\tau_w = -\mu(\frac{\partial u}{\partial y})_{y=0} \quad , \tag{31}$$

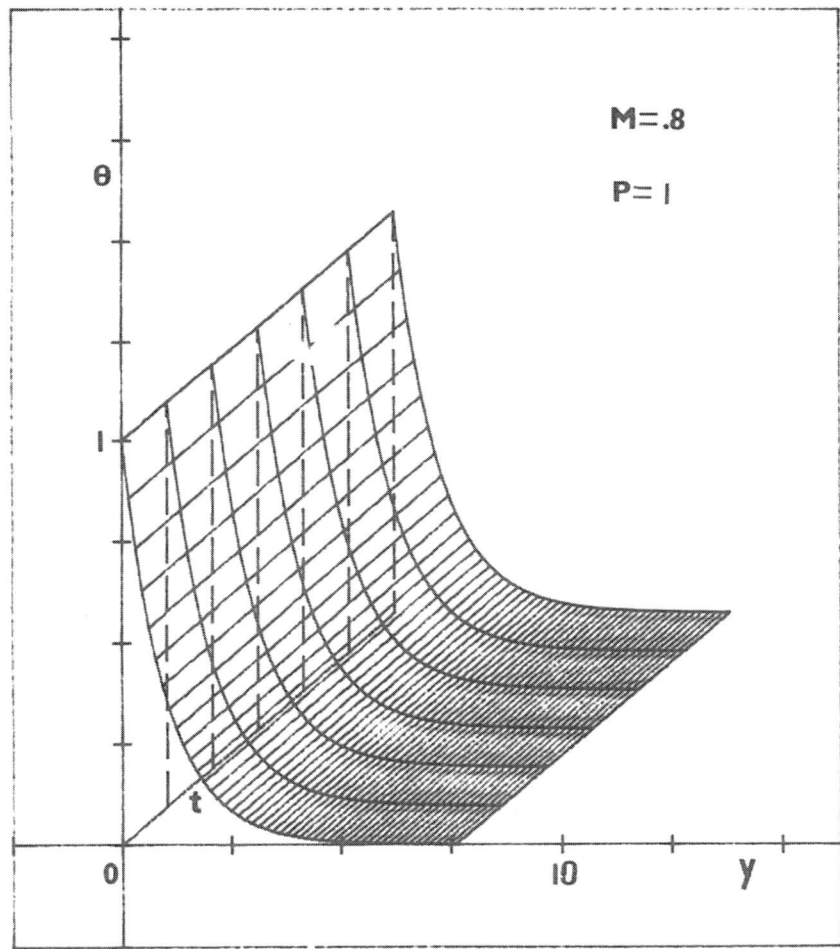

Fig. 4. Three-dimensional presentation of the temperature θ as function of y and t for M=0.8 and P=1.

in which μ is the dynamic viscosity coefficient. Therefore, the calculation of τ_w requires the computation of the quantity ∂u/∂y for y=0 for every instant of time. Since the values of the fluid velocity u are given from the numerical procedure already described, the value of its partial derivative with respect to y can be computed by some interpolation method, e.g. the polynomial method of Lagrange. The application of this method using a fourth order polynomial and hence the values of u at five points gives that

$$\frac{\partial u}{\partial y}\Big|_{y=0} = \pm \frac{1}{12h}(25u_1 - 48u_2 + 36u_3 - 16u_4 + 3u_5). \qquad (32)$$

The calculation of the skin friction follows immediately.

SOME NUMERICAL RESULTS

The general computer programme compiled, after the aforementioned tests, was used to calculate the viscous, thermal and magnetic boundary layers for the following cases:

i. Stationary plate and free-stream velocity

$$U(t) = 1 + \varepsilon e^{i\omega t} .$$

ii. Accelerating plate and stationary fluid.

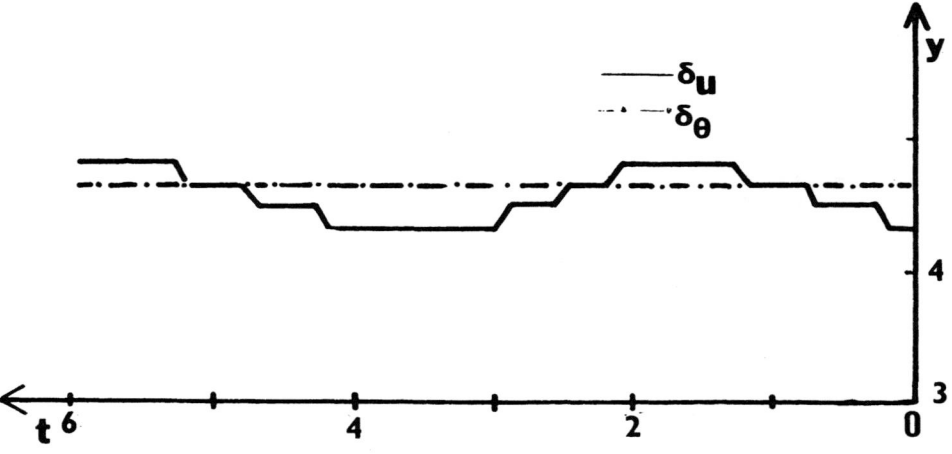

Fig. 5. Variation with the time t of the viscous and thermal boundary layer thickness δ_u and δ_θ for M=0 and P=1.

For case i the numerical values of the parameters were taken as follows:

$$a_o = b_o = c_o = 4, \quad \text{or} \quad c_o = 4/7 ,$$

k = 0.001,   h = 0.154919,   E = ±0.002,

P = 1, or P = 7, P_m = 1, ω = $\pi/2$,

M = 0.0   or 0.4,   or 0.8, ε = 0.2 .

Examples of the results obtained for this case are given in Figs. 1,2 and 3. Fig. 1 shows the variation of the velocity u against the distance y from the plate and the time t, for P=1, P_m=1 and M=0. The boundary condition u=0 on the plate is obviously fulfilled, while the effect of the periodic variation of the free-stream velocity is apparent. The viscous and the thermal boundary layers for this case are shown in Fig. 5. We observe that the two boundary layer thicknesses are roughly equal as predicted by theory and that the oscillatory motion of the free-stream influences the viscous boundary layer thickness, which varies with the period of the free-stream, whereas the thermal boundary layer thickness does not exhibit periodic changes.

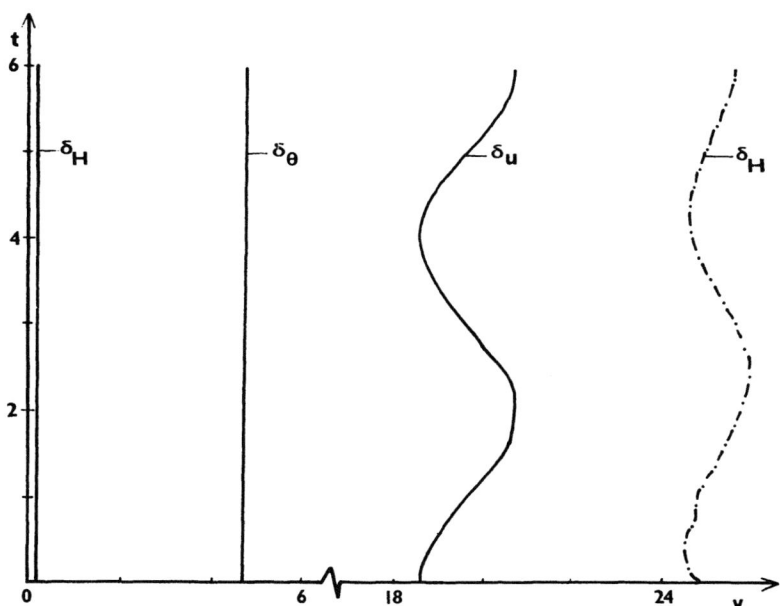

Fig. 6. Variation with the time t of the viscous, thermal and magnetic boundary layer thickness δ_u, δ_θ and δ_H for M=0.8 and P=1.

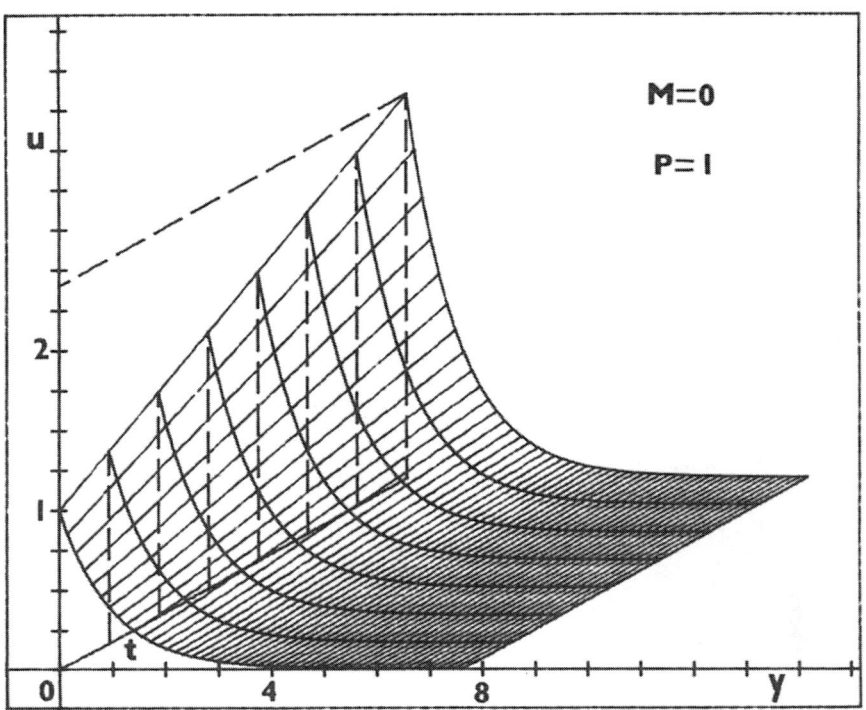

Fig. 7. Three-dimensional presentation of the velocity u of the fluid
as function of y and t for M=0 and P=1. The plate is accelerating and
the free-stream velocity zero.

In Fig. 2 we present the velocity distribution of the fluid as
function of y and t for P=1 and M=0.8. The presence of the magnetic
field affects considerably the velocity distribution and hence the size
of the viscous boundary layer thickness, which in this case (M=0.8) is
found to be almost three times the size corresponding to M=0. This is
illustrated in Fig. 6 where we plot the viscous boundary layer δ_u as
function of time. Comparison between Figs. 5 and 6 shows how serious is
the effect of the magnetic field.

In Fig. 3 we present the induced magnetic field H (this is the
component of the magnetic field along the Ox-axis) as function of y and
t for P=1 and M=0.8. The magnetic boundary layer thickness δ_H is given
in Fig. 6. We observe that the magnetically active region starts from a
small distance δ_H^1 from the plate (which is non-conducting and, in this
case, at rest) and terminates at a distance δ_H^2. Also, that the magnetic
boundary layer is thicker than the viscous one.

In Fig. 4 we present the temperature of the fluid for P=1 and M=0.8
as function of y and t. The temperature on the plate is constant and as
the distance y from the plate increases it drops rather fast to become

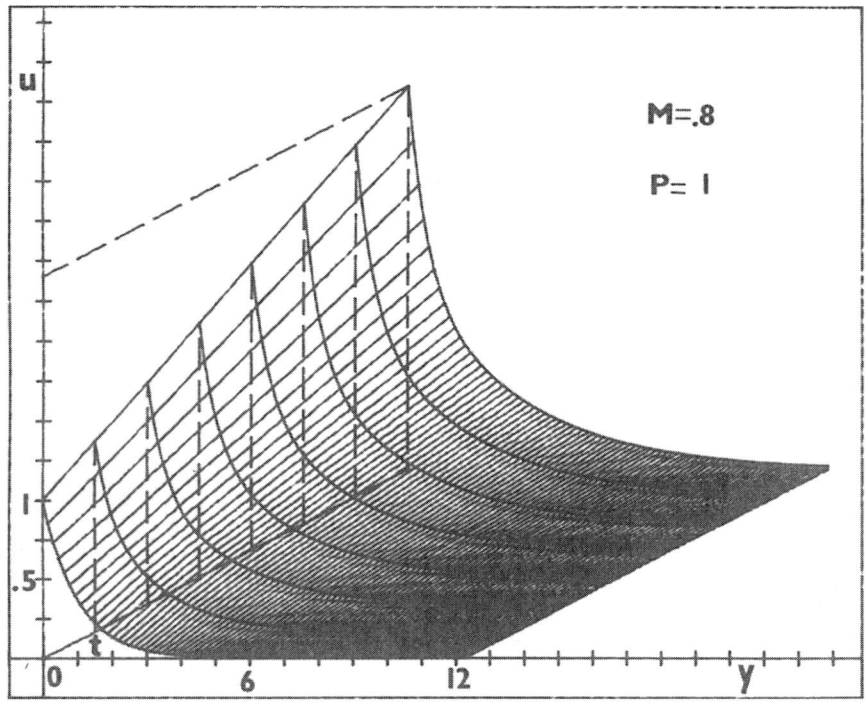

Fig. 8. Three-dimensional presentation of the velocity u of the fluid as function of y and t for M=0.8 and P=1.

practically equal to the temperature of the fluid at the free-stream region, i.e. T_∞. The thermal boundary layer thickness δ_T as function of the time for this case is given in Fig. 6. It is interesting to note that the presence of the magnetic field is reducing the size of the thermal boundary layer. This becomes obvious from a comparison between Figs. 5 and 6.

Case ii corresponds to an accelerating plate and a stationary fluid. The velocity of the fluid on the plate was taken to be

$$u'(0,t') = u_0 + \gamma t',$$

whereas for $t'=0$

$$u'(y',0) = u_0 e^{-\frac{v_0}{v} y'},$$

or in dimensionless form

$$u(0,t) = 1 + \frac{4v\gamma t}{u_0 v_0^2},$$

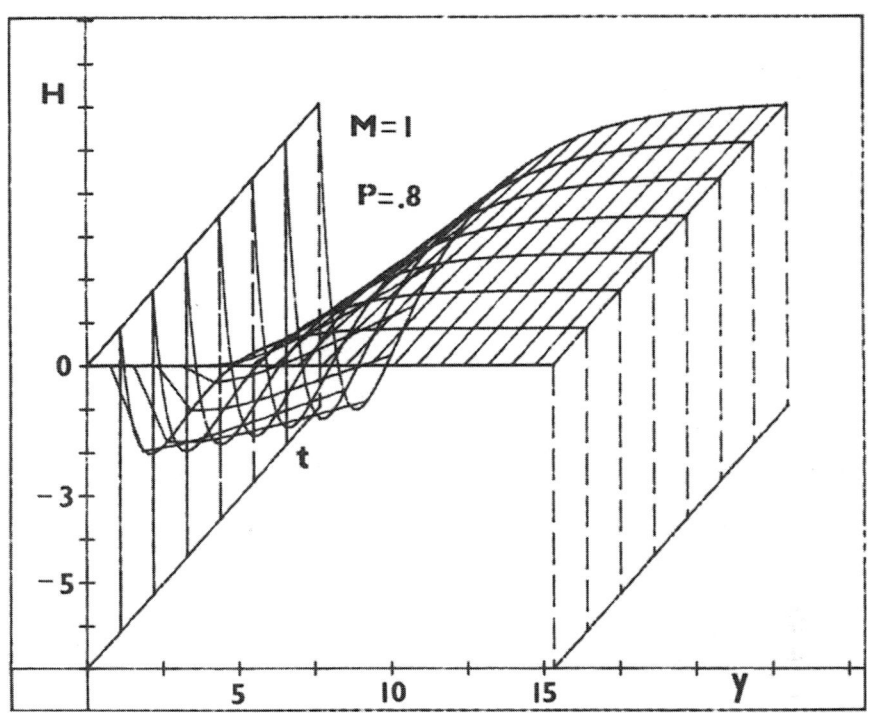

Fig. 9. Three-dimensional presentation of the temperature θ as function of y and t for M=0.8 and P=1.

$$u(y,0) = e^{-y}.$$

The initial and boundary values of the variables u, H, θ were taken to be

$$u(y,0) = e^{-y} \, , \quad H(y,0) = 0 \, , \quad \theta(y,0) = 0 \, ,$$

$$u(0,t) = 1 + \frac{4\nu\gamma t}{u_o v_o^2} \, , \quad \theta(0,t) = 1, \quad H(0,t) = 0 \, ,$$

$$u(\infty,t) = 0 \, , \quad \theta(\infty,t) = 0 \, , \quad H(\infty,t) = 0 \, .$$

The calculations presented in Figs. 7-10 correspond to the following values of the constants involved:

$$\nu = 1.827 \cdot 10^{-5} \ m^2/sec$$

$$v_o = 2.7 \cdot 10^{-2} m/sec$$

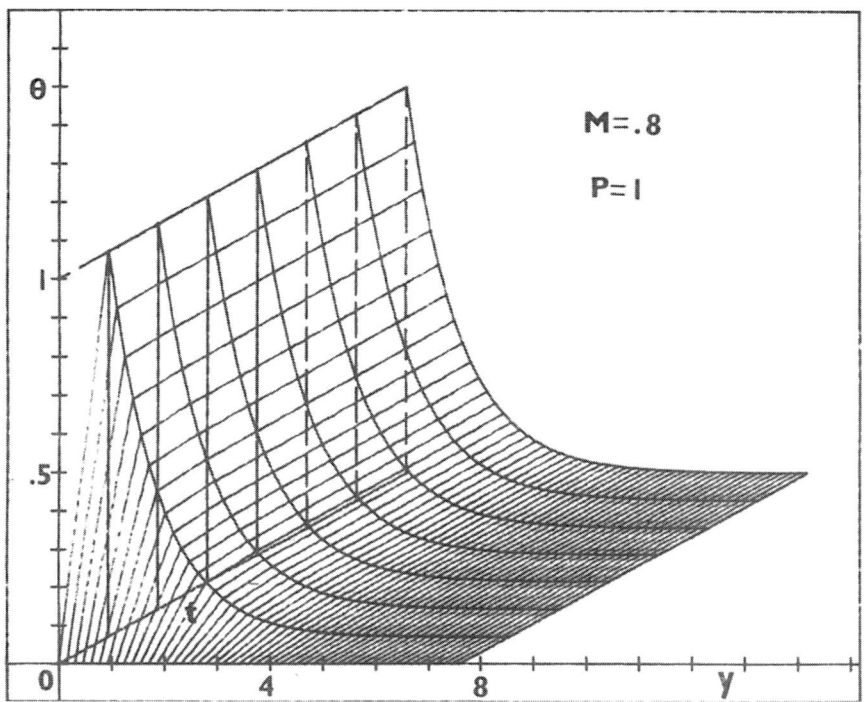

Fig. 10. Three-dimensional presentation of the induced field intensity H as function of y and t for M=0.8 and P=1.

$$\gamma = 2 \ m/sec^2$$

$$u_o = 1 \ m/sec .$$

The rest of the parameters are the same as in case i.

In Figs. 7 and 8 we present the velocity u as function of y and t without (M=0) and with (M=0.8) the effect of magnetic field, while in Figs. 9 and 10 we give only for M=0.8 the induced magnetic field of y and t.

The complete set of results and their discussion is given in reference number 4.

Acknowledgement

This research has been supported in part by the Scientific Research and Technology Agency, Ministry of Coordination, Greece.

REFERENCES

Cebeci, T., Stewartson, K., and Williams, P.G.: 1979, SIAM J. Appl.
    Math. 36, 190.
Der, J. and Raetz, G.S.: 1962, Inst. Aeron. Sci. paper No. 62-70.
Der, J.: 1971, AIAAJ, 9, 1249.
Drymonitou, M.: 1981, Doctoral Thesis, University of Patras.
Dukowicz, J.K., Matthias, C.S., and de Leeuw, J.H.: 1968, CASI Trans.
    1, 45.
Dwyer, M.A.: 1968, AIAAJ 6, 1336.
Hall, M.G.: 1967, Royal Aircraft Establishment TR 6714.
Hasimoto, H.: 1964, Rept. Aeronaut. Res. Inst. Univ. Tokyo, 29,145.
Ingham, D.B.: 1979, Nuclear Eng. Design, 52,325.
Kafousias, N.: 1976, Doctoral Thesis, University of Patras.
Khare, S.: 1977, Indian J. Pure Appl. Math. 8, 808.
Narayana, C.L. and Ramamoorthy, P.: 1972, AIAAJ 10, 1085.
Nath, G.: 1973, AIAAJ, 11,1429.
Richtmyer, R.D. and Morton, K.W.: 1967, Difference Methods for Initial
    Value Problems. Interscience Publ. Inc. New York.
Riley, N.: 1975, SIAM Rev., 17,274.
Rubbert, P.E.: 1965, Ph.D. Thesis, MIT, Mass.
Rubbert, P.E. and Landahl, M.T.: 1967, AIAAJ 5, 470.
Rubbert, P.E. and Landahl, M.T.: 1967, The Physics of Fluids 10, 831.
Sears, W.R. and Telionis, D.P.: 1975, SIAM J. Appl. Math. 28, 215.
Soundalgekar, V.M. and Takhar, H.S.: 1977, Nuclear Eng. Design. 42, 233.
Takhar, H.S. and Soundalgekar, V.M.: 1976, Z. Flugwiss. 24, 75.
Tan, C.W. and Dibiano, R.: 1972, AIAAJ 10, 923.
Yakovlev, M.N.: 1965, NASA Report TTF-254.

RAY TRACING AND CAUSTICS FROM LARGE REFLECTORS USED IN SPACECRAFTS

P.S. Theocaris
Fellow of the National Academy of Athens
Athens, Greece

SUMMARY

The general equations of caustic surfaces formed by reflections of a point-light source on the reflecting surfaces of ellipsoid, paraboloid and hyperboloid form were derived for the case when the light source is placed at an arbitrary distance and angle from the axis of symmetry of the reflector. A study of the evolution of the shape and form of the caustics was undertaken for each type of reflector. It was found that the position of the caustic relatively to the position of the reflector and its shape depend on the type of the reflector, as well as on the relative position of the light-source and the reflector. Interesting results were derived concerning the dependence of the shape the position and the aperture of the mirror, as well as the relative distance of the reflector and the light-source. All these properties are useful for the selection of systems of wide-angle reflectors having the possibility of collecting large amounts of light energy and presenting the advantage of minimization of third-order errors.

INTRODUCTION

The necessity of studying the optical fields created by the reflector light-waves on surfaces of different shape and form was dictated by the wide use of reflectors in modern optical systems. Indeed, the reflecting systems present great advantages against refracting systems constituting the typical elements of conventional optical systems. These advantages may be classified in the following: i) The reflecting systems may be used for any wavelength covering the whole wave spectrum from x-rays up to radio waves, whereas the refracting systems may be used only for wavelengths lying in the visible band of the spectrum. This possibility has a great significance for modern optical systems which are used for observations and recording of phenomena outside the atmosphere of the earth, which may be extended from the x-ray band to several centimeters waveband. ii) With a relatively limited number of reflecting optical elements it

415

E. G. Mariolopoulos et al. (eds.), Compendium in Astronomy, 415–445.
Copyright © 1982 by D. Reidel Publishing Company.

is possible to achieve observations of large areas, whereas for the
covering of such regions with refractive systems a large number of thick
lenses is required which complicates enormously the optical system.
iii) The reflecting optical systems, in which use is made of reflectors
corresponding to conic surfaces of revolution, are free of certain
chromatic aberrations for certain conditions of application, as well as
from monochromatic third-order aberrations, fact which cannot happen in
refractive systems.

The study of the optical fields created by the incidence of light-
waves on reflectors of axisymmetric form begun with Bartkowski (1961,1962)
who formulated the equations of the caustic surfaces in a parametric form.
These caustic surfaces were created by illuminating   different mirrors
by light-sources placed at different distances and angles from the axis
of symmetry of the mirror and they represent surfaces of concentration
of light rays reflected from the mirror. He gave also some relationships
for the caustic surfaces formed by refracted light rays from transparent
surfaces. Bartkowski tried also to solve the inverse problem, that is
the definition of the shape of a reflecting surface from the study of
the caustics formed during the illumination of the surface by a certain
light-source. This problem was studied only for the case of axisymmetric
surfaces and for special cases of conic mirrors.

The special case of a caustic surface formed by illuminating a
parabolic cylindrical mirror with a point-light source lying outside the
axis of symmetry of the mirror was studied by Scarborough (1964). He
considered only reflected light rays and he has established the parametric
equations of the caustic by considering it as the envelope of the
trajectories of the reflected light-rays. Stalzer (1965), in a review of
the paper by Scarborough, gave the parametric equations of the caustic
for the more general case of a parabolic mirror, whereas, for a special
case of a reflector, he had found the canonic equation of the caustic.

Especially interesting for the evaluation of the energy distribution
of the light-rays reflected or refracted from a surface of arbitrary
shape was the contribution by Shealy and Burkhard (1973a). They gave the
analytic expression of the light-energy density, which is defined as the
energy per surface and time unit of the light-rays reflected or
refracted from a curved surface of arbitrary shape and concentrated on
an equally arbitrary surface. The variation of the light-energy density
was expressed in terms of the geometric characteristic properties of the
deflecting surface. These properties were related to the three types of
curvatures of a deflector that is the Gaussian, the average and the
normal curvatures.

Shealy and Burkhard have used the previously developed general
principles, in their first paper, in a series of publications where they
have studied certain concrete optical problems. Thus, in their papers
(Burkhard and Shealy, 1973; Shealy and Burkhard, 1973b), they gave the
expressions of the caustic surfaces created from light rays deflected

from an ellipsoid, an elliptic paraboloid and an elliptic cone. For this purpose they calculated and gave analytic expressions for the Gaussian and the normal curvatures of the three types of deflectors and they have introduced these quantities in the general expression for the light-energy density. In all cases studied by them, the light-source used was assumed to be a point-light source.

Finally concerning the subject of caustics one should mention two scientific workers who contributed significantly in the understanding of the theory describing *"one of the few things in geometrical optics which has any physical reality"*, that is caustics. Both of them were and are of Greek descent and both worked or are working abroad. The one is Franciscus Maurolycus who as early as in 1543 in Sicily discovered the phenomenon of caustics and had described accurately some of the most interesting properties (Theocaris, 1978). The other is Orestis Stavroudis (1972) who, with his excellent book, revitalized the subjects concerning geometrical optics and gave a unified mathematical treatment in subjects already known otherwise, which was consistent when covering the subjects treated in this book.

In this paper we have studied the special case of rotationally symmetric reflectors corresponding either to ellipsoid or to hyperboloid-paraboloid forms which were considered as illuminated by point light sources lying along axes parallel to the principal axis of symmetry of the reflector at different distances apart from it. The technique used in this paper is an extension of a previously developed technique by the author (1971) for the study of caustic surfaces created in initially flat plates containing some points of stress singularities or stress concentrations and which when they are subjected to a stress field either in their plane or out of it they deform laterally and create types of smooth deflectors. Later on, Theocaris and Gdoutos (1976) have extended the method of caustics for the study of the most general type of smooth surfaces. They have developed a technique, based on the method of caustics, for the study of the geometry of an arbitrary surface from the respective caustics created when this surface is illuminated by a parallel, divergent or convergent light beam. The same authors (1977) have initiated a simple and versatile method for the evaluation of the distance between two points in space by using the caustics created by an ellipsoid mirror placed at the one point when at the other point a point-light source is illuminating the mirror.

Theocaris (1977a,b) has derived the general equations of the caustic surfaces which are formed by illuminating an ellipsoid, hyperboloid or paraboloid reflector with a point-light source in the case when this source is always placed along the axis of symmetry of the reflector. The influence of the shape and form of the reflector as well as of the distance between reflector and light-source are extensively studied.

THE EQUATIONS OF CAUSTICS

Consider a generic rotationally symmetric surface described by an equation of the form:

$$z = f(r) \qquad (1)$$

and a point-light source placed along the z-axis of symmetry of the surface at a distance A from the Oxy-plane indicated in Fig.1. If a reference screen Sc is placed at a distance z_0 from the Oxy-plane the deviation of the reflected light beam from the reflector originated from point P(r) of its surface on the reference screen is expressed by:

$$W = (z-z_0)\tan(2\alpha+\varphi) \qquad (2)$$

with:

$$\tan\alpha = \frac{dz}{dr} \quad \text{and} \quad \tan\varphi = \frac{r}{A+z} \qquad (3)$$

If we express now the deviation of the light-ray to the origin of the $O^-x^-y^-$-system which represents a parallel projection of the Oxy-system on the reference screen (Fig.1) we obtain:

$$r^- = r+(z-z_0)\tan(2\alpha+\varphi) \qquad (4)$$

Relation (4) represents the mapping of each point P(r) of the rotationally symmetric surface of the reflector to the respective point $P^-(r^-)$ on the screen. The necessary and sufficient condition for the points $P^-(r^-)$ to belong to one curve of the screen Sc is the zeroing of the derivative

$$dr^-/dr = 0 \qquad (5)$$

which corresponds to the respective zeroing of the Jacobian of the transformation.

Relation (5) defines for every case of an arbitrary rotationally symmetric surface the generatrix curve whereas the system of equations (4) and (5) defines the respective caustic.

If we consider now a reflector of an elliptic cross section by an arbitrary plane containing the principal axis of symmetry of the reflector and if the dimensions of semi-axes of the ellipse of this cross-section are z=a and r=b the equation expressing the form of the surface of the reflector is given by:

$$\frac{z^2}{a^2} + \frac{r^2}{b^2} = 1 \qquad (6)$$

from which it may be derived that:

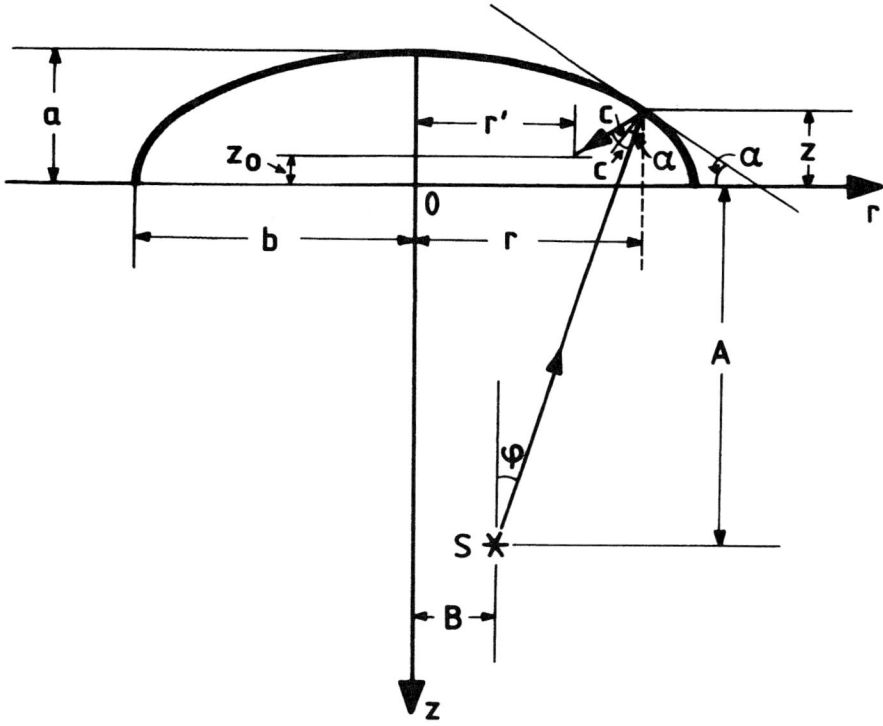

Figure 1. Geometry of the elliptical cross section of an
axisymmetric mirror illuminated by a point-light
source S.

$$z = f(r) = \frac{a}{b}(b^2-r^2)^{\frac{1}{2}} \tag{7}$$

For a point-light source S illuminating the reflector of Fig.1, which
is placed at a distance A from the Or-axis and a distance B from the
Oz-axis, angle φ, subtended by the light ray SP and a parallel to Oz-
axis passing through S, is expressed by:

$$\tan\varphi = \frac{r-B}{A+z} \tag{8}$$

Introducing relation (7) into Eq.(8) we obtain:

$$\tan\varphi = \frac{b(r-B)}{Ab+a(b^2-r^2)^{\frac{1}{2}}} \tag{9}$$

But:

$$\tan\alpha = \frac{dz}{dr} = -\frac{a}{b}\frac{r}{(b^2-r^2)^{\frac{1}{2}}} \tag{10}$$

then it may be derived that:

$$\tan 2\alpha = -\frac{2abr(b^2-r^2)^{\frac{1}{2}}}{b^4-r^2(a^2+b^2)} \tag{11}$$

Based on relations (8) and (11) we can derive an expression for the $\tan(2\alpha+\varphi)$. Introducing this expression into Eqs.(4) and (5) we derive the following parametric equations for the caustic under the form $z_0=z_0(r)$ and $r^-=r^-(r)$:

$$\frac{z_0}{b} = \frac{\Delta_1^2 \cdot \sqrt{1-(r/b)^2} + \left(\frac{a}{b}\right)\left[-\left(\frac{r}{b}\right)\Delta_1\Delta_2+\Delta_1\Delta_3+\Delta_2\Delta_4\right]}{\Delta_1\Delta_5+\Delta_2\Delta_6} \tag{12}$$

and

$$\frac{r^-}{b} = \frac{(\Delta_1\Delta_5+\Delta_2\Delta_6)\left[\left(\frac{r}{b}\right)\Delta_1+\left(\frac{a}{b}\right)\Delta_2\sqrt{1-\left(\frac{r}{b}\right)^2}\right]-\Delta_2\left[\Delta_1^2\cdot\sqrt{1-\left(\frac{r}{b}\right)^2}+\left(\frac{a}{b}\right)\left[-\left(\frac{r}{b}\right)\Delta_1\Delta_2+\Delta_1\Delta_3+\Delta_2\Delta_4\right]\right]}{\Delta_1(\Delta_1\Delta_5+\Delta_2\Delta_6)} \tag{13}$$

with:

$$\Delta_1 = \left(\frac{a}{b}\right)\left[1-\left[\left(\frac{a}{b}\right)^2-1\right]\left(\frac{r}{b}\right)^2-2\left(\frac{r}{b}\right)\left(\frac{B}{b}\right)\right]\cdot\sqrt{1-(r/b)^2}+\left(\frac{A}{b}\right)\left[1-\left[\left(\frac{a}{b}\right)^2+1\right]\left(\frac{r}{b}\right)^2\right]$$

$$\Delta_2 = -2\left(\frac{a}{b}\right)\left(\frac{r}{b}\right)\left(\frac{A}{b}\right)\cdot\sqrt{1-(r/b)^2} -2\left(\frac{a}{b}\right)^2\left(\frac{r}{b}\right)+\left(\frac{a}{b}\right)^2\left(\frac{r}{b}\right)^3-\left(\frac{r}{b}\right)-\left(\frac{r}{b}\right)^3-\left(\frac{B}{b}\right)\left[1-\left[\left(\frac{a}{b}\right)^2+1\right]\left(\frac{r}{b}\right)^2\right]$$

$$\Delta_3 = -2\left(\frac{a}{b}\right)\left(\frac{A}{b}\right)\left[1-\left(\frac{r}{b}\right)^2\right]\cdot\sqrt{1-(r/b)^2} +2\left(\frac{a}{b}\right)\left(\frac{A}{b}\right)\left(\frac{r}{b}\right)^2\sqrt{1-(r/b)^2} +$$
$$+\left[1-\left(\frac{r}{b}\right)^2\right]\cdot\left[-2\left(\frac{a}{b}\right)^2+1+3\left(\frac{r}{b}\right)^2\left[\left(\frac{a}{b}\right)^2-1\right]+2\left(\frac{r}{b}\right)\left[\left(\frac{a}{b}\right)^2+1\right]\left(\frac{B}{b}\right)\right]$$

$$\Delta_4 = 2\cdot\left(\frac{a}{b}\right)\left[\left[\left(\frac{a}{b}\right)^2-1\right]\left(\frac{r}{b}\right)+\left(\frac{B}{b}\right)\right]\left[1-\left(\frac{r}{b}\right)^2\right]\sqrt{1-(r/b)^2} +\left(\frac{a}{b}\right)\left(\frac{r}{b}\right)\left[1-\left[\left(\frac{a}{b}\right)^2-1\right]\left(\frac{r}{b}\right)^2\right] -$$
$$-2\left(\frac{r}{b}\right)\left(\frac{B}{b}\right)\sqrt{1-(r/b)^2} + 2\left(\frac{r}{b}\right)\left(\frac{A}{b}\right)\left[\left(\frac{a}{b}\right)^2+1\right]\left[1-\left(\frac{r}{b}\right)^2\right]$$

$$\Delta_5 = -2\left(\frac{a}{b}\right)\left(\frac{A}{b}\right)\left[1-\left(\frac{r}{b}\right)^2\right]+2\left(\frac{a}{b}\right)\left(\frac{A}{b}\right)\left(\frac{r}{b}\right)^2+\sqrt{1-(r/b)^2}\left[-2\left(\frac{a}{b}\right)^2+1+\right.$$
$$\left.+3\left(\frac{r}{b}\right)^2\left[\left(\frac{a}{b}\right)^2-1\right]+2\left(\frac{r}{b}\right)\left[\left(\frac{a}{b}\right)^2+1\right]\left(\frac{B}{b}\right)\right]$$

$$\Delta_6 = 2\left(\frac{a}{b}\right)\left[\left[\left(\frac{a}{b}\right)^2 - 1\right]\left(\frac{r}{b}\right) + \left(\frac{B}{b}\right)\right]\left[1 - \left(\frac{r}{b}\right)^2\right] + \left(\frac{a}{b}\right)\left(\frac{r}{b}\right)\left[1 - \left[\left(\frac{a}{b}\right)^2 - 1\right]\left(\frac{r}{b}\right)^2 - \right.$$

$$\left. - 2\left(\frac{r}{b}\right)\left(\frac{B}{b}\right)\right] + 2\left(\frac{r}{b}\right)\left(\frac{A}{b}\right)\left[\left(\frac{a}{b}\right)^2 + 1\right]\right] \sqrt{1 - (r/b)^2}$$

By the same procedure we obtain the parametric equations of the caustics for the cases of hyperboloid and paraboloid reflectors, the only difference being that Eq.(6) is now replaced either by:

$$\frac{z^2}{a^2} - \frac{r^2}{b^2} = 1$$

for the case of the hyperboloid reflector, or by:

$$z = \frac{r^2}{4a}$$

for the case of a paraboloid one.

PROPERTIES OF CAUSTICS

We shall now examine the variation of the characteristic properties of the caustics created by the reflected light-rays from either one of these three types of reflectors when it was illuminated by a point light source placed at a position $S(A,B)$ in front of each minor. We assume, further, that the light source is moving along axes parallel to the principal axis of the reflector so that distance B remains constant. For reasons of symmetry we consider such axes to lie always on the one side of the mirror where r/b are positive. In order to study also the influence of the eccentricity of the position of the light source on the shape of the respective caustic we shall consider also different such secondary axes of the mirror, parallel to the principal axis of symmetry and displaced at different distances B from it. Along each secondary axis the displacement of the light source will yield also the other parameter of influence of the shape of caustics.

Finally we shall limit ourselves to examine the shapes of caustics for reflectors with aspect ratios $(a/b)=0.25$ and 0.50 that is in both cases we are considering shallow reflectors.

a) Caustics From Ellipsoid Reflectors:

From the parametric equations of the caustics formed when an ellipsoid reflector is illuminated by a point-light source it may be derived that the form and shape of caustics depends on the aspect ratio of the mirror (a/b), as well as on the position of the point light source (coordinates A, B).

   Figs 2a to 2d and 3a to 3d represent the caustics formed by
illuminating ellipsoid reflectors with (a/b)=0.25 and 0.50 respectively
for a progressive displacement of the light source along different
secondary axes. Thus, in Figs 2a to 2d the point-light source moves
along the 'axes (B/b)=0,0.5,1.0 and 5.0 whereas in Figs 3a to 3d the
light source moves along the secondary axes (B/b)=0,0.5,1.0 and 3.0
respectively.Figs 2 correspond to an ellipsoid mirror with aspect ratio
(a/b)=0.25 and Figs 3 correspond to an ellipsoid with (a/b)=0.50.
Positive values for (A/b) mean that the point-light source is lying
beyond the plane z=0 (z>0), which is tangent to the lips of the
reflector, whereas positive sense for the (r/b)-axis means the side of
the mirror where the light source is placed.

   Figs 4a to 4b and 5a to 5b represent the shapes of caustics formed
by illuminating a hyperboloid reflector with aspect ratios (a/b)=0.25
and 0.50 respectively, for a progressive movement of the point-light
source along secondary axes. Thus, in Figs 4a to 4b the light source is
lying on the secondary axes with (B/b)=0 and 1.0 respectively, whereas
in Figs 5a to 5b this source is lying on the secondary axes with
(B/b)=0 and 2.0 respectively.

   Finally, Figs 6a to 6d represent the caustics formed by illuminating
a paraboloid reflector in the cases where the point-light source runs
along the secondary axes with (B/b)=0,0.5,1.0 and 2.0 respectively.

   From the above figures and other similar figures traced in a
digital computer and not shown in this paper for the shake of brevity
the laws of dependence of the position, the size and the shape of the
respective caustics on the type of the reflector and the relative
position of the reflector and the light source were found.

   Thus, for the ellipsoid reflectors we can derive the following
results:
i) The caustics formed are always non-symmetric relatively to the
principal axis of the reflector except for the case when the light
source lies along this axis, or at infinity (parallel light beam). This
asymmetry of the caustics becomes more severe as the reflector becomes
more shallow. Relatively to the position of the point-light source the
asymmetry of the caustics is increasing when the normalized distance
(B/b) of the light source from the principal axis becomes larger.
Similar phenomena appear if for the same distance (B/b) of the light
source from the principal axis, its distance (A/b) is diminishing that
is when the light source approaches the mirror.
ii) All caustics present cusp points where there is an intensive
concentration of the light-rays and where the larger part of the energy
of light is concentrated. As we recede from the cusp point to the tails
of each caustic the light intensity is progressively diminishing and
at both extremities of each caustic the intensity is minimum. Thus,
each caustic appears always as a curve (or a surface, in space) with
two branches in both sides of its cusp. The only caustic, which does
not present a cusp is the caustic formed when the light source lies on

a point of the mirror. In this case the caustic almost coincides with
the surface of the reflector and the light intensity along the caustic
is the less non-uniform from all other caustics.
iii) For small values of the (B/b)-distance the caustics formed
present always a cusp point, whereas for large distances B/b of the
light source from the principal axis this cusp point disappears and
only the one branch or a part of it appears illuminated. The remaining
part of the caustic which is eliminated could be formed from parts of
the reflector not considered and which are impractical. The limiting
(B/b)-distance for which the entity of the one branch of the caustic
appears,as well as its cusp point,is increasing with increasing
(A/b)-distances of the light source from the reflector.
iv) Relatively to the position of the cusps we may observe that these
points lie either in the opposite side of the space of the reflector
than that in which the light source exists relatively to its principal
axis of symmetry, or on the same side depending on the positive or
negative sign of the (A/b)-distances. Moreover, these points are
receding from the principal axis of the reflector for shallower mirrors
and for smaller values (positive or negative) of the (A/b)-distances,
that is as the light source approaches the (z=0)-plane.
v) The caustics are approaching each other more and more as the light
source is receding from the center of the reflector. Moreover, for an
increase of the (B/b)-distance of the secondary axis, on which the
light source lies, the caustics approach each-other and simultaneously
they approach the reflector. Thus, for a large (B/b)-distance, that is
a large eccentricity of the light source relatively to the axis of the
mirror, almost all caustics occupy the same space near the mirror with
the exception of the caustic corresponding to a parallel light beam.
vi) Another important observation is that the caustics become larger as
the aspect ratio of the reflector is increasing and the mirror becomes
shallower. The same phenomenon appears when the light source is either
approaching the mirror or is receding on the opposite side, that is the
lengths of the caustics increase for a continuous algebraic diminution
of the (A/b)-distance.

b) Caustics From Hyperboloid Reflectors:

     In the case of hyperboloid reflectors the following general results
may be derived:
i) The caustics formed by the reflected rays on hyperboloid mirrors
are again non-symmetric relatively to the principal axis of symmetry
of the reflector with the only exceptions of the caustics formed by a
point-light source placed either on the surface of the mirror, or at
infinity. The asymmetry of the caustics related to the type of
reflector becomes more and more intense as the mirror becomes more and
more shallow. With respect to the position of the point light source
the asymmetry of the caustics is increasing with an increase of the
(B/b)-distance, that is the eccentricity of the light source, and the
decrease of the (A/b) distance, that is the approaching of the light

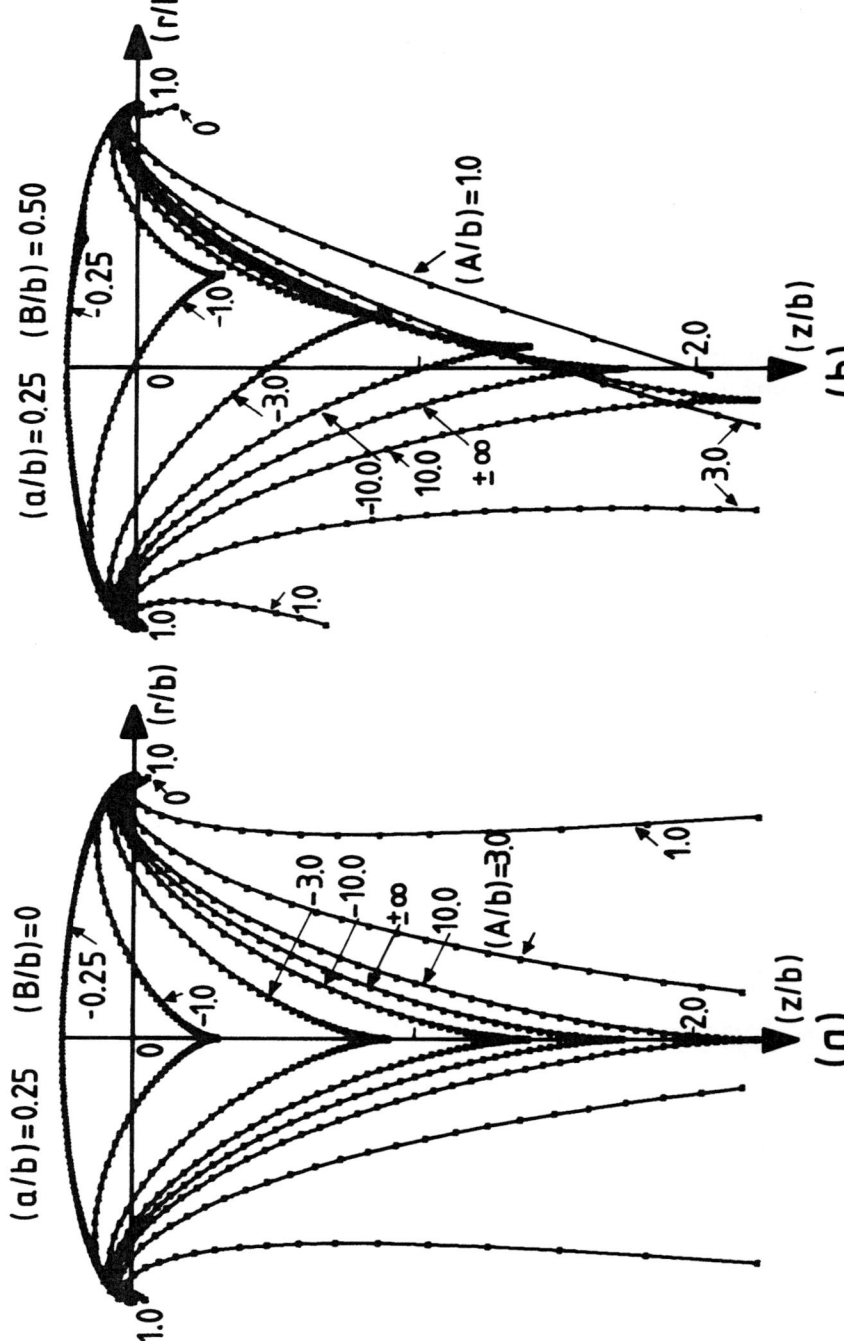

Figure 2. Caustics obtained from a shallow ellipsoid reflector with (a/b)=0.25 illuminated by a point-light source which runs along the secondary axes with (B/b)=0(a) and 0.5(b).

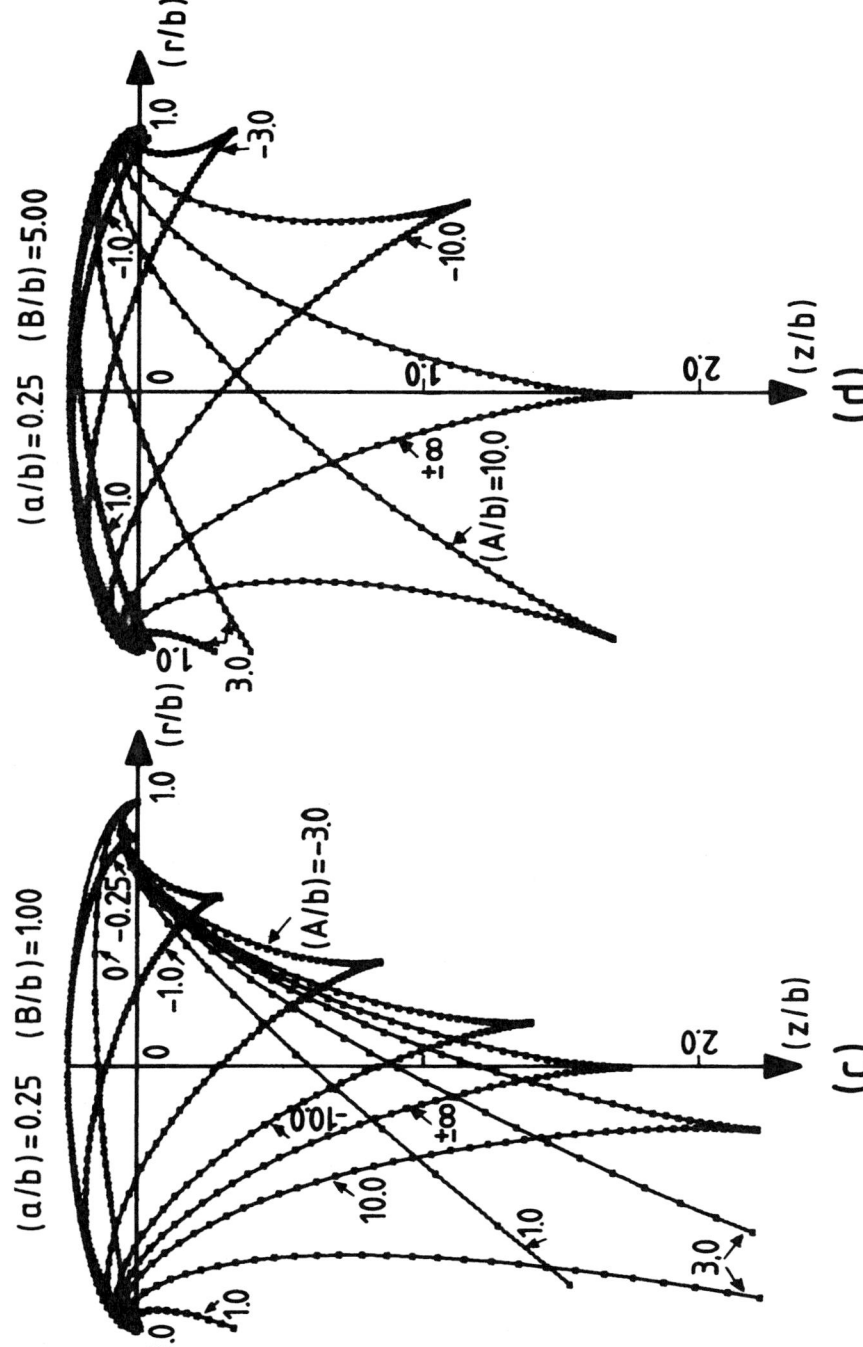

Figure 2. Caustics obtained from a shallow ellipsoid reflector with (a/b)=0.25 illuminated by a point-light source which runs along the secondary axes with (B/b)=1.0(c) and 5.0(d).

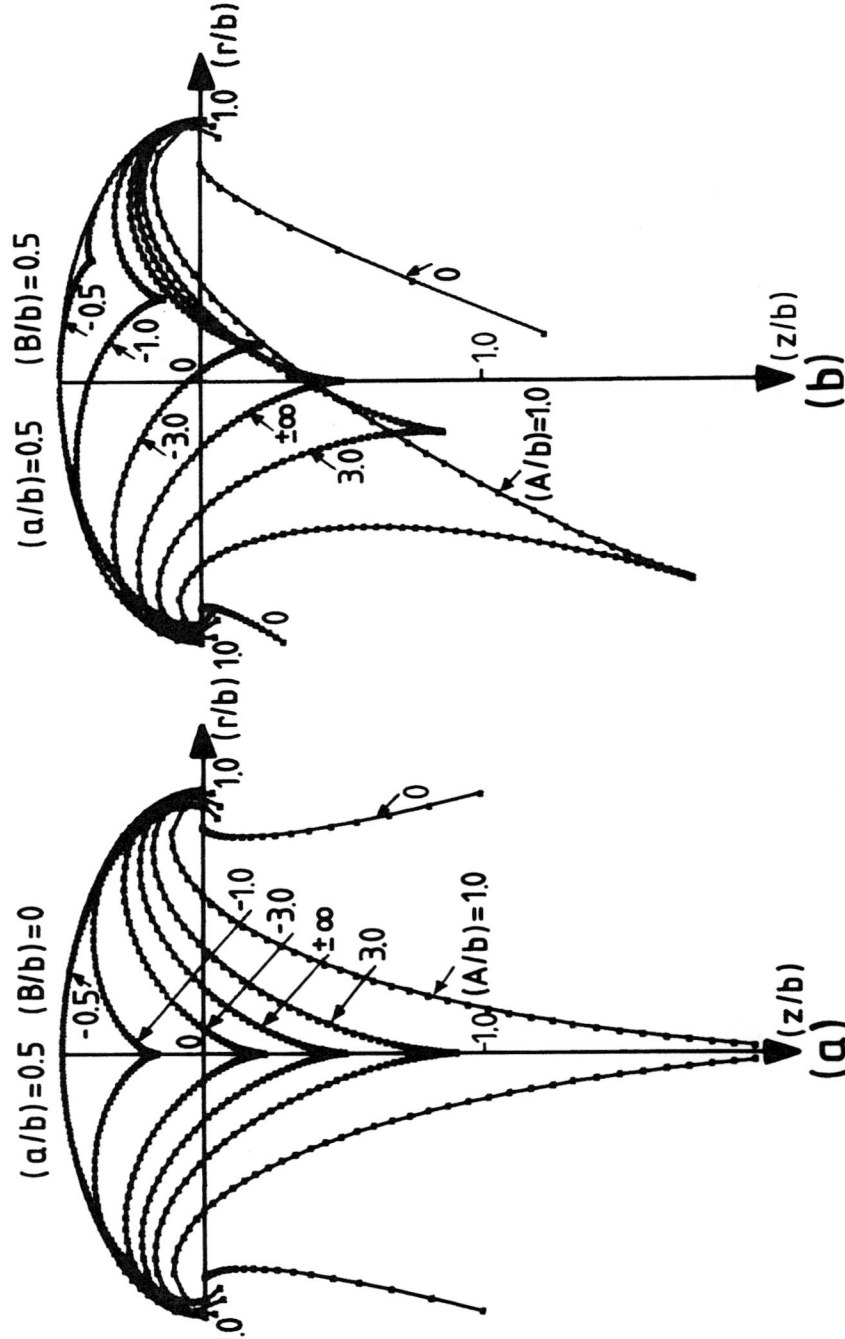

Figure 3. Caustics obtained from a shallow ellipsoid reflector with (a/b)=0.25 illuminated by a point-light source which runs along the secondary axes with (B/b)=0(a) and 0.5(b).

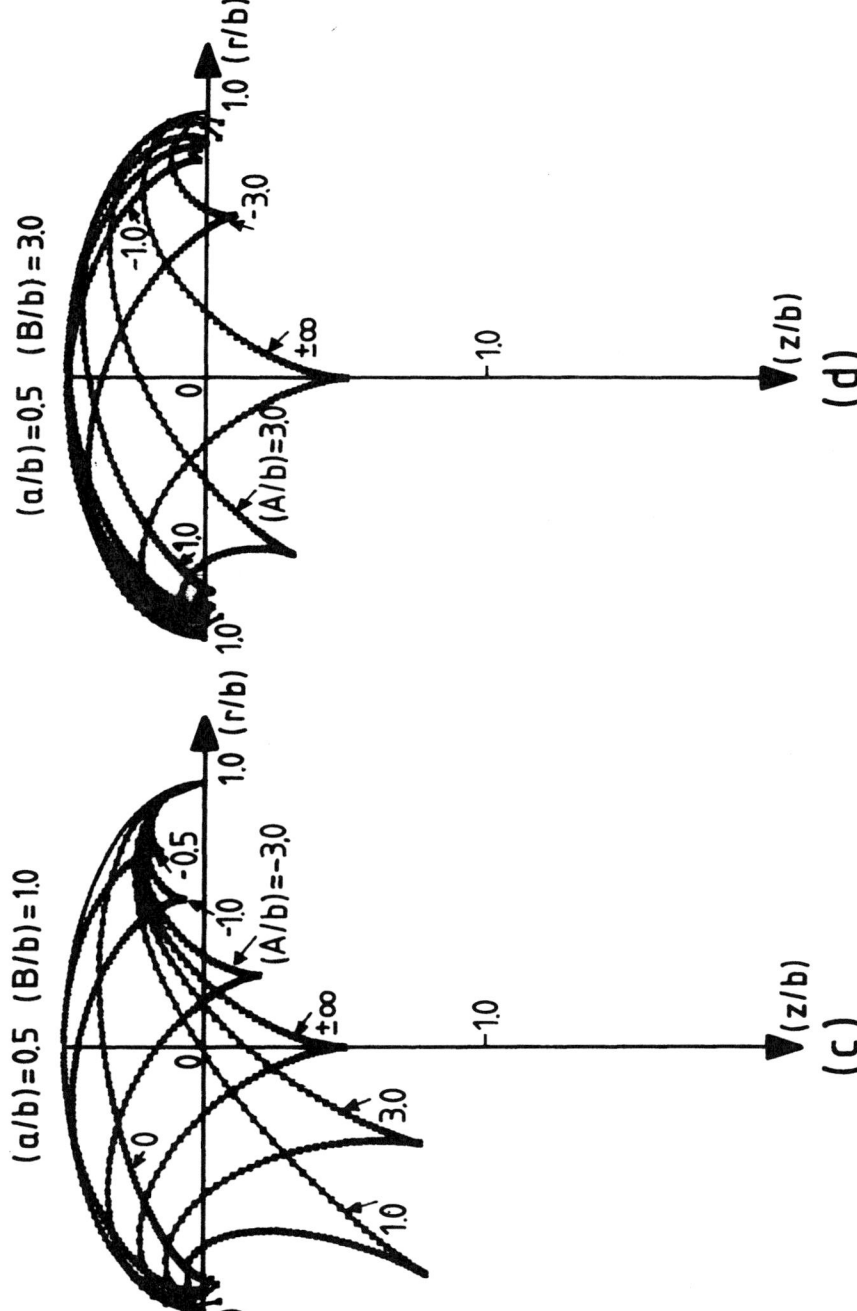

Figure 3. Caustics obtained from a shallow ellipsoid reflector with (a/b)=0.25 illuminated by a point-light source which runs along the secondary axes with (B/b)=1.0(c) and 3.0(d).

source to the mirror.
ii) All caustics present a cusp point, which is a point of high
concentration of light intensity. In the neighboring zone of the
cusp the largest part of the energy of light is concentrated so that
this part may be considered as the most important part in calculating
optical systems.

All caustics present two branches in either part of their cusp.
Again, the only caustic without a cusp is the caustic created by a
light source lying on the reflective surface of the mirror. This
caustic has also the smoothest variation of the distribution of the
energy of light along its length.
iii) For small values of the (B/b)-distance of the light source from the
principal axis of symmetry the caustics formed present always cusps,
whereas for larger values of the (B/b)-distance the cusps disappear
and the caustics become continuous curves. The limiting value for the
(B/b)-distance for which starts to appear a cusp point in the caustic
is increasing with increasing absolute values of the (A/b)-distances
of the light-source from the reflector in both sides of the point of
inversion of the orientation of the caustics.
iv) Concerning the position of the cusp points we can observe that
these points lie always either on the opposite or on the same side of
the space where the light source lies relatively to the position of the
principal axis of symmetry of the reflector, in the case when the
(A/b)-distance is either (a/b)<(A/b)<+∞ or -∞<(A/b)≦(a/b) respectively.
Moreover, these points become more and more remote from the principal
axis as the reflector becomes more and more shallow and the point-light
source approaches more and more the reflector from the positive or
negative values of the (A/b)-distance.
v) The caustics are approaching more and more between them as the
light source recedes from the center of the reflector. Moreover, for
an increase of the (B/b)-distance of the light-source from the principal
axis of symmetry the caustics are moving toward the reflector. Thus,
for a large (B/b)-distance almost all caustics occupy the same region
close to the mirror with the exception of the caustic corresponding to
a parallel light beam.
vi) There exists a position of the light source along the principal axis
of symmetry for which the respective caustic is degenerated into a
point. The orientation of the caustics and their cusps in both sides of
this point, corresponding to a zero-caustic, is opposite with both groups
of caustics pointing toward this characteristic point.
vii) Finally, we may observe that the sizes of caustics are progressively
increasing as these curves are receding from this zero-caustic point.

c) Caustics Formed by Paraboloid Reflectors

Finally, the case of paraboloid reflectors resembles closely to
the hyperboloid reflectors and yields the following results:
i) The caustics are, again, always asymmetric relatively to the
principal axis of symmetry of the reflector with the only exception the

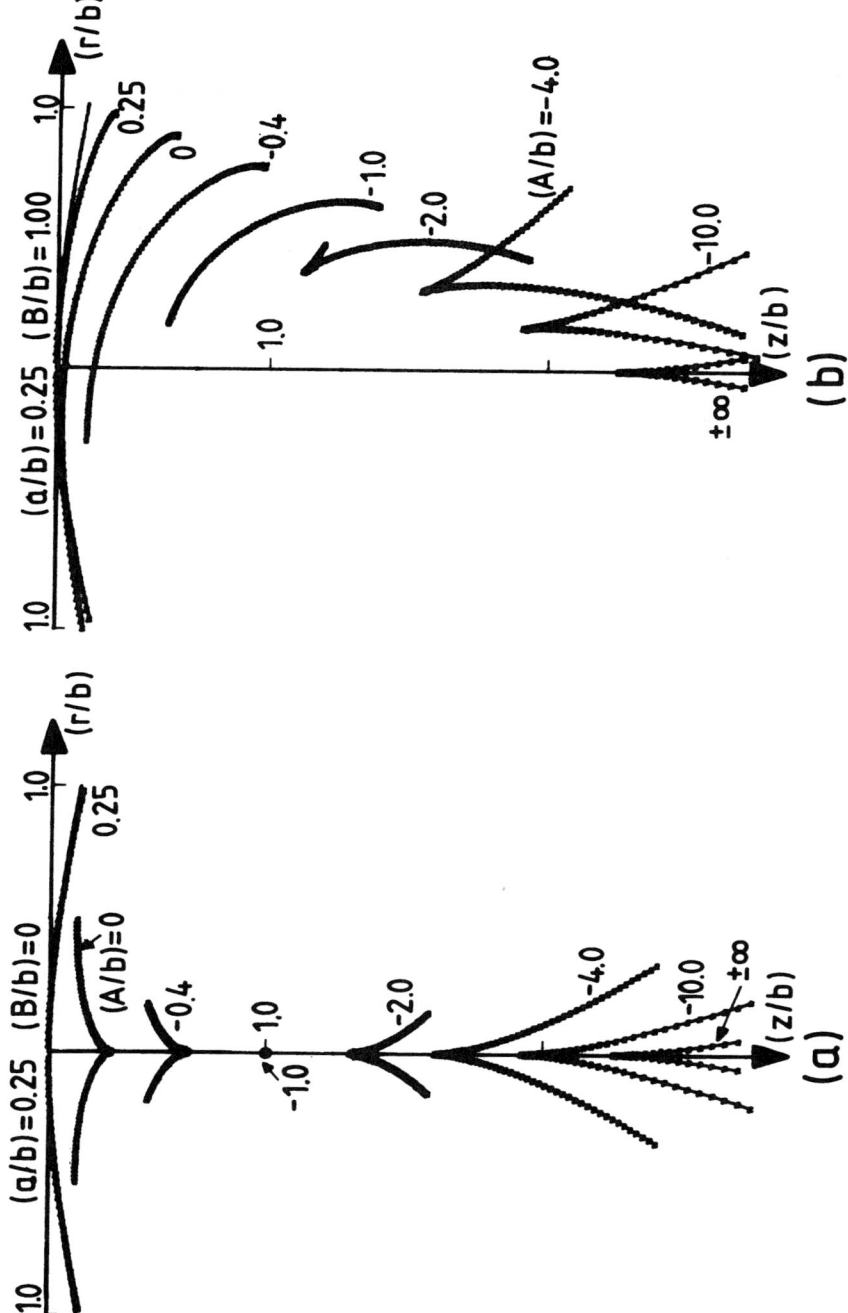

Figure 4. Caustics obtained from a hyperboloid reflector with (a/b)=0.25 and aperture equal to (r/b)=1.00 illuminated by a point-light source which runs along the secondary axes with (B/b)=0(a) and 1.0(b).

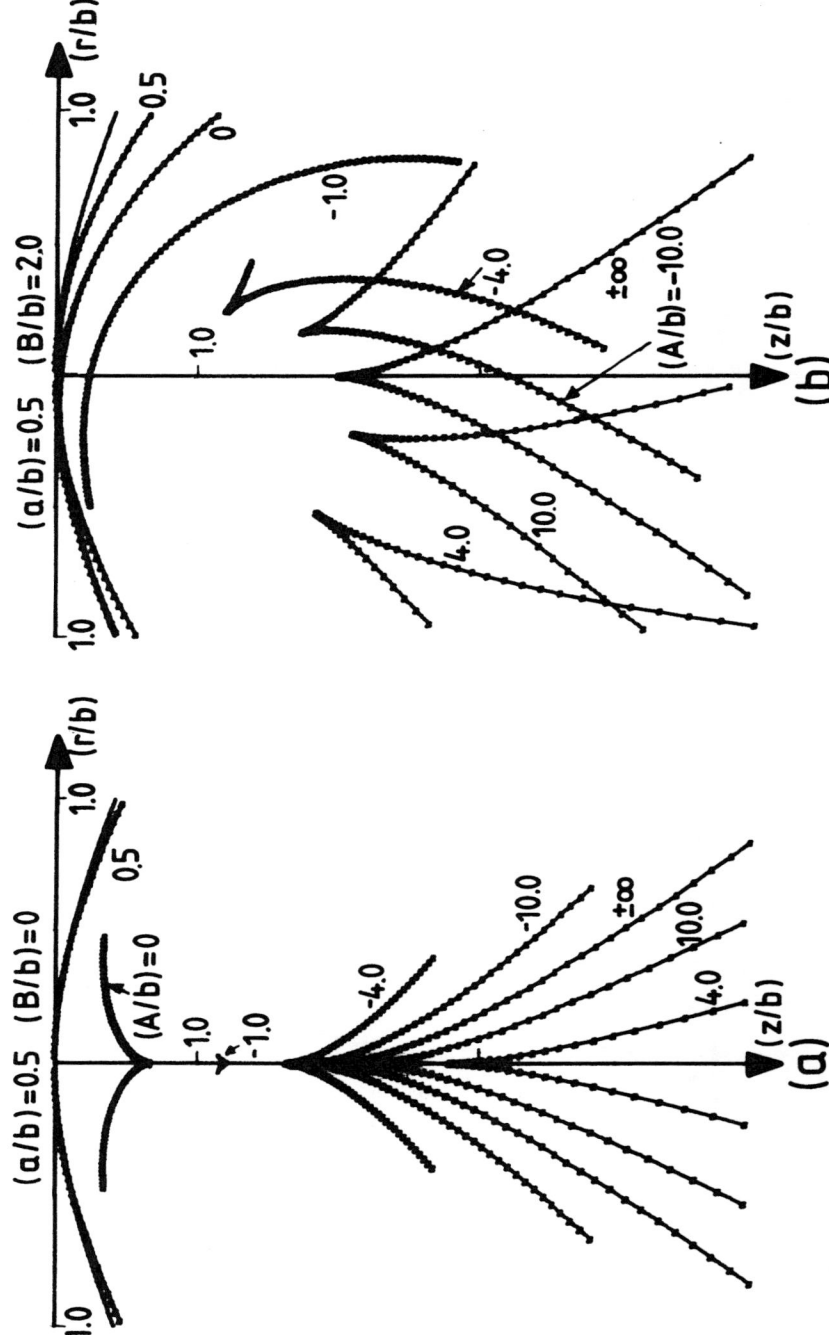

Figure 5. Caustics obtained from a hyperboloid reflector with (a/b)=0.5 and aperture equal to (r/b)=1.0 illuminated by a point-light source which runs along the secondary axes with (B/b)=0(a) and 2.0(b).

caustics formed by light sources lying either on the principal axis of the mirror, or at infinity. Relatively to the influence of the position of the light source on this asymmetry, it can be, again, derived that this asymmetry increases for increasing (B/b)-distances or for decreasing (A/b)-distances.

ii) All caustics present cusp points where the reflected light energy is concentrated. This energy distribution diminishes progressively as we recede from the cusp-point zone so that the tails of each caustic contain only a small fraction of the light energy. Thus, all caustics contain two branches in both sides of their cusp, the only exception being the caustic created by a light source lying on the surface of the mirror.

iii) There exist two positions of the point-light source on the principal axis of symmetry of the mirror for which the respective caustic is degenerated into a point. The orientation of the caustics on both sides of thie zero-caustic is different in the way that all caustics between this critical point and the mirror are oriented with their cusps looking outwards and all caustics lying between the zero-caustic point and infinity are oriented in the opposite direction with their cusps looking toward the mirror.

iv) For small values of the (B/b)-distance from the principal axis of symmetry of the reflector the caustics formed present cusp points, whereas for larger values these caustics are continuous curves without cusps. The limiting (B/b)-value for a cuspoid caustic is increasing for increasing absolute values of the distance $|(A/b)|$.

v) Relatively to the position of the cusp points we may observe that, again, these points lie always in the opposite or the same side with the point-light source depending on the distances (A/b). Thus for $-\infty<(A/b)\leqq0$ the cusp points lie on the opposite side, whereas for $0<(A/b)<\infty$ they lie on the same side. Moreover, the cusp points are more and more receding from the principal axis of the reflector as the light source apprpaches more and more the mirror from positive or negative values of the (A/b)-distance.

vi) The caustics are approaching closely to each other as the light source recedes from the center of the mirror. Moreover, for an increase of the (B/b)-distance the caustics are approaching the reflector. Thus, for a large value of the (B/b)-distance almost all caustics occupy the same zone of the space close to the mirror with the exception of the caustic corresponding to a parallel light beam.

vii) Finally, we observe that all caustics increase in size as they recede from the characteristic zero-caustic point.

REFLECTORS WITH VARIABLE APERTURE

In the previous discussion we have studied the caustics formed by illuminating with a point-light source rotationally symmetric conic reflectors of full size that is the reflector was occupying the surface of a half-section of the respective conic solid. However, in various cases of practical applications the reflectors used are not complete but

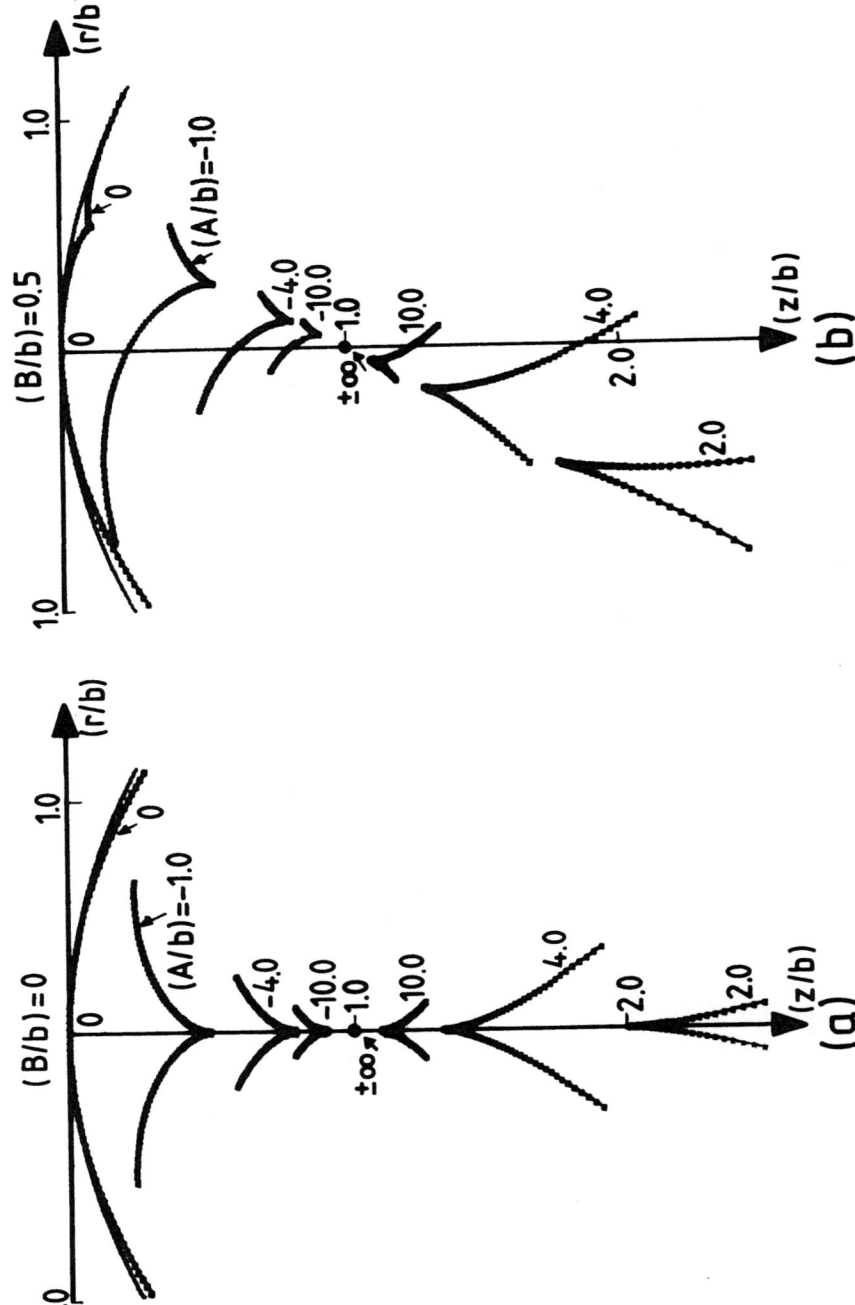

Figure 6. Caustics obtained from a paraboloid reflector with aperture equal to (r/b)=1.0 illuminated by a point-light source which runs along the secondary axes with (B/b)=0(a) and 0.5(b).

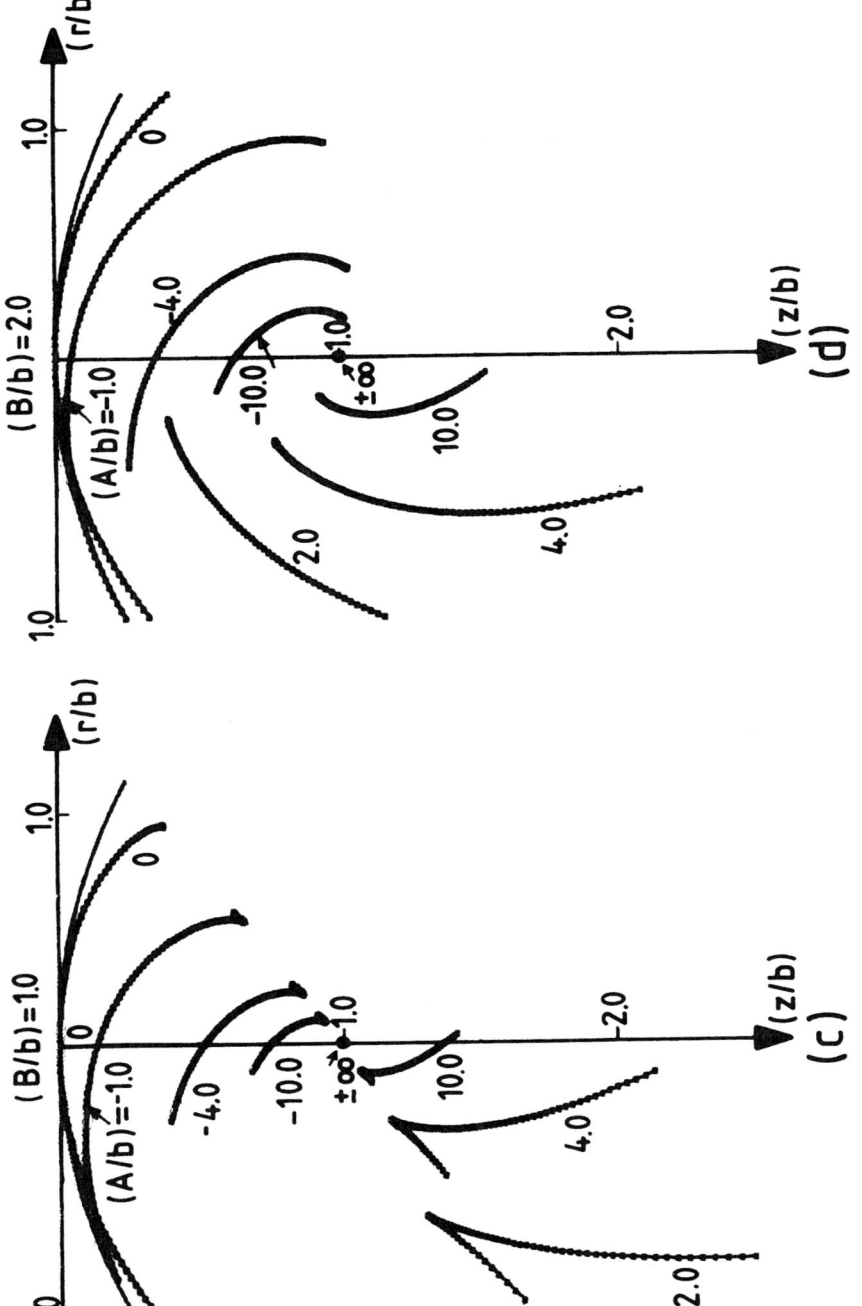

Figure 6. Caustics obtained from a paraboloid reflector with aperture equal to (r/b)=1.0 illuminated by a point-light source which runs along the secondary axes with (B/b)=1.0(c) and 2.0(d).

they are consisting of only parts of the whole form of the respective
conic surface.

Of special interest in this case is the size of the caustic which
is included by its extreme points defined from reflections from the
limits of the mirror: The study of the definition of the extreme points
of the caustics may be achieved by defining the characteristic curves
along which the extremities of each caustic lie in the case when the
light source is displaced along secondary axes, parallel to the
principal axis of symmetry for which the (B/b)-distance is remaining
constant. In this way the equations of caustics must be expressed in
terms of the parameter the distance (A/b) of the light source from the
(z=0)-plane, whereas in these equations the variable (r/b) must be
replaced by its respective value corresponding to the respective
aperture of the mirror.

The shapes of these loci are expressed in general by curves of
second degree which depend on the particular value of the (B/b)-distance.
Fig.7 presents, as an example the loci of the extremities of both
branches of caustics created from reflections of the light rays emanating
from a point-light source moving along a secondary axis with (B/b)=1.7
for various parametric values of the (A/b)-distance. Both curves (dushed
lines) are of the second degree resembling to ellipses.

The loci of the extremities of the caustics for the case of an
ellipsoid reflector with a/b=0.25 and for a displacement of the point-
light source along the secondary axes with parameters (B/b)=0,0.50,1.00
and 5.00 are given in Figs 8a to 8d respectively. Similar loci for the
case of an ellipsoid reflector with an aspect ratio (a/b)=0.50 and for
the cases where the point light source is moving along the secondary
axes with (B/b)=0,0.5,1.0 and 3.0 are represented in Figs 9a to 9d.
Similarly, the loci of the extremities of the caustics for the case of
an hyperboloid reflector with aspect ratio (a/b)=0.25 are given in
Figs 10a to 10b for the cases when the point light sourve is moving
along the secondary axes with parameters (B/b)=0 and 1.0. The respective
loci of the extremities of the caustics for an hyperboloid reflector
with aspect ratio (a/b)=0.50 and for a movement of the point-light source
along the secondary axes with (B/b)=0 and 2.0 are given in Figs 11a to
11b.

Finally, the loci of extremities of the caustics for a paraboloid
reflector and for a movement of the point-light source along the
secondary axes with parameters (B/b)=0,0.5,1.0 and 2.0 appear in Figs
12a to 12d.

In all the previous cases studied the regions of the mirrors lying
between their vertices with (r/b)=0 and the points defined by the values
(r/b)=-1.00,-0.75,-0.50,0.50,0.75 and 1.00 define the zones of the
reflector from which the respective loci have been traced in all
previous figures. The full-line tracings in these figures correspond to

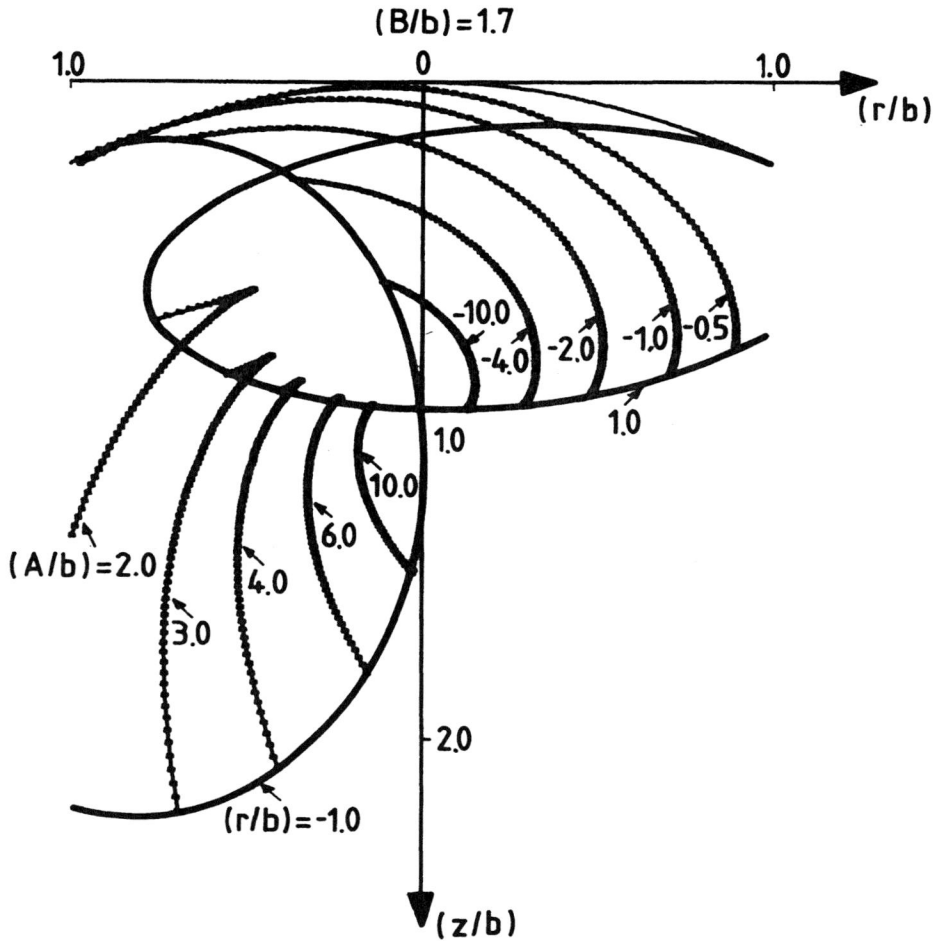

Fig.7. Caustics obtained from a paraboloid reflector with an
aperture equal to (r/b)=1.00 illuminated by a point-
light source moving along a secondary axis with
(B/b)=1.70 and the loci of the extremities of the
caustics for apertures equal to (r/b)=±1.00.

positions of the light source lying in the zone -20.0<(A/b)<20.0 whereas
the dashed-line parts correspond to respective positions of the light-
source lying outside the previous zone that is in the zone with
|(A/b)|>20.0.

    It is interesting to remark that in all these loci the dashed-line
parts of these curves occupy only a small portion of the whole locus.
This means that all caustics corresponding to positions of the light-

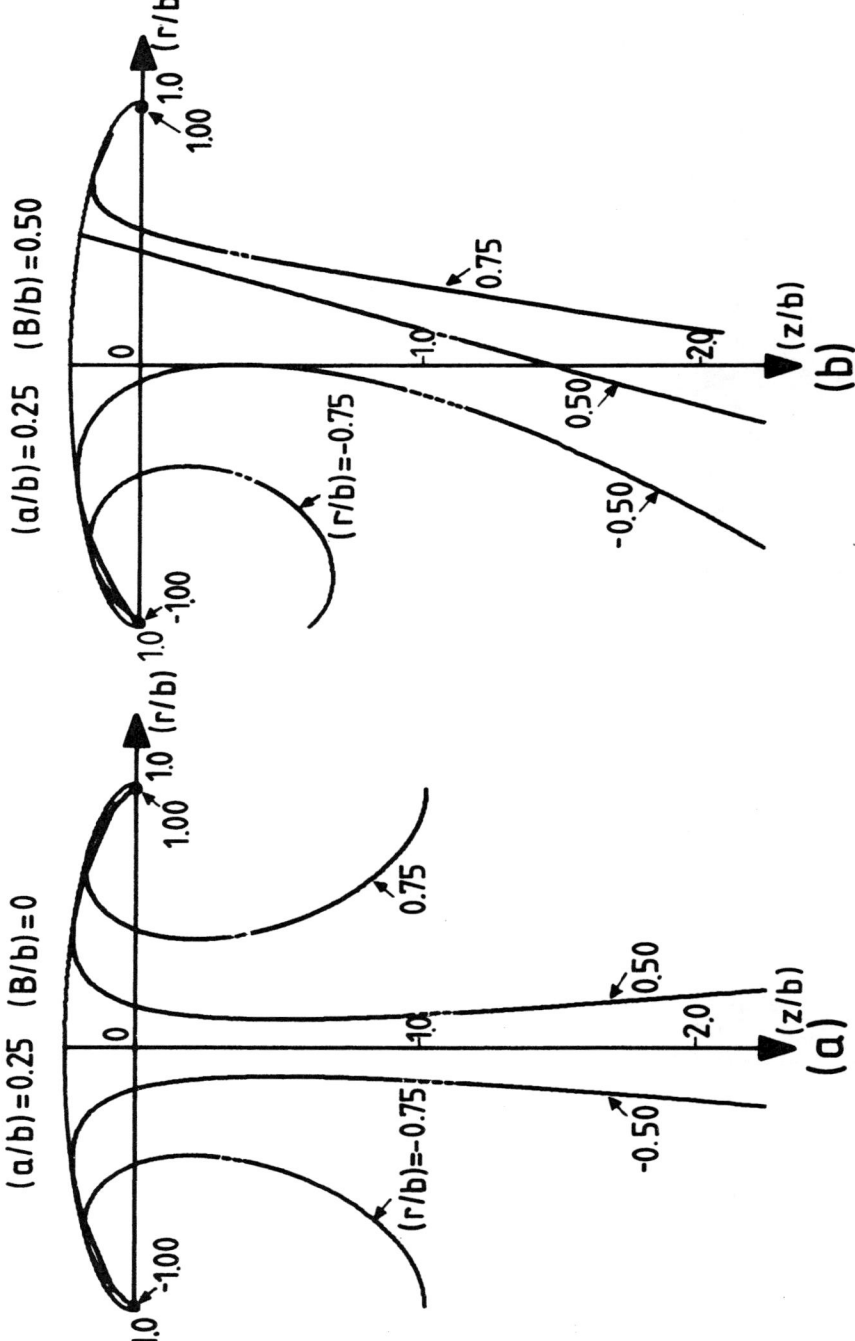

Figure 8. Loci of the extremities of the caustics corresponding to an ellipsoid reflector with (a/b)=0.25 and apertures equal to (r/b)=-1.00, -0.75, -0.50,0.50,0.75,1.00. The point-light source runs along the secondary axes with (B/b)=0(a) and 0.50(b).

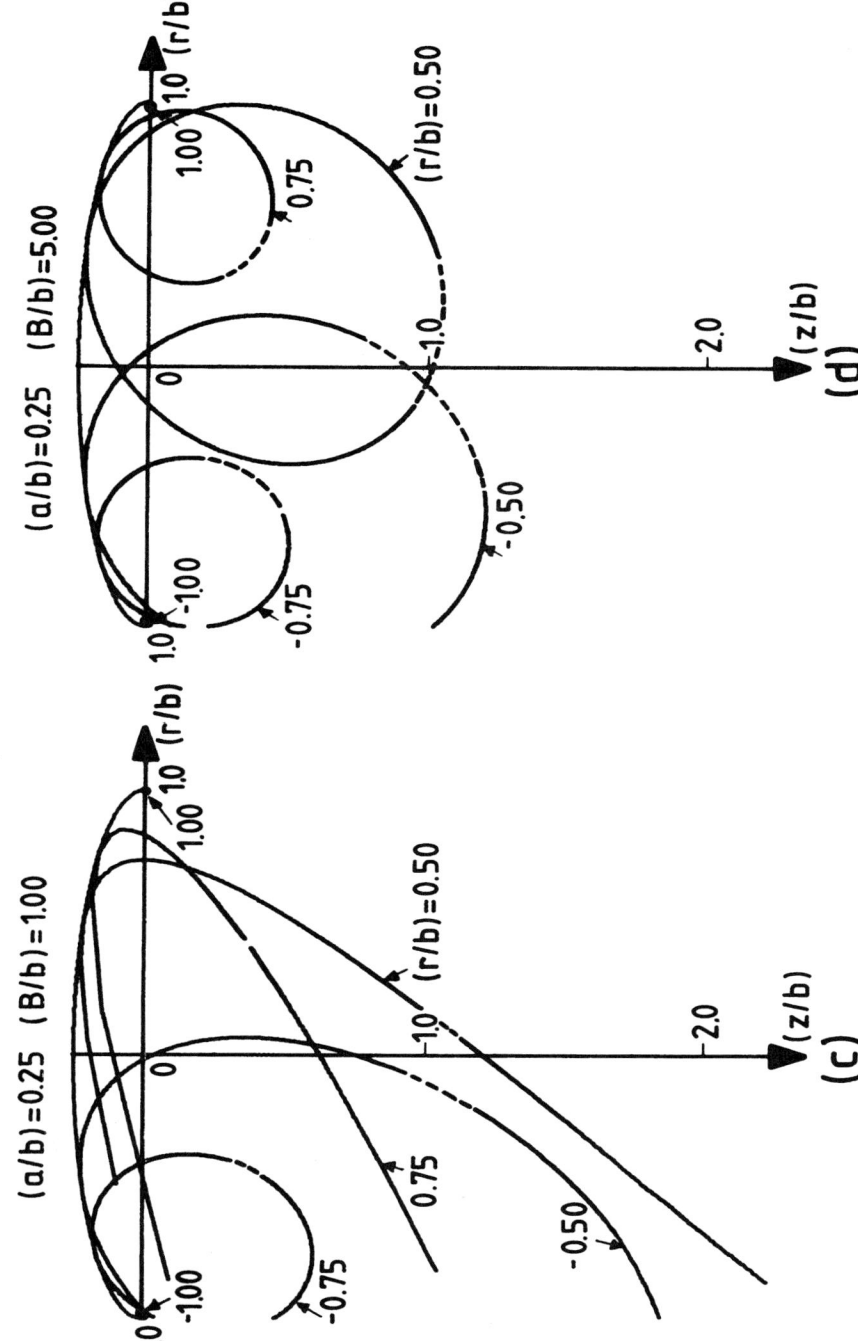

Figure 8. Loci of the extremities of the caustics corresponding to an ellipsoid reflector with (a/b)=0.25 and apertures equal to (r/b)=-1.00,-0.75,-0.50,0.50,0.75,1.00. The point-light source runs along the secondary axes with (B/b)=1.00(c) and 5.00(d).

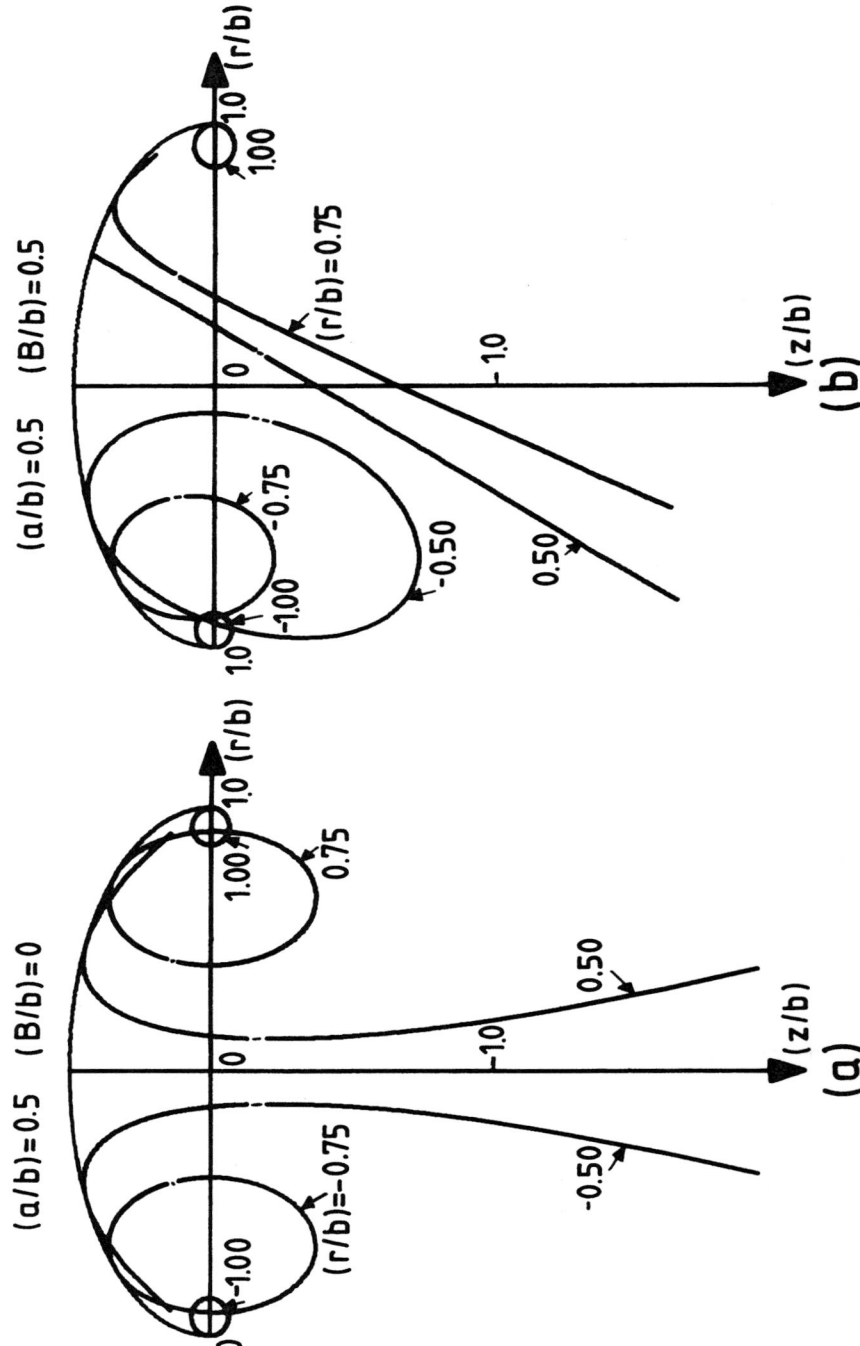

Figure 9. Loci of the extremities of the caustics corresponding to an ellipsoid reflector with (a/b)=0.50 and apertures equal to (r/b)=-1.00,-0.75,-0.50,0.50,0.75,1.00. The point-light source runs along the secondary axes with (B/b)=0(a) and 0.50(b).

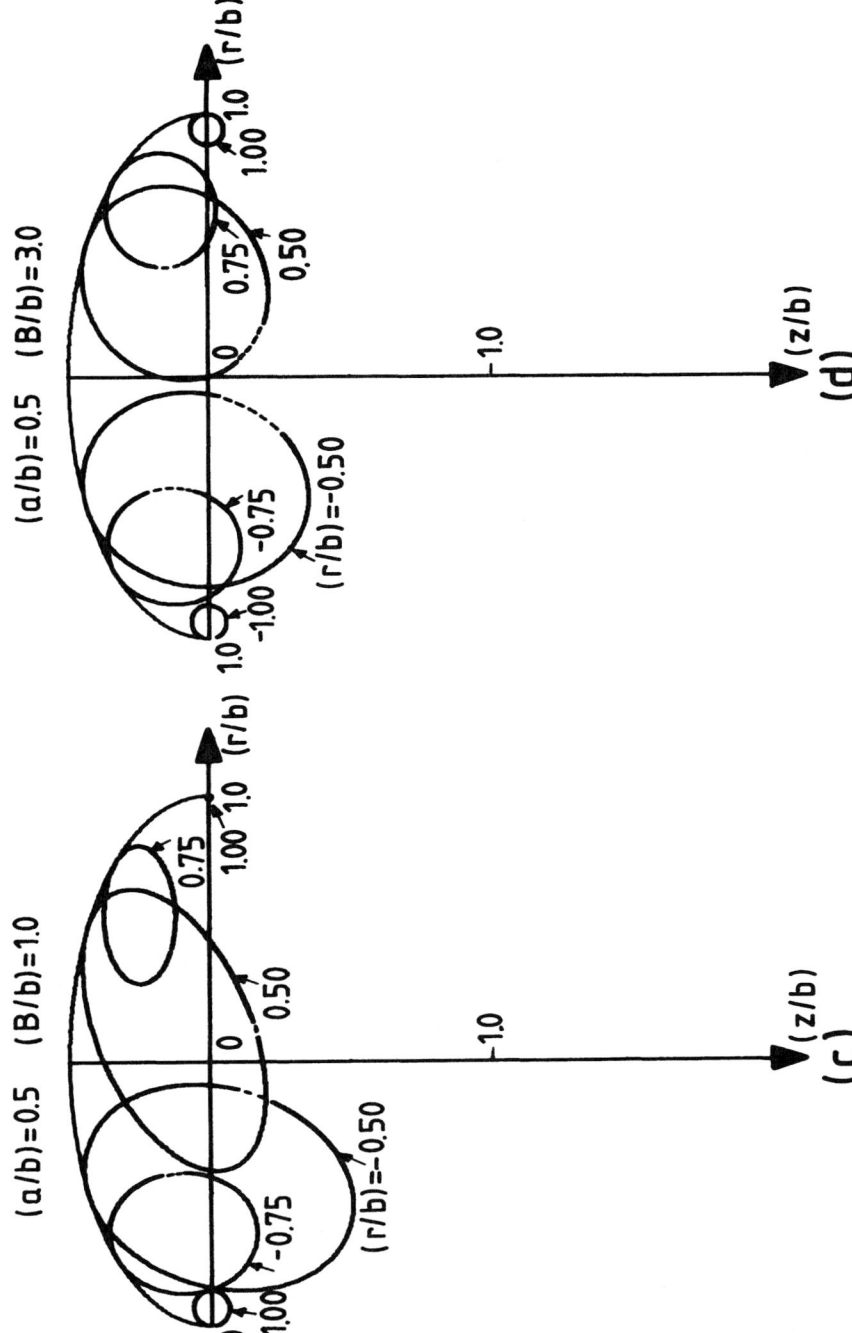

Figure 9. Loci of the extremities of the caustics corresponding to an ellipsoid reflector with (a/b)=0.50 and apertures equal to (r/b)=-1.00,-0.75,-0.50,0.50,0.75,1.00. The point-light source runs along the secondary axes with (B/b)=1.00(c) and 3.00(d).

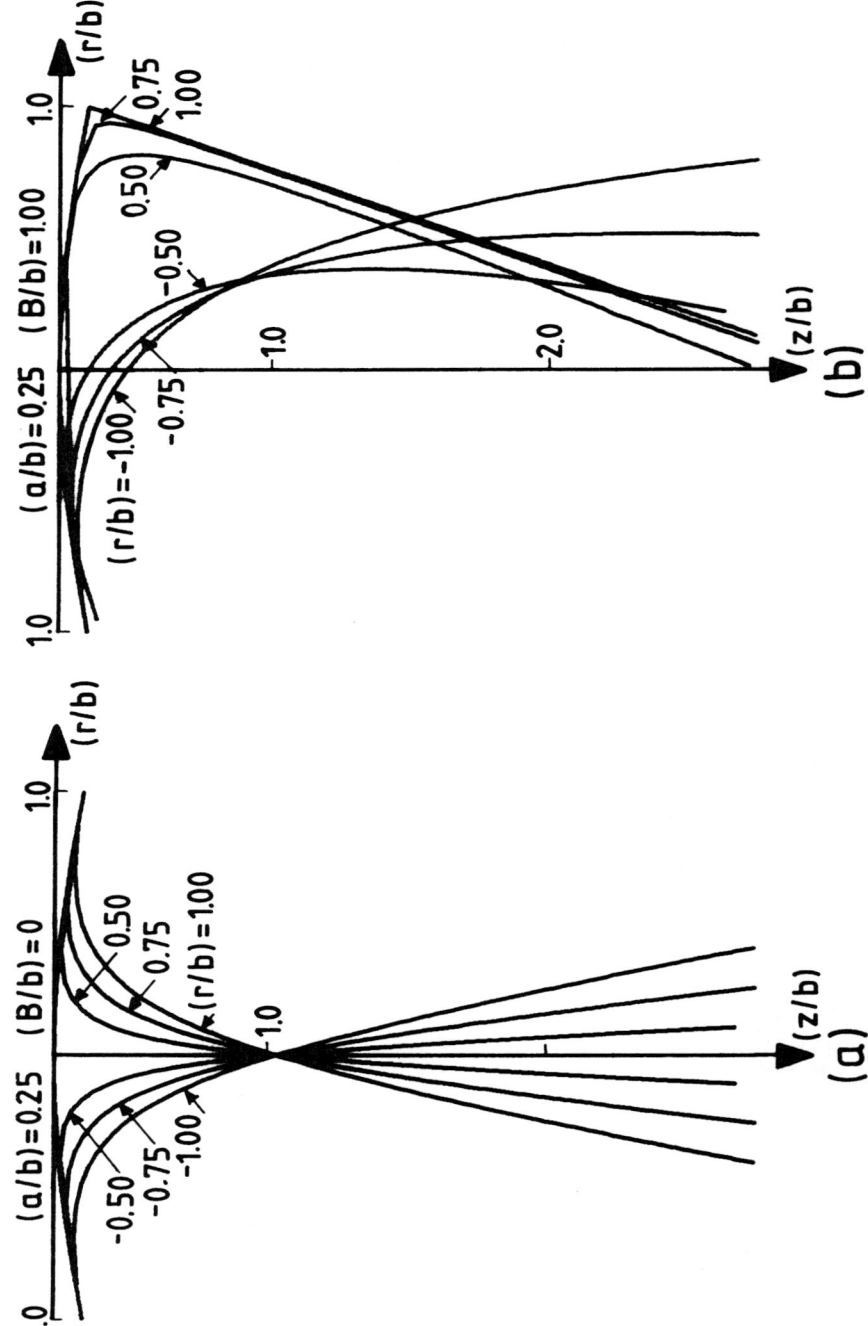

Figure 10. Loci of the extremities of the caustics corresponding to a hyperboloid reflector with (a/b)=0.25 and apertures equal to (r/b)=-1.00,-0.75,-0.50,0.50,0.75,1.00. The point-light source runs along the secondary axes with (B/b)=0(a) and 1.00(b).

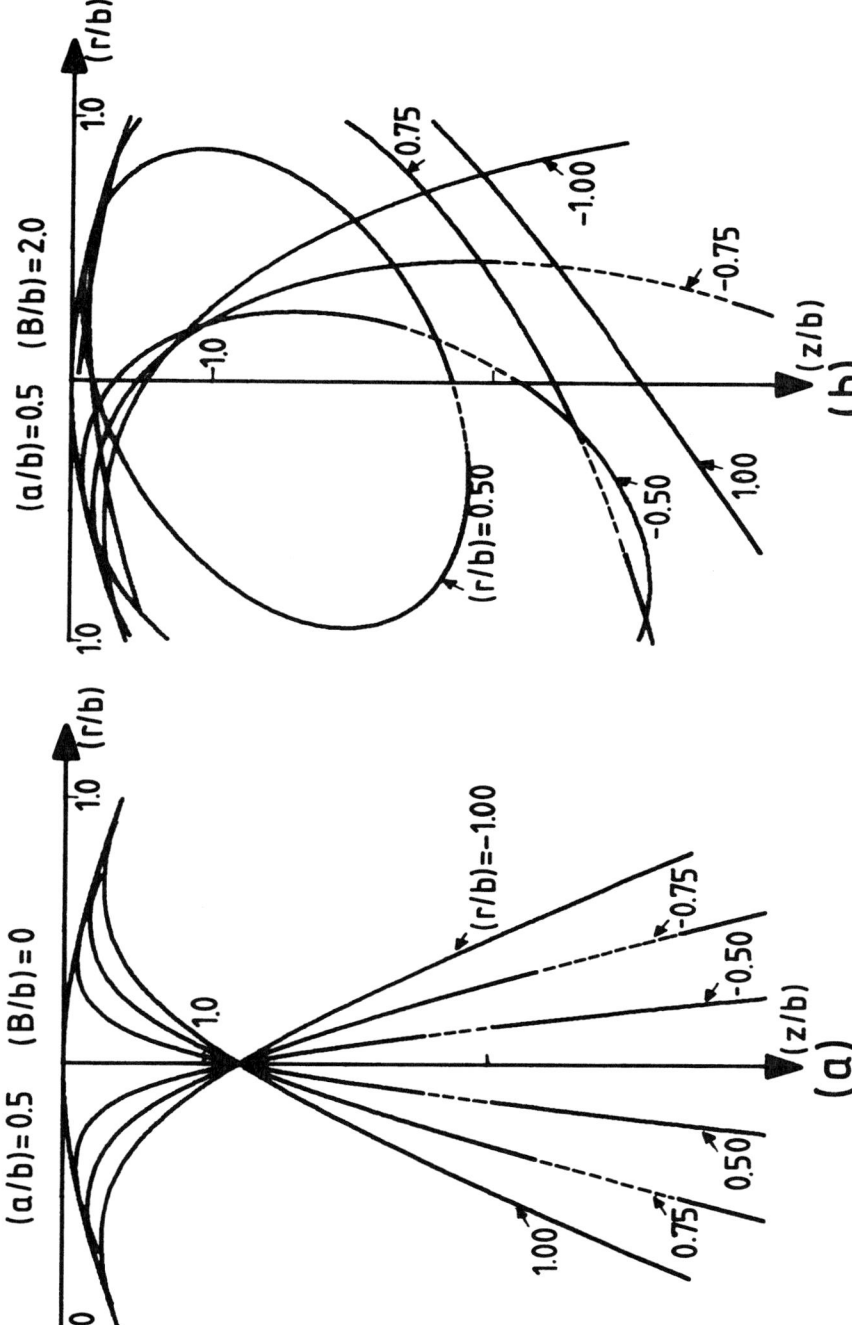

Figure 11. Loci of the extremities of the caustics corresponding to a hyperboloid reflector with (a/b)=0.50 and apertures equal to (r/b)=-1.00,-0.75,-0.50,0.50,0.75,1.00. The point-light source runs along the secondary axes with (B/b)=0(a) and 2.00(b).

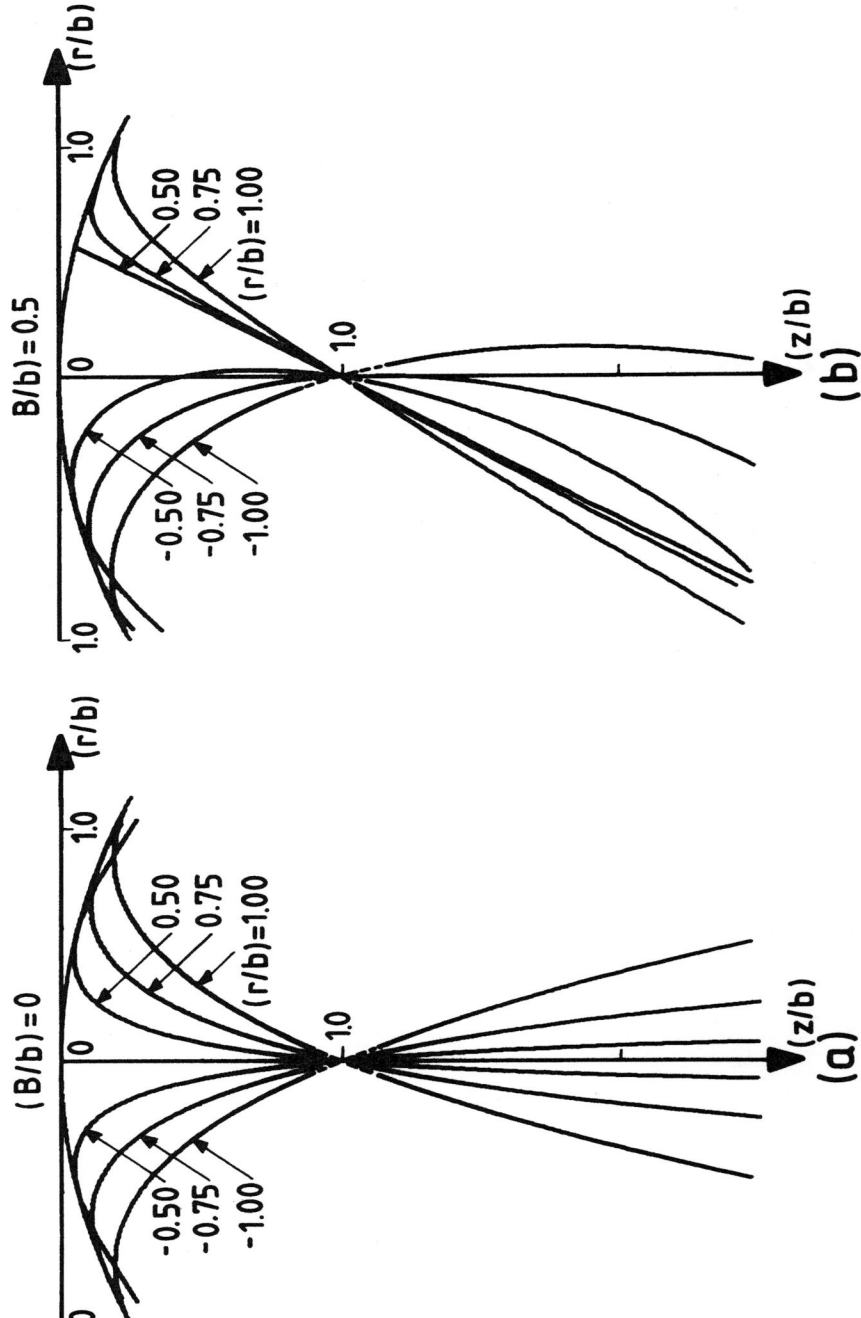

Figure 12. Loci of the extremities of the caustics corresponding to a paraboloid reflector and apertures equal to (r/b)=-1.00,-0.75,-0.50,0.50,0.75,1.00. The point-light source runs along the secondary axes with (B/b)=0(a) and 0.50(b).

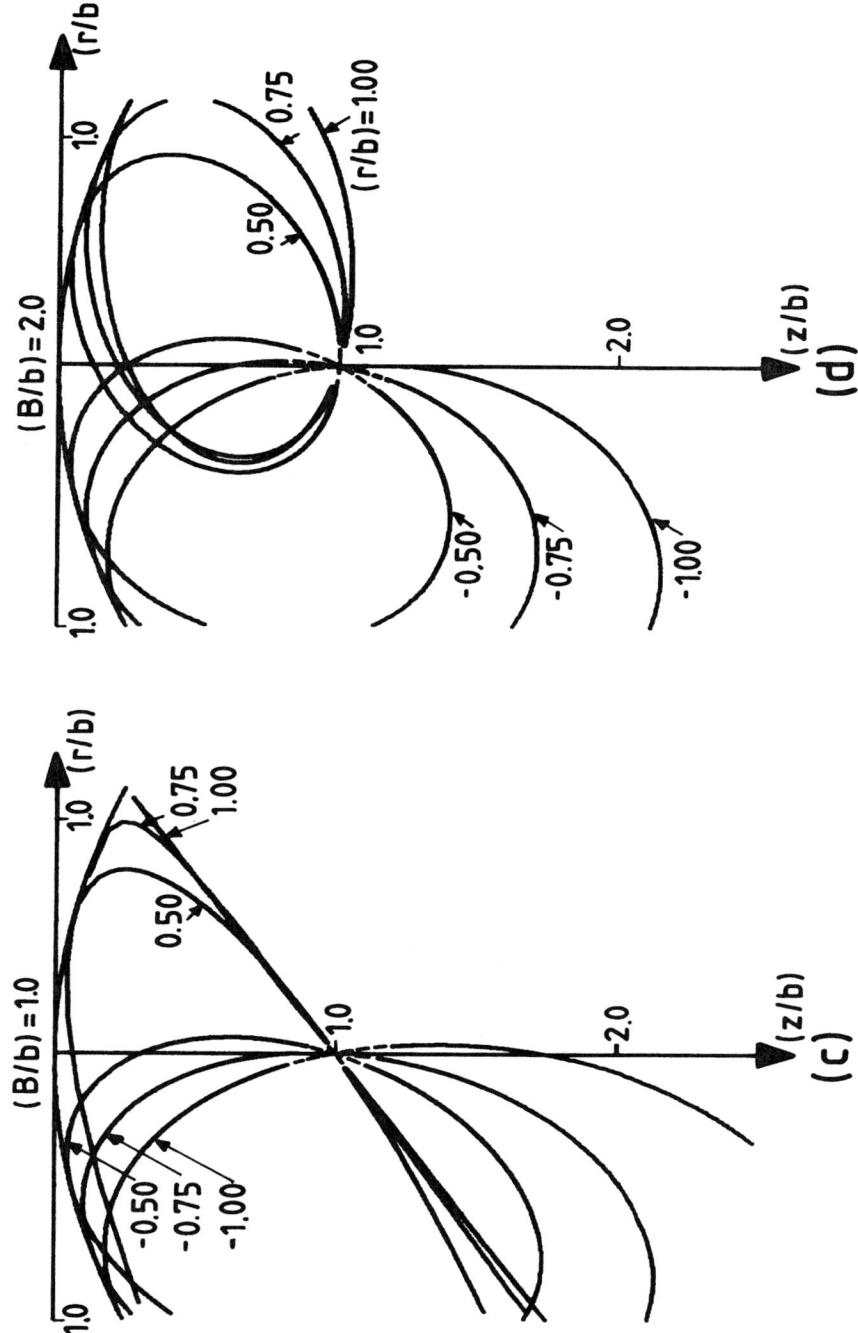

Figure 12. Loci of the extremities of the caustics corresponding to a paraboloid reflector and apertures equal to (r/b)=-1.00,-0.75,-0.50,0.50,0.75,1.00. The point-light source runs along the secondary axes with (B/b)=1.00(c) and 2.00(d).

source with $|A/b| > 20.0$ have their extremities lying close together. On the other hand, while for small values of the (B/b)-distance the dashed line parts are very small, as this parameter increases, that is as the eccentricity of the point-light source increases these parts are increasing becoming after a while significant.

Furthermore, it is worthwhile pointing out that all previous Figures of the loci of the extremities of the caustics and similar others not shown here, corresponding to different aspect ratios of the reflectors and different positions of the point-light source, are curves of the second degree (ellipses, hyperbolas and parabolas according to the particular case studied) which tend to become ellipses as the distance (B/b) expressing the eccentricity of the point light source is increasing beyond a certain limit.

In the above numerical study of the form, the shape and position of caustics created from reflected light rays impinging on rotationally symmetric conic mirrors from point-light sources placed anywhere in front of the reflector only the meridional caustics, that is the curves representing the sections of the caustic surfaces by a meridional plane passing through the light source were given.

In order to complete the analysis it is necessary to give at least another section of these caustic surfaces yielding the respective sagittal caustic lying on a plane normal to the plane of the respective meridional caustic. However, this tracing is a three-dimensional one and consitutes a formidable task with insuperable difficulties. The sagittal caustics for such types of reflectors in the case when the light source lies always along the principal axis of symmetry of the mirror were given in a previous paper (Theocaris, 1977b). The study of the form of the sagittal caustics in the case when the light-source is eccentric will constitute the subject of a companion paper.

Finally, it is interesting noting that the distribution of light energy along the caustics is of primary importance for the design of complex optical systems with a series of mirrors. This distribution of energy may be evaluated either graphically with the caustics or analytically.

ACKNOWLEDGMENT

The research work contained in this paper was financially supported by the University Research Fund. The author expresses his gratitude to his assistant Mr. C. Thireos for helping him in calculating and plotting in the computer all caustics presented in this paper and many more necessary for the complete study of this problem.

The paper is dedicated by the author to his very esteemed Colleague in the National Academy of Athens and Friend, Professor John Xanthakis, Secretary of the Proceedings, on the occasion of his 25-year anniversary in continuous and fruitful service at the Academy.

REFERENCES

Bartkowski, Z.: 1961, Optik 18, 22.
Bartkowski, Z.: 1962, Optik 19, 226.
Burkhard, D.G. and Shealy, D.L.: 1973, J. Opt. Soc. Am. 63, 299.
Scarborough, J.B.: 1964, Appl. Opt. 3, 1445.
Shealy, D.L. and Burkhard, D.G.: 1973a, Opt. Acta 20, 287.
Shealy, D.L. and Burkhard, D.G.: 1973b,Appl. Opt. 12, 2955.
Stalzer, H.J.: 1965, Appl. Opt. 4, 1205.
Stavroudis, O.N.: 1972, The Optics of Rays, Wavefronts and Caustics,
      Academic Press, New York.
Theocaris, P.S.: 1971, Appl. Opt. 10, 2240.
Theocaris, P.S.: 1977a, Appl. Opt. 16, 1705.
Theocaris, P.S.: 1977b, Trans. Natl. Acad. Athens 40, 1.
Theocaris, P.S.: 1978, Trans. Natl. Acad. Athens 53, 110.
Theocaris, P.S. and Gdoutos, E.E.: 1976, Appl. Opt. 15, 1629.
Theocaris, P.S. and Gdoutos, E.E.: 1977, Appl. Opt. 16, 722.

---

* The symbol ff. means that the corresponding subject occurs in more than two pages from the page given until the end of the corresponding paper.